Air, Water and Soil Science and Technology Series

Environmental and Regional Air Pollution

Air, Water and Soil Science and Technology Series

Trends in Air Pollution Research
James, V. Livingston (Editor)
2005. ISBN: 1-59454-326-7

Agriculture and Soil Pollution: New Research
James, V. Livingston (Editor)
2005. ISBN: 1-59454-310-0

Water Pollution: New Research
A.R. Burk (Editor)
2008. ISBN: 1-59454-393-3

Air Pollution: New Research
James, V. Livingston (Editor)
2007. ISBN: 1-59454-569-3

Air Pollution Research Advances
Corin G. Bodine (Editor)
2007. ISBN: 1-60021-806-7

Marine Pollution: New Research
Tobias N. Hofer (Editor)
2008. ISBN: 978-1-60456-242-2

Complementary Approaches for Using Ecotoxicity Data in Soil Pollution Evaluation
M. D. Fernandez and J. V. Tarazona
2008. ISBN: 978-1-60692-105-0

Lake Pollution Research Progress
Franko R. Miranda and Luc M. Bernard
2008. ISBN: 978-1-60692-106-7

Lake Pollution Research Progress
(Online Book)
Franko R. Miranda and Luc M. Bernard (Editors)
2008. ISBN: 978-1-60741-905-1

River Pollution Research Progress
Mattia N. Gallo and Marco H. Ferrari (Editors)
2009. ISBN: 978-1-60456-643-7

Heavy Metal Pollution
Samuel E. Brown and William C. Welton (Editors)
2008. ISBN: 978-1-60456-899-8

Cruise Ship Pollution
Oliver G. Krenshaw (Editor)
2009. ISBN: 978-1-60692-655-0

Industrial Pollution including Oil Spills
Harry Newbury and William De Lorne (Editors)
2009. ISBN: 978-1-60456-917-9

Environmental and Regional Air Pollution
Dean Gallo and Richard Mancini
2009. ISBN: 978-1-60692-893-6

Air, Water and Soil Science and Technology Series

ENVIRONMENTAL AND REGIONAL AIR POLLUTION

DEAN GALLO
AND
RICHARD MANCINI
EDITORS

Nova Science Publishers, Inc.
New York

For permission to use material from this book please contact us:
Telephone 631-231-7269; Fax 631-231-8175
Web Site: http://www.novapublishers.com

LIBRARY OF CONGRESS CATALOGING-IN-PUBLICATION DATA

Gallo, Dean.
Environmental and regional air pollution / Dean Gallo and Richard Mancini.
p. cm. -- (Air, water and soil pollution science and technology)
Includes bibliographical references and index.
ISBN 978-1-60692-893-6 (hardcover : alk. paper)
1. Air--Pollution. I. Mancini, Richard. II. Title.
TD883.G35 2009
628.5'3--dc22
2009009089

Published by Nova Science Publishers, Inc. ✢ New York

CONTENTS

PREFACE

The study of the environment includes interactions among the physical, chemical and biological components of the environment. The introduction of contaminants into the environment that cause instability, disorder, harm or discomfort to the physical systems or living organisms is a major concern that is addressed in this book. A simple field sampling/preconcentration strategy for determination and speciation of Cr(III) and Cr(VI) in seawater is described. Advantages of green methodologies over traditional techniques to extract pesticides from the environment are also presented. Also included in this book is a review of the most important sources of Hg into the environment and an evaluation of the use of different animal species as bioindicators of human mercury exposure. A variety of regional air pollution issues including the toxicity of smoke and the harmful effects of high surface ozone concentrations effects on human's health and crop production are examined.

Chapter 1 - Organochlorine pesticides have been used during decades as insecticides and for the control of vector-borne diseases. In 1980s these pesticides were banned in developed countries and were replaced by other kind of compounds like organophosphorus pesticides; however, they are usually more toxic than organochlorine pesticides. Nowadays these compounds are present in the environment; thus the long-term persistence and their metabolites are presented in water, sediments, soils and vegetables.

The methods usually used for the extraction of these compounds from environmental samples are often too time consuming, involve multi-step procedures that are prone to loss of analytes and use high amounts of toxic organic solvents.

Modern strategies in analytical chemistry have been developed towards the simplification of sample's preparation, as well as the minimization of organic solvent used on the extraction process. An alternative to organic solvents is the use of micellar systems as extractants of organic pollutants in liquid and solid environmental samples. Micellar media facilitate the solubilization of different compounds present in different kind of samples. Benefits obtained when micellar media are used, are lesser toxicity and cost than organic solvents.

Cloud-point extraction (CPE) process in liquid samples and microwave assisted micellar extraction (MAME) in solid samples, have been developed for the extraction of pesticides prior to their determination by high performance liquid chromatography. Moreover, during extraction process, many interfering components are usually co-extracted from solid samples together with target analytes. For increasing sensitivity and selectivity, MAME procedure can be coupled with other extraction and preconcentration techniques, introducing a clean-up step for removing substances that could interfere with the final determination of target analytes in

HPLC system. This coupling can be done with solid phase microextraction (SPME), called MAME-SPME, and with solid phase extraction (SPE), MAME-SPE, producing also an intensification of the analytes signals by means of extract clean-up and preconcentration.

In this study an overview of the use of micellar systems well alone, or well combined with the aforementioned extraction/preconcentration techniques, for the determination of organochlorine and organophosphorus pesticides in liquid and solid samples is presented. Fundaments, experimental, variables optimization, quantitative procedures and applications are described and discussed, making emphasis on the advantages of these green methodologies with regard to traditional extraction techniques.

Chapter 2 - For many contaminated sites *Natural Attenuation* (NA) of groundwater contaminants has become a common remediation strategy. The main process contributing to the NA of organic contaminants is their degradation by microorganisms inhabiting a groundwater aquifer, a process referred to as *in situ* biodegradation. As a consequence, the assessment of *in situ* contaminant biodegradation has gained increasing importance in recent years. Compound-specific stable isotope analysis (CSIA) is one of the few methods allowing for an evaluation of *in situ* biodegradation and this method is increasingly used for characterizing and quantifying microbial removal of groundwater contaminants. CSIA relies on the fractionation of stable isotopes during biodegradation, leading to an enrichment of heavier isotopes in the residual fraction of a contaminant. In many studies, the Rayleigh equation is used to obtain a quantification of contaminant biodegradation based on measured changes in isotope ratios. However, the standard Rayleigh equation considers only a single process affecting simultaneously the concentration and the isotope ratio of contaminants, a constraint often not fulfilled in the field. In many field cases, contaminant concentrations can also be affected by abiotic processes e.g. mixing, volatilization or sorption to the aquifer matrix. As the Rayleigh equation links changes in isotope ratios to concentration shifts, these abiotic processes need to be taken into account, even if they do not cause any isotope fractionation. The present study evaluates the applicability of the Rayleigh equation for the assessment of *in situ* biodegradation and shows the advantages and limitations of this concept. A review of studies in the literature addressing this issue is combined with the detailed discussion of a new modification of the Rayleigh equation. This modification allows considering abiotic concentrations changes in the quantification of *in situ* contaminant biodegradation based on stable isotope measurements. The applicability of this modification was tested against numerical modeling results using virtual flow path scenarios.

Meteorological drought results from a shortage of precipitation. Its definable characteristics include the magnitude of the precipitation deficit and how quickly it develops, duration, spatial extent, and timing. Climatological analyses of droughts can be obtained from various indices applicable at different time scales, usually 1, 3, 6, 9, 12 and 24 months, such as the Palmer Drought Severity Index (PDSI) and the Standard Precipitation Index (SPI). The PDSI is mainly a hydrological index. It is not used internationally because its empirical nature restricts its broad applicability. The SPI is based on widely measured precipitation data and can probabilistically describe precipitation shortages at any desired time scale, which have permitted its application around the world.

It has also been recently applied in the Iberian Peninsula, and in Europe. An alternative methodology, based solely in daily rainfall measures, can be derived from the distribution of the dry spells in relation to different daily rain amount thresholds, as will be shown.

According to Kiktev et al., for middle latitudes (30°-60°N) and the period 1950-1995, most of Europe and East Asia are characterized by a positive trend in the annual maximum of consecutive days with daily precipitation below 1.0 mm. This general behavior becomes of larger relevancy for the Mediterranean area, where an increase of extreme daily rainfall has been observed during 1951-1995 in spite of a decrease in total amounts.

Thus, in the Mediterranean area, the prospect is for a larger frequency of drought periods, with associated impacts on agriculture, on the exploitation of water resources and also on socio-economic activities.

Chapter 4 - A simple field sampling/preconcentration strategy for determination and speciation of Cr(III) and Cr(VI) in seawater is described. Seawater was collected in-situ by pumping it through a minicolumn packed with a poly(aminophosphonic acid) chelating resin (PAPhA, 20 mesh) connected to a field flow preconcentration system (FFPS). This packing can isolate and retain the desired chromium species, Cr(III) directly and Cr(VI) previous reduction to Cr(III) by ascorbic acid as reducing reagent. Minicolumns loaded with the analytes are returned to the laboratory where are sequentially inserted into a flow injection system for the on-line chromium species elution with diluted hydrochloric acid and flame atomic absorption spectrometric detection (FAAS). Factorial designs have been used to optimize the FFPS and the flow injection elution process for chromium speciation. A detailed study of the construction of these minicolumns and the stability of the chromium species retained in the packing is presented. With regard to the possible redox changes of the chromium species which can proceed after sampling before their determination, the field sampling/preconcentration technique was found to be optimal because it enables the separation of the species directly at a sampling site and thus avoids the risk of the conversion of Cr species during sample pretreatment and transfer to the laboratory. This procedure has been successfully applied to the determination of Cr(III) and Cr(VI) in seawater samples from Galicia (Spain).

Chapter 5 - Spatiotemporal distribution of soil nematode communities was investigated along a chronosequence of vegetation restoration in Horqin Sandy Land, Northeast China. Soil nematode communities in different seasons (spring, summer and fall) and soil depths (0-10 cm, 10-20 cm and 20-30 cm) were compared in an age sequence of 0, 5, 10 and 22-year-old *Caragana microphylla* plantations (treatments). The results showed that the total abundance of soil nematodes increased with increasing age of *C. microphylla* plantation, but decreased with depth in spring and fall. The relative abundances of plant parasites decreased, while those of omnivore-predators and fungivores increased with increasing age of *C. microphylla* plantation. Thirty-four genera were observed and *Acrobeles* was dominant genus in all treatments. Significant differences in the numbers of total nematodes, bacterivores, fungivores, plant parasites, omnivores-predators, *Acrobeles* and *Eudorylaimus* were observed among seasons, treatments, and depths. The evenness (J') and generic diversity (H') all increased with increasing age of *C. microphylla* plantation and varied significantly among the different seasons and treatments ($P < 0.05$). The dominance (λ) was significantly higher in 22-year treatment than in 0-year and 5-year treatments. The distribution of soil nematode communities in spring had a similar trend with that in fall, but different with that in summer. The soil nematode community structure and its ecological indices along different chronosequence of vegetation restoration reflected the different restoration stage and degree of the sandy land ecosystem. Vegetation rehabilitation played key roles for restoring the Horqin desertified ecosystem.

Chapter 6 - Although mercury (Hg) is one of the most toxic elements to living organisms, emission of Hg from natural sources worldwide has only a limited impact on life. However, over the past 150 years, anthropogenic Hg emissions have more than doubled the global Hg input from natural Hg sources (Hylander and Meili, 2003).

There have been sporadic episodes of human Hg poisoning from discharge of Hg-containing wastes to soil and water systems. Nevertheless, human activity still causes Hg releases into the environment. The main Hg pollution sources are atmospheric Hg discharges from coal-burning power plants, mining operations and incineration of urban and medical wastes (Morel et al., 1998).

The effect of Hg discharge into soils and aquatic systems is exacerbated by the transformation of inorganic Hg forms into methylmercury, which accumulates in living cells and tissues. It constitutes the greatest fraction of total mercury in living organisms and is quickly biomagnified in the aquatic and terrestrial food chains. Humans and wildlife can be exposed to methylmercury through ingestion of contaminated fish, which tend to accumulate high Hg concentrations in contaminated areas. Therefore, communities dependent on Hg-laden fish as a main food source frequently have elevated Hg levels in blood (Veiga and Hinton, 2002). Methylmercury is considered the sixth most toxic compound of a total of six million such toxins known to humankind (Malm, 2001).

The concept of using plants for environmental remediation has been intensively researched over the last 15 years. Phytoremediation embraces an array of low-cost plant-based technologies that could be potentially advantageous for remediation of Hg-polluted soils, especially in developing countries, where artisanal and small-scale mining for gold extraction has left a legacy of Hg pollution. Potentially, phytoremediation could remove Hg from the soil by growing successive crops of Hg-accumulating plants (phytoextraction). The Hg in the biomass could then be recovered, or stored in an area where it does not pose a risk to the environment. Alternatively, plants could be used to reduce the Hg flux in contaminated sites (phytostabilisation). This role relies on plants stabilising the soil surface, fixing Hg in the rootzone and returning water to the atmosphere via evapotranspiration. The reduction of Hg flux requires that the plants do not translocate large amounts of Hg into the shoots, where it may be consumed by herbivores.

Mercury in most soils is unavailable for uptake by plant roots. However, soil amendments can be used to increase the bioavailablility of Hg to plants. In this chapter, the authors examine the conditions by which the Hg concentration in the soil solution can be enhanced for plant uptake for three metal-contaminated mine sites where Hg pollution caused by anthropogenic or ASM activities are of concern.

In this chapter, the authors review the most important sources of Hg into the environment, the behaviour of Hg in soils, and describe plant Hg uptake experiments, with a view to elucidating the most viable strategies for the phytoremediation of Hg-contaminated sites worldwide. The authors describe laboratory and greenhouse experiments where the effect of total Hg concentrations, plant species, and humic acids was investigated on induced Hg solubilisation and plant-Hg accumulation. The results of a field-scale phytoextraction experiment in situ is also presented. Based on the soil Hg chemistry, thiol-containing ligands were tested for their potential to enhance Hg solubility for subsequent plant uptake. Research was conducted with mine wastes collected from Hg-contaminated sites in New Zealand, Brazil, and China.

The possibility of Hg (0) emissions to the atmosphere and/or Hg leaching to groundwater during the course of phytoextraction represents a threat to the health of humans and ecosystems and should not be overlooked. Therefore, the authors evaluated the generation of Hg-containing leachates and Hg vapour emissions from plant pots enclosed in gas-tight volatilisation chambers. These experiments provided a picture of Hg fate in the plant-soil-atmosphere continuum.

Finally, the authors discuss the role of phytostabilisation to reduce the flux of Hg in the environment.

Chapter 7 - Mercury is a highly mobile environmental pollutant, and small environmental releases of mercury can result in exposure of animals and human foodstuffs large enough to result in toxic effects. Point discharges, particularly coal burning power stations, but also other industrial plants and emissions from waste incinerators and crematoria, are major sources of environmental mercury and can result in marked spatial variation in mercury deposition and subsequent uptake by biota. Although chemical analysis of air, soil or water provides information on mercury environmental contamination, these analyses alone can be inadequate to assess bioavailability or predict subsequent toxicity to animals and humans. Animals in these environments, however, integrate exposure spatially, temporally and across media, and consequently can be expected to exhibit a wide amplitude of exposure and biological response. In this chapter, the use of different animal species as bioindicators of human mercury exposure are evaluated.

Chapter 8 - A theoretical analysis of heat transfer and pressures drop during the condensation of pure HFC refrigerants (and HCFC22) inside a horizontal smooth tube is developed. The tube wall is subject to a constant heat flux density. Heat and mass transfer through the liquid are described by classical equations of continuity, momentum and energy. An overall good agreement was obtained with the experimental data of local heat transfer coefficient and pressure drop in the literature.

Chapter 9 - This paper had the fundamental objective of assessing product stewardship initiatives designed by a plethora of chemical industries in Germany. The paper justifies the need for product stewardship in Germany's chemical industry's and presents the results of a survey of some chemical industries in Germany in June 2003. A total of 122 questionnaires were administered to environmental managers in the selected industries using systematic random sampling. The return rate was 51% corresponding to 66 industries. Survey results revealed that despite the inadequacy in the implementation of some product stewardship tools like emission reduction (particularly solid waste emission), implementation of product stewardship tools by the sampled industries was advanced. It was concluded that most of the attempts made thus far by industries to address product stewardship have been major successes. However, it was emphasized that further improvement could be achieved if incentives for product stewardship implementation are promoted and the merits of strategic research in the area of reduction in solid waste emission is further developed and encouraged.

Chapter 10 - In the first phase of the research an analytical method was developed in order to determine three groups of priority organic pollutants simultaneously such as polycyclic aromatic hydrocarbons (PAHs), polychlorinated biphenyls (PCBs) and organochlorine pesticides (OCPs) in atmosphere quantitatively by chromatographic techniques. In the second phase of the research, PAHs, PCBs and OCPs were monitored in gas and particle phases of urban air.

In determining particle phase compounds, the best extraction efficiencies were obtained by ultrasonic extraction method using 25 mL of n-hexane/petroleum ether (1/1, v/v) solvent mixture three times for 15 minutes. For determining compounds in gas phase of air samples soxhlet extraction method was carried out for 16 hours with 150 mL diethyl ether/n-hexane (1/9, v/v) solvent mixture. 10 gram 5% deactivated silica gel column was used for cleaning up the interfering substances while simultaneously fractionation of target compounds prior to GC analyses. Elutions of compounds were carried out as follow: Fraction 1: 70 mL n-hexane, Fraction 2: 3 x 20 mL n-hexane/ethyl acetate (1/1, v/v). PAHs and PCBs compounds were determined with GC/MS systems while OCPs compounds were determined using GC/μ-ECD system. Surrogate and standard spiked filters were analyzed after optimized method. Recovery were between 86%±2 and 104%±3 for PCBs with Relative Standard Deviation (RSD)< ±4, between 66%±6 and 94%±5 for PAHs with RSD<±8, between 67%±5 and 120%±10 for OCPs with RSD<10. Spiked PUF plugs were also analyzed following the optimized method. Recovery ratios obtained from PUF plugs were between 94%±9 and 118%±8 for PCBs with RSD%±9 while the recovery ratios changed from 88%±1 to 120%±2 for PAHs with RSD<±8. Recovery ratios of OCP compounds were between 65%±4 and 108%±6 with RSD<±9. In order to determine the matrix effect on optimized method, air sampled filters and PUF plugs were spiked with standards and surrogate standards and analyzed according to the optimized method. Recovery ratios of air sampled and spiked filters and PUF plugs were not statistically significantly different from the recoveries of non air sampled spiked filters and PUF plugs at 95% level.

Gas phase and particle phase PAHs, PCBs and OCPs contents were analyzed in air samples taken from five different locations selected considering residential, industrial and traffic areas between August 2006 and May 2007. Mean PAHs concentrations were between 30 ng m^{-3} and 214 ng m^{-3} while mean PCBs and OCPs were between 0.04 ng m^{-3} and 0.14 ng m^{-3} and 0.81 ng m^{-3} and 6.15 ng m^{-3}, respectively. Total concentration ratios found in gas phase were 61%-63% for PAHs, 63%-89% for PCBs and 45%-70% for OCPs. With increasing in temperature concentration ratios in gas phase were increased while concentration ratios in particle phase decreased for all compounds. The higher concentrations of PAHs were determined in winter seasons while the lower concentrations of PAHs were determined in summer. Results from seasonal monitoring of PCBs in urban air of Konya show that the highest PCBs concentrations were determined in September while the lowest PCBs concentrations in April. The highest total OCPs concentrations were found in autumn seasons that in these months, lands were prepared for planting by farmers. Considering the risk characterization of these compounds determined in Konya urban air it could be concluded that these compounds pose some health risks for human.

Chapter 11 - The northeastern coast of the Mediterranean Sea is under the influence of sporadic dust transport from North African (The Great Sahara) and Middle Eastern (Arabian Peninsula) deserts, particularly during spring and autumn. Mineral dust from these arid regions typically has a characteristic reddish yellow - reddish brown color owing to its high iron oxide content. Precipitation samples collected at Mersin, on the eastern Mediterranean coast of Turkey, during the period of December 2003 – May 2005 have been analyzed with respect to their red dust contents, and in relation to backward air mass trajectories corresponding to the rainy days to determine potential source regions. Classification of trajectories revealed local sources from Anatolian mainland and in the Cilician Basin, in addition to remote sources in North Africa and the Middle East. Approximately, one fourth of the precipitation samples can be classified as "*red rain*", annually. The mean electrical conductivity and the volume weighted mean (VWM) Aluminium (Al) concentration of red rain were found to be more than four and eight times higher than normal rain, respectively; while the dissolution of calcite in dust was found to result in an order of magnitude decrease in the VWM hydrogen ion concentration of red rain, by neutralizing (buffering) the acidity of rainwater. Due to dissolution of atmospheric CO_2 in rainwater, natural precipitation is slightly acidic (pH ≈ 5.6). The VWM pH values were calculated to be 6.10 for normal rain and 7.21 for red rain. The mean bicarbonate concentration of red rain, estimated from the theoretical relation between pH and HCO_3^-, was sixteen times higher than that of normal rain. Similarly, abundance of the major anions (Cl^-, NO_3^-, SO_4^{2-}) and major cations (Na^+, K^+, Ca^{2+}, Mg^{2+}, NH_4^+) measured in red rain samples were all high, relative to the values observed in normal rain samples, including formaldehyde (HCHO), the only organic parameter measured in Mersin precipitation. Similar enhancements were also observed in soluble (Al_s) and particulate (Al_p) aluminium species. Based on these results, red rain can be identified as a rich source of all major ions, formaldehyde and aluminium suggesting mineral dust particles to act as host for all measured species in the northeastern Mediterranean atmosphere.

Chapter 12 - The objective of this research was to develop and evaluate models for predicting hourly ozone concentrations, ozone exceedances and hourly air quality index (AQI) in Dayton, OH. As the hourly ozone concentrations are closely related to the meteorological conditions, three variables - temperature, wind speed, and dew point temperature were chosen for this study. The ozone data were extracted from the EPA's Aerometric Information Retrieval System (AIRS) database for the period between 1996 and 2003. The meteorological data were taken from the National Climatic Data Center (NCDC) for the same period.

An analysis of variations in hourly ozone concentrations and ozone episode occurrences was carried out for the period Apr.-Oct. for the years 1996-1999. Also, analysis of the long-term trends in annual means of ozone concentrations, temperature, wind speed, and dew point temperature was performed using the same data set. Based on this analysis, the ozone data was divided into pre-summer (Apr.-Jul.) and post-summer (Aug.-Oct.) seasons, to account for seasonal variations, and each season was further divided into three regimes, namely, stable period (hours: 1-8), ascent period (hours: 9-16), and descent period (hours: 17-24). The Kolmogorov – Zurbenko (KZ) filter technique was used to reduce the scatter in the time series, and models were developed for the three regimes for each season by regression, using the corresponding independent parameter values.

A total of twelve models were developed to predict ozone concentrations for pre-summer and post-summer periods. Six models considered temperature, wind speed, and dew point temperature as the independent variables (three-parameter models), and the other six considered temperature, and wind speed as variables (two-parameter models). Also, three models each for pre-summer and post-summer season were developed for predicting the ozone exceedances. The performance of the models was evaluated using statistical indicators in three ways:

a. Initial evaluation (or validation) of the models was conducted using 2002 data.

b. The effectiveness of these models was further evaluated using available MM5 (a mesoscale meteorological forecasting model) real-time forecasts from the Ohio State University for the months of Aug.-Oct., 2003.
c. Finally, the performance of the three-parameter models was compared with that of the two-parameter models.

The study shows that the forecasts of hourly ozone concentrations made by the models based on KZ filters are reliable only to a limited extent. However, the models performed well in predicting AQI values reported by the EPA. Also, the three-parameter models performed better in predicting the peak concentrations when compared to the two-parameter models.

Chapter 16 - When the relative magnitudes of air pollutants are arranged on a map, the local distribution of the each pollutant becomes apparent. This approach will be clues revealing the emission sources, behaviors and impacts of pollutants. Of ambient VOCs, passive air sampler will be useful device for the sampling because of highly portable, sensitive, and selective feature. The spatial distribution patterns in an industrial area were quite different from each targeted VOC, indicating that those emission sources are specific for each VOC. Similar trend has been observed in the distribution of the ambient PAHs in an industrial area. It indicates that the high-polluted distributions were apparently contributed to the industrial plants such as paper-making factories. On the other hand, the local distributions of ambient PAHs in urban city showed that the highly polluted areas were spread along the busy roads of the outer city, with an enclosed area of lower pollution in the central city. In certain industrial area, the lifetime cancer risks of exposure to ambient VOCs were significantly higher in winter than in summer, which were derived mainly from benzene, carbon tetrachloride, formaldehyde, and acetaldehyde. The carcinogenicity of ambient PAH mixture was occupied (~77%) by BaP, DBahA, and Nap in both summer and winter.

Chapter 17 - In this chapter, the environmental quality (EQ) of three rural communities in connection with two dams in Kumasi, Ghana, is overhauled over four distinct years. The study is organized into two main parts: (i) an empirical analysis that aims at investigating the retrospective EQ of three riparian communities, and (ii) model development and prediction (backcasting/hindcasting) of the EQ of the communities, in one year lag. The authors use discontinuous, irregularly spaced, multivariate geostatistical, limited time series data consisting of sets of predictors and predictands. Missing values in the datasets are estimated using kriging technique. A multivariate empirical orthogonal function (EOF) analysis is performed on the datasets out of which the first three principal components (PCs) are retained for the study. The construction of the operational activity indices (OAIs) and EQ indices (EQIs) is based on the scores of the first three PCs of the predictor and predictand datasets, respectively. This is followed by performing dependency measures typically, zero order and lagged cross correlation analyses between the two sets of indices to monitor the EQ of the communities over the four years. Generally, the results from the retrospective environmental analysis (REA) show moderate to strong relationships between the EQIs and OAIs, a signal of profound influences exerted by the dam(s). The most important signals are observed in the years when the two dams are operated in tandem. The potential forecasting (hindcasting) abilities of the OAIs determined from cross-validations, give us the confidence to incorporate the REA results in the stochastic and temporal environmental autoregressive AR (1) model developments for performing retrospective backcasts. A comparison between the parametric and non-parametric stochastic model results shows that the latter outperforms the former, and thus, better fits the data. The AR (1) model results theoretically signal progressive deteriorations in the backcast EQ for three out of the four lagged years. However, the remaining one shows no significant difference from the preceding year`s, an indication of environmental stability and non-responsiveness to the operational activities of the dam(s). Lack of observational data posed a limitation in comparing our hindcast model results with real life situation. Notwithstanding this, the model biases and forecast (hindcast) errors seem to be very reasonable. These suggest beyond doubts that the findings can be relied upon for decision support system. Remediation of the environmental impacts of the dams proposed does not differ from that of our previous studies.

Short Communication - Poly-γ-glutamic acid (γ-PGA) is a water soluble, polyanionic, extracellular polymer produced by several members of the genus *Bacillus*, the most notable among them *Bacillus subtillis* and *Bacillus licheniformis.* This polymer consists of D- and L-glutamic acid repeating units that are linked between the α-amino and γ-carboxylic acid functionalities. A peculiar characteristic of γ-PGA formed by *B. licheniformis* is that the polymer stereochemistry can be controlled by the composition of the medium used in the polymer production in the laboratory. γ-PGA is water soluble, biodegradable, edible and non-toxic to humans and the environment. Therefore, the potential applications of this polymer and its derivatives have been of growing interest in the past few years in a broad range of industrial fields such as food, cosmetics, medicine and water treatment.

Most of the reports published on the microbial or enzymatic biodegradation of microbial γ-PGA are only concerned with γ-D-PGA. In the present work, γ-L-PGA was biosynthesized using *B. licheniformis* NCIMB 11709, and the biodegradation of the different enantiomeric forms of PGA, as well as the α and γ forms (γ-L-PGA, γ-D-PGA, γ-D,L-PGA, α-L-PGA and α-D-PGA) of this bacterial strain, were studied. γ-PGA with L-units content of 70% was produced with a Mw of 1 million g/mole. *B. licheniformis* NCIMB 11709 did not grow when the α-forms of the polymer were used as sole carbon source, independently of the D or L contents. However, a decrease in the molecular weight of the α-L-PGA in the bacterial culture was observed, which was not due to hydrolytic degradation. On the other hand, this strain could degrade and use the polymer when it was in the γ-form, regardless of the D or L contents.

This finding can have an implication in defining bacterial factors involved in the degradation of the polymer. Moreover, the generated knowledge of the degradation of the different forms of the poly(glutamic acid) will facilitate selection of the better form according to the intended application.

In: Environmental and Regional Air Pollution
Editors: D. Gallo, R. Mancini

ISBN: 978-1-60692-893-6

Chapter 1

ENVIRONMENTAL FRIENDLY METHODS FOR THE EXTRACTION AND DETERMINATION OF ORGANOPHOSPHORUS AND ORGANOCHLORINE PESTICIDES

J. J. Santana Rodríguez*[*], *Z. Sosa Ferrera and D. Vega Moreno
Department of Chemistry, Faculty of Marine Sciences,
University of Las Palmas de Gran Canaria, 35017,
Las Palmas de Gran Canaria, Spain

ABSTRACT

Organochlorine pesticides have been used during decades as insecticides and for the control of vector-borne diseases. In 1980s these pesticides were banned in developed countries and were replaced by other kind of compounds like organophosphorus pesticides; however, they are usually more toxic than organochlorine pesticides. Nowadays these compounds are present in the environment; thus the long-term persistence and their metabolites are presented in water, sediments, soils and vegetables.

The methods usually used for the extraction of these compounds from environmental samples are often too time consuming, involve multi-step procedures that are prone to loss of analytes and use high amounts of toxic organic solvents.

Modern strategies in analytical chemistry have been developed towards the simplification of sample's preparation, as well as the minimization of organic solvent used on the extraction process. An alternative to organic solvents is the use of micellar systems as extractants of organic pollutants in liquid and solid environmental samples. Micellar media facilitate the solubilization of different compounds present in different kind of samples. Benefits obtained when micellar media are used, are lesser toxicity and cost than organic solvents.

Cloud-point extraction (CPE) process in liquid samples and microwave assisted micellar extraction (MAME) in solid samples, have been developed for the extraction of pesticides prior to their determination by high performance liquid chromatography.

[*] E-mail: jsantana@dqui.ulpgc.es

Moreover, during extraction process, many interfering components are usually co-extracted from solid samples together with target analytes. For increasing sensitivity and selectivity, MAME procedure can be coupled with other extraction and preconcentration techniques, introducing a clean-up step for removing substances that could interfere with the final determination of target analytes in HPLC system. This coupling can be done with solid phase microextraction (SPME), called MAME-SPME, and with solid phase extraction (SPE), MAME-SPE, producing also an intensification of the analytes signals by means of extract clean-up and preconcentration.

In this study an overview of the use of micellar systems well alone, or well combined with the aforementioned extraction/preconcentration techniques, for the determination of organochlorine and organophosphorus pesticides in liquid and solid samples is presented. Fundaments, experimental, variables optimization, quantitative procedures and applications are described and discussed, making emphasis on the advantages of these green methodologies with regard to traditional extraction techniques.

ABBREVIATIONS

ASE	Accelerated Solvent Extraction
CMC	Critical Micellar Concentration
CPE	Cloud Point Extraction
DDT	Dichlorodiphenyltrichloroethane
EPA	Environmental Protection Agency
HPLC	High Pressure Liquid Chromatography
LLE	Liquid-Liquid extraction
LOD	Limit of detection
MAE	Microwave Assisted Extraction
MAME	Microwave Assisted Micellar Extraction
PLE	Pressurized Liquid Extraction
POLE	Polyoxyethylene 10 Lauryl Ether
RSD	Relative Standard Desviation
SFE	Supercritical Fluid Extraction
SPE	Solid Phase Extraction
SPME	Solid Phase Microextraction

INTRODUCTION

Organochlorine and Organophophorus Pesticides

Organochlorine pesticides have been used during decades in agricultural as insecticide and for the control of vector-borne diseases. Although these pesticides have banned in developed countries since 1970s and 1980s, some of them have continued to use DDT, especially for public health applications such as malaria control [1, 2]. Despite not being used by many countries for some years they are present in the environment; in this way, the long-term persistence of DDT and its metabolites in soil have been reported [3-6].

The lipophilic nature, hydrophobicity and low chemical and biological degradation rates of organochlorine pesticides have led their accumulation in biological tissues and subsequent magnification of concentrations in organisms progressing up the food chain with consequences for the humans [7, 8]; for these reasons are listed as US Environmental Protection Agency (EPA) priority pollutants [9].

Organophosphorus pesticides, which may be considered as second generation of synthetic pesticides, after the use of organochlorine pesticides, have become more popular, as they are generally of low persistence, biodegradable and do not tend to accumulate in the food chain. However they are generally more toxic (potent cholinesterase inhibitors) than organochlorine or carbamate compounds [10].

Over recent decades, research into developing new pesticides and analytical methodology has been severely curtailed in favour of alternate means of pest control. Nevertheless, the use of pesticides continues to be our major weapon for controlling pests. Large amounts of pesticide residues remain in agricultural and forest soils after their application [11]. They leach into groundwater and are also found in run-off water. Therefore, the development of new and improved techniques for analysing these agents and their breakdown products is mandatory.

Tradional techniques for analysis of organochlorine and organophosphorus pesticides are based in gas-chromatographic multiresidue procedures [12-14]. However, these methods are often too time consuming, due to the slow chromatographic elution and the intensive pre-analytical clean-up procedures [15-17] followed, in some cases, by a derivatization [17, 18].

They usually involve multi-step procedures, are prone to loss of analytes, and use toxic organic solvents [19, 20]. Consequently, most of the analysis time is spent on the sample preparation steps alone. Furthermore, some pesticides are thermolabile and therefore difficult to volatilize without decomposition [21].

New methods therefore have been developed in order to supplement or even replace existing gas-chromatographic methods. These altenative methodologies are based in a high performance liquid chromatography determination combined with a previous micellar extraction: cloud point extraction (CPE) procedure for liquid samples and microwave assisted micellar extraction (MAME) for solid samples. Specifically, the organochlorine and organophosphorus pesticides listed in Table 1 have been studied and determined using these green methods.

Table 1. Organochlorine and Organophosphorus pesticides studied

Organochlorine Pesticides	Organophosphorus Pesticides
	DIMETHOATE
4,4'-Dichlorodiphenyldichloroethane (4,4'-DDD)	Methidation
Dieldrin	Parathion methyl
4,4'-Dichlorodiphenyltrichloroethane (4,4'-DDT)	Malathion
2,4'-Dichlorodiphenyltrichloroethane (2,4'-DDT)	Ethoprophos
4,4'- Dichlorodiphenyldichloroethylene (4,4'-DDE)	Parathion ethyl
Aldrin	Diazinon
	Chlorpyrifos

MAIN CONVENTIONAL EXTRACTION TECHNIQUES FOR LIQUID AND SOLID SAMPLES

The need to determine low concentrations of pollutants in environmental samples, as well as the matrix complexity, means that analytical methods with high sensitivity, selectivity and resolution have to be applied to soil, sediment, water and biological samples. Elaborate sample preparation involving analyte isolation and enrichment is generally necessary before the final analysis, which is usually performed using gas or liquid chromatography.

The most frecuently used methods for the extraction and preconcentration of organochlorine and organophosphorus pesticides in water samples are liquid-liquid extraction (LLE) and solid-phase extraction (SPE) [22,23]. Solvent extraction methods have the disadvantage of poor recoveries, which vary from compound to compound and use a large amount of organic solvent. There is therefore a trend to replace solvent extraction procedure in order to minimised sample manipulation, analyte losses and the use of toxic solvents [24]. So there is an increasing tendency to replace LLE by solid-phase extraction (SPE). SPE is widely accepted as an alternative extraction/cleanup method to LLE in order to determine pollutants in liquid samples [25]. However, SPE have certain disadvantages for water analysis: the cross-sectional area is small, sample processing rates are slow, the particle blockage tolerance and adsorbed matrix components is low and channelling reduces the capacity to retain analytes. Another drawback in SPE, which is the same for LLE, is the considerable amount of time needed and the manual operations involved [26]. Other methodologies have been developed with a view to eliminating or, at least minimising, the use of organic solvents. Solid-phase microextraction (SPME) is a recent technique for the extraction of organic compounds from environmental samples. SPME is based on the partition equilibrium of analytes between a polymeric stationary phase, and the sample matrix. The desorption step is usually carried out by placing the fiber in a hit injector of a gas chromatograph [27] or it can also be performed in an HPLC system by introducing an interface [28]. An inherent disadvantage of SPME is that quantitative work is still rather laborious because severe carry-over may occur between samples. Moreover, a relatively long equilibration time for the absorption of analytes is needed.

An alternative extraction technique for liquid samples is cloud point methodology or cloud point extraction (CPE) [29]. It lets extraction and preconcentration steps based on phase separation using non ionic surfactants as extractants. It offers a convenient alternative to more conventional extraction procedures. From an analytical point of view, one of the most important properties of surfactants is their good capacity to solubilise solutes of different type and nature due to their organised structure and their anphipatic character.

Table 2 gives a description of the aforementioned techniques and lists their advantages and their disadvantages for liquid samples. The extraction of organochlorine and organophosphorus pesticides from solid samples requires the use of organic solvent, which competes in the releasing of the analytes retained owing to the high activity of the matrix. Traditional methods employ large volumes of solvent under aggressive shaking and/or temperature conditions. The most frequently used method for the extraction of organic compound from soils is Soxhlet extraction or the use of an ultrasonic bath.

Table 2. Extraction techniques for liquid samples [from ref. 54]

	LLE	SPE	SPME	CPE
Brief description	Analyte is partitioned between two immiscible solvents	Analyte retained on a solid adsorbent	Partition equilibrium of analytes between a polymeric stationary phase and the sample matrix	Analyte is portioned between two isotropic phase 'surfactant-rich phase' and 'aqueous phase'
Extraction Time	Up to 24 h	20-30 min	Up to 60 min	20 min
Solvent Usage	Organic Solvent	Organic Solvent	None	Surfactant solution
Cost	Low	Relatively low	Relatively low	Low
Ease of operation	Easy	Relatively easy	Relatively easy	Easy
Disadvantages	Large consumption of solvent and concentration of sample required after extraction	Insufficient retention of very polar compounds and limited selectivity	Choice of fiber and optimisation of procedure required	Optimisation of operating conditions

Table 3. Extraction techniques for solid samples [from ref. 54]

	Soxhlet	Sonication	SFE	PLE	MAE
Brief description	Sample is placed in an extraction thimble and leached with hot solvent	Sample is placed in a ultrasonic bath with an organic solvent, with shaking	Sample is placed in a high-pressure cartridge and extracted with a supercritical fluid	Sample is placed in an extraction vessel and pressurized with a solvent heated	Sample is placed immersed in solvent, and heated with microwave energy
Extraction Time	Up to 24- 48 h	30-60 min	30-60 min	15 min	20-30 min
Solvent usage	Organic solvent	Organic solvent	Solvent in its supercritical state	Organic solvent at high temperature and pressure	Organic solvent heated with microwave energy
Cost	Low	Low	High	High	Moderate
Ease of operation	Easy	Easy	Relatively easy	Relatively easy	Relatively easy
Dis-advantages	Large amount of solvent and evaporation of extract are neccesary	Large amount of solvent and filtration are required	Limited sample size, and dependence of matrix and analyte type	Dependence on matrix	Clean-up procedure is required

Soxhlet extraction, the most conventional of all methods, is particularly suitable for organic pollutants strongly absorbed in soil matrices but it requires long extraction times (from 24 to 48 hours) and the use of large volumes of frequently toxic organic solvents (300 to 500 ml per sample) [30, 31]. Sonication extraction is faster than Soxhlet extraction (30 to 60 min per sample) and allows the extraction of large amounts of sample, but it still uses about as much solvent as the Soxhlet extraction [32].

The newer extraction techniques such as supercritical fluid extraction (SFE), based in the extraction of target compounds with solvents near supercritical conditions [33]; accelerated solvent extraction or pressured liquid extraction (ASE or PLE), which extracts target compounds using solvents at high pressures and high temperatures [34] and microwave assisted solvent extraction (MAE), based in the extraction by microwave energy and temperature [35,36]. All of them are very attractive because they are faster, use much smaller amounts of solvents and are more environmentally friendly techniques than traditional ones [37].

In the last few years, the number of procedures using extraction of organic compounds from environmental matrices by microwave energy has increased [38, 39]. Microwave assisted extraction (MAE) allows rapid extraction of solutes from solid matrices by employing microwave energy as a heat source, with an extraction efficiency comparable to that of the classical techniques. It exhibits many substantial improvements in analytical sample preparation, as it requires much lower volume of organic solvent (30 ml in MAE versus 300 - 500 ml in Soxhlet extraction), reduces extraction time (20 to 30 min per sample) and lets prepare multiple samples in one step [40-42]. Optimisation of MAE conditions is rather easy, as there are few parameters (nature of solvent, time and power) and it is cheaper than other modern techniques, such as supercritical fluid extraction. The main drawbacks of this system are: extraction solvent has to be able to absorb microwaves, a clean-up procedure may be required before analysis and the vessels need to be cooled to room temperature after extraction and before they can be opened, thus increasing the overall extraction time. This methodology has been applied to the extraction of different organic compounds, such as organochlorine insecticides [43], polynuclear aromatic hydrocarbons, [44], arochlors [45], etc. with good results. In all these studies, the extractant has always been an organic solvent. Table 3 provides a summarised description of these techniques and show their advantages and disadvantages for solid samples.

SURFACTANTS

An alternative to the use of organic solvents would be the use of micellar systems (surfactants) as extractants.

Organic solvents are toxic compounds which are dangerous to researchers who work with them and also for environment. For this reason it is necessary look for new extraction solvents that will be less toxic for environment and work's atmosphere. In this sense, surfactants can be used as extractants in the analysis of organic pollutants like organochlorine and organophosphorus pesticides in different kind of matrices.

Surfactants are known as tensoactives, whose main characteristic is their anphipatic structure; the presence in a same molecule of two or more groups with anthagonic properties respect the solubility of them [46]. A monomer of surfactant has an structure R-X where R is a hydrocarbon chain between 8 and 18 carbon of hydrophobic characteristics and X is a polar group of ionic or non-ionic nature (Figure 1).

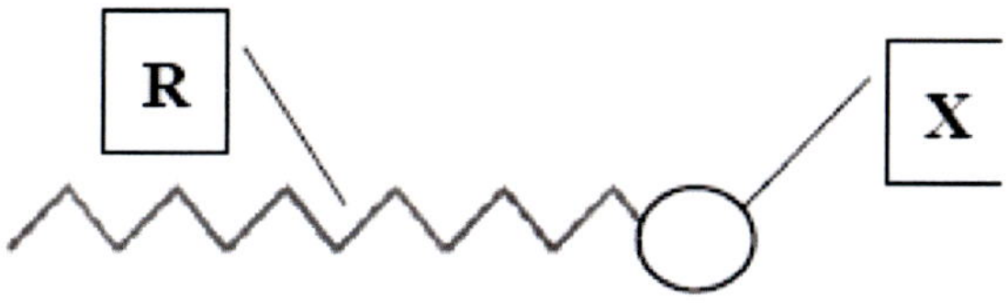

Figure 1. Scheme of surfactant's monomer molecular structure.

Depending on the nature of hydrophilic head, surfactants are classified in four groups: Cationic, Anionic, Zwitterionic (positive and negative charge) and Non-ionic [47]:

Due to the anphipatic character of surfactants, their molecules in presence of water are absorbed in interfaces, decreasing interfacial pressure and forming organized structures, called 'micelles' when surfactant concentration are higher than critical micellar concentration (c.m.c.).

Main property of surfactants, related with their capacity to micelle formation, it is analytes solubilization. Micelles, formed in aqueous solution (direct micelle) or in non aqueous solution (reverse micelle) can solubilize polar and non polar molecules. In next scheme it is showed a micellar aggregated depending on their nature, polar, semipolar or non polar (Figure 2).

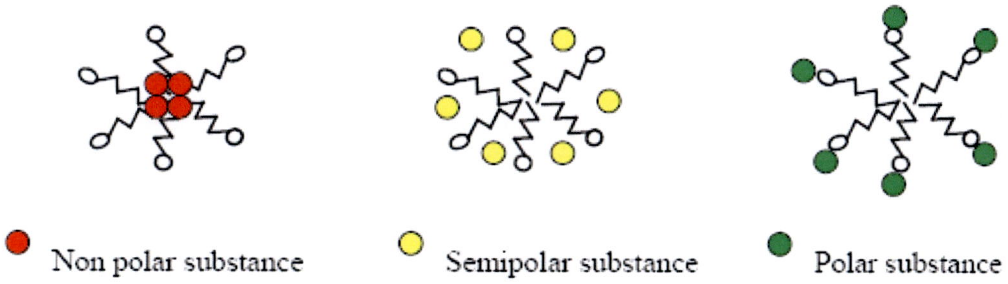

Figure 2. Micellar aggregated depending on substance nature.

Solubilization is located in different points depending on the nature of solubilized substance, but also depending on surfactant. In non-ionic surfactants the solubilization is generally in hydrophilic heads [47]. This capacity for solubilization produces that micellar solutions have important applications in fields like chemistry, pharmacological, cosmetics, etc. [48]

From the analytical point of view, this capacity to solubise solutes of different types and nature as well as their compatibility with the mobile phase in HPLC are the most important properties. Moreover, other advantages of surfactants are safety, low cost and no toxicological effects due to the biodegradability [49, 50].

For liquid samples, the use of extraction and preconcentration steps based on phase separation using the cloud point methodology offers a convenient alternative to more conventional extraction systems. Aqueous solutions of some surfactants have been used in cloud point extraction (CPE) of different species prior to their determination by several techniques [51-53]. The small volume of the surfactant-rich phase obtained in CPE process allows to extract and preconcentrate the analytes in one step, prior to gas or liquid chromatographic analysis (Figure 3).

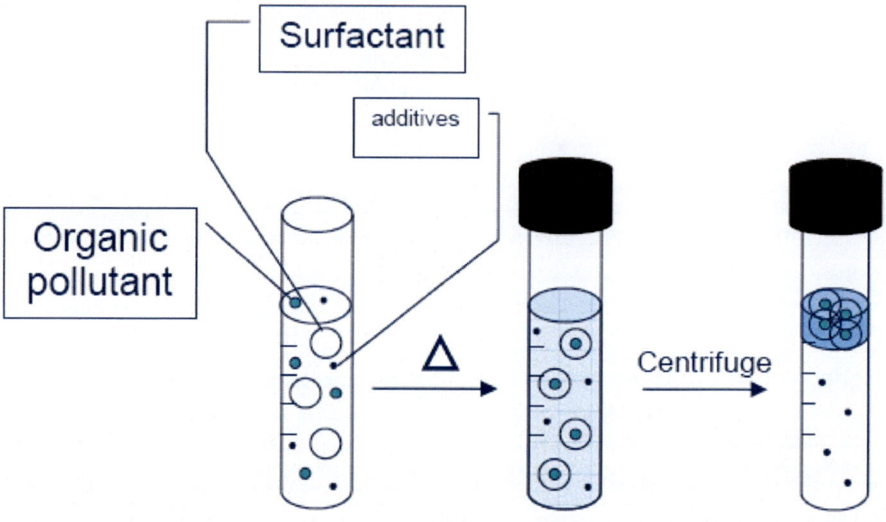

Figure 3. Squeme of the Cloud-Point process.

For solid samples, using the combination of the surfactant capacity to solubilise different species and the energy supplied by microwave assisted extraction, hereinafter microwave assisted micellar extraction (MAME), it can be applied to the extraction of different compounds present in solid samples [54], which always offers the benefit of low cost, easy handling, reduced toxic effects and compatibility with aqueous-organic mobile phase in HPLC (Figure 4).

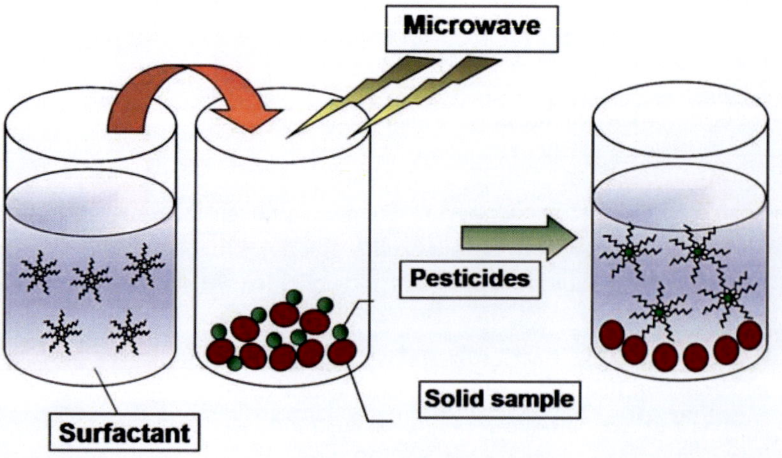

Figure 4. Squeme of Microwave Assisted Micellar Extraction process.

The MAME procedure have been applied to the extraction of several compounds as phenols [55], organochlorine and organophosphorous pesticides [56,57] present in different solid environmental samples such as marine sediments, soils and biological samples.

DETERMINATION OF PESTICIDES BY USING CLOUD POINT EXTRACTION (CPE)

Introduction

When a micellar solution of a non-ionic surfactant is heated, it become turbid over a narrow temperature range, which is referred to as its cloud point temperature [58]. Above the cloud point temperature, such solutions separate into two isotropic phases, one of which will contain a surfactant-rich phase with a small amount of water, "surfactant rich phase", which is separated from the bulk aqueous solution, "aqueous phase", in which the surfactant concentration will be approximately equal to the critical micelle concentration of the non-ionic surfactant present. The phenomenon is reversible and, upon cooling, a single phase is again obtained. Table 4 shows the main characteristics of the non-ionic surfactants most used in CPE.

Table 4. Characteristics of non-ionic surfactants [from ref. 54]

Surfactant	Commercial name	CMC[a]	Cloud point (°C)[b]
Polyoxyethylene 4 lauryl ether	Brij 30 ($C_{12}E_4$)	0.064	2.0
Polyoxyethylene 6 lauryl ether	$C_{12}E_6$	0.068	50.3
Polyoxyethylene 9 lauryl ether	Polidocanol ($C_{12}E_9$)	0.069	86.6
Polyoxyethylene 10 lauryl ether	Pole ($C_{12}E_{10}$)	0.090	77.0
Polyoxyethylene 8 dodecanol	Genapol X80 ($C_{12}E_8$)	0.050	75.0
Polyoxyethylene 7.5 nonylphenyl ether	Ponpe 7.5	0.085	5.0
Polyoxyethylene 9.5 octylphenyl ether	Triton X-100	0.240	64.0
Polyoxyethylene 10 cetyl ether	(Brij 56) $C_{16}E_{10}$	0.001	69.0
Polyoxyethylene 10 stearyl ether	Brij 97 ($C_{18}E_{10}$)	0.002	72.0

a) Critical micellar concentration (CMC) values given in mM/L.

b) Values given are the critical temperatures for micellar solutions in which the surfactants concentration is 1% (w/v).

The solubilisation of organic material in micelles and subsequent cloud point extraction technique thus offers a convenient alternative to conventional liquid-liquid extraction that uses organic solvents. The extraction and preconcentation of organic compounds by CPE prior to their analysis by gas or liquid chromatography has been used for different authors [51-53].

The cloud point temperature depends on the structure of the surfactant and on its concentration. For a homologous series of polyoxyethylene non-ionic surfactant, the cloud point temperature increases with decreasing length of the hydrocarbon chain or increasing length of the ethylene oxide moiety. This temperature can be modified by the presence of salts, alkalis, acids, polymers, urea and other surfactants [58]. The ionic strength can alter the cloud point temperature as well as facilitate the separation of the two phases by altering the density of the bulk aqueous phase. It has been shown, for example, that for Genapol X-080, the cloud point temperature obtained when NaCl is added in a 5% w/v concentration, is lesser than the temperature obtained in the absence of this salt [59].

In CPE, extraction needs to be carried out under conditions in which the preconcentration factor will be the maximum or the extraction yield will be 100%. The preconcentration factor is defined by the expression $F_c = C_s/C_{in}$, where C_s is the analyte concentration in the surfactant-rich phase after phase separation, and C_{in} is the analyte concentration in the initial solution, before the preconcentration step. This depends on the phase relationship, on the distribution constant of the analyte between the phase and on the surfactant concentration used.

Moreover, the ratio between the volume of aqueous solution to be preconcentrated and the surfactant-rich phase volume (V_w/V_s) increases with the decrease in the surfactant concentration. This shows that the smaller the surfactant concentration, the higher the preconcentration factor; but when the volume of surfactant-rich phase is small, the extraction process become more difficult, and the accuracy and reproducibility probably suffer [29,60]. In this way, a compromise must be reached so that the surfactant concentration will allow a high phase ratio and a manageable volume of surfactant-rich phase.

The extraction process can be also altered by different factors such as equilibration time, pH, concentration and nature of the surfactant and addition of salt [61]. The effect of these factors on the percentage extraction of the analytes studied therefore needs to be established.

Optimization of Variables

Surfactant Concentration

The effect of the surfactant concentration is related with the surfactant capacity for the extraction of target pesticides [62]. It depends on the kind of surfactant used and on the concentration of pesticides in the sample. The range of surfactant concentration can vary between 1-10% (v/v) [56,62]. As an example, for a mixture of organophosphorus pesticides, it was determined by measuring a series of aqueous solutions containing the mixture of target pesticides (listed in Table 5) and different surfactant concentrations ranging from 2 to 7% (v/v) for each of the surfactants under study, POLE and Genapol (X-080).

Table 5. Organophosphorus Pesticides Studied

Identification Number	*Compound*
1	Dimethoate
2	Methidation
3	Parathion methyl
4	Malathion
5	Ethoprophos
6	Parathion ethyl
7	Diazinon
8	Chlorpyrifos

Figure 6 shows how the recovery percentages with POLE remain practically constant up to 4% (v/v), from which point onwards they drop sharply, before stabilising again round 5% (v/v). This indicates that pesticides are not extracted as efficiently at high surfactant concentrations. The Genapol X-080 concentration can therefore be considered practically not to influence on the extraction of these analytes in the range studied.

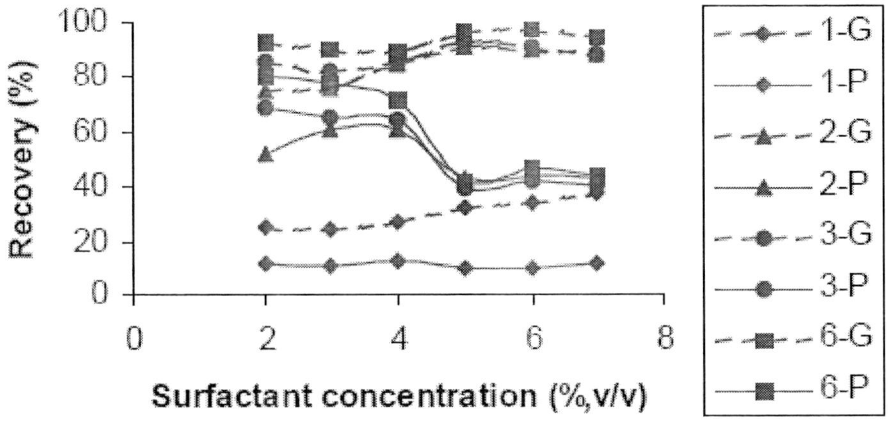

Figure 6. Effect of surfactant concentration on recovery. Dashed line, Genapol X-080 (G); continuos line, POLE (P). CPE conditions: 95°C, 15 min. Identification number in Table 5. [from ref. 56].

In view of these results and taking into account that the surfactant concentration has a proportionally inverse impact on the preconcentration factor [63], the lowest possible surfactant concentrations must be used, but within a range that allowed the samples to be adequately handled.

Equilibration Time

The recovery percentage depends on the time that the analytes have to interact with the micelles and get into their core. It has been reported that longer equilibration times (more than 30 min) do not have any significant effect on the extraction parameters and that an equilibration time of 20 min is sufficient to obtain a good extraction [59,60]. It has likewise been demonstrated that the analyte preconcentration factor and percent recovery in CPE process increases as the equilibration temperature for phase separation is progressively increased above that of the cloud point temperature of the system [64]. Likewise, an increase in the cloud point temperature leads to a slight decrease in the volume of the surfactant-rich phase [61]. As the temperature, or the equilibration time, increases, the amount of water in a surfactant-rich phase decreases and hence the volume of that phase decreases. This can be interpreted in terms of the fact that as the temperature is increased, the hydrogen bonds are disrupted and dehydratation occurs.

Equilibration time studies normally range between 5 and 25 min, higher extraction times than 20 min generally don't produce better recoveries.

Figure 7 shows the results obtain for the optimization of this variable on the recovery of a mixture of organophosphorus pesticides. For this case, this variable practically does not affect the recovery of the analytes, in the case of POLE and Genapol X-080 surfactants [56].

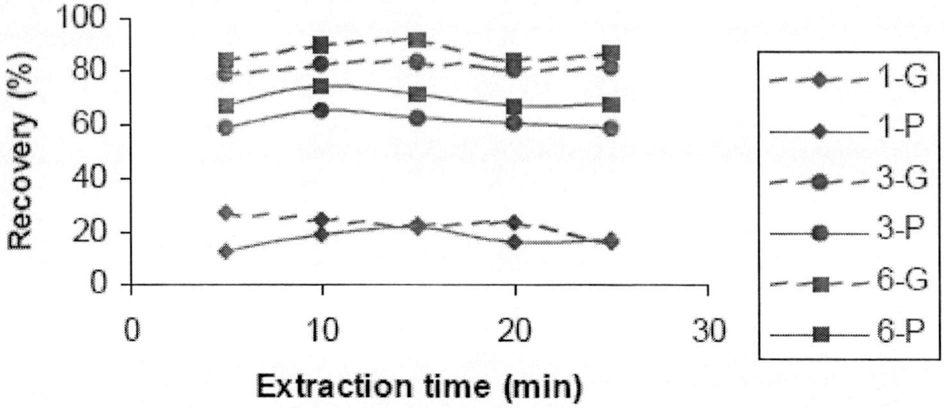

Figure 7. Effect of the extraction time on the recovery when using 2% (v/v) Genapol X-080 and 3% (v/v) POLE with 3 and 4% (w/v) of NaCl, respectively. Dashed line, Genapol X-080 (G); continous line, POLE (P). CPE conditions: 95°C, 15 min. Identification number in Table 5. [from ref. 56].

pH

The ionic form of a neutral molecule normally does not interact with the micellar aggregate as strongly as its neutral form does and a lesser amount of the analyte is therefore extracted [61, 65]. Thus, the pH should be adjusted to ensure that the neutral molecular form of the analytes is present prior to performing the CPE step. The influence of the pH on the recovery percentages is not important for those compounds that do not present ionic forms, such as organochlorine or organophosphorus pesticides. However, in the case of the phenolic derivatives (can appear as degradation products of some pesticides like 2,4-D), when working with pH modified by the addition of 1% acetic acid, the percentage extraction is better for most solutes and is the most noticeable change for more polar compounds [59].

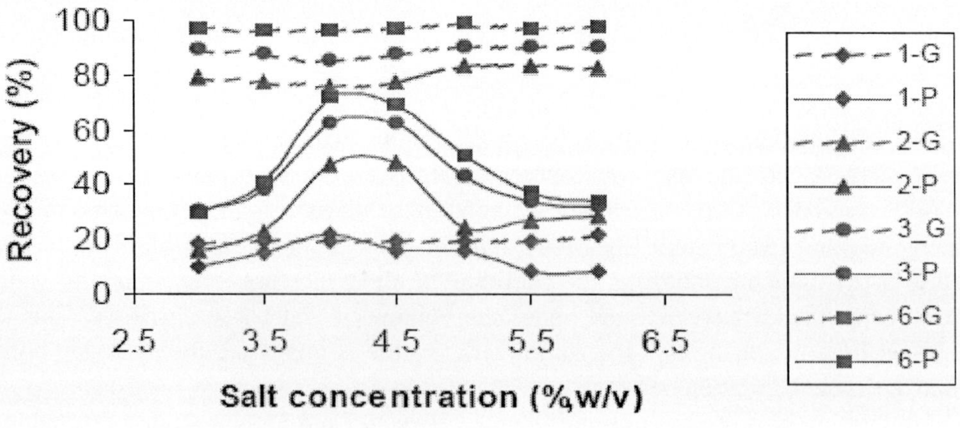

Figure 8. Effect of salt addition on the recovery when using 2% (v/v) Genapol X-080 and 3% (v/v) POLE. Dashed line, Genapol X-080 (G); continous line, POLE (P). CPE conditions: 95°C, 15 min. Identification number in Table 5. [from ref. 56].

Ionic Strength

The addition of an inert salt can facilitate the phase separation process for some non-ionic surfactant systems since it increases the density of the bulk aqueous phase [58,66]. When the salt concentration is increased, the micelle size and the aggregation number are increased and the critical micellar concentration remains constant [67]. In addition, non-polar analytes may become less soluble in the solution at higher salt concentrations and thus contribute to higher recoveries. The results obtained indicate that the addition of salt produces an increase in the extraction of the more polar solutes while the recoveries of the less polar compounds is not affected [65,68].

In the case of organophosphorus pesticides the maximum extraction for surfactant POLE is obtained with 4% (w/v) NaCl, while with Genapol X-80 practically this parameter does not affect on the recovery. It can be observed in Figure 8.

Analytical Parameters and Applications

Most analytical applications of CPE methodology for the extraction of organic compounds make use of reversed-phase high performance liquid chromatography (HPLC). The surfactant-rich phase obtained in the extraction process is compatible with the hydro-organic phase usually employed in this chromatographic mode. From this, another important step is the optimization of the chromatographic conditions. There are two important factors to be taken into account in liquid chromatography separations: the analysis time and a good separation of the analytes.

The CPE methodology has been successfully exploited for the preconcentration of pesticides as a step prior to their determination by HPLC using non-ionic surfactants [56]. Figure 9, shows a chromatogram obtained for the mixture of the organophosphrus pesticides in Genapol X-080 solution, after CPE process was applied.

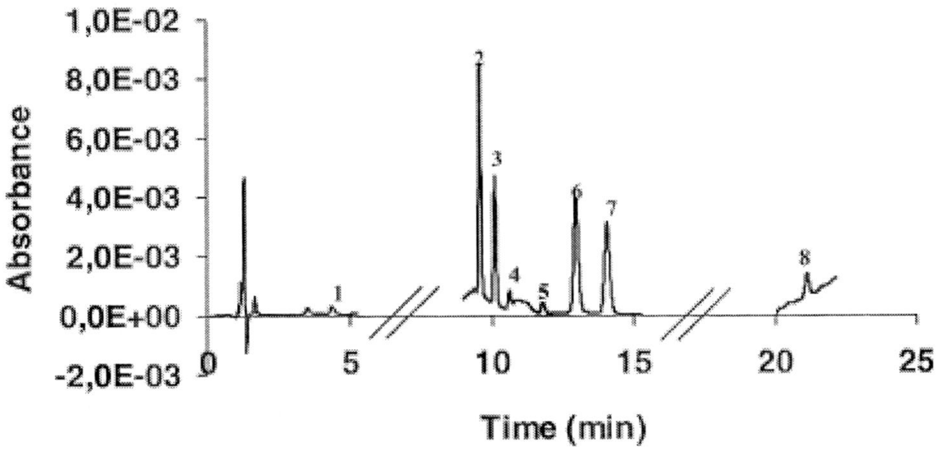

Figure 9. Chromatogram of the organophosphorus pesticides in the presence of 5% (v/v) Genapol X-080. Pesticide concentration: 2mg L-1; Detection wavelength: 247 nm. Identification number in Table 5. [from ref. 56].

Relative standard deviation values are listed in Table 6, where values ≤5.5 and 2.6% were obtained for POLE and Genapol X-080 respectively, in the case of organophosphorus pesticides extraction. The limits of detection were also calculated for each analyte using a signal-to-noise ratio (s/n = 3) [69] once the CPE method was fully applied. The results obtained are also listed in Table 6.

Table 6. Analytical parameters for the cloud point extraction method of organophosphorous pesticides [from ref. 56]

Compounds	POLE		Genapol X-80	
	R.S.D[a] (%)	L.O.D.[b]($ngml^{-1}$)	R.S.D[a] (%)	L.O.D.[b]($ngml^{-1}$)
Dimethoate	1.3	1.86	0.6	0.61
Methidathion	2.5	1.65	0.8	1.37
Parathion methyl	3.2	1.86	2.1	0.03
Malation	0.5	28.45	0.7	1.37
Ethoprophos	5.5	0.88	1.4	1.84
Parathion ethyl	1.6	2.03	2.6	0.36
Diazinon	2.8	2.96	1.0	0.47
Chlorpyrifos	2.4	3.54	2.2	2.81

a) Relative standard desviation (n = 6).
b) Limit of detection (s/n = 3).

In order to corroborate its efficiency, the method was applied to a certified standard containing the pesticides: Parathion-methyl, Parathion-ethyl, Malathion and Diazinon (Pesticide Mix 1, Dr. Ehrenstorfer). The results (Table 7) show the satisfactory recoveries obtained using the optimised CPE procedure.

Table 7. Recoveries obtained on the application of the CPE method using Genapol X-080 to a certified standard pesticide mixture [from ref. 56]

Compound	Recovery (%)[a]	
	Added amount (10 $ng{\cdot}ml^{-1}$)	Added amount (100 $ng{\cdot}ml^{-1}$)
Parathion methyl	81.7 ± 2.3	80.5 ± 1.1
Malation	75.0 ± 0.0	66.0 ± 0.7
Parathion ethyl	80.4 ± 1.6	84.5 ± 1.1

a) Mean ± S.D. (n = 3)

The proposed method can be applied to real samples for the determination of pesticide's concentration in them. As example, it was applied to various spiked aqueous samples from different locations in Gran Canaria island (Canary Islands, Spain). Samples of depurated waste water and groundwater were employed in this study. After checking the absence of pesticides in the samples, they were spiked with the analytes under study (6–30 $ng.ml^{-1}$) and the established method with Genapol X-080 was applied to them. The results obtained for all the samples analysed are set out in Table 8. Recoveries between 70 and 100% were obtained for the majority of cases. Only Dimethoate and Ethoprophos were extracted with recoveries <50%.

Table 8. Recoveries obtained on the application of the CPE method using Genapol X-080 to pesticides added to real aqueous samples [from ref. 56]

Compound	Added amount (ng ml^{-1})	RECOVERY (%)^{A}	
		Depurated waste water	Groundwater
Dimethoate	10	27 ± 0.5	29 ± 1.5
Methidathion	10	86 ± 2.2	85 ± 3.1
Parathion methyl	6	92 ± 2.0	83 ± 1.2
Malation	10	87 ± 1.1	81 ± 0.6
Ethoprophos	30	43 ± 1.8	47 ± 2.3
Parathion ethyl	6	105 ± 1.2	100 ± 1.6
Diazinon	6	89 ± 1.0	84 ± 0.0
Chlorpyrifos	30	99 ± 3.6	99 ± 5.0

a) Mean ± S.D. (n = 3)

Determination of Pesticides by Using Microwave Assisted Micellar Extraction (MAME)

Introduction

In recent years, the use of microwave technology for sample preparation has been widely accepted. The reasons for the large dissemination of microwave-assisted extraction (MAE) relate to its clear advantages over more traditional technology:

- a shorter digestion time
- a supposed better recovery of volatile elements and compounds
- lower contamination levels
- minimal volumes of reagent being required
- more reproducible procedures
- better working environment

Ganzler et al. [70] were the first researchers ever to use a microwave oven to extract organic compounds from contamined soil. Since then microwave digestion methods have been developed for different types of matrices to extract organic contaminants, such as PAHs [23, 71], PCBs [72,73], pesticides [73,74], phenols [75, 76], etc. using an organic solvent as extractant.

It has been reported that the combination of MAE with micellar media as extractants (MAME) allows the extraction of different organic compounds from solid samples [77-80]. The extraction process starts with the loading of the sample into the extraction vessel, followed by surfactant addition and closing of the vessel. Microwave radiation is applied and an extraction step is initiated in order to heat the extractant to set values. When the extraction is concluded the samples are allowed to cool down to a temperature that is reasonable to handle. Prior to the analysis, a clean-up step might be needed. In most applications, simple filtration of the extract using a membrane syringe filter is enough.

Optimization of MAME Process

There are different parameters that can influence the extraction process when MAME is used. The most commonly studied parameters are: nature and volume of surfactant, extraction temperature or extraction power and duration of extraction. Factorial, central composite and orthogonal array can be used to find optimal conditions.

Nature and Volume of Surfactant

A correct choice of kind of surfactant is fundamental for obtaining an optimal extraction process. When selecting the extractant, consideration should be given to its interaction with the matrix, as well as the solubility of the analyte. Another important aspect is the compatibility of the extraction surfactant with the analytical method used for the final analysis step. The non-ionic surfactant solutions should be effective in desorbing and extracting organic compounds from solid matrices due to their power of solubilisation. In addition, the surfactant phase is compatible with the hydro-organic mobile phases usually employed in HPLC.

In some cases, extractant solvent volume may be an important parameter for efficient extraction. The surfactant volume must be sufficient to ensure that the entire sample is immersed, especially when having a matrix that will swell during the extraction process. Research has led to the conclusion that the proportion of sample in the extraction solution should not exceed 30-34% (w/v) sediment/solvent ratio. In conventional extraction techniques, a higher volume of solvent will generally increase the recovery, but in the case of MAME, a higher surfactant volume does not influence the extraction efficiency. Therefore, if the sample is full immersed in the extractant, an increase of the surfactant volume does not seem to have any influence on the extraction percentages [77,81]. Normally at extraction volumes lower than 5 ml irreproducible data are obtained (due to the insufficient covering of the extractant) and at volumes higher than 20 ml some evaporation losses take place (due to the high temperatures reached). This last effect maybe due to the high capacity of the aqueous surfactant solutions in absorbing the microwave radiation and transforming it into heat [82].

As example, for MAME extraction of organophosphorus pesticides using POLE and Genapol X-080 surfactants, the optimum extraction volume is around 10 ml [78] and for organochlorine pesticides using POLE, 8 ml [79], a very similar value in both cases.

Surfactant Concentration

Surfactant concentration is one of the most important parameters that needs to be optimized in MAME process. Surfactant concentration have to secure that the amount of micelles is enough for a efficient interaction with analytes and a correct solubilization, but it should be not too high due to possible viscosity problems.

Optimum surfactant concentration can be determined using sequential optimization steps or applying factorial designs.

For example, to study the effect of surfactant concentration on the recovery percentage of organophosphorus pesticides, several samples containing different POLE and Genapol X-080 concentrations, 1, 3, 5 and 7% (v/v), were analyzed using sequential mode. The pesticides from enriched samples were extracted with both surfactant solutions in the microwave system at 300 W during 5 min [78]. The recoveries obtained for some of the compounds studied can be seen in Figure 10.

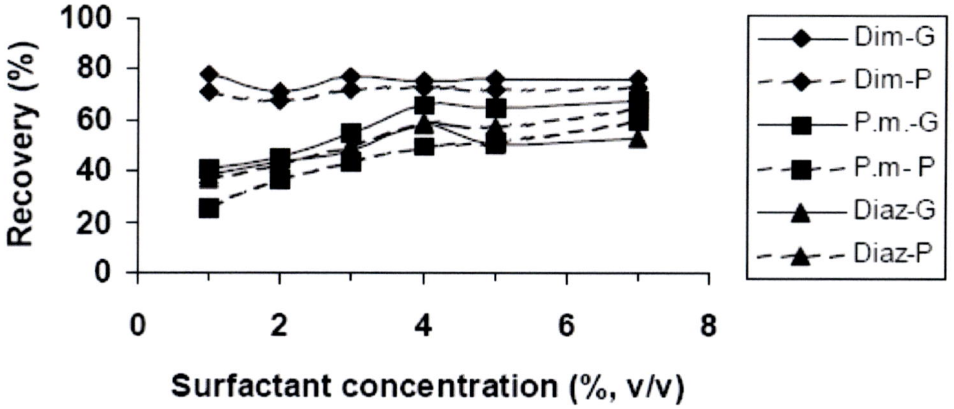

Figure 10. Effect of surfactant concentration on the recovery of the pesticides under study. Continous line (—): Genapol X-080 (G); dashed line (- - -): POLE (P). Dim: Dimethoate, P.m.:parathion-methyl and Diaz:diazinon. [from ref. 78]

For the extraction of organochlorine pesticides, the results of the experiment design applied and the correlation between variables with the recovery and each other show that the surfactant concentration is the variable which mostly affect to the recovery and is correlated with the microwave power. The results obtained for the case of 4,4'-DDT using a mixture of surfactants POLE and Polyoxyethylene 10 Stearyl Ether (Stearyl mixture) are shown in Figure 11 [79].

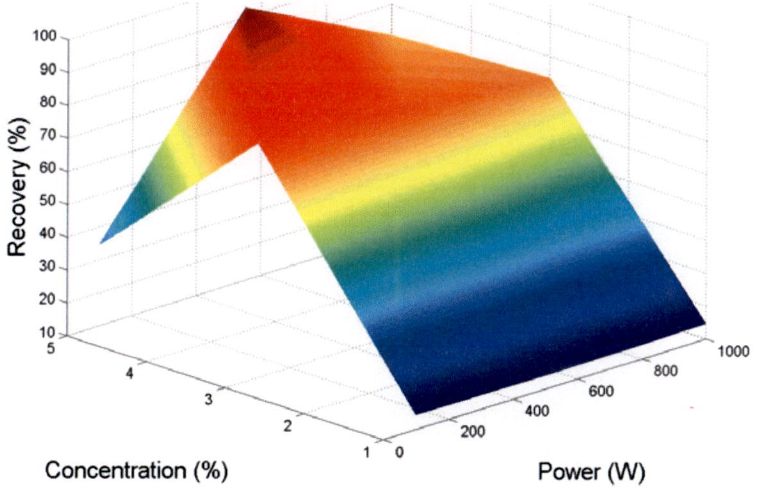

Figure 11. Response surface for the effect of surfactant concentration and power on the extraction of 4,4'-DDT using Stearyl mixture. [from ref. 79].

Extraction Temperature and Extraction Power

The most researched MAME parameter is the extraction temperature or extraction power, as an increase in the power involves a corresponding increase in heat. When MAME is conducted in closed vessels, the temperature may rise well above the boiling point of the solvent at atmospheric pressure, thus enhancing both extraction speed and efficiency. These elevated temperatures result in improved extraction efficiencies, since desertion of analytes from active sites in the matrix will increase. However, the effect of the temperature depend on the type of analyte as well as the type of matrix.

As in other extraction techniques, time is another parameter whose influence needs to be taken in account. Extraction times in MAME are very short compared to conventional techniques where extraction procedures require as much as 24 hours. 30 minutes are often sufficient to obtain good results in MAME. With thermolabile compounds, long extraction times may result in degradation, obtaining lower recovery percentages when increase extraction time [80].

A central composite design can be used in order to study the effect on the recovery, it allows the direct evaluation of the considered variables [83]. The values normally vary between 200 to 800 W during 2 to 14 min [78]. The optimization of both variables can be done using a 3^2 factorial design, where the recovery is the dependent variable. The experiment involves 10 run by duplicate and other variables involved in the extraction process were kept constant in their optimized values.

Figure 12a shows the response surface for 4,4'-DDT using Cetyl mixture (a mixture of POLE and Polyoxyethylene 10 Cetyl ether) as extractant, where we can observe that from 750 W for microwave power and during 14 minutes the best recoveries are obtained; for Stearyl mixture (Figure 12b) the optimum conditions are 775 W and 8 minutes.

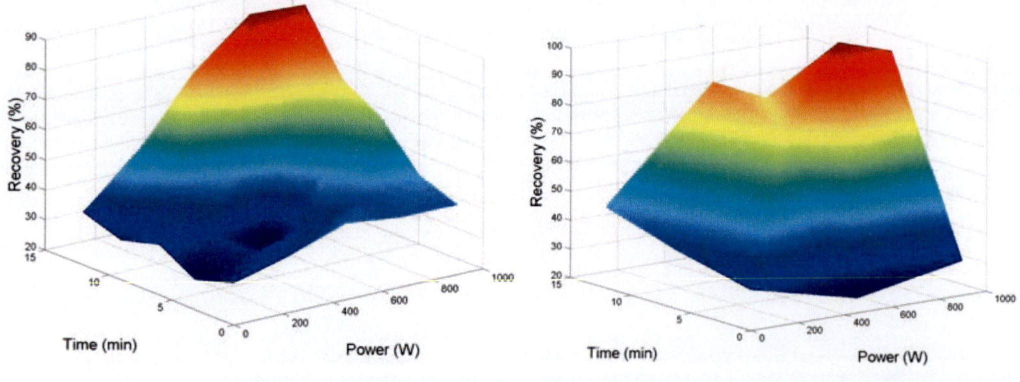

Figure 12. Response surface for the effect of power and time on the extraction of 4,4'-DDT using Cetyl mixture (a-left) and Stearyl mixture (b-right) [from ref. 79].

Matrix Characteristic

The nature of the matrix in which the analytes of interest are bound can have a profound effect on the recoveries of the compounds. The different interactions that the organic pollutants present with the organic and inorganic matter of the matrix can play different roles in their solubility or diffusivity into the micellar medium.

The pH, conductivity and organic content of the sample can vary the recoveries obtained. Moreover, the texture of a soil is extremely important in the sorption process. When the particles are small they present a high superficial area thus increasing the adsorption, although this adsorption process depends on the nature of the analytes. In this sense, a high variability can be observed in the recoveries obtained with the different granulometry of the soil samples. In general, it can be said that the proposed method is applicable in the sense that it enables the extraction of the pesticides under study even if the recovery efficiency depends on the characteristics of the soil as well as the surfactant to be used.

Appling the method to different soil samples containing a mixture of organochlorine pesticides can be observed that recoveries vary depending on the soil characteristics (Table 9) and the surfactant used as extractant [79]. The results obtained are shown in Tables 10a and 10b.

Table 9. Soil characteristics (from ref. [79])

Soil	Agricultural Uses	Particle size (%) 0.3 mm	0.2 mm	0.15mm	≤0.1 mm	pH	Organic matter (%)	Conductivity (μS/cm)
Tafira	Garden	44.4	17.2	14.6	23.8	8.3	4.8	292
San Roque	Bananas	52.1	20.0	13.2	14.6	7.86	12.54	3266
Sta Brigida	Pine Forest	56.5	17.8	12.8	12.8	5.9	3.9	483
Valleseco I	Potatoes	36.33	16.43	15.19	32.06	4.84	4.4	202
Valleseco II	Potatoes	25.16	13.17	14.38	47.29	3.91	6.21	100

Table 10a. Recoveries obtained after MAME procedure for organochlorine pesticides in five different soils using Cetyl mixture*. [from ref. 79]

Soil	4,4'-DDD	Dieldrin	4,4'-DDT	2,4'-DDT	4,4'-DDE	Aldrin
Tafira	99.1	66.2	67.7	58.6	65.0	63.0
San Roque	69.4	44.7	78.0	59.4	67.1	52.6
Sta Brigida	96.8	80.7	84.0	75.1	73.1	63.8
Valleseco I	78.3	73.3	76.5	66.3	65.3	64.4
Valleseco II	65.2	72.3	61.7	59.4	56.1	60.2

* Mean of three determinations.

Table 10b. Recoveries obtained after MAME procedure for organochlorine pesticides in five different soils using Stearyl mixture*. [from ref. 79]

Soil	4,4'-DDD	Dieldrin	4,4'-DDT	2,4'-DDT	4,4'-DDE	Aldrin
Tafira	103.6	78.6	101.8	62.8	71.8	79.4
San Roque	101.3	46.1	94.4	86.2	76.0	80.0
Sta Brigida	84.7	80.8	78.8	62.4	66.0	65.6
Valleseco I	55.2	56.3	100.7	83.5	44.8	57.5
Valleseco II	54.3	50.2	98.3	68.4	49.2	72.8

* Mean of three determinations.

The soils with high organic matter have good recoveries because the surfactant can also extract humic substances which are linked to the pesticides [84]. These recoveries decreases when the temperature is too high (observed in the most extremely conditions in the experimental design).

As we commented before, texture of a soil is quite important in the sorption process [85]. Clay is by far the most adsorbent of the three main soil textures (clay, silt and sand) due to its small particle size, high surface area, and high surface charge [86]. This has a relevant effect on the recovery; the higher clay percent a soil has, the lower the recovery is so; in this case, the Valleseco II soil has a higher amount of particles with smaller size and therefore, lower recoveries are obtained.

The effect of ageing the samples on the recovery of the analytes is a known phenomenon [87]. Decreasing recoveries resulting from aging of matrices is frecuent [88,89]. The analytes present in recent soil samples are more easily extracted than those that have had a longer contant time. This can be explained according to whether the analytes are incorporated by adsorption (short periods) [90] or by sequestration (long periods). The former phenomenon occurs at the early stages of sorption, where H-bonding and Van der Waals forces prevail. On the other hand, sequestration involves sorption at remote microsites within the soil matrix [91]. As the contact time increases between the analyte and the matrix the general recuperation of all the analytes under study was reduced which could be explained by the greater interaction between them [92]. This effect could also be caused by a process of degradation because as is well known the degradability of this family of compounds is far greater than that of organochlorine pesticides. Among the different pathways of organophosphates decomposition (hydrolysis, photolytic oxidation, microbial transformations, etc.) [93,94], hydrolysis is the most common degradation process which can occur at several reactive centres in a given organophosphorus pesticide molecule [95], but also can occur different degradation mechanisms including the presence of humic acids, metal oxides and the soil structure [94-97].

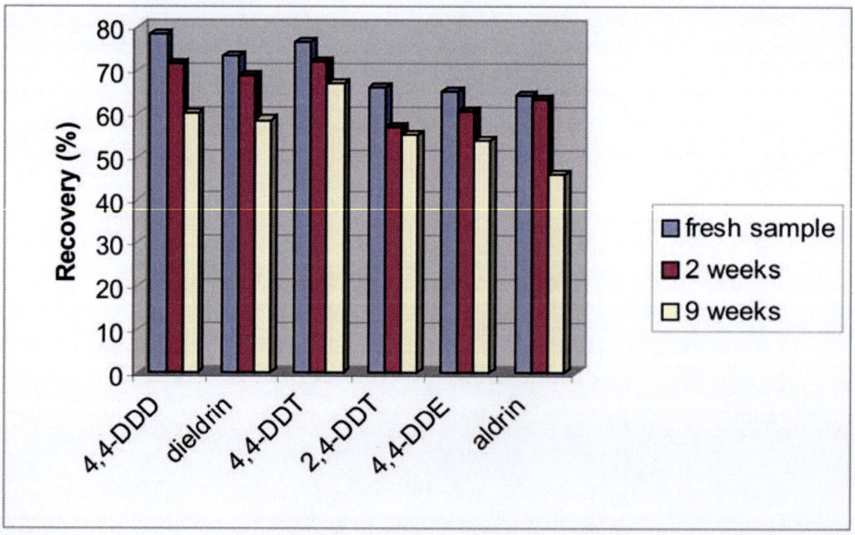

Figure 13 (Continued).

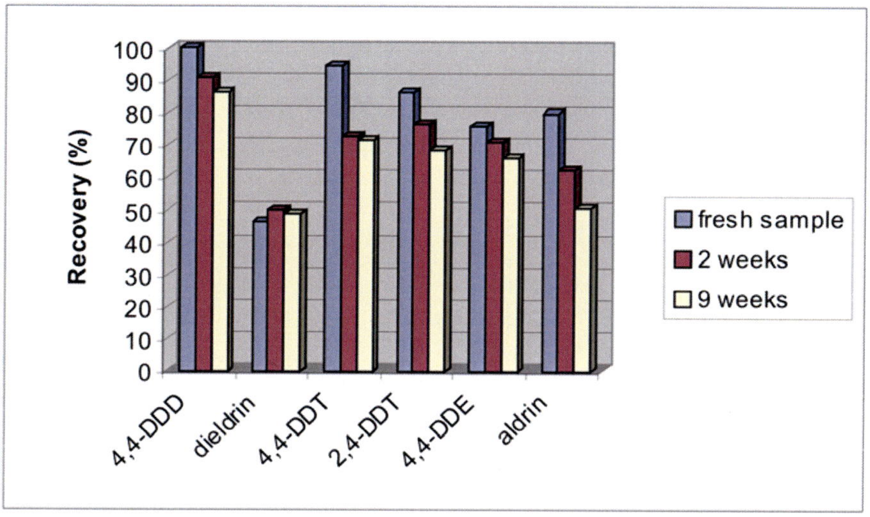

Figure 13. Recoveries obtained after MAME procedure for organochlorine pesticides for three different aging times for Valleseco I soil using Cetyl mixture (a-left) and San Roque soil using Stearyl mixture (b-right) [from ref. 79].

In the case of the extraction of organochlorine pesticides, MAME procedure was applied to four soils for different time periods after conditioning. Figure 13a shows the recoveries obtained for the different pesticides in the case of Valleseco I soil using Cetyl mixture with the time aging and Figure 13b shows the results obtained in the case of San Roque soil using Stearyl mixture with the same aging time.

It can be seen that the recoveries decrease with the aging time for all compounds. Only in the case of dieldrin using Stearyl mixture, the recovery remains practically constant with the time. Moreover, none of the metabolites of DDT increases with the time then we can assert that degrading processes were not happen in these aging time. These results indicate that the developed method leads determine the presence of pesticides in real samples after a contact time and show the viability of the method as alternative to the conventional ones.

Analytical Parameters and Validation

Separation and detection of pesticides have been developed using MAME procedure coupled with HPLC. The calibration curves can be obtained by injecting standard solutions containing a known concentration of the pesticide mixture and the surfactant chosen into the chromatographic system. This method provides optimum detection limits and reproducibility in soil matrices. For organophosphorus pesticides, on the application of this process a linear relationship is obtained in the interval 100–2500 ng ml^{-1}, when POLE or Genapol X-80 are used as extractants, with high correlation coefficients (0.999) for all pesticides.

Relative standard deviation lower than 2.1 and 2.6% for Genapol-X080 and POLE repectively and LODs lower than 0.2 in both cases were obtained (Table 11). Figure 14 shows the chromatogram of the pesticides mixture extracted from a soil sample under the optimized conditions for POLE.

Table 11. Analytical parameters for the determination of organophosphorus pesticides under study using MAME procedure [from ref. 78]

	POLE		*Genapol X-080*	
Compound	RSD[a] (%)	LOD[b] (ng·ml^{-1})	RSD[a] (%)	LOD[b] (ng·ml^{-1})
Dimethoate	0.4	0.2	1.8	2.4
Methidathion	2.2	4.5	2.1	3.0
Parathion/methyl	1.1	2.1	1.2	4.8
Malathion	2.2	14.3	1.1	11.3
Ethoprophos	2.6	95.0	1.0	0.8
Parathion-ethyl	1.6	0.7	1.6	1.8
Diazinon	1.2	2.5	0.8	0.9
Chlorpyrifos	0.3	6.7	1.1	0.2

a) Relative standard deviation (n = 6).

b) Limit of detection.

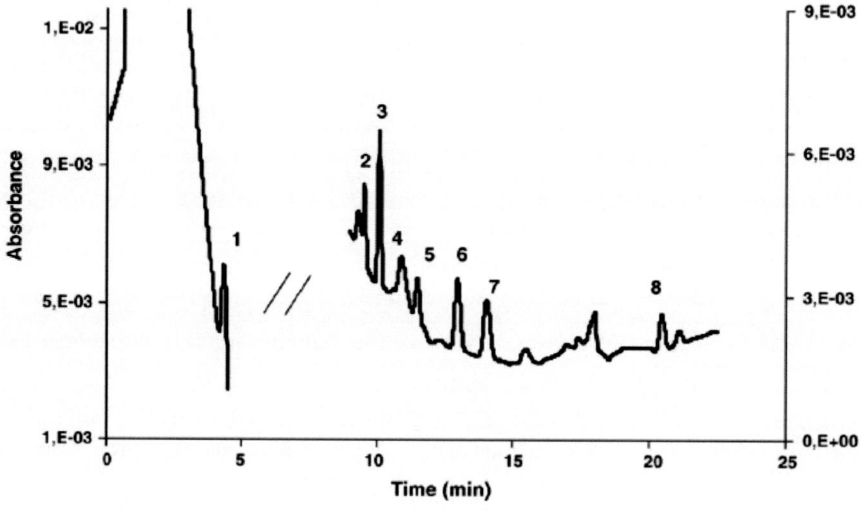

Figure 14. Chromatogram of the organophosphorous pesticide mixture extracted from soil sample n° 1 with POLE at the optimized conditions. Identification number in Table 5 [from ref. 78].

In the case of organochlorine pesticides, a good success was obtained appling MAME-HPLC method with POLE-Cetyl and POLE-Stearyl mixtures as extractants, with RSDs values between 5.2-8.4% for POLE-Cetyl and 5.2-9.7 for POLE-Stearyl and LODs of 86.4-806.4 ng·ml^{-1} and 150.8-734.0 ng·ml^{-1}, respectively [79]. In Figure 15 we can observe the chromatogram obtained for the mixture of organochlorine pesticides extracted of a spiked soil using Steary mixture as extractant. Recoveries obtained with spiked compounds may not be representative of those found with native compounds. Spiked analytes are generally lightly coated on the surface of the matrix whereas native ones can be strongly adsorbed inside the porous matrix. This can be explained by the diffusional and the kinetic limitations of the sorption process, and the several interactions which may have been simultaneously established between native analyte and the matrix [98]. This is the reason why it is necessary to validate the extraction procedure with certified reference matrices.

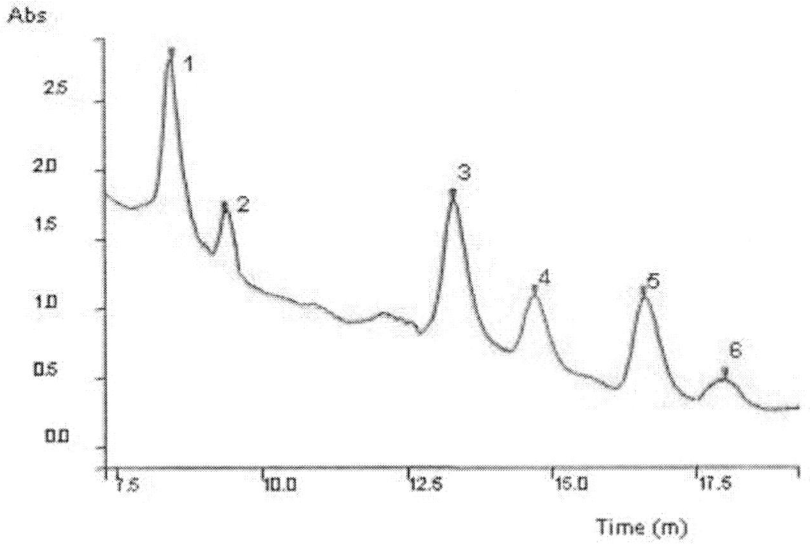

Figure 15. Chromatogram of an extract of a spiked Tafira soil after MAME procedure. (1- 4,4'-DDD; 2- Dieldrin; 3- 4,4'-DDT; 4- 2,4'-DDT; 5- 4,4'-DDE; 6- Aldrin) [from ref. 79].

Due to the lack of a certified material available containing target organophosphorus pesticides, the developed extraction method was applied in a soil sample enriched with a certified mixture, Pesticide Mix 1, EPA Method 914, and the results obtained are shown in Table 12.These results were compared with those found using the traditional Soxhlet extraction procedure as proposed by the EPA in the 3540 Method [82], finding an important similarity of results in both methods.

Table 12. Application of MAME procedure and Soxhlet extraction to a soil sample containing a certified mixture of organophosphorous pesticides, (Pesticide Mix 1, EPA Method 914)* [from ref. 78]

	MAME		*Soxhlet*	
Compound	Added ($\mu g\ g^{-1}$)	Found ($\mu g\ g^{-1}$)	Added ($\mu g\ g^{-1}$)	Found ($\mu g\ g^{-1}$)
Parathion/methyl	1.50	1.39 ± 0.02	2.00	2.04 ± 0.02
Malathion	1.50	1.27 ± 0.04	2.00	1.89 ± 0.11
Parathion-ethyl	1.50	1.49 ± 0.12	2.00	1.85 ± 0.02
Diazinon	1.50	1.46 ± 0.02	2.00	1.64 ± 0.05

*) Mean of three determinations.

For organochlorine pesticides, a contaminated certified reference soil from an agricultural region of the Western United States (CRM804-050) was used. Table 13 reports the pesticides present in the certified soil, the corresponding certificated values, the confidence intervals and the concentration of each analyte that was obtained using the MAME procedure with Stearyl mixture. The concentrations obtained, with the proposed method, falling within the certified range for all compounds analyzed, indicate that the proposed extraction procedure is suitable for this kind of matrix.

Table 13. Extraction of organochlorine pesticides compounds using MAME with Stearyl mixture in a certified reference soil[a]. [from ref. 79]

Compound	Reference Value[b]	RSD (%)	Confidence Interval	Concentration found
4,4'-DDD	1.531	0.476	1.294 – 1.767	1.733 ± 0.009
Dieldrin	1.863	0.655	1.539 – 2.186	2.242 ± 0.084
4,4'-DDT	1.060	0.275	0.926 – 1.195	1.079 ± 0.015
4,4'-DDE	1.520	0.410	1.325 – 1.715	1.502 ± 0.022

[a]All values are expresed in Rg·g-[1].

[b]The pesticides values in the sample were certified by USEPA SW846 (3rd edition). Extraction Methods 3540A/3541 (soxhlet), 3550 (sonication) and Analysis Method 8081.

Determination of Pesticides by Using Coupling MAME-SPME and MAME-SPE

Coupling MAME with Solid Phase Microextraction (MAME-SPME)

Solid phase microextraction (SPME) is a new variation of extraction techniques which has been used mainly for the analysis of pollutants in environmental water samples [99], but this methodology can be coupled with other techniques, as the microwave extraction, for the analysis of soil samples [100]. SPME allows the simultaneous extraction and preconcentration of analytes from a sample. The principle of SPME is equilibration of the analytes between the sample matrix and the organic polymeric phase usually coating a fused-silica fibre. The amount of the analyte absorbed by the fibre is proportional to the initial concentration. Figure 16 shows the scheme of SPME process.

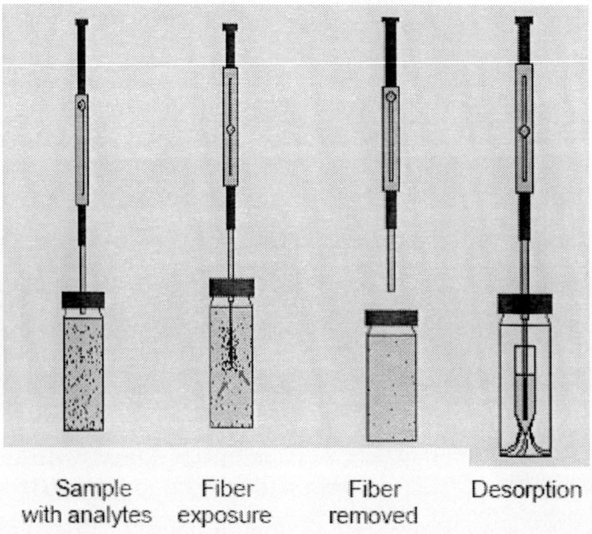

Figure 16. Scheme of the solid phase microextraction (SPME) process.

Normally the application of SPME has been performed in combination with gas chromatography, however, it is possible coupled with HPLC [101]. The advantage using HPLC is that compounds with low volatility or thermal labile compounds can be analysed. An organic solvent (static desorption) or the mobile phase (dynamic mode) is used to disorb the analytes from the SPME fiber.

One application of solid phase microextraction (SPME) procedure can be a clean-up and preconcentration steps applied to extracts obtained with other extraction systems such as microwave assisted micellar extraction (MAME). Using HPLC-UV as determination system, the coupling MAME-SPME has been used for the analysis of five organochlorine pesticides (4,4'-DDD, 4,4'-DDT, 2,4'-DDT, 4,4'-DDE, Dieldrin and Aldrin) in agricultural soil samples [57,102]. A non ionic surfactant, POLE, was used, and the different variables for the optimization of MAME and SPME procedures were studied. This method was successfully applied to the determination of these pesticides in several kinds of agricultural soil samples, obtaining lower detection limits than determining MAME extract directly.

Microwave assisted micellar extraction

Once the soil sample is transferred to the vessel, the optimized volume of surfactant is added and the solid sample is subjected to the MAME process [15, 78]. The vessels are closed for avoiding the loss of extract during the extraction process at microwave power and time optimized; after this, the vessels are allowed to cool first before being opened.

Optimization of SPME Procedure

For MAME-SPME extraction, a fraction of MAME extract is diluted with water and introduced in a vial for the SPME extraction under the optimized conditions [103]. The desorption of the compounds is done in a glass vial with a small volume of desorption solvent (100 μl of methanol) during several minutes.

The efficiency of analytes extraction by SPME can vary widely depending upon the matrix nature, the time period of absorption, temperature, salt addition to the sample and pH [104,105]. Figure 17 shows different variables optimized for the analytes 4,4'-DDT, 4,4'-DDD and 4,4'-DDE as example [57]. One of the most important steps is the determination of the time needed until it reaches equilibrium between the sample matrix and the coating of the fiber. Most of organochlorine pesticides reach the equilibrium after about 40 min; with this time, it is obtained a good reproducibility and an acceptable analysis time (Figure 17a).

Desorption is done normally using static mode by immersion of the fiber into the vial during enough time for producing the efficienct analytes desorption. This time normally is studied between 1 and 15 min, longer times don't give better recoveries. In the studied case, it was proved that a time of 4 min was enough for obtaining a desorption of target analytes (Figure 17b).

Extraction temperature can play an important role on the absorption of analytes, because it influences the mass transfer rates and the partition coefficients of the analytes. A high temperature could increase the diffusion rate of the analytes, however, absorption is generally an exothermic process and the amount of analytes absorbed could decrease with a temperature increase [106]. For that, it is necessary to search the optimum extraction temperature. In the case of organochlorine pesticides, for temperatures lower to 40 °C extraction efficiency is constant, but to higher temperatures, it decrease due to the exothermic process (Figure 17c).

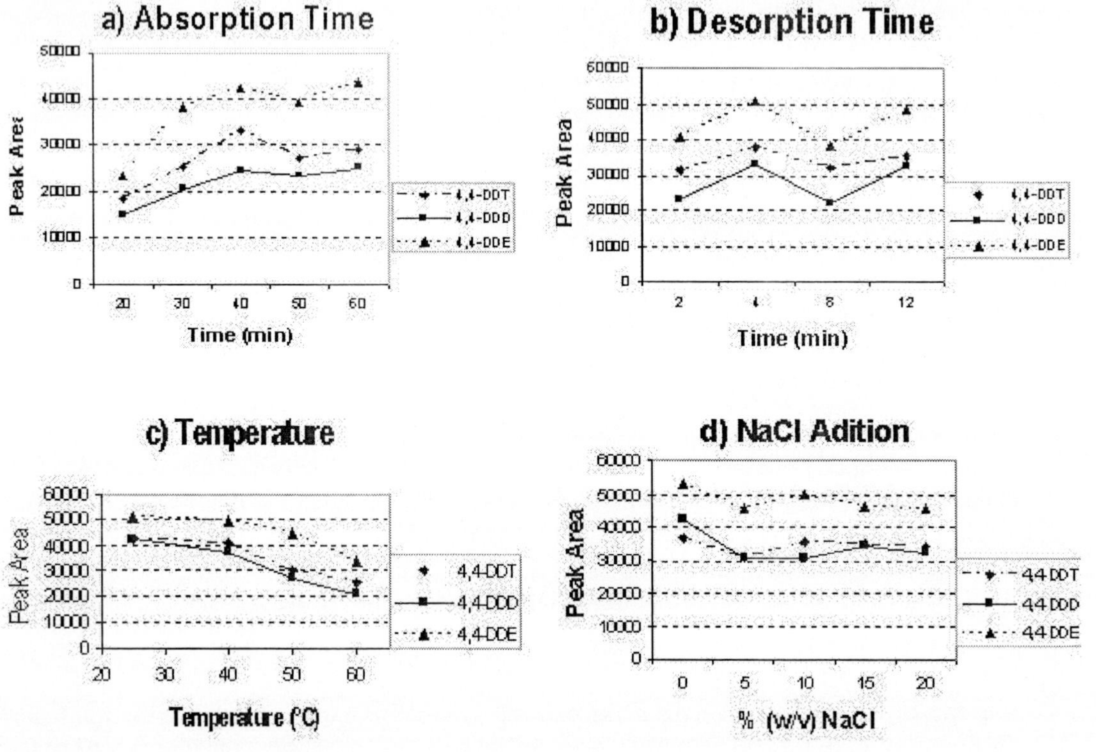

Figure 17. Optimized variables for SPME extraction using POLE coupling with PDMS/DVB fiber. [from ref. 57].

Another extraction parameter which has a well-establish effect in the extraction procedure is the ionic strength of the sample by addition of salt. An increase of ionic strength can improve extraction efficency due to the reduction of the solubility of least polar analytes. However, an increase of salt concentration can generate damage on the fiber by salt deposition. Figure 17d shows the effect of ionic strength on the peak area of target analytes, between 0% and 20% (w/v) NaCl concentrations. It can be seen that for these compounds salt addition doesn't improve the analyte absorption.

Analytical Parameters and Validation

Liquid chromatography with UV detection can be used in the analysis of organochlorine pesticides after their MAME-SPME extraction. Limits of detection reached using this process are in the range 56-96 ng·g^{-1} for all compounds studied and relative standard desviations are lower than 9% in all cases (Table 14).

When this process MAME-SPME is applied to the determination of a mixture of pesticides 4,4'-DDD, 4,4'-DDT, 4,4'-DDE and Dieldrin present in a certified soil sample (CRM804-050), recoveries obtained for all compounds analyzed, falling within the certificated range.

Table 14. Analytical parameters for the determination of organochlorine pesticides using MAME-SPME procedure [from ref. 57]

Compound	RSD (%)*	LOD (ng· g^{-1})
4,4'-DDD	7.4	68
Dieldrin	5.5	96
4,4'DDT	7.3	72
2,4'DDT	8.4	84
4,4'-DDE	7.9	56

* n=6.

Applications

Soil characteristics could also have a great influence on the recovery. However, if we applied the optimized method MAME-SPME-HPLC to the extraction and determination of organochlorine pesticides mixture in different types of agricultural soils, with different characterisitics, extraction efficiency is satisfactory for most compounds in study, independently of the characteristics of the matrix (Figure 18).

This method has been also applied to different mud samples with satisfactories results [107]. Therefore, the versatility and less complex sample's preparation make MAME-SPME a suitable process for solid samples.

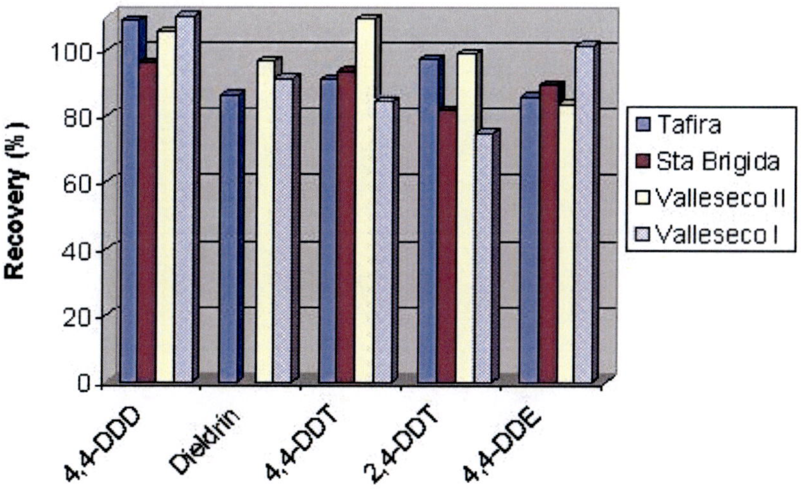

Figure 18. Recoveries obtained after MAME-SPME procedure for organochlorine pesticides for four different soils using POLE as extractant [from ref. 57].

Coupling MAME with Solid Phase Extraction (MAME-SPE)

During extraction step, many interfering components are co-extracted from solid samples together with target analytes which could interfere on the final quantification. Removal of interfering substances can be made by SPME [100] as we studied before, and by SPE [108] process, producing also an intensification of the analytes signal by means of extract clean-up and preconcentration.

With the aim to obtain cleaner extracts and lower detection limits, it can be applied the coupling between MAME procedure and solid phase extraction (SPE). This coupling has a simply handling and allows simultaneous analysis. MAME-SPE procedure has been applied to the extraction and determination of organochlorine pesticides from seaweed [109] and mollusc samples [110] with very satisfactories results.

MAME-SPE extraction can be divided in two steps: microwave assisted micellar extraction (MAME) to obtain MAME extract under the optimum conditions (optimized before) and the application of solid phase extraction (SPE) clean up procedure to this extract.

Solid phase extraction (SPE) is widely used as preparation process based on the selective retention of interest's compounds in a solid sorbent and then eluted with an organic solvent. But also can be coupled with other extraction process as clean-up step (see Figure 19).

In SPE procedure, before the extraction it is necessary conditioning the cartridge. After absorption, a wash step is done for reducing surfactant remains with ultra-pure water. The retained analytes are eluted with methanol. Finally, 50 µl of extract were injected into chromatographic system.

There are different parameters that may influence in the clean-up and preconcentration step such as type of sorbent, sample volume, composition and volume of washing solution and type and volume of desorption solution.

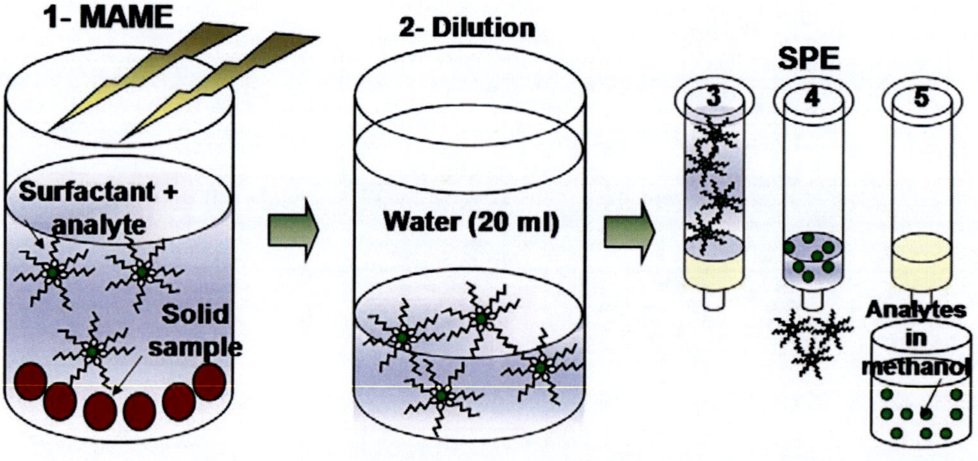

Figure 19. Scheme of the coupling MAME-SPE (1- MAME extract, 2- Extract's Dilution, 3- SPE Absorption, 4- SPE Clean step, 5- SPE Desorption). [from ref. 110].

Optimization of MAME-SPE procedure

It is necessary to optimize and compare different SPE cartridges for obtaining the best extraction efficiencies [111]. Several papers reported on the evaluation of a number of stationary phases that can be used based on compound's characteristics. For example, in the case of organochlorine pesticides, results obtained (as peak area) for different SPE cartridges show that the best recoveries were obtained with C_{18} and Envirelut-Pesticides cartridges. However, the last one was selected because it gives a better baseline in the chromatogram.

For MAME-SPE procedure it is necessary to study the amount and concentration of MAME extract which are going to be use in SPE step [112]. Sample must be diluted with water to facility the pass of the sample through the solid phase and minimize the competition for analytes between surfactant and cartridge sorbent.

Figure 20 shows a comparison between relative peak areas obtained using 5 ml of MAME extract (5%, v/v of surfactant) without and with dilution with different volumes of ultra-pure water. It can be observed that, results were quite similar for each analyte using different final volume samples. However, from a volume of 25 ml the ratio signal-noise obtained in the chromatogram was better and this volume was selected as the optimum for further experiments.

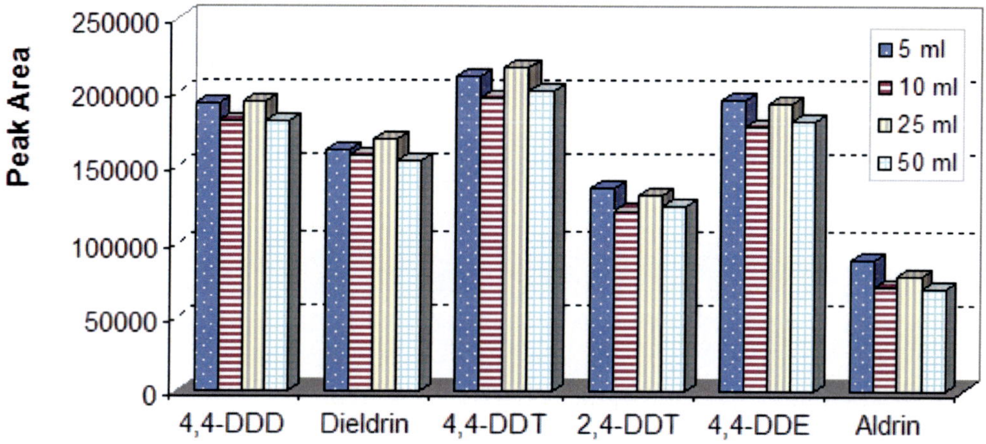

Figure 20. Study of absorption volume, represented as final volume, used in MAME-SPE process. [from ref. 109].

In order to increase preconcentration factor in MAME-SPE procedure, methanol on the desorption volume can be reduced until a value which would allow high sensitivity but with an acceptable reproducibility of the method [109,110].

Analytical Parameters

After MAME-SPE extraction, analytes can be determined by HPLC. This procedure has been applied to the determination of organochlorine pesticides in seaweeds [109] and molluscs [110]. In both cases, average recoveries were in the range between 78.8-107.7% for seaweeds and 68.2-100.4% for molluscs. Precision was determined by reproducibility studies, expressed by relative standard desviation (RSD). Reproducibility was measured by analysis of six spiked seaweed samples under optimum conditions. RSD values were below 10% and limits of detection of all studied compounds were calculated [69], obtaining values ranged from 2 to 38 $ng \cdot g^{-1}$ for seaweeds and from 6 to 38 $ng \cdot g^{-1}$ for molluscs. Results obtained for seaweeds are reported in Table 15.

Characteristic chromatograms for MAME-SPE of spiked (4 $\mu g \cdot g^{-1}$ for 4,4'-DDD, 4,4'-DDT, 2,4'-DDT and 4,4'-DDE and 8 $\mu g \cdot g^{-1}$ for Aldrin and Dieldrin) (a) and non-spiked (b) seaweed samples are shown in Figure 21.

Table 15. Analytical parameters in the determination of organochlorine pesticides using MAME-SPE [from ref. 109]

Compound	Recovery (%)	RSD[a] (%)	LOD[b] (ng· g^{-1})
4,4'-DDD	92.9	3.5	12
Dieldrin	100.5	3.3	22
4,4'DDT	92.6	2.8	2
2,4'DDT	91.2	3.6	6
4,4'-DDE	107.7	5.3	2
Aldrin	78.8	4.6	38

a) Relative Standard Desviation (n=6).

b) Limit of Detection.

c) Concentration of spiked seaweed: 4 μg/g for 4,4'-DDD; 4,4'-DDT; 2,4'-DDT and 4,4'-DDE, and 8 μg/g for Aldrin and Dieldrin.

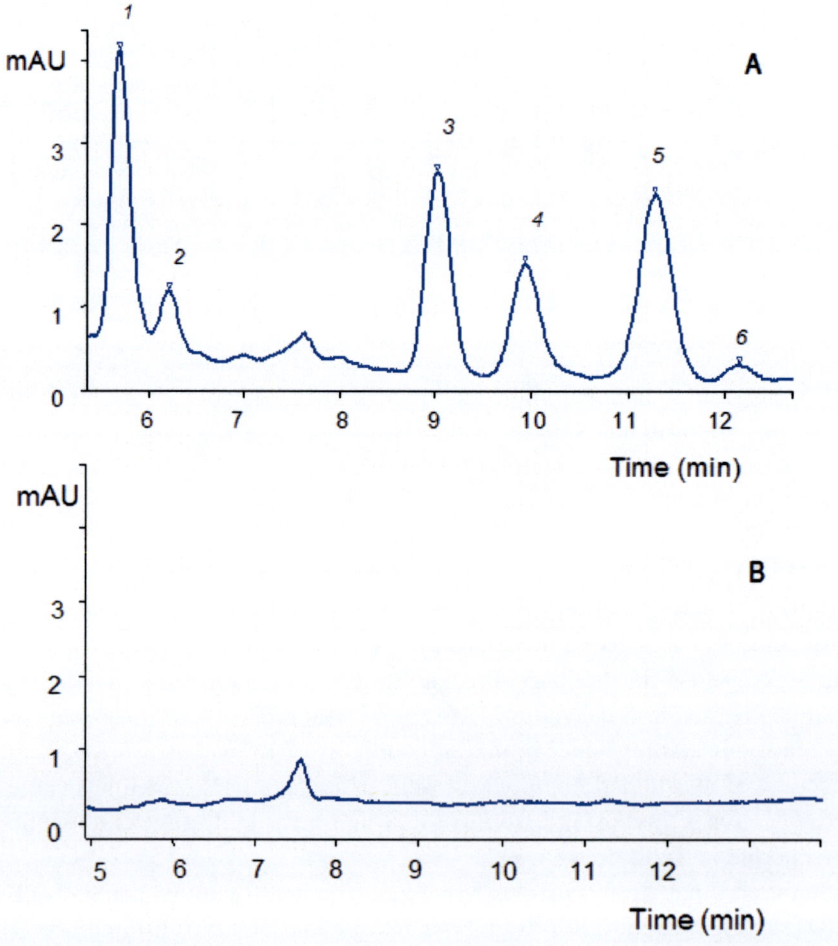

Figure 21. Chromatogram of an extract of a spiked (a) and non-spiked (b) Ulva seaweed sample after MAME-SPE procedure. λ=238 nm. (1- 4,4'-DDD; 2- Dieldrin; 3- 4,4'-DDT; 4- 2,4'-DDT; 5- 4,4'-DDE; 6- Aldrin). [from ref. 109].

Applications and Validation

When this method is applied to several seaweed samples or different kind of molluscs, recoveries obtained in all cases were equal or higher than 73.9 and 68.3%, respectively. It is demonstrated that the proposed method MAME-SPE-HPLC can be an alternative to the determination of organochlorine pesticides in complex solid samples [109,110].

If we compared traditional Soxhlet extraction [113] with MAME-SPE process developed, results obtained by both methods are comparable (Table 16). A clean-up of Soxhlet extracted was not done because it was only used to validate our method, and with direct injection was enough for obtaining recovery's results but with bad baseline and higher detection limits than in MAME-SPE. Moreover, MAME-SPE offers as advantages that is a faster process and it needs much less organic solvent.

Table 16. Application of MAME-SPE procedure and Soxhlet extraction to a seaweed sample containing a certified mixture of pesticides[a,b] [from ref. 109]

Compound	MAME-SPE ($\mu g \cdot g^{-1}$)	Soxhlet ($\mu g \cdot g^{-1}$)
4,4'-DDD	3.72 ± 0.08	4.12 ± 0.24
Dieldrin	4.04 ± 0.14	3.94 ± 0.30
4,4'DDT	3.82 ± 0.10	4.04 ± 0.18
2,4'DDT	3.52 ± 0.18	3.44 ± 0.22
4,4'-DDE	3.78 ± 0.14	3.76 ± 0.15

[a)] Concentration of spiked seaweed: 4 Rg/g for all compounds.

[b)] Mean of two determinations.

General Conclusions and Future Trends

New analytical methods are continually being developed for sample preparation. Although LLE continues to be widely used, CPE methodology is a good alternative extraction technique for liquid samples and offers a series of highly interesting advantages from an analytical point of view, such as the possibility of extracting and preconcentrating the analytes of different polarities in only one step.

The pre-concentration factor can be optimised by modifying the type and the concentration of surfactant as well as the experimental conditions under which extraction and phase separation are carried out. Surfactants are lesser toxic and cheaper than the extractants used in LLE. The most commonly used surfactants are commercially available and, since it is not necessary to evaporate off the solvents, no analyte is lost in the process. The experimental operations involved in CPE methodology are very simple and the surfactant-rich phase is compatible with the mobile phases used in HPLC.

MAME procedure could replace the conventional Soxhlet extraction, offering several advantages over traditional extraction techniques applied to solid samples. The major benefits are:

- a reduction in extraction time (typically 10–30 min)
- a reduction in solvent consumption (around 25–50 ml, reduced by 5–10-fold compared to traditional extractions); and,

- the opportunity to perform multiple extractions.

MAME procedure is easy to use and is cheaper than other modern techniques. Furthermore, the use of micellar medium reduces exposure to solvent vapours and wastes. When considering economical and practical aspects, the use of micellar systems for the extraction of organic compounds in solid matrices offers strong competition to other extraction techniques.

This competitiveness may be improved in the future in favour of two-directional micellar extraction. It may also be applied to a great number of types of solid real samples (plants, marine organisms, etc.), using other types of surfactants, as well as a mixture of them as extractants. In addition, combining the use of surfactants with other extraction techniques and with highly sensitive chromatographic techniques, such as LC–MS and GC–MS or GC-ECD, may be of great interest for the extraction and the determination of organochlorine and organophosphorus pesticides that, in many occasions, are found in the environment in concentrations too small to be detected using conventional techniques.

Finally, the combination of microwave assisted micellar extraction (MAME) with SPME or SPE as a clean-up and preconcentration step of MAME extract, provided sensitive and selective methods for the extraction of organochlorine and organophosphorus pesticides residues in environmental samples. Specifically, the results obtained for MAME-SPE-HPLC indicate that this technique is more sensitive for the extraction and determination of this kind of compounds where the LODs are from 10 to 100 times lower than those obtained using MAME-SPME-HPLC.

REFERENCES

[1] Kusvuran, E. and Erbatur, O. (2004) Degradation of aldrin in adsorbed system using advanced oxidation processes: comparison of the treatment methods. *Journal of Hazardous Materials,* 106B, 115-125.

[2] Stockholm Convention on Pops: *A Global Public Health Treaty*. (2001) Sweden.

[3] Shivaramaiah, H.M., Odeh, I.O.A., Kennedy, I.V. and Skerritt, J.H. (2002) Mapping the distribution of DDT residues as DDE in the soils of the irrigated regions of Northern New South Wales, Australia using ELISA and GIS. *Journal of Agricultural and Food Chemistry,* 50, 5360-5367.

[4] Chen, W., Zhang, L., Xu, L., Wang, X., Hong, L. and Hong, H. (2002) Residue levels of HCHs, DDTs and PCBs in shellfish from coastal areas of east Xiamen Island and Minjiang Estuary, China. *Marine Pollution Bulletin*, 45, 385-390.

[5] Nawab, A., Aleem, A. and Malik, A. (2003) Determination of organochlorine pesticides in agricultural soil with special reference to gamma-HCH degratation by Pseudomonas strains. *Biosesource Technology*, 88, 41-46.

[6] Megharaj, M., Kantachote, D., Singleton, I. and Naidu, R. (2000) Effects of long-term contamination of DDT on soil microflora with special reference to soil algae and algal transformation of DDT. *Environmental Pollution*, 109, 35-42.

[7] Mwevura, H., Othman, O. and Mhehe, G.L. (2002) Organochlorine pesticide residues in sediments and biota from the coastal area of Dar es Salaam city, Tanzania. Marine *Pollution Bulletin*, 45, 262-267.

[8] Maroni, M., Colosio, C., Ferioli, A. and Fait, A. (2000) *Organochlorine pesticides. Toxicology*, 143, 61-75.

[9] U.S. Environmental Protection Agency. *Toxic Substance Control Act*, US EPA. (1979) Washington DC.

[10] Anson Moye, H. (Ed.) Analysis of Pesticide Residues (1981) vol 58. John Wiley and Sons, New York/Chichester/Brisbane/Toronto. EEUU.

[11] Woodwell, G.M., Craig, P.P. and Johnson, H.A. (1971) DDT in the Biosphere: Where Does It Go?. *Science*, 174, 1101-1107.

[12] Hernández, F., Pitarch, E., Beltran, J. and López, F.J. (2002) Headspace solid-phase microextraction in combination with gas chromatography and tandem mass spectrometry for the determination of organochlorine and organophosphorus pesticides in whole human blood. *Journal of Chromatography* B, 769, 65-77.

[13] Zulin, Z., Huasheng, H., Xinhong, W., Jianqing, L., Weiqui, C. and Li, X. (2002) Determination and load of organophosphorus and organochlorine pesticides at water from Jiulong River Estuary, China. *Marine Pollution Bulletin*, 45, 397-402.

[14] Viñas, P., Campillo, N., López-García, I., Aguinaga, N. and Hernández-Córdoba, M. (2002) Determination of pesticides in waters by capillary gas chromatography with atomic emission detection. *Journal of Chromatography* A, 978, 249-256.

[15] Torres, C.M. and Mañes, Y.P.J. (1996) Determination of pesticide residues in fruit and vegetables. *Journal of Chromatography* A, 754, 301-331.

[16] Tadeo, J.L., Sánchez-Brunete, C., Pérez, R.A. and Fernández, M.D. (2000) Analysis of herbicide residues in cereals, fruits and vegetables. *Journal of Chromatography* A, 882, 175-191.

[17] Carabias-Martínez, R., Rodríguez-Gonzalo, E., Paniagua-Marcos, P.H. and Hernández-Méndez, J. (2000) Analysis of pesticide residues in matrices with high lipid contents by membrane separation coupled on-line to a high-performance liquid chromatography system. *Journal of Chromatography* A, 869, 427-439.

[18] Caballo-López, A. and Luque de Castro, M.D. (2003) Continuous ultrasound-assisted extraction coupled to on line filtration–solid-phase extraction–column liquid chromatography–post column derivatisation–fluorescence detection for the determination of N-methylcarbamates in soil and food. *Journal of Chromatography* A, 998, 51-59.

[19] Rastrelli, L., Totaro, K. and De Simone, F. (2002) Determination of organophosphorus pesticide residues in Cilento (Campania, Italy) virgin olive oil by capillary gas chromatography. *Food Chemistry*, 79, 303-305.

[20] Russo, M.V., Campanella, L. and Avino, P. (2002) Determination of organophosphorus pesticide residues in human tissues by capillary gas chromatography-negative chemical ionization mass spectrometry analysis. *Journal of Chromatography* B, 780, 431-441.

[21] Jeong, M.L. and Chesney, D.J. (1999) Determination of *N*-methylcarbamate pesticides from spiked matrices using supercritical fluid extraction. *Analytica Chimica Acta*, 389, 53-57.

[22] Hennion, M.C., Cau-Dit-Coumes, C. and Pichon, V. (1998) Trace analysis of polar organic pollutants in aqueous samples: Tools for the rapid prediction and optimisation of the solid-phase extraction parameters. *Journal of Chromatography* A, 823, 147-191.

[23] Hennion, M.C. (1999) Solid-phase extraction: method development, sorbents, and coupling with liquid chromatography. *Journal of Chromatography* A, 856, 3-54.

[24] Chiron, S., Fernández Alba, A. and Barceló, D. (1993) Comparison of on-line solid-phase disk extraction to liquid-liquid extraction for monitoring selected pesticides in environmental waters. *Environmental Science Technology*, 27, 2352-2359.

[25] Ahmed, F.E. (2001) Analyses of pesticides and their metabolites in foods and drinks. *Trends in Analytical Chemistry*, 20, 649-661.

[26] Rodriguez, I., Llompart, M.P. and Cela, R. (2000) Solid-phase extraction of phenols. *Journal of Chromatography* A, 885, 291-304.

[27] Lores, M., Llopart, M., González-García, R., González-Barreiro, C. and Cela, R. (2002) On-fibre photodegradation studies of polychlorinated biphenyls using SPME-GC-MS-MS: a new approach. *Chemosphere*, 47, 607-615.

[28] Eisert, R. and Pawliszyn, J. (1997) Automated In-Tube Solid-Phase Microextraction Coupled to High-Performance Liquid Chromatography. *Analytical Chemistry*, 69, 3140-3147.

[29] Moreno Cordero, B., Pérez Pavón, J.L., García Pinto, C. and Fernández Laespada, E. (1993) Cloud point methodology: A new approach for preconcentration and separation in hydrodynamic systems of analysis. *Talanta,* 40, 1703-1710.

[30] Heemke, O.P., Norbert, T. and Wenclawiak, B.W. (1997) Comparison of ASE and SFE with Soxhlet, Sonication, and Methanolic Saponification Extractions for the Determination of Organic Micropollutants in Marine Particulate Matter. *Analytical Chemistry*, 69, 2171-2180.

[31] Snyder, J., Grob, R., McNally, M. and Gostdyk, T. (1992) Comparison of supercritical fluid extraction with classical sonication and Soxhlet extractions for selected pesticides. *Analytical Chemistry*, 64, 1940-1946.

[32] Fernández, P., Grifoll, M., Solanas, A.M., Bayona, J. and Albaigés, J. (1992) Bioassay-directed chemical analysis of genotoxic components in coastal sediments. *Environmental Science Technology*, 26, 817-829.

[33] Punín Crespo, M. and Lage Yusty, M. (2006) Comparison of supercritical fluid extraction and Soxhlet extraction for the determination of aliphatic hydrocarbons in seaweed samples. *Ecotoxicology and Environmental Safety*, 64, 400-405.

[34] Li, K.; Landrialult, M.; Fingas, M. and Llompart, M. (2003) Accelerated solvent extraction (ASE) of environmental organic compounds in soils using a modified supercritical fluid extractor. *Journal of Hazardous Materials*, 102, 93-104.

[35] Zuloaga, O., Etxebarria, N., Fernández, L.A. and Madariaga, J.M. (2000) *Fresenius Journal Analytical Chemistry*, 367, 733-737.

[36] Sanghi, R. and Kannamkumarath, S. (2004) Comparison of extraction methods by soxhlet, sonicator, and microwave in the screening of pesticides residues from solid matrices. *Journal of Analytical Chemistry*, 59, 1032-1036.

[37] Sporring, S.; Bowadt, S.; Svensmark, B. and Björklund, E. (2005) Comprehensive comparison of classic Soxhlet extraction with Soxtec extraction, ultrasonication extraction, supercritical fluid extraction, microwave assisted extraction and accelerated

solvent extraction for the determination of polychlorinated biphenyls in soil. *Journal of Chromatography* A, 1090, 1-9.

[38] Pastor, A., Vazquez, E., Ciscar, R. and de la Guardia, M. (1997) Efficiency of the microwave-assisted extraction of hydrocarbons and pesticides from sediments. *Analytica Chimica Acta*, 344, 241-249.

[39] Shu, Y.Y., Tey, S.Y. and Wu, D.K.S. (2003) Analysis of polycyclic aromatic hydrocarbons in airborne particles using open-vessel focused microwave-assisted extraction. *Analytica Chimica Acta*, 495, 99-108.

[40] López-Ávila, V. and Young, R. (1994) Microwave-assisted extraction of organic compounds from standard reference soils and sediments. *Analytical Chemistry*, 66, 1097-1106.

[41] Esteve-Turrillas, F.A., Aman, C.S., Pastor, A. and de la Guardia, M. (2004) Microwave-assisted extraction of pyrethroid insecticides from soil. *Analytica Chimica Acta,* 522, 73-78.

[42] Barriada-Pereira, M., Concha-Graña, E., González-Castro, M.J., Muniategui-Lorenzo, S., López-Mahía, P., Prada-Rodríguez, D. and Fernández-Fernández, E. (2003) Microwave-assisted extraction versus Soxhlet extraction in the analysis of 21 organochlorine pesticides in plants. *Journal of Chromatography* A, 1008, 115-122.

[43] Onuska, F.I. and Terry, K.A. (1993) Extraction of pesticides from sediments using a microwave technique. *Chromatographia*, 36, 191-194.

[44] Letellier, M., Budzinski, H. and Garrigues, P. (1999) Focused microwave-assisted extraction of polycyclic aromatic hydrocarbons. *LC-GC International*, 12, 222-225.

[45] López-Avila, V., Young, R., Benedicto, J., Ho, P., Kim, R. and Beckert, W.F. (1995) Extraction of organic pollutants from soild samples using microwave energy. *Analytical Chemistry,* 67, 2096.

[46] Rosen, M.J. (1986) *Surfactants and Interfacial phenomena*, John Wiley and sons. USA.

[47] Pelizzetti, E. and Pramauro, E. (1985) Analytical applications of organized molecular assemblies. *Analytica Chimica Acta*, 169, 1-29.

[48] Uhrich, K.E. (1999) *Chemical Review*, 99, 3181-3198.

[49] Goto, T., Ito, Y., Oka, H., Saito, I., Matsumoto, H. and Nakazawa, H. (2003) Simple and rapid determination of N-methylcarbamate pesticides in citrus fruits by electrospray ionization tandem mass spectrometry. *Analytica Chimica Acta*, 487, 201-209.

[50] Namera, A., Yashiki, M., Liu, J., Okajima, K., Hara, K., Imamura, T. and Kojima, T. (2000) Simple and simultaneous analysis of fenfluramine, amphetamine and methamphetamine in whole blood by gas chromatography–mass spectrometry after headspace–solid phase microextraction and derivatization. *Forensic Science International,* 109, 215-223.

[51] Paleologos, E. K., Giokas, D. L. and Karayannis, M. I. (2005) Micelle-mediated separation and cloud point extraction. *Trends in Analytical Chemistry*, 24, 426-436.

[52] Faria, A. M., Dardengo, R. P. Lima, C. F., Neves, A. and Queiroz, M. E. (2007) Determination of disulfoton in surface water samples by cloud-point extraction and gas chromatography. *International Journal of Environmental Analytical Chemistry*, 87, 249-258.

[53] Shen, J. and Shao, X. (2006) Determination of tobacco alkaloids by gas chromatography-mass spectrometry using cloud point extraction as a preconcentration step. *Analytica Chimica Acta*, 561, 83-87.

[54] Sosa Ferrera, Z., Padrón Sanz, C., Mahugo Santana, C. and Santana Rodríguez, J.J. (2004) The use of micellar systems in the extraction and pre-concentration of organic pollutants in environmental samples. *Trends in Analytical Chemistry*, 23 (7), 469-479.

[55] Mahugo Santana, C., Sosa Ferrera, Z. and Santana Rodríguez, J.J. (2004) Use of polyoxyethlylene-6-lauryl ether and microwave-assisted extraction for the determination of chlorophenols in marine sediments. *Analytica Chimica Acta*, 524, 133-139.

[56] Padrón Sanz, C., Halko, R., Sosa Ferrera, Z. and Santana Rodríguez, J.J. (2004) Micellar extraction of organophosphorus pesticides and their determination by liquid chromatography. *Analytica Chimica Acta*, 524, 265-270.

[57] Vega Moreno, D., Sosa Ferrera, Z. and Santana Rodríguez, J.J. (2006) Microwave assisted micellar extraction coupled with solid phase microextraction for the determination of organochlorine pesticides in soil simples. *Analytica Chimica Acta,* 571, 51-57.

[58] Hinze, W.L. and Pramauro, E. (1993) A critical review of surfactant-mediated phase separations (cloud point extractions): theory and applications. *Critical Reviews in Analytical Chemistry*, 24(2), 133-177.

[59] Mahugo Santana, C., Sosa Ferrera, Z. and Santana Rodríguez, J.J., (2002) Use of non-ionic surfactant solutions for the extraction and preconcentration of phenolic compounds in water prior to their HPLC-UV detection. *The Analyst*, 127, 1031-1037.

[60] Eiguren Fernández, A., Sosa Ferrera, Z. and Santana Rodríguez, J.J., (1997) *Química Analítica*, 16(2), 283.

[61] Frankewich, R.P. and Hinze, W. (1994) Evaluation and Optimization of the Factors Affecting Nonionic Surfactant-Mediated Phase Separations. *Analytical Chemistry*, 66, 944-954.

[62] Carabias-Martínez, R., Rodríguez-Gonzalo, E., Domínguez-Álvarez, J., García Pinto, C., Hernández-Méndez, J. (2003) Prediction of the behaviour of organic pollutants using cloud point extraction. *Journal of Chromatography* A, 1005 (1-2), 23-34.

[63] Carabias, R., Rodríguez, E., Moreno, B., Pérez, J.L., García, C. and Fernández, E. (2000) Surfactant cloud point extraction and preconcentration of organic compounds prior to chromatography and capillary electrophoresis. *Journal of Chromatography* A, 902, 251-265.

[64] Silva, M.F., Fernández, L., Olsina, R.A. and Stacchiola, D. (1997) Cloud point extraction, preconcentration and spectrophotometric determination of erbium(III)-2-(3,5-dichloro-2-pyridylazo)-5-dimethylaminophenol. *Analytica Chimica Acta*, 342, 229-238.

[65] Eiguren Fernández, A., Sosa Ferrera, Z. and Santana Rodríguez, J.J. (1998) Determination of polychlorinated biphenyls by liquid chromatography following cloud-point extraction. *Analytica Chimica Acta*, 358, 145-155.

[66] Faria, A.M., Dardengo, R.P., Lima, C.F., Neves, A.A., Queiroz, M.E.L.R. (2007) Determination of disulfoton in surface water samples by cloud-point extraction and gas chromatography. *International Journal of Environmental Analytical Chemistry* 87 (4), 249-258

[67] Froschl, B., Stangl, G. and Niessner, R. (1997) Combination of micellar extraction and GC-ECD for the determination of polychlorinated biphenyls (PCBs) in water. *Fresenius Journal Analytical Chemistry*, 357, 743-746.

[68] Eiguren Fernández, A., Sosa Ferrera, Z. and Santana Rodríguez, J.J. (1999) Application of cloud-point methodology to the determination of polychlorinated dibenzofurans in seawater by high-performance liquid chromatography. *Analyst*, 24, 487-491.

[69] Lindsay, S. (1992) High Performance Liquid Chromatography, John Wiley and Sons: New York, pp. 337.

[70] Ganzler, K., Salgo, A. and Lebensm-Unter, Z. (1987) *Forsch*, 184, 274.

[71] Pineiro Iglesias, M., López Mahía, P., Vázquez Blanco, E., Muniategui Lorenzo, S., Prada Rodríguez, D. and Fernández Fernández, E. (2000) Microwave assisted extraction of polycyclic aromatic hydrocarbons from atmospheric particulate samples. *Fresenius Journal Analytical Chemistry*, 367, 29-34.

[72] López-Avila, V., Benedicto, J., Charan, C. and Young, R. (1995) Determination of PCBs in soils/sediments by microwave-assisted extraction and GC/ECD or ELISA. *Environmental Science Technology*, 29, 2709-2712.

[73] Hummert, K., Vetter, W. and Luckas, B. (1996) Fast and effective sample preparation for determination of organochlorine compounds in fatty tissue of marine mammals using microwave extraction. *Chromatographia*, 42, 300-304.

[74] Pastor, A., Vázquez, E., Ciscar, R. and de la Guardia, M. (1997) Efficiency of the microwave-assisted extraction of hydrocarbons and pesticides from sediments. *Analytica Chimica Acta*, 344, 241-249.

[75] Egizabal, A., Zuloaga, O., Etxebarria, N., Fernández, L.A. and Madariaga, J.M. (1998) Comparison of microwave-assisted extraction and Soxhlet extraction for phenols in soil samples using experimental designs. *The Analyst*, 123, 1679-1684.

[76] Llompart, M.P., Lorenzo, R.A., Cela, R., Li, K., Belanger, J.M.R. and Pare, J.R.J. (1997) Evaluation of supercritical fluid extraction, microwave-assisted extraction and sonication in the determination of some phenolic compounds from various soil matrices. *Journal of Chromatography* A, 774, 243-251.

[77] Padrón Sanz, C., Sosa Ferrera, Z. and Santana Rodríguez, J.J. (2004) Extraction of PCDDs from marine sediments using polyoxyethylene 10 lauryl ether and oligoethylene glycol monoalkyl ether surfactants. *Analytical Letters*, 37, 1385-1399.

[78] Padron Sanz, C., Halko, R., Sosa Ferrera, Z. and Santana Rodríguez, J.J. (2005) Combination of microwave assisted micellar extraction and liquid chromatography for the determination of organophosphorous pesticides in soil simples. *Journal of Chromatography* A, 1078, 13-21

[79] Vega Moreno, D., Sosa Ferrera, Z. and Santana Rodríguez, J.J. (2006) Use of polyoxyethylene surfactants for the extraction of organochlorine pesticides from agricultural soils. *Journal of Chromatography* A, 1104, 11-17.

[80] Padrón Sanz, C., Eiguren Fernández, A., Sosa Ferrera, Z. and Santana Rodríguez, J.J. (2002) Determination of Organochlorinated Compounds in Marine Organisms by Microwave-Assisted Extraction with Molecular Organized Systems and Liquid Chromatography with Fluorescence Detection. *Journal of AOAC Internacional*, 85(1), 44-49.

[81] Bianco Prevot, A., Gulmini, M., Zelano, V. and Pramauro, E. (2001) Microwave-Assisted Extraction of Polycyclic Aromatic Hydrocarbons from Marine Sediments Using Nonionic Surfactant Solutions. *Analytical Chemistry*, 73, 3790-3795.

[82] U.S. Environmental Protection Agency (1996), *Method* 3540 C, Washington, DC.

[83] Llompart, M.P., Lorenzo, R.A. and Par´e, J.R. (1997) Optimization of a Microwave-assisted Extraction Method for Phenol and Methylphenol Isomers in Soil Samples Using a Central Composite Design. *Analyst,* 122, 133-137.
[84] Huang, W., Peng, P., Yu, Z. and Fu, J. (2003) Effects of organic matter heterogeneity on sorption and desorption of organic contaminants by soils and sediments. *Applied Geochemistry,* 18, 955-972.
[85] Thiele-Bruhn, S., Seibicke, T., Schuelten, H.R. and Leinweber, P. (2004) Sorption of Sulfonamide Pharmaceutical Antibiotics on Whole Soils and Particle-Size Fractions. *Journal of Environmental Quality*, 33, 1331-1342.
[86] Aquino, A. J., Tunega, D., Haberhauer, G., Gerzabek, M. and Lischka, H. (2003) Adsorption of Organic Substances on Broken Clay Surfaces: A Quantum Chemical Study. *Journal of Computational Chemistry*, 24(15), 1853-1863.
[87] Shu, Y.Y., Lai, T.L., Lin, H.S., Yang, T.C. and Chang, C.P. (2003) Study of factors affecting on the extraction efficiency of polycyclic aromatic hydrocarbons from soils using open-vessel focused microwave-assisted extraction. *Chemosphere*, 52, 1667-1676.
[88] Kiflom, G., Wandiga, S.O., Ng'ang'a, P.K. and Kamau, G.N. (1999) Variation of plant p,p'-DDT uptake with age and soil type and dependence of dissipation on temperature. *Environmental Internacional*, 25(4), 479-487.
[89] Hawthorne, S.B., Björklund, E., Bowadt, S. and Mathiasson, L. (1999) Determining PCB sorption/desorption behavior on sediments using selective supercritical fluid extraction. 3. Sorption from water. *Environmental Science Technology*, 33, 3152-3159.
[90] Dec, J. and Bollag, J. (1997) Determination of covalent and noncovalent binding interactions between xenobiotic chemicals and soil. *Soil Science*, 162, 858-874.
[91] Kopinke, F., Pörschmann, J. and Stottmeister, U. (1995) Sorption of organic pollutants on anthropogenic humic matter. *Environmental Science Technology*, 29, 941-950.
[92] Stevenson, F.J. (1982) Humus Chemistry, John Wiley and Sons Inc., USA.
[93] Racke, K.D. in: Chambers, J.E. and Levi, P.E. (Eds.) (1992) *Organophosphates: Chemistry*, Fate, and Effects. Academic Press, San Diego, pp. 47–73.
[94] Pehkonen, S.O. and Zhang, Q. (2002) The degradation of organophosphorus pesticides in natural waters: A critical review. *Critical Review Environmental Science Technology,* 32, 17-72.
[95] Dannenberg, A. and Pehkonnen, S.O. (1998) Investigation of the Heterogeneously Catalyzed Hydrolysis of Organophosphorus Pesticidas. *Journal of Agricultural and Food Chemistry*, 46, 325-334.
[96] Dowling, K.C. and Lemley, A.T. (1995) Organophosphate insecticide degradation by non-amended and cupric ion-amended Fenton's reagent in aqueous solution. *Journal Environmental Science Health* B, 30 (5), 585-604.
[97] Noblet, J.A., Smith, L.A. and Suffet, I.H. (1996) Influence of natural dissolved organic matter, temperature, and mixing on the abiotic hydrolysis of triazine and organophosphate pesticidas. *Journal of Agricultural and Food Chemistry*, 44, 3685-3693.
[98] Dupeyron, S., Dudermel, P.M. and Counturier, D. (1997) Focused microwave assisted extraction (FMAE) of polynuclear aromatic hydrocarbons from contaminated soil: Role of acetone and water content impact on microwave efficiency. *Analusis,* 25, 286-292.

[99] Peñalver, A., Pocurull, E., Borrull, F. and Marcé, R.M. (1999) Trends in solid-phase microextraction for determining organic pollutants in environmental samples. *Trends in Analytical Chemistry*, 18(8), 557-567.

[100] Ho, W. and Hsieh, S. (2001) Solid phase microextraction associated with microwave assisted extraction of organochlorine pesticides in medicinal plants. *Analytica Chimica Acta,* 428, 111-120.

[101] Aulakh, J. and Malik, A. (2005) A review on solid phase microextraction-high performance liquid chromatography (SPME-HPLC) analysis of pesticides. *Critical Reviews in Analytical Chemistry,* 35, 71-85.

[102] Vega Moreno, D., Sosa Ferrera, Z. and Santana Rodríguez, J.J. (2006) Sample extraction method combining micellar extraction-SPME and HPLC for the determination of organochlorine pesticides in agricultural soils. *Journal of Agricultural and Food Chemistry*, 54, 7747-7752.

[103] Beltrán, J., López, F. and Hernández, F. (2000) Solid-phase microextraction in pesticides residue analysis. *Journal of Chromatography* A, 885, 389-404.

[104] Krutz, L., Senseman, S. and Sciumbato, A. (2003) Solid-phase microextraction for herbicide determination in environmental samples. *Journal of Chromatography* A, 999, 103-121.

[105] Hernández, F., Beltrán, J., López, F. and Gaspar, (2000) Use of solid-phase microextraction for the quantitative determiantion of herbicides in soil and water samples. *Journal Analytical Chemistry*, 72, 2313-2322.

[106] Arthur, C.L., Killiam, L.M., Buchholz, K.D. and Pawliszyn, J. (1992) Automation and optimization of solid-phase microextraction. *Journal Analytical Chemistry*, 64, 1960-1966.

[107] Vega Moreno D, Sosa Ferrera Z, Santana Rodríguez JJ (2008) Application of microwave assisted micellar extraction combined with solid phase microextraction and HPLC-UV for the determination of organochlorine pesticides in diferents mud samples. *International Journal on Environmental Analytical Chemistry*, in press.

[108] Mutavdzic, D., Horvat, A., Babic, S. and Kastelan-Macan, M. (2005) SPE- Microwave-assisted extraction coupled system for the extraction of pesticides from water samples. *Journal of Separation Science*, 28, 1485-1492.

[109] Vega Moreno, D., Sosa Ferrera, Z. and Santana Rodríguez, J.J. (2007) SPME and SPE comparative study for coupling with microwave-assisted micellar extraction in the analysis of organochlorine pesticides residues in seaweed samples. *Microchemical Journal,* 87, 139-146.

[110] Vega Moreno D, Sosa Ferrera Z, Santana Rodríguez JJ (2008) Analysis of organochlorine pesticides in mollusc samples by HPLC after microwave assisted micellar extraction coupled with solid phase extraction. *Journal of Liquid Chromatography and Related Technologies*, in press.

[111] Deger, A., Gremm, T. and Frimmel, F. (2000) Problems and solutions in pesticides analysis using solid-phase extraction (SPE) and gas chromatography ion-trap mass spectrometry detection (GC-MS). *Acta Hydrochim. Hydrobiol.*, 28, 292-299.

[112] Papadakis, E. and Papadopoulou-Mourkidou, E. (2006) LC-UV determination of atrazine and its prinicipal conversion products in soil after combined microwave-assisted and solid-phase extraction. *International Journal on Environmental Analytical Chemistry,* 86, 573-582.

[113] AOAC Official Method 970.52, AOAC Official Methods of Analysis, *Pesticide and Industrial*.

In: Environmental and Regional Air Pollution
Editors: D. Gallo, R. Mancini
ISBN: 978-1-60692-893-6

Chapter 2

CHARACTERIZATION AND QUANTIFICATION OF *IN SITU* BIODEGRADATION OF GROUNDWATER CONTAMINANTS USING STABLE ISOTOPE FRACTIONATION ANALYSIS: ADVANTAGES AND LIMITATIONS

***Martin Thullner*[*]*, Hans-Hermann Richnow*[+] *and Anko Fischer*[+,#]**
[*]Department of Environmental Microbiology, UFZ - Helmholtz Centre for Environmental Research, Leipzig, Germany
[+]Department of Isotope Biogeochemistry, UFZ - Helmholtz Centre for Environmental Research, Leipzig, Germany
[#]Isodetect - Company for Isotope Monitoring, Neuherberg/Oberschleißheim, Germany

ABSTRACT

For many contaminated sites *Natural Attenuation* (NA) of groundwater contaminants has become a common remediation strategy. The main process contributing to the NA of organic contaminants is their degradation by microorganisms inhabiting a groundwater aquifer, a process referred to as *in situ* biodegradation. As a consequence, the assessment of *in situ* contaminant biodegradation has gained increasing importance in recent years. Compound-specific stable isotope analysis (CSIA) is one of the few methods allowing for an evaluation of *in situ* biodegradation and this method is increasingly used for characterizing and quantifying microbial removal of groundwater contaminants. CSIA relies on the fractionation of stable isotopes during biodegradation, leading to an enrichment of heavier isotopes in the residual fraction of a contaminant. In many studies, the Rayleigh equation is used to obtain a quantification of contaminant biodegradation based on measured changes in isotope ratios. However, the standard Rayleigh equation considers only a single process affecting simultaneously the concentration and the isotope ratio of contaminants, a constraint often not fulfilled in the field. In many field cases, contaminant concentrations can also be affected by abiotic processes e.g. mixing, volatilization or sorption to the aquifer matrix. As the Rayleigh equation links changes in isotope ratios to concentration shifts, these abiotic processes need to be taken into account, even if they do not cause any isotope fractionation. The present study evaluates the applicability of the Rayleigh equation for the assessment of *in situ* biodegradation and shows the advantages and limitations of this concept. A review of studies in the literature addressing this issue is combined with the detailed discussion of a new modification of the Rayleigh equation. This modification allows considering abiotic concentrations changes in the quantification of *in situ* contaminant biodegradation based on stable isotope measurements. The applicability of this modification was tested against numerical modeling results using virtual flow path scenarios.

INTRODUCTION

Due to the increasing world population, the demand of water is steadily rising. Ground water is used to a major part for irrigation and as a source for drinking water. Since the industrialization in the middle of 19th century, groundwater reservoirs have been contaminated by organic chemicals. Worldwide, chlorinated and monoaromatic hydrocarbons belong to the most frequent contaminants in aquifer systems [1,2] and many of them are known for their toxicity [3-7].

Since the organic contaminants represent a risk for humans and the environment, it is important to understand their behavior in aquifer systems. In groundwater environments, abiotic processes such as sorption, advection, dispersion, and volatilization cause an immobilization, spreading or displacement of organic contaminants into other environmental compartments. Only biodegradation leads to a sustainable reduction of contaminant concentrations associated with an effective mass reduction of the contaminants.

To use *Natural Attenuation* (NA) processes to control organic contaminants in aquifers, the evaluation of *in situ* contaminant biodegradation is essential. Thus, for the implementation of *Natural Attenuation* (NA) or *Enhanced Natural Attenuation* (ENA) as groundwater management strategies [8,9] the characterization and assessment of *in situ* biodegradation are critical factors.

Geochemical and biological concepts have been suggested to assess NA processes [10,11]. However, a direct assessment of *in situ* degradation is often not possible or very time consuming and expensive and, thus, the quantification of *in situ* degradation is still a challenge.

Isotope fractionation of contaminants during biodegradation is a perspective to characterize *in situ* degradation [12-14]. Thus, in recent years, compound-specific stable isotope analysis (CSIA) has gained more and more attention for assessing *in situ* biodegradation of organic contaminants in contaminated aquifers [15-20]. This concept relies on the observation of shifts in stable isotope ratios of carbon, hydrogen or other elements that are involved in the breakage or generation of chemical bonds during biodegradation of organic contaminants. The detection of a significant enrichment of the heavier stable isotope in the residual fraction of a contaminant gives indications for the biodegradation of this substance. These changes in stable isotope ratios of contaminants can be used to decipher biodegradation pathways and to calculate the extent of biodegradation in contaminated aquifers [17,21-23].

In many studies, it has been shown that CSIA can provide more reliable and easily ascertainable information about *in situ* contaminant biodegradation compared to other approaches such as electron acceptor - electron donor balances, field tracer tests, microcosm studies or column experiments. Therefore, in the future CSIA might be a promising tool in environmental chemistry and contaminant hydrology. The purpose of this contribution is to give an overview about CSIA, its potential for characterizing and quantifying *in situ* biodegradation, and to discuss the limitations of this concept encountered at the field scale.

COMPOUND-SPECIFIC STABLE ISOTOPE ANALYSIS (CSIA)

Isotopes of an element have the same amount of protons but differ in the number of neutrons, and therefore, they have different masses. Organic contaminants mainly consist of carbon and hydrogen and can additionally contain chlorine, oxygen, sulfur and nitrogen in their functional groups. All these elements exhibit at least two stable isotopes. The quotient between the heavy and the light isotope is called isotope ratio or isotope signature (e.g. $^{13}C/^{12}C$, $^{2}H/^{1}H$). Due to biodegradation the isotope ratio of a substrate changes since molecules consisting of only light isotopes are more preferentially transformed during an enzymatic reaction compared to molecules bearing heavy isotopes. Hence, an enrichment of heavy isotopes occurs within the non-biodegraded residual substrate fraction. The extent of the change in isotope ratios (isotope fractionation) depends among other factors on the relative mass difference of the isotopes of an element. The higher the difference the higher is the expected isotope fractionation. The relative mass difference of hydrogen for protonium (^{1}H) and deuterium (^{2}H) is 100 % and that for ^{12}C- and ^{13}C-carbon is approximately 8 %. Thus, the ascertainment of biodegradation should be easier detectable by the analysis of hydrogen isotope fractionation. However, currently the error and required amount of analyte for hydrogen isotope analysis is higher than for carbon isotope analysis. Therefore, carbon isotope analysis is used more often than hydrogen isotope analysis. Methods for compound specific chlorine isotope measurements of organic contaminants are time consuming and expensive and routine on–line-techniques for chlorine are not established yet. Only a few examples for nitrogen isotope fractionation analyses in the laboratory currently exist and the potential for nitrogen isotope fractionation to assess biodegradation in field studies is not explored, yet. This also accounts for oxygen isotopes.

Isotope Measurements of Contaminants

To measure compound-specific stable isotope ratios of contaminants gas chromatography isotope ratio mass spectrometry (GC IRMS) systems are used [24]. The separation of analytes is performed by a gas chromatograph (GC) which is connected via an interface by the isotope ratio mass spectrometer (IRMS) (Figure 1). For carbon isotope measurements, the interface consists of a combustion device where the separated analytes are oxidized to CO_2 [25]. The isotope signatures of the analytes are determined by the online-connected IRMS as CO_2-specific molecule masses and are given in comparison to a reference gas (CO_2) with known isotope ratio calibrated by an international standard.

For the hydrogen isotope analyses, the interface consists of a graphite-covered ceramic oven where the analytes are pyrolized at 1400 °C to hydrogen [26,27].

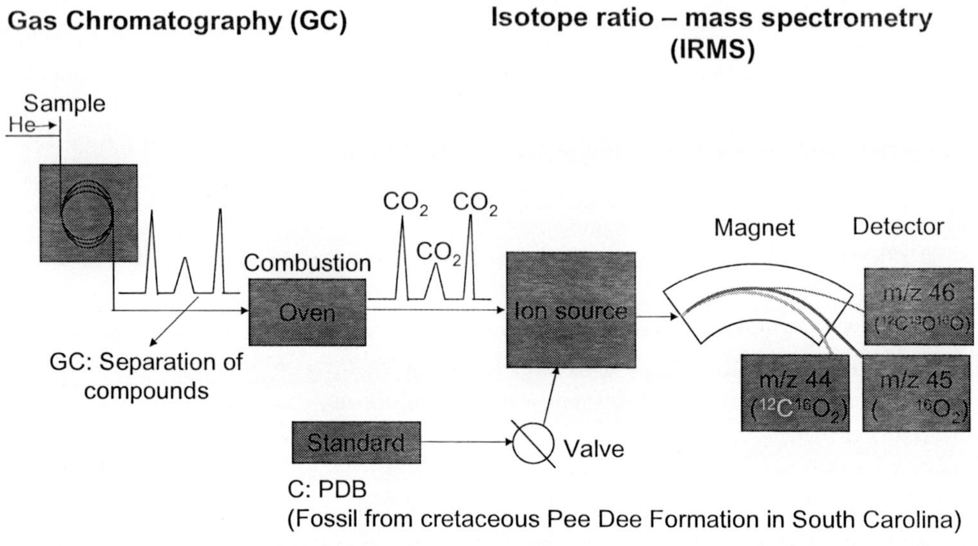

Figure 1. Scheme of gas chromatography isotope ratio mass spectrometry (GC IRMS) system for compound-specific carbon isotope ratio analysis.

Alternatively, organic compounds can be catalytically reduced by chrome to hydrogen by a temperature of 900 °C [28,29]. The hydrogen isotope ratio is measured for the mass traces 2 and 3 ($^1H_2$, $^2H^1H$) by the online-connected IRMS in comparison to a reference gas (H_2) calibrated by an international standard.

Isotope ratios are conventionally expressed in the delta notation ($\delta^{13}C$ and δ^2H) in per mil [‰] relative to an international standard according to:

$$\delta^{13}C_{sample} \quad \text{or} \quad \delta^2H_{sample}\,[‰] = \left(\frac{R_{sample} - R_{standard}}{R_{standard}}\right) \cdot 1000 \qquad (1)$$

In Eq. 1, R_{sample} and $R_{standard}$ are the $^{13}C/^{12}C$-ratios or $^2H/^1H$-ratios of the sample and an international standard, respectively. Vienna Pee Dee Belemnite (VPDB, $^{13}C/^{12}C = (11237{,}2 \pm 2{,}9) \times 10^{-6}$) is used as the standard for the analysis of carbon isotope signature and Vienna Standard Mean Ocean Water (VSMOW, $^2H/^1H = (155.76 \pm 0.05) \times 10^{-6}$) is used as the standard for the detection of hydrogen isotope ratios (*30-33*). Note that the delta notation also allows expressing changes in isotope ratios as:

$$\frac{R_1}{R_2} = \frac{(\delta_1 + 1000)}{(\delta_2 + 1000)} \qquad (2)$$

The reproducibility of $\delta^{13}C$ values and δ^2H values is under good experimental conditions smaller than 0.5 ‰ and 10 ‰, respectively. The linearity of most isotope mass spectrometers is expected to have a dynamic range limited to three orders of magnitude, which implies that analytes must be adjusted to an optimal amount for isotope measurements. Contaminant mixture can require multiple measurements if the species of interest are present at highly different concentrations. The current limits for carbon and hydrogen isotope measurements are about 1 and 10 nM, respectively, which is poor compared to GC-MS techniques. The relatively low sensitivity is a challenge for extraction and separation methods considering the measurements of trace contaminants in the environment (see below).

Isotope measurements depending on the pyrolysis or combustion of analytes to simpler products such as H_2, CO, CO_2 or N_2 require highly effective separation techniques. A base line separation of the analytes is necessary which can be a challenge for the gas chromatography and pose certain limits on the investigation of complex mixtures.

To measure isotope ratios of contaminants dissolved in groundwater target compounds have to be extracted from the water sample and concentrated to an amount sufficient for isotope measurements. This extraction of analytes is a further challenge considering the relatively low detection limits required to meet demands of environmental regulations. The method of extraction is dependent on the properties of the analytes. Headspace, purge and trap, solvent extraction and solid phase micro extraction (SPME) are used most frequently [16,18].

Headspace and SPME-approach are mostly applied for isotope analysis of MTBE, chlorinated hydrocarbons and BTEX. In general, extraction methods should be checked for isotope effects in reference experiments. For the headspace-technique, no significant isotope effects are known for the transfer of most nonpolar and halogenated contaminants from the water into the gas phase [34,35]. In contrast, undesirable isotope effects can occur for chlorinated methanes using the headspace-approach [36]. SPME-techniques can sometimes produce isotope effects, too. If there is no adequate study about isotope effects of the compound of interest, experiments should be performed to find an appropriate SPME-fiber and to estimate a correction factor, respectively [35,37]. Systematic studies about isotope effects for the SPME-approach are only known for a few contaminants [35-38]. For many target compounds, the requirements and boundary conditions for using headspace and SPME-approaches have not been studied thoroughly with respect e.g., to concentration limits or effects of the sample matrix. Therefore, often a new technique has to be developed to meet specific analytical demands.

Currently, it is hard to evaluate the general capability of these techniques based on a few available studies, only. Thus, isotope signatures of analytes might be determined also for lower concentrations than mentioned in the literature. For carbon and hydrogen isotope analysis of MTBE, reliable results were measured for concentrations of approximately 5 and 50 mg/L, respectively, using the headspace-approach [39]. Using the same technique, hydrogen isotope analysis of toluene could be performed for a concentration of 2 mg/L [40]. The headspace-approach was used to determine carbon isotope ratios of BTEX, chlorinated ethenes and methanes in a concentration range between 0.1 and 0.5 mg/L [34,41,42]. Using a dynamic headspace method where a GC-IRMS system was modified to include a headspace extraction system followed by a cryogenic trap, carbon isotope values of dissolved chlorinated ethenes could be measured at µg/L concentrations [43]. The SPME-technique allowed to perform isotope analysis for contaminant concentrations from 0.01 to 0.35 mg/L for carbon isotope signatures and 1 mg/L for hydrogen isotope signatures [35-39].

The solvent extraction is a well established method for CSIA. Generally, there are no significant isotope effects know for this technique [44]. To measure isotope signatures for samples with a low concentration at least one liter should be extracted with a few milliliter of the solvent [13,14]. The optimal duration of extraction for certain contaminants should be assessed by preliminary tests. A sufficient extraction of BTEX was achieved by shaking the groundwater samples for 12 hours using *n*-pentane or *n*-hexane as a solvent [45,46]. Carbon and hydrogen isotope ratios of BTEX were determined for concentrations of approximately 0.02 and 0.4 mg/L by the extraction of one liter groundwater with 2 mL *n*-pentane [21]. The efficiency of the solvent extraction is depended on the water solubility and polarity of the compound of interest. A lowering of the extraction efficiency can be caused by groundwater ingredients such as tensides and humic substances. The solvent extraction may be increased by an optimal adjustment of the pH-value or the addition of salt.

The purge and trap approach allows isotope analysis for tiny contaminant concentrations. The carbon isotope ratio of volatile organic compounds was measured for concentrations smaller than 0.2 μg/L [37]. The time for purge must be adjusted during method development experiments to avoid isotope effects. However, no significant isotope effects were observed for this technique if purge times are optimized [37,47-49].

Isotope Fractionation Due to Microbial Degradation

The microbially induced isotope fractionation of a substrate depends on rate differences due to the difference in activation energy in cleavage of chemical bonds substituted by compounds only consisting of light isotopes (^{12}C, ^{1}H) or bearing a heavy isotope (^{13}C, ^{2}H). Due to a preferential transformation of the lighter isotope species the heavy isotopes will accumulate in the non-biodegraded residual fraction. The mass-dependent stability of a chemical bond leads to higher reaction rates of light isotope species in a (bio)chemical reaction and is mainly affected by the atoms directly involved in the cleavage reaction (primary isotope effect).

Atoms of a substrate, which are not directly involved in the initial step of the biochemical reaction mechanism, cause only insignificant or low isotope fractionation due to secondary isotope effects. The reaction type governs the extent of isotope fractionation and may allow for determination of the biochemical reaction pathway [22,50-56].

Theoretical Description of (Microbial) Isotope Fractionation

Microbial degradation of organic contaminants is constrained by the reaction kinetics of the degradation process. The quantification of such biodegradation processes using stable isotope analysis is based on the work of Mariotti et al. (1981) [57] who adapted the theoretical framework presented by Rayleigh (1896) [58] for liquid distillation to allow for the description of microbial isotope fractionation. This so-called Rayleigh (distillation) model considers a closed system where the degradation of a substance A is the only process taking place.

The kinetic isotope fractionation factor refers to irreversible chemical reactions of the type

$$A + (\text{reaction partners}) \xrightarrow{r} B + (\text{further products}) \qquad (3)$$

where the progress of the reaction is controlled by the reaction rate r. It is considered that an arbitrary element X with two stable isotopes is present in species A and B. The ratio

$$R = {}^{h}X / {}^{l}X \tag{4}$$

between the concentration of the heavy, ${}^{h}X$, and the light isotope, ${}^{l}X$, may differ between species A and B. If in species A the binding of the heavier isotope is stronger than the binding of the lighter isotope the breaking of such chemical bonding occurs slower if the heavier isotope is involved. This kinetic isotope effect (KIE) results in a slower consumption of the fraction of A containing the heavier isotope and as a consequence also a slower production of the corresponding fraction of B. As a consequence, the isotopic composition of A and B changes throughout the course of the reaction leading to an enrichment of the heavier isotope in the remaining reactant. Note that the normal kinetic isotope fractionation describes the slower consumption of the heavier isotope compared to lighter species which is commonly the case. An inverse isotope effect i.e. faster transformation of heavier isotope species is rare but can occur in complex reactions as well [59-62].

To quantify the kinetic isotope fractionation an isotope fractionation factor α can be defined as:

$$\alpha = \frac{{}^{h}r / {}^{h}X_A}{{}^{l}r / {}^{l}X_A} \tag{5}$$

with ${}^{h}X_A$ and ${}^{l}X_A$ as the concentration of the reactant containing the heavy and the light isotope and ${}^{h}r$ and ${}^{l}r$ describing their consumption rate, respectively. Note that the total concentration of A is given by ${}^{h}X_A + {}^{l}X_A$ and that ${}^{h}r + {}^{l}r$ represents the total rate of the reaction. Other species involved in the reaction are considered as isotopically neutral. However, the magnitude of the kinetic isotope fractionation factor varies with the reaction partner of the reactant of interest and depends, therefore, on the specific reaction (e.g., aerobic vs. anaerobic degradation). Thus, deriving values for α from known KIE requires detailed knowledge on the proceeding reaction (see [22] for a detailed review). However, for a given reaction the isotope fractionation factor is commonly considered as a constant (see sections below for a discussion of some limitations of this assumption).

Considering that ${}^{h}r = -d{}^{h}X_A / dt$ and ${}^{l}r = -d{}^{l}X_A / dt$, and rearranging Eq. 5 leads to

$$\frac{d{}^{h}X_A / dt}{{}^{h}X_A} = \alpha \frac{d{}^{l}X_A / dt}{{}^{l}X_A} \tag{6}$$

which can be integrated leading to

$$\ln\left({}^{h}X_{A} \big/ {}^{h}X_{A,0}\right) = \alpha \cdot \ln\left({}^{l}X_{A} \big/ {}^{l}X_{A,0}\right) \tag{7}$$

and

$$\frac{{}^{h}X_{A}}{{}^{h}X_{A,0}} = \left(\frac{{}^{l}X_{A}}{{}^{l}X_{A,0}}\right)^{\alpha} \tag{8}$$

where ${}^{h}X_{A,0}$ and ${}^{l}X_{A,0}$ are the initial concentrations of the reactant containing the heavy and the light isotope, respectively.

Furthermore, the fraction f of the initial amount of the reactant left after a certain amount of time is defined as

$$f = \frac{{}^{h}X_{A} + {}^{l}X_{A}}{{}^{h}X_{A,0} + {}^{l}X_{A,0}} \tag{9}$$

which can be transformed to

$$f = \frac{R \cdot {}^{l}X_{A} + {}^{l}X_{A}}{R_{0} \cdot {}^{l}X_{A,0} + {}^{l}X_{A,0}} = \frac{{}^{l}X_{A}}{{}^{l}X_{A,0}} \cdot \frac{R_{A} + 1}{R_{A,0} + 1} \tag{10}$$

with R_A and $R_{A,0}$ as the ratio between the heavy and the light isotope (as defined by Eq. 4) in the reactant and its initial value, respectively. Combining Eq. 8 with Eq. 10 then leads to

$$\frac{R_{A}}{R_{A,0}} = \left(\frac{f}{\dfrac{R_{A}+1}{R_{A,0}+1}}\right)^{\alpha-1} \text{ or } f = \frac{R_{A}+1}{R_{A,0}+1} \cdot \left(\frac{R_{A}}{R_{A,0}}\right)^{\frac{1}{\alpha-1}} \tag{11}$$

which links the change in the isotopic composition of a reactant to its total consumption.

Eq. 11 is generally valid for irreversible reactions in closed systems (no further processes taking place) not implying any restrictions on the kinetics of the reactions. A further simplification of Eq. 11 can be achieved if the heavy isotope is much less abundant than the lighter one (${}^{h}X_{A} << {}^{l}X_{A}$), which is commonly the case for the natural abundance of many stable isotopes (e.g., H, C, N, O, S, Cl). In this case the approximation $R_A+1 \approx 1$ is valid and Eq. 11 transfers to

$$\frac{R_{A}}{R_{A,0}} = (f)^{\alpha-1} \text{ or } f = \left(\frac{R_{A}}{R_{A,0}}\right)^{\frac{1}{\alpha-1}} \tag{12}$$

which is called Rayleigh equation in analogy to the formalism describing the fractional distillation of a mixed liquid introduced by Rayleigh (1896) [58]. In the remainder of this text the term "Rayleigh equation" always refers to Eq. 12 whereas the term "extended Rayleigh equation" refers to Eq. 11.

Commonly, kinetic isotope fractionation effects are relatively small, which results in fractionation factors close to 1. It is therefore convenient to introduce an enrichment factor ε (given in ‰) with

$$\varepsilon = (\alpha - 1) \cdot 1000 \tag{13}$$

Converting the Rayleigh equation into a logarithmic form:

$$\ln\left(\frac{R_A}{R_{A,0}}\right) = (\alpha - 1) \cdot \ln(f) = \varepsilon/1000 \cdot \ln(f) \tag{14}$$

shows a simple linear relation between $\ln(R_A/R_{A,0})$ and $\ln(f)$ (Figure 2). This so-called Rayleigh plot provides the opportunity for testing the validity of the Rayleigh equation and/or determining values for α or ε using experimental data (see sections below).

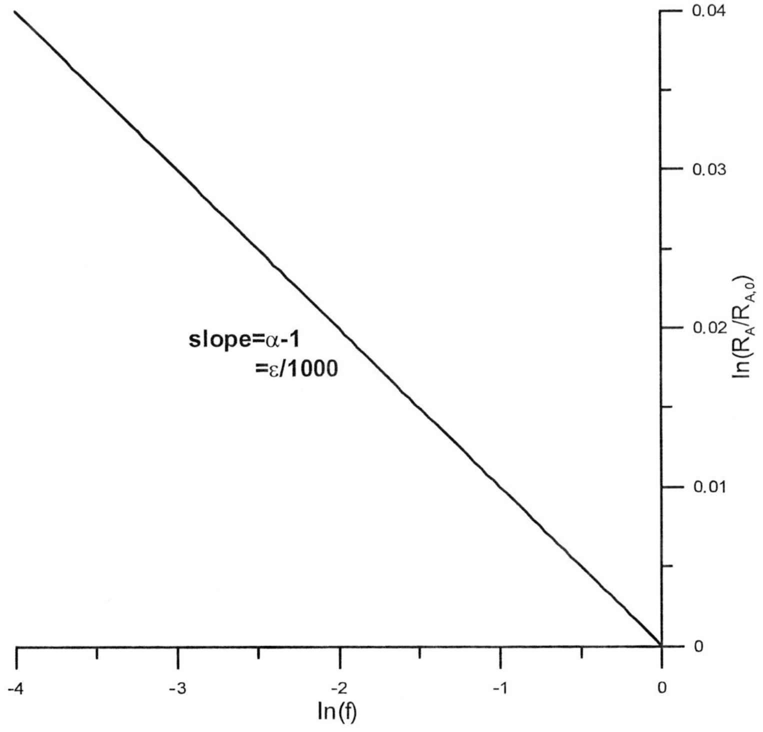

Figure 2. Hypothetical Rayleigh plot - double logarithmic graphical representation of the Rayleigh equation (Eq. 14) - for the degradation of an arbitrary substrate with ε = -10‰ (α = 0.990).

The Rayleigh equation predicts not only the change in the isotopic composition of the degraded substrate *A* but also allows the calculation of the isotopic composition of the product *B*. At any given time the isotope ratio R_B^* of the newly produced reaction product is given by the ratio

$$R_B^* = \frac{d^hX_B / dt}{d^lX_B / dt} = \frac{{}^hr}{{}^lr} \qquad (15)$$

using Eq. 5, Eq. 15 can be transformed to

$$R_B^* = \alpha \cdot R_A \qquad (16)$$

which provides another definition of the kinetic isotope fractionation factor, α (*57*).

Furthermore, the isotope composition of the accumulated reaction product, R_B, is given as

$$R_B = \frac{{}^hX_B}{{}^lX_B} = R_{A,0} \cdot \frac{1-f^\alpha}{1-f} \qquad (17)$$

if $R_A+1\approx 1$ can be assumed and if the initial product concentration is 0. These related changes of R_A, R_B^* and R_B are illustrated in Figure 3.

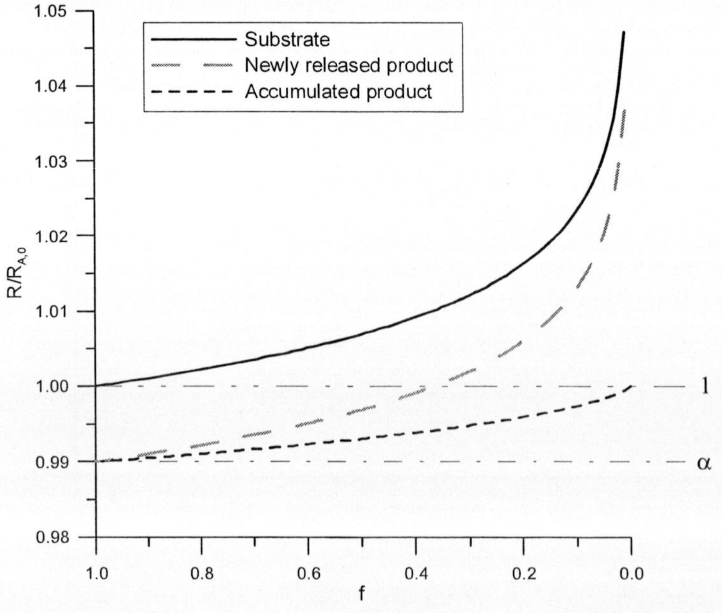

Figure 3. Hypothetical changes in isotopic composition during the kinetically controlled conversion of a substrate *A* into a reaction product (according to Eq. 15, 16 and 17) for an isotope fractionation factor α=*0.990*.

Isotope fractionation, and enrichment factors, are obtained experimentally from controlled laboratory experiments for which the changes in concentrations and isotope ratios of the substrate are only affected by the biodegradation activity of the microorganisms. From such experimental data α and ε are derived using the logarithmic form of the Rayleigh equation (Eq. 14) (if $R_A+1\approx 1$) and plotting $\ln(C_t/C_0) = \ln(f)$ versus $\ln(R_t/R_0)$ in a Rayleigh plot (Figure 4). The slope of the linear regression, m, allows to determine α and ε according to:

$$\alpha = m + 1 \tag{18}$$

$$\varepsilon = m \cdot 1000 \tag{19}$$

Further methods and modifications of this approach are discussed in the literature (*63,64*).

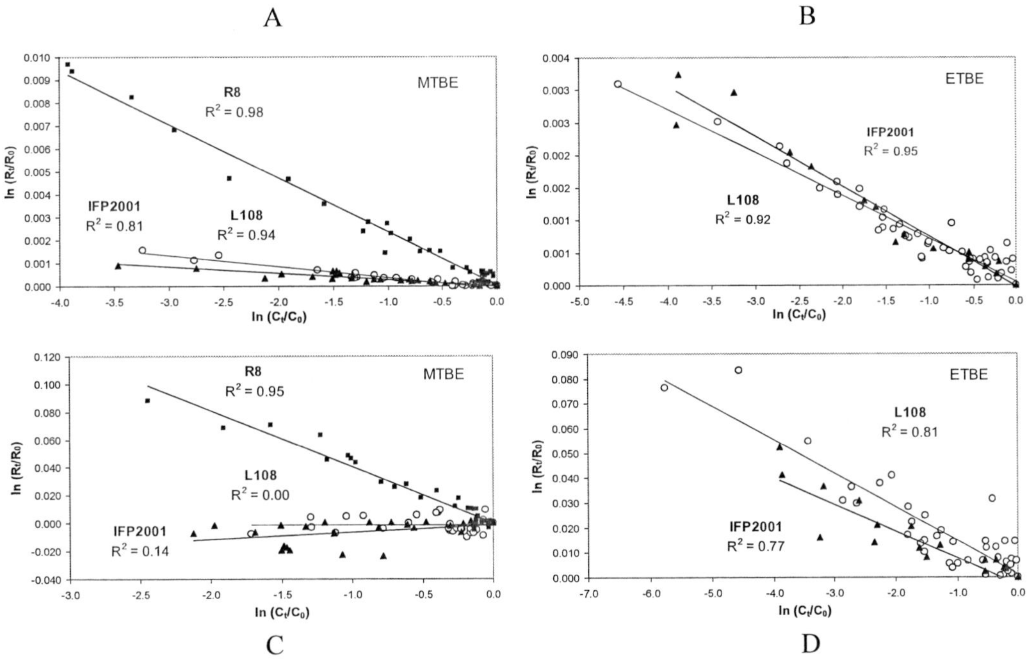

Figure 4. Double logarithmic plot according to the Rayleigh equation of the isotopic composition versus the residual concentration of substances (A) Carbon isotope fractionation of MTBE (B) Carbon isotope fractionation of ETBE (C) Hydrogen isotope fractionation of MTBE and (D) Hydrogen isotope fractionation of ETBE; by strain L108 (open circles); strain R8 (solid squares) and IFP 2001 (solid triangles). The lines correspond to a linear regression model [80].

It should be taken into account that for microbial fractionation of isotopically labelled substrates $R_A+1\approx 1$ is often not fulfilled, and therefore, the Rayleigh equation (Eq. 12) cannot used a basis for the estimation of α and ε. In this case the "extended Rayleigh equation" (Eq. 11) has to be applied instead [65].

An overview on enrichment factors for contaminant biodegradation under several redox conditions is given in the literature [16,18] or in data bases in the internet [66]. It is obvious that microbial isotope fractionation is less pronounced for molecules with a high number of carbon atoms. Thus, enrichment factors for biodegradation of BTEX, naphthalene or methylnaphthalene are low compared to values obtained for the transformation of chlorinated ethenes [16,18]. The reason for less pronounced microbial isotope fractionation in the degradation of larger molecules is that only the isotopic composition of atoms at the reaction position has a considerable impact on isotope fractionation. The isotopic composition of the remainder of the molecule does not have a significant influence. Therefore, changes of the isotope signature due to an isotope effect at the reaction position will be diluted by the rest of the non-reacting part of the molecule. Thus, the larger the molecule the smaller is the bulk isotope fractionation effect.

Hence, using stable carbon isotope analysis to proof *in situ* biodegradation is more a common approach for small molecules than for larger molecules, and it is assumed that carbon isotope fractionation during biodegradation of higher polyaromatic hydrocarbons (PAH) is too low to be analyzed by CSIA [67,68]. In this case, it might be helpful to measure hydrogen isotope fractionation because this process is more sensitive to biodegradation reactions. However, the fact that the isotope signature of higher weight molecules is not or only weakly influenced by biodegradation allows a straightforward identification of the source of product compounds using CSIA [69-71].

Field Applications of Compound-Specific Stable Isotope Analysis (CSIA)

In laboratory or column experiments isotope fractionation has been reported for biodegradation of environmental contaminants, such as petroleum hydrocarbons [14,35,40,41,72-77], fuel oxygenates [39,78-82] and chlorinated compounds [83-92] and may serve as an indicator for *in situ* biodegradation. Using CSIA, biodegradation pathways of these contaminants were deciphered within contaminated aquifers [21,79,89,93,94]. Moreover, there is a large number of publications showing the application of the Rayleigh equation-based approach for quantification of *in situ* contaminant biodegradation [12-14,20,45,46,79,94-97]. Thus, CSIA seems to be an appropriate tool for monitoring contaminant biodegradation at the field scale. In the following sections advantages and limitations of this approach will be discussed in more detail.

Analysis of Degradation Pathways

Isotope fractionation patterns have potential to study degradation processes. On one hand, the extent of isotope fractionation can be used to analyze the transition state of a chemical reaction which reflects the reaction mechanism of the chemical bond cleavage to some extent [50]. On the other hand, the extent of isotope fractionation can be used to characterize rate limitations of a biologically catalyzed reaction [50]. The rate limitation on preceding reaction steps of the overall biological reaction such as transport to and binding to an enzyme can lower the extent of isotope fractionation which will complicate the interpretation of isotope effects in biological systems [50]. The tendency of the biological catalyzed reaction to continue forward and becoming irreversible was described as "forward commitment to catalysis" [50]. Depending on the "forward commitment to catalysis" isotope fractionation can significantly vary on the chemical mechanism of the bond cleavage as well as on non isotope sensitive steps of a particular biodegradation pathway [90].

For a mechanistic interpretation of the observed isotope discrimination the enrichment factors have to be converted into the apparent kinetic isotopic effect (AKIE) of the bond cleavage taking into the account the effects of nonreacting positions and intramolecular isotopic competition [22]. The AKIE describes the isotope discrimination of the bond cleavage and can be used to compare isotope fractionation of different compounds. According to Elsner et al. (2005), the enrichment factor calculated from isotopic changes of the chemical (ε_{bulk}) is be used to calculate the enrichment factor specific for the reactive position ($\varepsilon_{reactive\ position}$):

$$\varepsilon_{reactive\ position} \approx \frac{n}{x \cdot \varepsilon_{bulk}} \tag{20}$$

where n is the number of the atoms of an element considered of which x are located at the reactive site [22]. Using $\varepsilon_{reactive\ position}$, an AKIE can be calculated by:

$$AKIE = \frac{1}{1 + z \cdot \varepsilon_{reactive\ position} / 1000} \tag{21}$$

where z is number of the atoms of an element in indistinguishable, reactive positions between which intermolecular isotopic competition exists. Re-evaluation of existing ε_{bulk} from literature values results in consistent ranges of AKIE (subscripts C and H indicate values for carbon and hydrogen): $AKIE_C = 1.01\text{-}1.03$ and $AKIE_H = 2\text{-}23$ for oxidation of C-H bonds; $AKIE_C = 1.03\text{-}1.07$ for S_N2-reactions; $AKIE_C = 1.02\text{-}1.03$ for reductive cleavage of C-Cl bonds; $AKIE_C = 1.00\text{-}1.01$ for C=C bond epoxidation; $AKIE_C = 1.02\text{-}1.03$ for C=C bond oxidation by permanganate [22].

Depending on the influence of the commitment to catalysis, isotope fractionation of an element can significantly vary for a given biodegradation pathway. So different enrichment factors were observed for aerobic toluene biodegradation related to the growth temperature as well as effects of trace element concentrations and for reductive dechlorination of tetrachloroethene [74,90,98]. Hence, the characterization and assessment of biodegradation using stable isotope fractionation analysis of only one element can lead to misinterpretation. Since commitment to catalysis affects elements involved in the bond conversion similarly, two-dimensional isotope fractionation analysis is a valuable tool for determining biodegradation pathways or conditions [22].

Using this technique, it could be determined for a BTEX contaminated aquifer that benzene is mainly biodegraded under anoxic conditions [21]. Therefore, theoretical carbon and hydrogen isotope ratios (δ_t) were calculated by means of the Rayleigh equation using arbitrarily chosen changes in benzene concentrations (C_t/C_0), highest and lowest known enrichment factors (ε_C, ε_H) for aerobic and anaerobic benzene biodegradation [35,41], and measured carbon and hydrogen isotope ratios of the benzene source (δ_0). The theoretical ranges of isotope signature patterns obtained for aerobic and anaerobic benzene degradation were plotted together with the isotope ratios detected for the sampling wells of the investigated transect (Figure 5). Most of the measured data exhibited carbon and hydrogen isotope signatures which are indicative of anaerobic benzene biodegradation.

Two-dimensional isotope fractionation analysis was also applied to other contaminants. As a result, conditions under which methyl *tert*-butyl ether (MTBE) was biodegraded in contaminated aquifers could be identified by measuring carbon and hydrogen isotope signatures [49,79,94]. Furthermore, the evaluation of defined laboratory studies as reference show the potential to study pathways for benzene, toluene and MTBE biodegradation using two-dimensional isotope fractionation analysis [35,41,80].

In the future, it is envisioned that CSIA will be available for measuring oxygen, chlorine, sulfur or nitrogen isotope signatures of contaminants, routinely. Hence, it should be possible to enlarge the two-dimensional isotope fractionation analysis to multi-dimensional isotope fractionation analysis. This gives the opportunity for a detailed and precise deciphering of degradation pathways of contaminants comprising fuel oxygenates, explosives, heterocyclic aromatic and aliphatic compounds, as well as chlorinated organic substances.

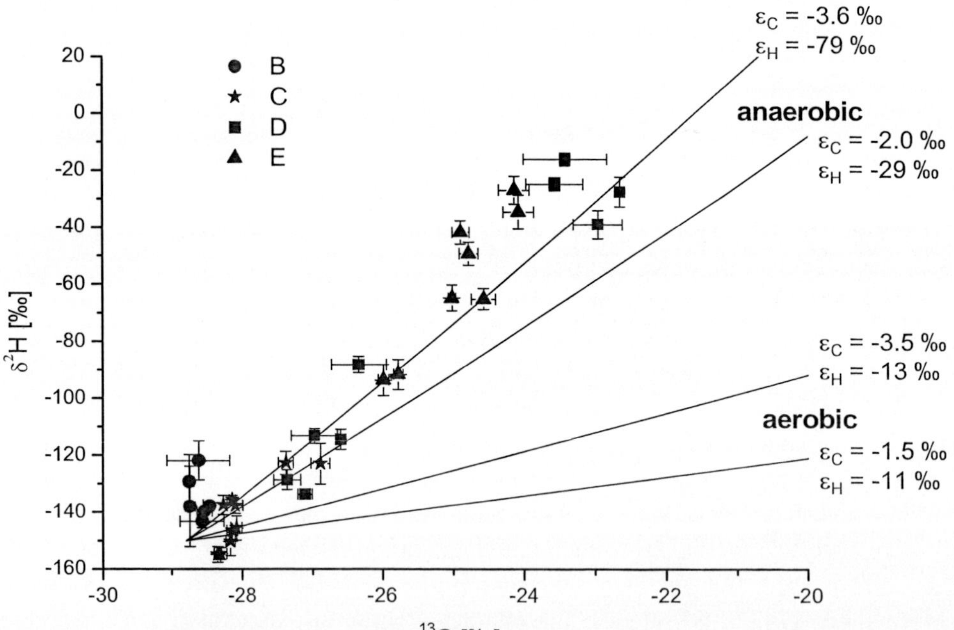

Figure 5. Concurrent carbon and hydrogen isotope ratios of benzene measured at the various sampling depths of wells B - E together with isotope patterns for aerobic and anaerobic benzene degradation calculated from published enrichment factors for carbon (ε_C) and hydrogen (ε_H) as well as the isotope signature of the contaminant source using the Rayleigh equation. The reproducibilities of $\delta^{13}C$- and δ^2H-values are given as error bars and were always smaller than 0.5 ‰ and 10 ‰, respectively. Error bars which are smaller than 0.08 ‰ for carbon and 2.0 ‰ for hydrogen are not visible due to the size of the symbols [21].

Quantification of *In Situ* Biodegradation

Applying the formalism described in the section on “Theoretical description of (microbial) isotope fractionation“ to the quantification of the biodegradation of contaminants in the environment is straightforward when considering species *A* as the contaminant degraded by the microorganisms. The extent of contaminant biodegradation is commonly expresses as percentage, *B*[%], of the initial contaminant concentration removed due to biodegradation:

$$B[\%] = (1 - f) \cdot 100 \tag{23}$$

Combining this expression with the Rayleigh equation allows quantifying contaminant biodegradation within a time interval or along a migration path of the contaminant in a given environment using measured isotope signatures, only. Required data are the initial isotope ratio of the contaminant at a starting point in time or in space and the isotope ratio of the remaining contaminant at a temporal or spatial observation point. The amount of contaminants degraded between the initial and observation point is then given by

$$B[\%] = (1 - f) \cdot 100 = \left(1 - \left(\frac{R_A}{R_{A,0}} \right)^{\frac{1000}{\varepsilon}} \right) \cdot 100 \tag{24}$$

using the Rayleigh equation (Eq. 12) or by an analogous expression in case the extended Rayleigh equation (Eq. 11) needs to be used instead.

In several studies the Rayleigh equation based approach was used to quantify the biodegradation of contaminants in contaminated aquifers [12-14,20,45,46,79,94-97]. Appropriate enrichment factors from laboratory experiments and field measurements of changes in isotope signatures of the contaminant were used to calculate the decrease in the contaminant concentration due to biodegradation along a flow path between a contaminant source and a sampling well.

Furthermore, apparent *in situ* first-order biodegradation rate constants (λ) can be estimated by changes in isotope ratios between a contaminant source and a sampling well using a Rayleigh equation based approach [99].

$$\lambda = \frac{-1000}{\varepsilon \times t} \ln \left(\frac{R_t}{R_0} \right) \tag{25}$$

Time (t) can be assessed by the groundwater flow velocity (v) and the distance between the sampling well and the contaminant source (s):

$$t = \frac{s}{v} \tag{26}$$

Limitations of Rayleigh Equation Approaches for the Quantification of *In Situ* Biodegradation

The standard Rayleigh equation considers only a single process affecting simultaneously the concentration and the isotope ratio of a contaminant, a constraint often not fulfilled in the field. In many field cases, contaminant concentrations can also be affected by non-isotope-fractionating abiotic processes such as dilution, volatilization or sorption to the aquifer matrix. As the Rayleigh equation links changes in isotope ratios to concentration shifts, these abiotic processes need to be taken into account, even if they do not change isotope ratios [20,21,100-102].

Furthermore, the interpretation of data obtained in the field is subject to inherent uncertainties due to the complexity of groundwater systems. Besides errors introduced by the sampling or analytical technique [14,20,45,103], uncertainties include the proper identification of groundwater flow paths, which might be required to determine the initial isotope ratio of the contaminants at sites affected by contaminants sources with different isotopic compositions [95,104]. In addition, using isotope enrichment factors measured in the laboratory to describe the *in situ* processes taking place in the field also represents a potential errors source as apparent enrichment factors in the field may depend on specific microbial degradation pathways and on environmental factors not fully represented in the laboratory system used as a reference [102].

The following sections address first the errors in the quantification of biodegradation caused by improperly constrained input parameters of the Rayleigh equation. Thereafter, the impact of additional transport and abiotic reactive processes on the applicability of the Rayleigh equation to field cases is discussed.

Sensitivity to Input Parameters

The sensitivity of a term T towards variations or uncertainties in a parameter p used for its calculations is given as [105]

$$\Delta T = \frac{\partial T}{\partial p} \cdot \Delta p \tag{27}$$

where ΔT is the change in T caused by varying p by Δp.

Applying this concept to the calculation of $B[\%]$ using the Rayleigh equation, errors ($\Delta B[\%]$) for the quantification of $B[\%]$ due to uncertainties in ε, or measured isotope ratios at an observation point (R_x) or at the contaminant source (R_0) can be determined:

$$\frac{\Delta B[\%]}{B[\%]} = \frac{100 - B[\%]}{B[\%]} \cdot \ln\left(\frac{100 - B[\%]}{100}\right) \cdot \frac{\Delta \varepsilon}{\varepsilon} \tag{28}$$

$$\frac{\Delta B[\%]}{B[\%]} = -\frac{1000}{\varepsilon}\cdot\frac{100-B[\%]}{B[\%]}\cdot\frac{\Delta R_x}{R_x} \approx -\frac{100-B[\%]}{B[\%]}\cdot\frac{\Delta\delta_x}{\varepsilon} \tag{29}$$

$$\frac{\Delta B[\%]}{B[\%]} = \frac{1000}{\varepsilon}\cdot\frac{100-B[\%]}{B[\%]}\cdot\frac{\Delta R_0}{R_0} \approx \frac{100-B[\%]}{B[\%]}\cdot\frac{\Delta\delta_0}{\varepsilon}. \tag{30}$$

These expressions show that in each case the relative accuracy of $B[\%]$ is given by the product of a term depending on $B[\%]$ itself and a term representing the uncertainty of the input parameter in relation to the isotope enrichment factor. For a given ratio between $\Delta\delta_x$ or $\Delta\delta_0$ and ε, respectively, the relative accuracy of $B[\%]$ is lowest at low amounts of biodegradation and is monotonously increasing towards higher amounts of biodegradation, whereas the sensitivity to $\Delta\varepsilon$ is highest at intermediate amounts of biodegradation (see Tab. 1 for an example). The above equations also indicate that in the typical case of negative ε values a conservative estimate of $B[\%]$ is obtained when using the most negative potential value for ε. However, the use of the most negative potential value for ε can lead to a significant underestimation of $B[\%]$. Therefore, in recent studies it was tried to find an optimal method for the assessment of an appropriate value of ε [106,107].

In the field, each of the input parameters is affected by a certain uncertainty/error and the combined effect of these errors on $B[\%]$ is given as

$$|\Delta B| = \frac{1}{\varepsilon}\cdot(100-B[\%])\cdot\sqrt{\left(\ln\left(1-{}^{B[\%]}\!\big/_{100}\right)\right)^2\cdot(\Delta\varepsilon)^2+(\Delta\delta_x)^2+(\Delta\delta_0)^2} \tag{31}$$

These combined effects are illustrated using data on benzene degradation from a field study published by Mancini et al. (2002) [95]. There, three different source zones for benzene could be determined each exhibiting individual isotope ratios for carbon and hydrogen. In contrast to Mancini et al. (2002), who tried to allocate which source zone affected the results at each specific sampling well, we here use for demonstration purpose the average value and standard deviation of the source zone data obtained for carbon (R_0 = -28.1 ‰ ± 0.7 ‰) and for hydrogen (R_0 = -37.8 ‰ ± 4.0 ‰). At each sampling location typical uncertainty ranges for measured carbon and hydrogen isotope ratios were 0.2 ‰ and 2.0 ‰, which were taken as respective values for ΔR_x. Values for isotope enrichment factors during anaerobic benzene degradation vary for different redox conditions and without knowledge of the actual degradation conditions the most appropriate enrichment factor is not known. Mancini et al. (2003) determined enrichment factors for benzene degradation under different redox conditions (nitrate reduction, sulfate reduction, methanogenesis) [41] and for the example calculations here we used the average and standard deviation of the values for each condition to determine ε_C = *-2.6 ‰ ± 0.9 ‰* and ε_H = *-57.3 ‰ ± 23.1 ‰*. With these numbers values for *B[%]* and *ΔB[%]* were calculated using Eqs. 24 and 31 for arbitrary shifts in the isotope signals (Figure 6). Results show that for both, carbon and hydrogen isotope signatures, predicted amounts of biodegradation are subject to significant errors. However, errors do not exceed 15 %, and thus results still provide a reasonable estimate of *in situ* contaminant biodegradation at that site.

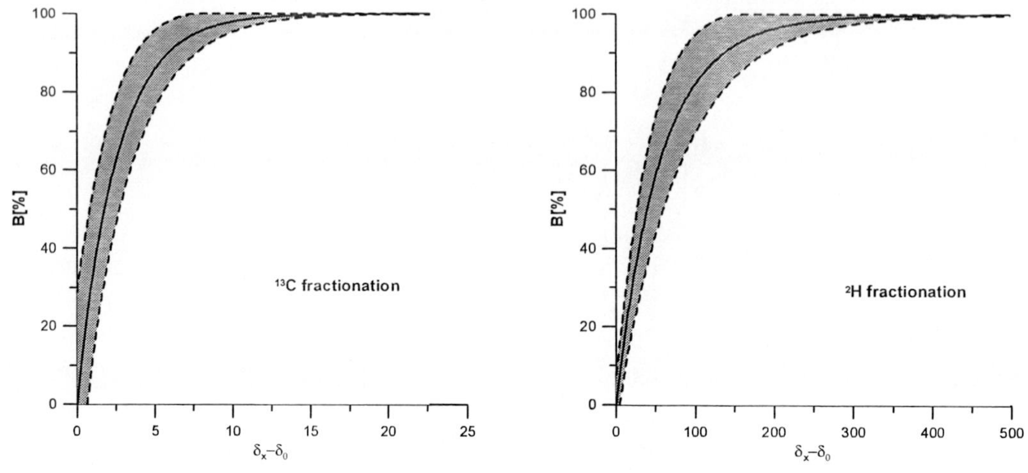

Figure 6: Uncertainty in B[%] due to limited knowledge on enrichment factors, isotopic composition of source zone and analytical errors for a field case example [*95*]. Plots show calculated values of B[%] using averaged input parameters (solid lines) and error margins (dashed lines), $\pm\Delta B$[%] calculated as combined effect of input parameters errors. Only data with $0 \leq B[\%] \leq 100\%$ are shown.

Nevertheless, the example also highlights the need to reduce uncertainties by e.g., using carbon and hydrogen isotopes in a combined approach as done by Mancini et al. (2002) for the shown example [95] or suggested by other authors [21,79,94]. Such a combined approach might also lead to an identification of dominant degradation pathways allowing to select the most appropriate enrichment factor for calculations (see section on "Analysis of degradation pathways"). However, the influence of environmental factors on the apparent isotope fractionation factor representative for *in situ* biodegradation still needs to be considered.

Sampling Related Problems

Heterogeneity of aquifers can result in varying extents of biodegradation within different groundwater flow paths which may affect the quantification by the Rayleigh equation approach and lead to an underestimation of biodegradation [14,20,45,103]. If different extents of biodegradation occur on distinct flow paths and if mixing of these flow paths takes place in at a sampling well, the isotope ratio of contaminants will be dominated by the fraction contributing the higher amount of contaminant, which is commonly the less biodegraded fraction. Groundwater contaminants often exhibit a vertical concentration gradient along the depth of an aquifer. Especially in the case of petroleum hydrocarbons, the assumption of a homogeneous vertical distribution of solutes within an aquifer is not justified. Most of these species are nonaqueous-phase liquids, which are lighter than water (LNAPLs), and thus, they float on the groundwater table where their dissolution into groundwater takes place. Dissolved petroleum hydrocarbons, therefore, often exhibit distinct vertical concentration profiles with maximum solute concentrations near the water table. In addition, aquifer heterogeneities may lead to different groundwater flow velocities with depth and, thus, to aquifer depths depending travel times of dissolved contaminants between their source and the sampling location. Such hydrogeological heterogeneities of the aquifer may also be combined with heterogeneities concerning the degradation kinetics of a contaminant. The latter can be caused e.g., by geochemical heterogeneities or degradation preferentially occurring at the fringe of a plume [108-111] where the contaminated groundwater is mixing with pristine groundwater containing electron acceptors. As a consequence, the vertical concentration distribution of the contaminant may exhibit a complex pattern not to be known a priori. Usually, measurements taken in fully screened observation wells represent the flux weighted average solute concentration for the screened section of the aquifer independently from the position of the pump [112]. These data contain no information on the vertical gradients of the measured parameters. Furthermore, isotope signals obtained for such samples only partially reflect the isotope fractionation processes in the aquifer, and thus, the biodegradation of the contaminant [103].

For an appropriate evaluation of samples from a fully screened well one would need the spatial distribution, $Q(s)$, of the influx along the length, s, of the well screen. With $Q_0 = \int_{s_{\min}}^{s_{\max}} Q(s)ds$ the relative flux distribution can be defined as $q(s)=Q(s)/Q_0$ with $\int_{s_{\min}}^{s_{\max}} q(s)ds = 1$. Furthermore, the distribution of contaminant concentration, $C(s)$, and isotopic composition, $R(s)$, would have to be obtained, too. Concentration, C_m, and isotope ratio, R_m, of the bulk sample are linked to the screen length related values via

$$C_m = \int_{s_{\min}}^{s_{\max}} q(s) \cdot c(s) ds \tag{32}$$

and

$$R_m = \frac{\int_{s_{\min}}^{s_{\max}} q(s) \cdot c(s) \cdot R(s) \cdot ds}{\int_{s_{\min}}^{s_{\max}} q(s) \cdot c(s) \cdot ds} \tag{33}$$

if $R+1 \approx 1$ can be assumed. This expression shows that the average isotope signal depends not only on the flux distribution along the well but also on the concentration distribution as additional weighting factor. Thus, those aquifer parts contributing most to the contaminant mass flux to the well are the most relevant for the measured average isotope signal. As a consequence, such averaged isotope signals are biased towards the signals indicating no or less biodegradation whereas high biodegradation rates in other parts of the aquifer are not well reflected in the mixed signal.

If the Rayleigh model describes the biodegradation along each individual flow path reaching the sampling well at a given depth in the aquifer $\left(\frac{R(s)}{R_0}=\left(\frac{C(s)}{C_0}\right)^{\alpha-1}\right)$ then the total amount biodegradation, $B[\%]_m$, is given as

$$B[\%]_m=\left(1-\int_{s_{\min}}^{s_{\max}} q(s)\cdot\frac{C(s)}{C_0}ds\right)\cdot 100$$
$$=\left(1-\int_{s_{\min}}^{s_{\max}} q(s)\cdot\left(\frac{R(s)}{R_0}\right)^{\frac{1}{\alpha-1}}ds\right)\cdot 100 \tag{34}$$

Obviously, $B[\%]_m\neq\left(1-\left(\frac{R_m}{R_0}\right)^{\frac{1}{\alpha-1}}\right)\cdot 100$ and thus, any quantification based on screen length averaged values, only, leads to a misinterpretation of the biodegradation at the investigated field site if contaminant concentration and/or stable isotope ratios exhibit vertical gradients.

To demonstrate the differences between the actual biodegradation taking place at a site and the apparent amount determined from average isotope ratios measured in the bulk sample, we here assume an example of an aquifer being subdivided into two distinct horizontal layers. The relative contributions of each layer to the total flux to the well are q and $(1-q)$, respectively. For this case, Eqs. 33 and 34 can be simplified to

$$R_m=\frac{q\cdot R_1^{\frac{\alpha}{\alpha-1}}+(1-q)\cdot R_2^{\frac{\alpha}{\alpha-1}}}{q\cdot R_1^{\frac{1}{\alpha-1}}+(1-q)\cdot R_2^{\frac{1}{\alpha-1}}} \tag{35}$$

and

$$B[\%]_m=\left(1-q\cdot\left(\frac{R_1}{R_0}\right)^{\frac{1}{\alpha-1}}-(1-q)\cdot\left(\frac{R_2}{R_0}\right)^{\frac{1}{\alpha-1}}\right)\cdot 100 \tag{36}$$

with subscript 1 and subscript 2 referring to layers 1 and 2, respectively. Even if both layers contribute equally to the flux into the well ($q = (1-q) = 1/2$) the isotopic signal of a sample taken from this well exhibits a non-trivial dependency on the isotope signals to be found in each layer (Figure 7a). Shifts in each of these isotope signals do indicate an increasing amount of biodegradation (Figure 7b) but a quantification of this amount by applying the Rayleigh equation to the average isotope signal of a bulk sample from the well does not lead to the same results (Figure 7c) and may result in predicting lower biodegradation at higher shifts in isotope signals. This leads to an underestimation of *in situ* contaminant biodegradation when using samples from such wells (Figure 7d). If the aquifer comprises more than only two layers and/or if the Rayleigh equation is not applicable to all of the aquifer layers an evaluation of such bulk samples is further challenged [103].

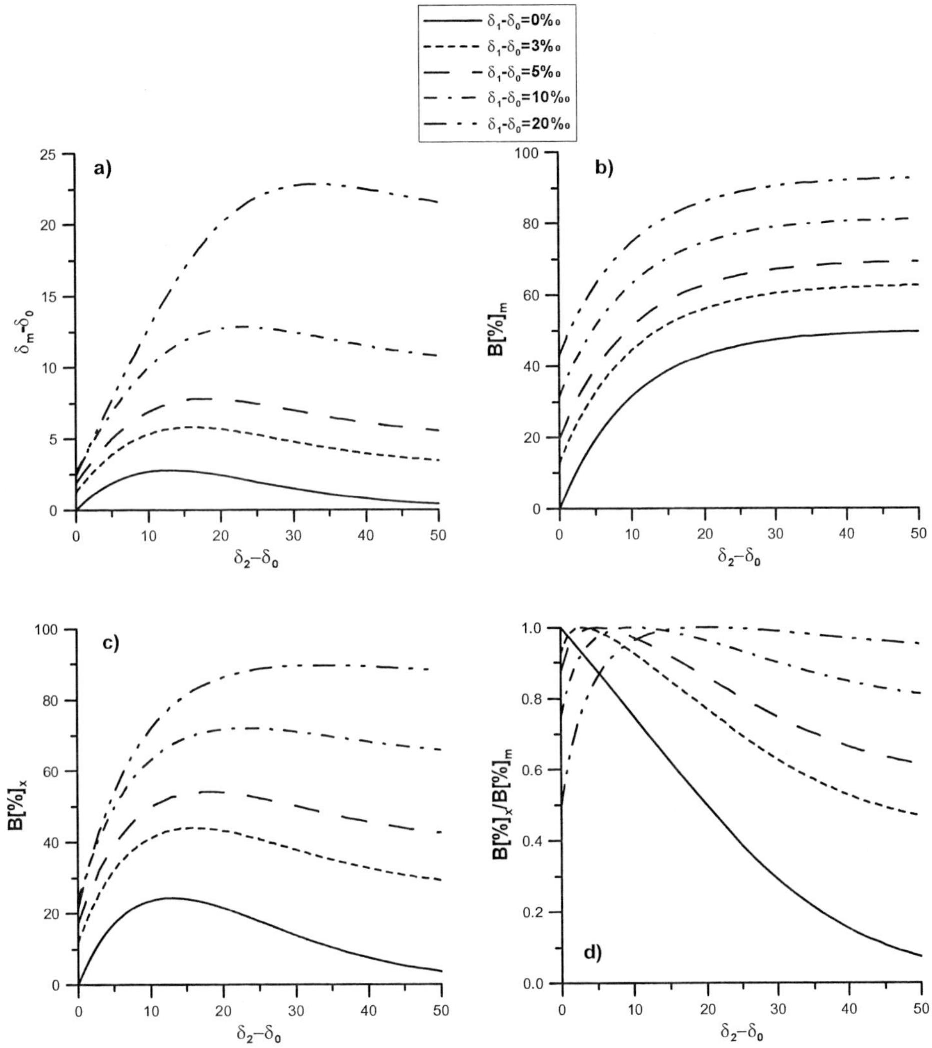

Figure 7. Contaminant isotope signatures and biodegradation amounts derived from changes in isotope signatures at a fully screened observation well in a two-layered aquifer. Both layers contribute equally to the flux into the well. Subscripts 1 and 2 refer to the respective layer, subscript 0 to contamination source. a) isotopic composition of a bulk sample from the well for given isotope signatures of each layer. b) actual biodegradation, $B[\%]_m$, taking place at the site for given isotope signatures of each layer. c) apparent biodegradation, $B[\%]_x$, as obtained when applying the Rayleigh equation to the average bulk isotope ratio. d) ratio between apparent and actual biodegradation, showing the fraction of actual biodegradation detected by the Rayleigh model when using the average bulk isotope ratio for calculation.

Isotope Effects Due to Abiotic Processes

Abiotic processes affecting the isotopic signature of the target compound have to be known in order to characterize or estimate biodegradation of organic contaminants in aquifers. No significant changes in the isotopic signature of contaminants occur during the dissolution with pristine water [34,40,44].

Volatilization leads to (usually insignificant) inverse isotope effects when a considerable amount of the compound of interest is transferred from the water to the gas phase [34,78,113-118]. In contrast to microbial isotope fractionation, inverse isotope effects due to volatilization cause an enrichment of light isotopes (^{12}C, ^{1}H) in the residual contaminant fraction dissolved in groundwater. If the aquifer is highly influenced by volatilization the microbial isotope fractionation would be diminished leading to an underestimation of *in situ* contaminant biodegradation by the Rayleigh equation approach. Since in most of the cases the volatilization of contaminants can be neglected within contaminated aquifers [11,119] this process does not play a significant role for the interpretation of isotope data.

Several laboratory studies show that sorption of contaminants causes no significant isotope fractionation, leading to the assumption that this process has no influence on the isotope signature of a contaminant dissolved in groundwater [120-122]. In contrast, a significant isotope effect for organic compounds due to consecutive sorption steps was observed during the transport through high performance liquid chromatography (HPLC) columns [103,123]. Sorption processes in aquifers are mainly affected by the concentration and quality of organic carbon of the aquifer matrix which can retard the migration of contaminants. Kopinke et al. (2005) predicted that for an aquifer where no biodegradation occurs sorption can lead to a significant isotope fractionation at the front of an expanding contaminant plume or a breakthrough curve occurring after a short contamination event [103]. In the case of a stationary contaminant plume with biodegradation activity, the degree of isotope fractionation due to sorption was predicted to be smaller and possibly insignificant [103]. In a field tracer test, the fate of ring-deuterated (d_5) and completely deuterium-labeled (d_8) toluene as reactive tracers was investigated within a BTEX-contaminated aquifer using bromide as a conservative tracer [124]. Given the average concentration of organic carbon of the aquifer matrix sorption was expected to alter the isotope composition of the migrating front. However, the fractionation of both isotopologues measured at the front of the tracer breakthrough curves was much lower than expected [124]. This indicates that reliable predictions using the organic carbon content to estimate isotope effects due to sorption processes in the aquifer are difficult to obtain. Longitudinal dispersion and preferential flow paths due to aquifer heterogeneities as well as the quality of organic matter with respect to sorption affect the chromatographic properties of an aquifer. Thus, further research is needed to clarify mechanisms and extent of isotope fractionation due to sorption at field sites.

Chemical reactions can lead to significant isotope effects. For the reductive dehalogenation of chlorinated methanes and ethenes by iron, a considerable carbon isotope fractionation was found [125-128]. Also for the hydrolysis of MTBE, carbon and hydrogen isotope fraction was observed [129]. However, abiotic dehalogenation and hydrolysis should be negligible in contaminated aquifers, therefore, no significant effect on contaminant isotope signature might be expected due to these processes.

Abiotic Non-Fractionating Changes of Concentrations

The fate of contaminants in the subsurface is determined by their microbial degradability but also by a variety of abiotic reactive as well as transport processes. Some of these processes can alter the isotopic composition and the concentration of a contaminant (see section above) but many of them have an impact on contaminant concentration only, leaving the isotope signature unchanged. As a consequence, the relation between concentration changes and isotope signatures predicted by the Rayleigh equation might not by valid in such cases and an application of the Rayleigh equation for the assessment of contaminant biodegradation may lead to a misquantification at a site. Of particular interest are abiotic processes which do not lead to a mass removal of the contaminant. For example, sorption and dispersive mixing in groundwater reduce the dissolved contaminant concentration, but in contrast to biodegradation these processes do not remove any contaminants from an aquifer and thus do not contribute to the remediation of a site [11]. Therefore, it is important to be able to distinguish between concentration changes due to biodegradation and due to other, abiotic processes.

In the following sections, we discuss the uncertainty caused by such non-fractionating processes in quantifying *in situ* biodegradation based on stable isotope fractionation.

Dispersion Along Flow Paths

Aquifers are characterized by heterogeneities at different length scales, ranging from the pore scale to regional scales [11]. As a consequence, velocities describing the movement of groundwater and chemical species dissolved in groundwater vary within and along flow paths connecting the source of a groundwater contamination and the location of a sampling well. This effect - called dispersion - phenomenologically results in a spreading of contaminant plumes or, regarded from a different perspective, in contaminant molecules exposed to different travel times along the same macroscopic flow path [130].

If biodegradation (or any other kinetically controlled consumption) takes place in the aquifer, the actual amount of degradation depends on the exposure time of the contaminant to the aquifer environment. Due to dispersion caused variability in travel times of the contaminants, the exposure time varies for contaminants traveling from their source to an observation well. Concentrations at the observation well represent, thus, a mixed signal showing the averaged concentration after different exposure times - or degrees of biodegradation. Therefore, isotope signatures reflecting the biodegradation process are also averaged at the sampling well [131]. Note that the mixing here is not caused by the sampling process itself as discussed above in section "Sampling related problems", but by the longitudinal and transversal dispersive mixing taking place in the aquifer itself.

To quantify this effect for aquifers of varying heterogeneities, numerical simulations of the biodegradation along a virtual flow path were performed using the BRNS simulation platform [132,133]. A 1-D flow path of 300 m length was simulated with a contaminant source represented by a fixed concentration at the inflow boundary of the flow path. Along the flow path, the contaminant is degraded following first order degradation kinetics, and the degradation is assumed to lead to an isotope fractionation with an isotope fractionation factor of $\alpha = 0.990$. Steady state simulations were performed for three different flow paths with varying heterogeneities obtaining the concentration profiles and isotopic composition of the contaminant along each flow paths. For each simulation, the advective transport velocity, v, and first order degradation rate parameter, k, were set to yield a Damköhler number, Da, of $Da = \frac{k \cdot x}{v} = 0.024 \cdot x$ with x as the distance between observation point and contaminant source. Aquifer heterogeneity is quantified by the (longitudinal) dispersivity, α_l, which was set to a value of 0.50 m, 5.00 m and 50.0 m for each of the three flow paths, respectively, covering the range of typical values for groundwater aquifers [134]. Simulations were performed using a spatial discretization of 0.20 m, which yields simulated concentrations and isotope ratios for 1500 "observation points" along the flow path. Data from these observation points were subsequently handled the same way as measurements from real field sites and the Rayleigh equation was applied to them. The simulated release of a degradation product was used to quantify the actual biodegradation along the virtual flow path.

Results show that for low heterogeneities (α_l = 0.5 m) the biodegradation calculated by the Rayleigh equation meets the actual values nearly exactly (Figure 8). However, with increasing dispersivity (and heterogeneity) predicted amounts of biodegradation derived from the Rayleigh equation tend to be smaller than the actual values. The relative importance of dispersive vs. advective transport can be expressed by the Peclet number, Pe, which is in the present example given as $Pe = \frac{x}{\alpha_l}$, with x again as the distance between the observation point and the source of the contamination (inflow boundary of the flow path). When plotting the ratio of Rayleigh derived and actual biodegradation vs. Pe results show that for small Pe (< 100) the use of the Rayleigh equation leads to a significant underestimation of the *in situ* contaminant biodegradation (Figure 9). Errors are highest for smallest Peclet numbers reaching more than 20 % at $Pe = 1$. These results are confirmed by Abe and Hunkeler (2006) who did a more rigorous analysis of such systems [131]. These authors also show that errors in the quantification of biodegradation tend to decrease with increasing Damköhler numbers, which explains the differences between the results from the individual flow paths in Figure 9 (in the present example $Da = \alpha_l \cdot \frac{k}{v} \cdot Pe$). Abe and Hunkeler (2006) also showed that with realistic ranges of Pe and Da the relative error in the quantification of biodegradation using the Rayleigh equation does not exceed 20 % even if the location of the sampling well is not positioned at the center line of the contaminant plume [131]. In all cases the Rayleigh equation underestimates the biodegradation at a site and thus provides a conservative estimate of the *in situ* biodegradation.

Mass Removal from a Flowpath

If contaminants are removed from the flow path by other processes than biodegradation, these processes need to be taken into account in the interpretation of measured isotope signatures also if these additional processes do not contribute to isotope fractionation [21].

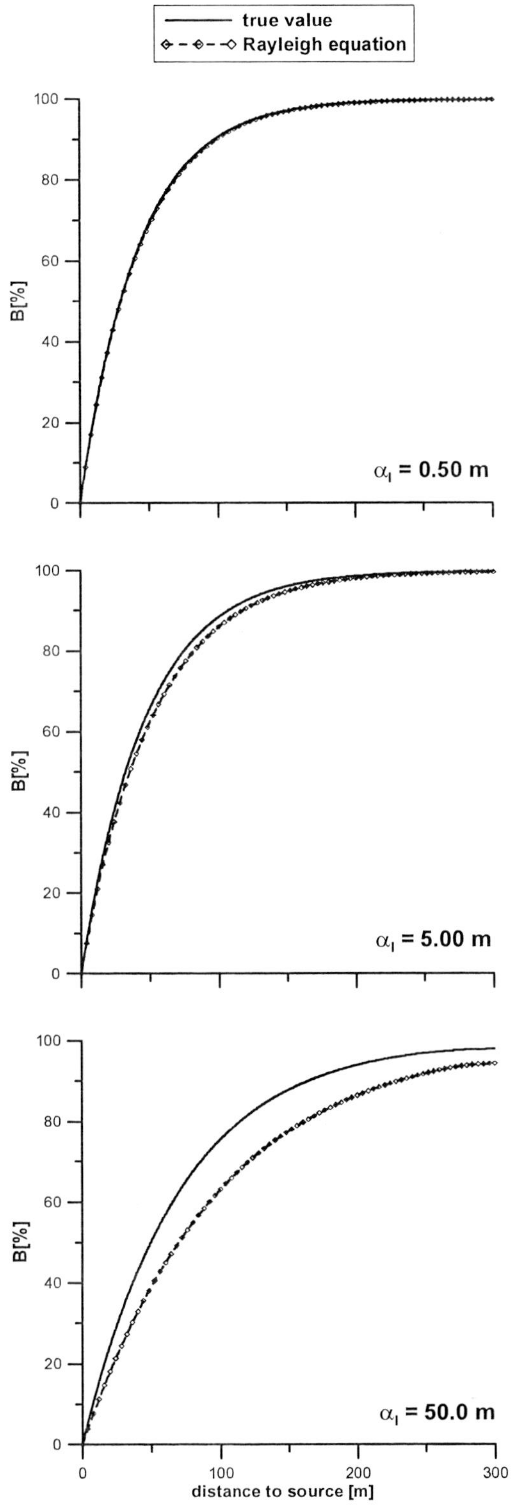

Figure 8. Comparison between actual amounts of biodegradation and estimates based on the Rayleigh equation along virtual groundwater flow paths with different heterogeneities (dispersivities).

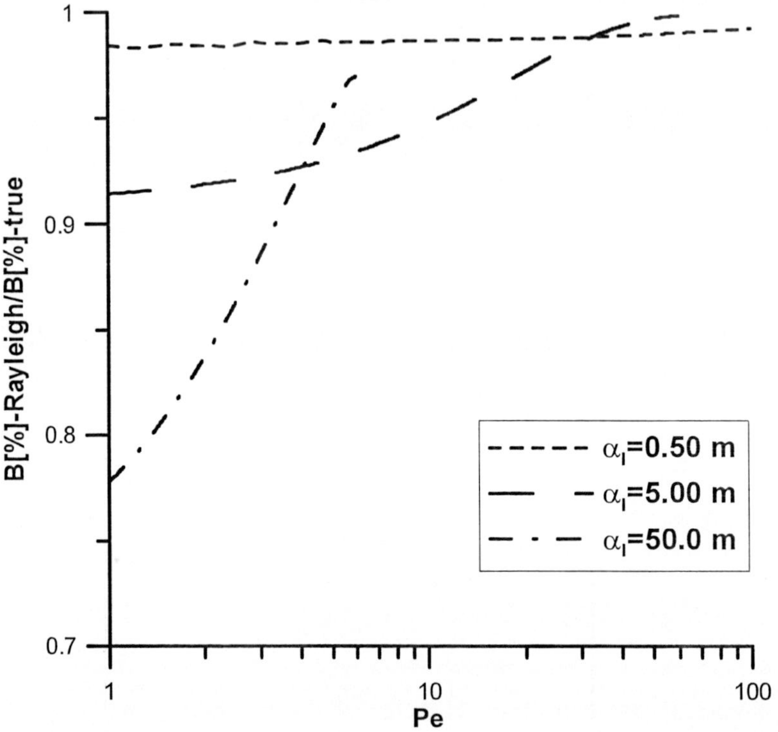

Figure 9. Relative accuracy of the Rayleigh equation for the quantification of biodegradation along virtual flow paths in heterogeneous aquifers.

Such processes might be irreversible sorption of the contaminant to the aquifer matrix, which results in an undesired accumulation of contaminants in the aquifer, or transversal dispersive mixing processes, which dilute the contaminated groundwater with pristine groundwater from outside the plume. Mixing processes occur mainly at the fringe of contaminant plumes and might be a prerequisite for biodegradation if electron acceptors are already depleted in the plume are left in the pristine water, only [108-111]. In addition and of main interest here is the effect that such mixing leads to a mass removal of contaminants from a flow path without removing the mass from the aquifer. The Rayleigh equation has been derived for closed systems where the degradation reaction is the only process taking place. Thus, any application of the Rayleigh equation for the quantification of biodegradation in the field is challenged by the presence of additional processes. Even if these processes do not contribute to isotope fractionation, they alter the relation between observed changes in concentration and shifts in the isotopic composition of a contaminant. Therefore, the precise quantification of biodegradation requires an expansion to the simple Rayleigh equation based approach [21] unless the aquifer is biogeochemically highly homogeneous. In the latter case it might be assumed that contaminants removed from a flowpath by mixing are biodegraded elsewhere in the aquifer at similar rates than along the flow path, which allows applying the simple Rayleigh equation based approach for the quantification of biodegradation [23].

In all other cases, quantification approaches need to consider the relative magnitude of these additional processes - e.g., mixing or irreversible sorption, and referred to as dilution in the remainder of this section - relative to biodegradation [135]. In addition, the impact of these processes on the quantification of biodegradation using the stable isotope signatures depends on whether these processes take place in parallel or sequentially to biodegradation. Biodegradation and dilution occurring strictly sequentially is not a realistic assumption for natural aquifer environments but highlights the range of possible amounts of biodegradation corresponding to a given change in isotope ratio (and concentration) of a contaminant. If a scenario is considered, where along a flowpath in a contaminated aquifer an initial phase of biodegradation is followed by a phase of dilution in a strictly sequential order, quantifying biodegradation with the simple Rayleigh equation (Eq. 24) would be an adequate approach. Vice versa, for a scenario with an initial phase of dilution and a subsequent phase of biodegradation, biodegradation in the aquifer can be quantified by

$$B[\%]=\left(\frac{\frac{c_x}{c_0}}{\left(\frac{R_x}{R_0}\right)^{\frac{1}{\alpha-1}}}\cdot\left(1-\left(\frac{R_x}{R_0}\right)^{\frac{1}{\alpha-1}}\right)\right)\cdot 100 \qquad (37)$$

where c_x and c_0 are the contaminant concentration at location x and the contaminant source, respectively, and R_x and R_0 the associated isotope ratios. Applying these two extreme scenarios to data on benzene biodegradation obtained from a field site reveals huge differences in the estimated amounts of biodegradation depending on the approach used [21]. For some observation points, the calculation using the Rayleigh equation (Eq. 24) (considering biodegradation preceding dilution) results in more than 80 % of the benzene being biodegraded whereas using Eq. 37 (considering dilution preceding biodegradation) predicts less than 1 % [21]. Although, the assumption of biodegradation and dilution occurring strictly sequential is not realistic these results indicate the significance of the dilution processes for the interpretation of measured stable isotope ratios. In particular, the application of the simple Rayleigh equation based quantification approach (Eq. 24) leads to reliable estimates only in the absence of any significant dilution processes or in the unlikely case that biodegradation precedes any major dilution processes taking place along a flowpath in a contaminated aquifer. In any other case, using Eq. 24 may lead to an overestimation of the biodegradation of a contaminant. These examples demonstrate the uncertainty in assessing biodegradation by the Rayleigh approach neglecting the hydrogeological situation.

To allow for a reliable quantification of biodegradation at a field site where dilution processes take place Fischer et al. (2007) proposed an alternative approach, where biodegradation and dilution occur simultaneously with a constant ratio between their rates [21]. For this case the authors derived a correction of the Rayleigh equation approach in case biodegradation and dilution follow first order rate laws each [21]. Here, we avoid the restriction to first order rate laws and require only that the rate of biodegradation, r_b, and the rate of dilution, r_d, have a constant ratio, $k = r_d/r_b$, along the entire flowpath. Thus, the decrease of total concentration, C, is given by

$$\frac{dC}{dt} = r_d + r_b = k \cdot r_b + r_b = (1+k) \cdot r_b . \tag{38}$$

The isotope fractionation factor of the biodegradation process is given as

$$\alpha = \frac{{}^h r_b / {}^h C}{{}^l r_b / {}^l C} = \frac{1}{R} \cdot \frac{{}^h r_b}{{}^l r_b} \tag{39}$$

which can be rearranged to

$${}^h r_b = \alpha \cdot R \cdot {}^l r_b \tag{40}$$

where the superscripts h and l denote rates and concentration of the heavier and lighter isotope, respectively. If no isotope fractionation is caused by the dilution processes an analogous expression for the latter process is

$${}^h r_d = R \cdot {}^l r_d . \tag{41}$$

Considering the low natural abundance of the heavier isotopes we again assume that $(R+1) \approx 1$, which allows to express the concentration changes of each isotope as

$$\frac{d^{13}C}{dt} = {}^{13}r_d + {}^{13}r_b \approx R \cdot r_d + R \cdot \alpha \cdot r_b = (\alpha + k) \cdot R \cdot r_b \tag{42}$$

and

$$\frac{d^{12}C}{dt} = {}^{12}r_d + {}^{12}r_b \approx r_d + r_b = (1+k) \cdot r_b . \tag{43}$$

With these expressions an apparent isotope fractionation factor α^* can be defined:

$$\alpha^* = \frac{\frac{d^{13}C}{dt} / {}^{13}C}{\frac{d^{12}C}{dt} / {}^{12}C} = \frac{(\alpha + k) \cdot R \cdot r_b / {}^{13}C}{(1+k) \cdot r_b / {}^{12}C} = \frac{\alpha + k}{1+k} \tag{44}$$

which is constant (if the assumption of a constant k is valid) and allows to link changes in contaminant concentration to shifts in the isotope ratio according to the Rayleigh equation:

$$f = \frac{C}{C_0} = \left(\frac{R}{R_0}\right)^{\frac{1}{\alpha^*-1}} = \left(\frac{R}{R_0}\right)^{\frac{1+k}{\alpha-1}} = \left(\left(\frac{R}{R_0}\right)^{\frac{1}{\alpha-1}}\right)^{1+k} \tag{45}$$

The amount of biodegradation along the flow path is then given as

$$\begin{aligned} B[\%] &= \left(\int r_b\right)\cdot 100 = \left(\int \frac{1}{1+k}\cdot\frac{dC}{dt}\right)\cdot 100 = \frac{1}{1+k}\cdot(1-f)\cdot 100 \\ &= \frac{1}{1+k}\cdot\left(1-\left(\left(\frac{R}{R_0}\right)^{\frac{1}{\alpha-1}}\right)^{1+k}\right)\cdot 100 \end{aligned} \tag{46}$$

which shows that the expression presented by Fischer et al. (2007) for first order kinetics is applicable for arbitrary kinetics, too [21]. These results indicate that if dilution has a significant influence on the concentration decrease along a groundwater flow path the dilution effect needs to be considered in order to obtain a reliable estimate of biodegradation based on stable isotope signatures. In such a case a simple Rayleigh equation based quantification leads to an overestimation of biodegradation and the stronger the impact of dilution or the larger the ratio k between the rates of dilution and biodegradation the larger the overestimation will be (Figure 10). In turn, using stable isotope signatures combined with measured contaminant concentrations allows for a quantification of biodegradation even in the presence of concentration changes caused by other processes than biodegradation.

If high dilution rates are caused by transversal dispersive mixing processes it is likely to assume that longitudinal dispersion is also high along the flow path i.e., measured contaminant concentrations and isotope signatures are exhibiting a combined effect of both types of dispersion. To test if the above statements also hold in such a case the virtual flow path simulations presented in section "Dispersion along flow paths" are repeated assuming an additional dilution of the contaminants. Two different magnitudes of dilution are considered: a) dilution and biodegradation following the same rate law ($k = 1$), and b) dilution being the dominant process for concentration decrease ($k = 5$) along a flowpath in a contaminated aquifer. The total rate of dilution and biodegradation is kept constant. Again simulated concentrations and isotope ratios were handled as measured data from a field site and the Rayleigh equation or its modification were applied to them. Results of the simulation confirm that in the presence of dilution the quantification of biodegradation solely based on the Rayleigh equation is overestimating the actual amount of biodegradation, although the overestimation is less severe than using measured concentrations, only (Figure 11). In turn, considering the dilution effects in a modified Rayleigh equation (Eq. 45) allows for an accurate estimate of the simulated biodegradation (Eq. 46). Estimation errors due to longitudinal dispersivities obtained for applying the Rayleigh equation in the absence of dilution (Figure 9) also describe the situation for applying the modified Rayleigh equation to dilution affected sites.

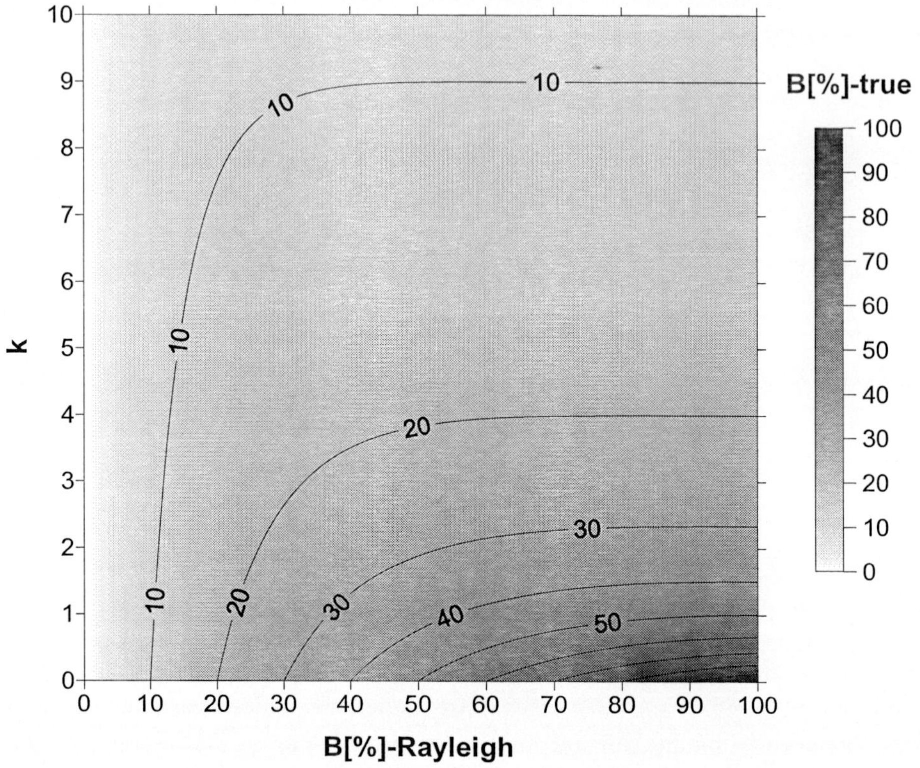

Figure 10. Relevance of abiotic non-fractionating processes for the quantification of biodegradation along a groundwater flow path (adapted from [*21*]). Gray scales indicate the actual or "true" amount of biodegradation (Eq. 46) taking place in contrast to values predicted by the simple Rayleigh equation based approach. The larger the ratio k between dilution and biodegradation rate the larger the difference between actual and Rayleigh equation based estimates. E.g., if the simple Rayleigh equation approach predicts a biodegradation of 70 % this value is only correct if $k = 0$ or if no dilution takes place. If both process take place at the same rate ($k = 1$) then the actual biodegradation would be less than 50 % and for $k = 4$ this value would drop to approx. 20 %.

Table 1. Errors in the quantification of biodegradation caused by weakly constrained input parameters of the Rayleigh equation

B[%]	$\|\Delta B[\%]\|$ if $\Delta\varepsilon = \pm\, 0.1 \cdot \varepsilon$	$\|\Delta\, B[\%]\|$ if $\Delta\delta_x$ or $\Delta\delta_0 = \pm\, 0.1 \cdot \varepsilon$
10	0.9	9.0
50	3.5	5.0
90	2.3	1.0
99	0.5	0.1

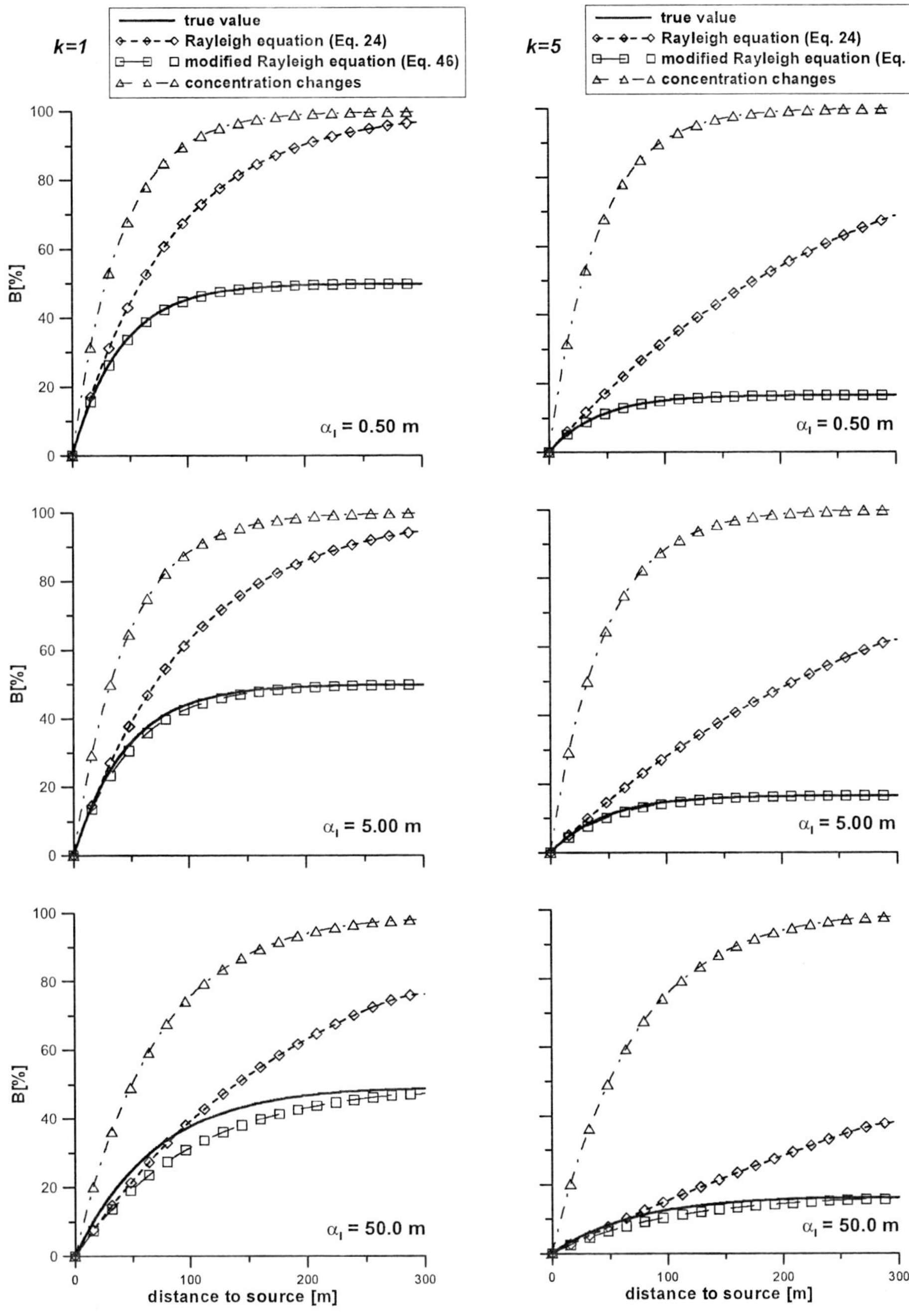

Figure 11. Comparison between actual amounts of biodegradation (*B [%]*) and estimates based on a measured concentration changes, a simple Rayleigh equation approach (Eq. 24), and a modified Rayleigh equation considering biodegradation and dilution taking place simultaneously (Eq. 46) along virtual groundwater flow paths with different heterogeneities (dispersivities). Left: Biodegradation and dilution having identical rates ($k = 1$). Right: Dilution rate five times higher than biodegradation rate ($k = 5$).

CONCLUSIONS

Compound-specific stable isotope analysis (CSIA) is a powerful method for the detection of contaminant biodegradation in an aquifer and can also be used for the determination of degradation pathways and the quantification of biodegradation. However, the applicability of the simple Rayleigh equation based approach for the quantification of biodegradation requires idealized conditions not always to be found in the field. There, additional processes (e.g., mixing or sorption) can lead to significant uncertainties using the Rayleigh equation based approach, only. These uncertainties can be reduced by using extensions of the Rayleigh equation based approach, which include additional information on processes in the aquifer into the quantification of biodegradation.

ACKNOWLEGEMENTS

This study was supported by the Helmholtz Association grant VG-NG-338 (GReaT MoDE). In addition, we are grateful for support by the SAFIRA Project of the Helmholtz Centre for Environmental Research – UFZ, and by the Helmholtz Enterprise Fund for the foundation of Isodetect GmbH- Company for Isotope Monitoring.

REFERENCES

[1] Arneth, J. D.; Milde, G.; Kerndorff, H. H.; Schleyer, R.; Waste deposit influences on groundwater quality as a tool for waste type and site selection for final storage quality. In: *The Landfill.- Lecture Notes in Earth Sciences*; Baccini, P., Ed.; Springer-Verlag: Berlin, 1989; Vol. 20.

[2] Plumb, R. H.; A comparison of groundwater monitoring data from CERCLA and RCRA Sites. *Ground Water Monit. R.* 1987, 7, 94-100.

[3] Smith, M. T.; Benzene, NQO1, and genetic susceptibility to cancer. *P. Natl. Acad. Sci. USA.* 1999, 96, 7624-7626.

[4] Snyder, R.; Xenobiotic metabolism and the mechanism(s) of benzene toxicity. *Drug Metab. Rev.* 2004, 36, 531-547.

[5] Korte, J. E.; Hertz-Picciotto, I.; Schulz, M. R.; Ball, L. M.; Duell, E. J.; The contribution of benzene to smoking-induced leukemia. *Environ Health Persp.* 2000, 108, 333-339.

[6] Fishbein, L.; Industrial mutagens and potential mutagens I. Halogenated aliphatic derivatives. *Mutat. Res.* 1976, 32, 267-307.

[7] Dean, B. J.; Recent findings on the genetic toxicology of benzene, toluene, xylenes and phenols. *Mutat. Res.* 1985, 154, 153-181.

[8] US-EPA; "Use of monitored natural attenuation at superfund, RCRA corrective action, and underground storage tank sites", United States Environmental Protection Agency, Office of Solid Waste and Emergency Response (OSWER), 1999.

[9] US-EPA; "Monitored natural attenuation of MTBE as a risk management option at leaking underground storage tank sites", National Risk Management Research

Laboratory, Office of Research and Development, U.S. Environmental Protection Agency, 2005.

[10] Christensen, T. H.; Bjerg, P. L.; Banwart, S. A.; Jakobsen, R.; Heron, G.; Albrechtsen, H. J.; Characterization of redox conditions in groundwater contaminant plumes. *J. Contam. Hydrol.* 2000, 45, 165-241.

[11] Wiedemeier, T. H., Rifai, H.S., Newell, C.J., Wilson, J.T.; Natural attenuation of fuels and chlorinated solvents in the subsurface; 1 ed.; John Wiley and Sons, INC.: New York, 1999.

[12] Sherwood Lollar, B.; Slater, G. F.; Sleep, B.; Witt, M.; Klecka, G. M.; Harkness, M.; Spivack, J.; Stable carbon isotope evidence for intrinsic bioremediation of tetrachloroethene and trichloroethene at area 6, Dover Air Force Base. *Environ. Sci. Technol.* 2001, 35, 261-269.

[13] Richnow, H. H.; Meckenstock, R. U.; Reitzel, L. A.; Baun, A.; Ledin, A.; Christensen, T. H.; *In situ*-biodegradation determined by carbon isotope fractionation of aromatic hydrocarbons in an anaerobic landfill leachate plume (Vejen, Denmark). *J. Contam. Hydrol.* 2003, 64, 59-72.

[14] Richnow, H. H.; Annweiler, E.; Michaelis, W.; Meckenstock, R. U.; Microbial *in situ*-degradation of aromatic hydrocarbons in a contaminated aquifer monitored by carbon isotope fractionation. *J. Contam. Hydrol.* 2003, 65, 101-120.

[15] Kästner, M.; Fischer, A.; Nijenhuis, I.; Geyer, R.; Stelzer, N.; Bombach, P.; Tebbe, C. C.; Richnow, H. H.; Assessment of microbial *in situ*-activity in contaminated aquifers. *Eng. Life Sci.* 2006, 6, 234-251.

[16] Meckenstock, R. U.; Morasch, B.; Griebler, C.; Richnow, H. H.; Stable isotope fractionation analysis as a tool to monitor biodegradation in contaminated aquifers. *J. Contam. Hydrol.* 2004, 75, 215-255.

[17] Rosell, M.; Häggblom, M. M.; Richnow, H. H.; Compound specific isotope analysis (CSIA) to characterise degradation pathways and to quantify in situ degradation of fuel oxygenates and other fuel derived contaminants. In: *Fuel Oxygenates*; Barceló, D., Ed.; Springer-Verlag: Berlin Heidelberg, 2007.

[18] Schmidt, T. C.; Zwank, L.; Elsner, M.; Berg, M.; Meckenstock, R. U.; Haderlein, S. B.; Compound-specific stable isotope analysis of organic contaminants in natural environments: a critical review of the state of the art, prospects, and future challenges. *Anal. Bioanal. Chem.* 2004, 378, 283-300.

[19] Slater, G. F.; Stable isotope forensics - When isotopes work. *Environ. Forensics.* 2003, 4, 13-23.

[20] Wilson, J. T.; Kaiser, P.; Adair, C.; Monitored MTBE biodegradation with stable isotope ratios. In: *Monitored natural attenuation of MTBE as a risk management option at leaking underground storage tank sites*; U.S. Environmental Protection Agency: Oklahoma, 2005; 31-36.

[21] Fischer, A.; Theuerkorn, K.; Stelzer, N.; Gehre, M.; Thullner, M.; Richnow, H. H.; Applicability of stable isotope fractionation analysis for the characterization of benzene biodegradation in a BTEX contaminated aquifer. *Environ. Sci. Technol.* 2007, 41, 3689-3696.

[22] Elsner, M.; Zwank, L.; Hunkeler, D.; Schwarzenbach, R. P.; A new concept linking observable stable isotope fractionation to transformation pathways of organic pollutants. *Environ. Sci. Technol.* 2005, 39, 6896-6916.

[23] van Breukelen, B. M.; Quantifying the degradation and dilution contribution to natural attenuation of contaminants by means of an open system Rayleigh equation. *Environ. Sci. Technol.* 2007, 41, 4980-4985.

[24] Matthews, D. E.; Hayes, J. M.; Isotope-Ratio-Monitoring Gas Chromatography-Mass Spectrometry. *Anal Chem.* 1978, 50, 1465-1473.

[25] Merritt, D. A.; Freeman, K. H.; Ricci, M. P.; Studley, S. A.; Hayes, J. M.; Performance and optimization of a combustion interface for isotope ratio monitoring gas-chromatography mass-spectrometry. *Anal. Chem.* 1995, 67, 2461-2473.

[26] Hilkert, A. W.; Douthitt, C. B.; Schluter, H. J.; Brand, W. A.; Isotope ratio monitoring gas chromatography mass spectrometry of D/H by high temperature conversion isotope ratio mass spectrometry. *Rapid Commun. Mass Sp.* 1999, 13, 1226-1230.

[27] Burgoyne, T. W.; Hayes, J. M.; Quantitative production of H_2 by pyrolysis of gas chromatographic effluents. *Anal. Chem.* 1998, 70, 5136-5141.

[28] Gehre, M.; Hoefling, R.; Kowski, P.; Strauch, G.; Sample preparation device for quantitative hydrogen isotope analysis using chromium metal. *Anal. Chem.* 1996, 68, 4414-4417.

[29] Gehre, M.; Hoefling, R.; Kowski, P.; Methodical studies for D/H-isotope analysis - A new technique for the direct coupling of sample preparation to an IRMS. *Isot. Environ. Healt. S.* 1996, 32, 335-340.

[30] Coplen, T. B.; Brand, W. A.; Gehre, M.; Groning, M.; Meijer, H. A. J.; Toman, B.; Verkouteren, R. M.; New guidelines for delta C-13 measurements. *Anal. Chem.* 2006, 78, 2439-2441.

[31] Craig, H.; Isotopic standards for carbon and oxygen and correction for mass-spectrometric analysis of carbon dioxide. *Geochim. Cosmochim. Ac.* 1957, 12, 133-149.

[32] Gonfiantini, R.; Stichler, W.; Rozanski, K.; Standards and intercomparison materials for stable isotopes of light elements; 1995; IAEA-TECDOC-825, Vienna, 13-29.

[33] Hagemann, R.; Nief, G.; Roth, E.; Absolute isotopic scale for deuterium analysis of natural waters. Absolute D/H ratio for SMOW. *Tellus.* 1970, 22, 712-715.

[34] Slater, G. F.; Dempster, H. S.; Sherwood Lollar, B.; Ahad, J.; Headspace analysis: A new application for isotopic characterization of dissolved organic contaminants. *Environ. Sci. Technol.* 1999, 33, 190-194.

[35] Hunkeler, D.; Anderson, N.; Aravena, R.; Bernasconi, S. M.; Butler, B. J.; Hydrogen and carbon isotope fractionation during aerobic biodegradation of benzene. *Environ. Sci. Technol.* 2001, 35, 3462-3467.

[36] Hunkeler, D.; Aravena, R.; Determination of compound-specific carbon isotope ratios of chlorinated methanes, ethanes, and ethenes in aqueous samples. *Environ. Sci. Technol.* 2000, 34, 2839-2844.

[37] Zwank, L.; Berg, M.; Schmidt, T. C.; Haderlein, S. B.; Compound-specific carbon isotope analysis of volatile organic compounds in the low-microgram per liter range. *Anal. Chem.* 2003, 75, 5575-5583.

[38] Dias, R. F.; Freeman, K. H.; Carbon isotope analyses of semivolatile organic compounds in aqueous media using solid-phase microextraction and isotope ratio monitoring GC/MS. *Anal. Chem.* 1997, 69, 944-950.

[39] Gray, J. R.; Lacrampe-Couloume, G.; Gandhi, D.; Scow, K. M.; Wilson, R. D.; Mackay, D. M.; Sherwood Lollar, B.; Carbon and hydrogen isotopic fractionation

during biodegradation of methyl *tert*-butyl ether. *Environ. Sci. Technol.* 2002, 36, 1931-1938.

[40] Ward, J. A. M.; Ahad, J. M. E.; Lacrampe-Couloume, G.; Slater, G. F.; Edwards, E. A.; Sherwood Lollar, B.; Hydrogen isotope fractionation during methanogenic degradation of toluene: Potential for direct verification of bioremediation. *Environ. Sci. Technol.* 2000, 34, 4577-4581.

[41] Mancini, S. A.; Ulrich, A. C.; Lacrampe-Couloume, G.; Sleep, B.; Edwards, E. A.; Sherwood Lollar, B.; Carbon and hydrogen isotopic fractionation during anaerobic biodegradation of benzene. *Appl. Environ. Microb.* 2003, 69, 191-198.

[42] Harper, D. B.; Kalin, R. M.; Hamilton, J. T. G.; Lamb, C.; Carbon isotope ratios for chloromethane of biological origin: Potential tool in determining biological emissions. *Environ. Sci. Technol.* 2001, 35, 3616-3619.

[43] Morrill, P. L.; Lacrampe-Couloume, G.; Sherwood Lollar, B.; Dynamic headspace: a single-step extraction for isotopic analysis of µg/L concentrations of dissolved chlorinated ethenes. *Rapid Commun. Mass Sp.* 2004, 18, 595-600.

[44] Dempster, H. S.; Sherwood Lollar, B.; Feenstra, S.; Tracing organic contaminants in groundwater: A new methodology using compound-specific isotopic analysis. *Environ. Sci. Technol.* 1997, 31, 3193-3197.

[45] Fischer, A.; Vieth, A.; Knöller, K.; Wachter, T.; Dahmke, A.; Richnow, H. H.; Charakterisierung des mikrobiellen Schadstoffabbaus mit Hilfe von isotopenchemischen Methoden. *Grundwasser.* 2004, 9, 159-172.

[46] Vieth, A.; Kästner, M.; Schirmer, M.; Weiss, H.; Gödeke, S.; Meckenstock, R. U.; Richnow, H. H.; Monitoring *in situ*-biodegradation of benzene and toluene by stable carbon isotope fractionation. *Environ. Toxicol. Chem.* 2005, 24, 51-60.

[47] Song, D. L.; Conrad, M. E.; Sorenson, K. S.; Alvarez-Cohen, L.; Stable carbon isotope fractionation during enhanced *in situ*-bioremediation of trichloroethene. *Environ. Sci. Technol.* 2002, 36, 2262-2268.

[48] Smallwood, B. J.; Philp, R. P.; Burgoyne, T. W.; Allen, J. D.; The use of stable isotopes to differentiate specific source markers for MTBE. *Environ. Forensics.* 2001, 2, 215-221.

[49] Kolhatkar, R.; Kuder, T.; Philp, P.; Allen, J.; Wilson, J. T.; Use of compound-specific stable carbon isotope analyses to demonstrate anaerobic biodegradation of MTBE in groundwater at a gasoline release site. *Environ. Sci. Technol.* 2002, 36, 5139-5146.

[50] Northrop, D. B.; The expression of isotope effects on enzyme-catalyzed reactions. *Annu. Rev. Biochem.* 1981, 50, 103-131.

[51] Hayes, J. M.; Fractionation of carbon and hydrogen isotopes in biosynthetic processes. In: Stable isotope geochemistry; 43 ed.; Valley, J. W., Cole, D. R., Eds.; The Mineralogical Society of America: Washington D.C., 2001; 225-277.

[52] Galimov, E. M.; Isotope organic geochemistry. *Org. Geochem.* 2006, 37, 1200-1262.

[53] Melander, L.; Saunders, W. H.; Reaction rates of isotopic molecules; Robert E. Krieger Publishing Company, Inc.: Malabar, Florida, 1987.

[54] Cook, P. F.; *Enzyme mechanism from isotope effects*; CRC Press: Boca Raton, Ann Arbor, Boston, London, 1991.

[55] Cleland, W. W.; The use of isotope effects to determine enzyme mechanisms. *Arch. Biochem. Biophys.* 2005, 433, 2-12.

[56] O'Leary, M. H.; Multiple isotope effects on enzyme-catalyzed reactions. *Annu. Rev. Biochem.* 1989, 58, 377-401.

[57] Mariotti, A.; Germon, J. C.; Hubert, P.; Kaiser, P.; Letolle, R.; Tardieux, A.; Tardieux, P.; Experimental-determination of nitrogen kinetic isotope fractionation - some principles - illustration for the denitrification and nitrification processes. *Plant Soil.* 1981, 62, 413-430.

[58] Rayleigh, J. W. S.; Theoretical considerations respecting the separation of gases by diffusion and similar processes. *Philos. Mag.* 1896, 42, 493-498.

[59] Mitchell, K. H.; Rogge, C. E.; Gierahn, T.; Fox, B. G.; Insight into the mechanism of aromatic hydroxylation by toluene 4-monooxygenase by use of specifically deuterated toluene and p-xylene. *P. Natl. Acad. Sci. USA.* 2003, 100, 3784-3789.

[60] de Visser, S. P.; Oh, K.; Han, A. R.; Nam, W.; Combined experimental and theoretical study on aromatic hydroxylation by mononuclear nonheme iron(IV)-oxo complexes. *Inorg. Chem.* 2007, 46, 4632-4641.

[61] Churchill, D. G.; Janak, K. E.; Wittenberg, J. S.; Parkin, G.; Normal and inverse primary kinetic deuterium isotope effects for C-H bond reductive elimination and oxidative addition reactions of molybdenocene and tungstenocene complexes: Evidence for benzene sigma-complex intermediates. *J. Am. Chem. Soc.* 2003, 125, 1403-1420.

[62] Marlier, J. F.; Multiple isotope effects on the acyl group transfer reactions of amides and esters. *Accounts Chem. Res.* 2001, 34, 283-290.

[63] Mahieu, K.; De Visscher, A.; Vanrolleghem, P. A.; Van Cleemput, O.; Carbon and hydrogen isotope fractionation by microbial methane oxidation: Improved determination. *Waste Manage.* 2006, 26, 389-398.

[64] Scott, K. M.; Lu, X.; Cavanaugh, C. M.; Liu, J. S.; Optimal methods for estimating kinetic isotope effects from different forms of the Rayleigh distillation equation. *Geochim. Cosmochim. Ac.* 2004, 68, 433-442.

[65] Hunkeler, D.; Quantification of isotope fractionation in experiments with deuterium-labeled substrate. *Appl. Environ. Microb.* 2002, 68, 5205-5206.

[66] www.isodetect.de.

[67] Mazeas, L.; Budzinski, H.; Raymond, N.; Absence of stable carbon isotope fractionation of saturated and polycyclic aromatic hydrocarbons during aerobic bacterial biodegradation. *Org. Geochem.* 2002, 33, 1259-1272.

[68] Trust, B. A.; Mueller, J. G.; Coffin, R. B.; Cifuentes, L. A.; The biodegradation of fluoranthrene as monitored using stable carbon isotopes. In: Monitoring and verification of bioremediation; Hinchee, R. E., Douglas, G. S., Ong, S. K., Eds.; Batelle Press: Columbus, 1995; 233– 239.

[69] Pond, K. L.; Huang, Y. S.; Wang, Y.; Kulpa, C. F.; Hydrogen isotopic composition of individual n-alkanes as an intrinsic tracer for bioremediation and source identification of petroleum contamination. *Environ Sci Technol.* 2002, 36, 724-728.

[70] Yan, B. Z.; Abrajano, T. A.; Bopp, R. F.; Benedict, L. A.; Chaky, D. A.; Perry, E.; Song, J.; Keane, D. P.; Combined application of delta C-13 and molecular ratios in sediment cores for PAH source apportionment in the New York/New Jersey harbor complex. *Org Geochem.* 2006, 37, 674-687.

[71] Mazeas, L.; Budzinski, H.; Molecular and stable carbon isotopic source identification of oil residues and oiled bird feathers sampled along the Atlantic coast of France after the Erika oil spill. *Environ Sci Technol.* 2002, 36, 130-137.

[72] Ahad, J. M. E.; Sherwood Lollar, B.; Edwards, E. A.; Slater, G. F.; Sleep, B. E.; Carbon isotope fractionation during anaerobic biodegradation of toluene: Implications for intrinsic bioremediation. *Environ Sci Technol.* 2000, 34, 892-896.

[73] Meckenstock, R. U.; Morasch, B.; Warthmann, R.; Schink, B.; Annweiler, E.; Michaelis, W.; Richnow, H. H.; C-13/C-12 isotope fractionation of aromatic hydrocarbons during microbial degradation. *Environ. Microbiol.* 1999, 1, 409-414.

[74] Morasch, B.; Richnow, H. H.; Schink, B.; Meckenstock, R. U.; Stable hydrogen and carbon isotope fractionation during microbial toluene degradation: Mechanistic and environmental aspects. *Appl. Environ. Microb.* 2001, 67, 4842-4849.

[75] Morasch, B.; Richnow, H. H.; Schink, B.; Vieth, A.; Meckenstock, R. U.; Carbon and hydrogen stable isotope fractionation during aerobic bacterial degradation of aromatic hydrocarbons. *Appl. Environ. Microb.* 2002, 68, 5191-5194.

[76] Morasch, B.; Richnow, H. H.; Vieth, A.; Schink, B.; Meckenstock, R. U.; Stable isotope fractionation caused by glycyl radical enzymes during bacterial degradation of aromatic compounds. *Appl Environ Microb.* 2004, 70, 2935-2940.

[77] Wilkes, H.; Boreham, C.; Harms, G.; Zengler, K.; Rabus, R.; Anaerobic degradation and carbon isotopic fractionation of alkylbenzenes in crude oil by sulphate-reducing bacteria. *Org Geochem.* 2000, 31, 101-115.

[78] Hunkeler, D.; Butler, B. J.; Aravena, R.; Barker, J. F.; Monitoring biodegradation of methyl *tert*-butyl ether (MTBE) using compound-specific carbon isotope analysis. *Environ. Sci. Technol.* 2001, 35, 676-681.

[79] Kuder, T.; Wilson, J. T.; Kaiser, P.; Kolhatkar, R.; Philp, P.; Allen, J.; Enrichment of stable carbon and hydrogen isotopes during anaerobic biodegradation of MTBE: Microcosm and field evidence. *Environ. Sci. Technol.* 2005, 39, 213-220.

[80] Rosell, M.; Barcelo, D.; Rohwerder, T.; Breuer, U.; Gehre, M.; Richnow, H. H.; Variations in C-13/C-12 and D/H enrichment factors of aerobic bacterial fuel oxygenate degradation. *Environ. Sci. Technol.* 2007, 41, 2036-2043.

[81] Somsamak, P.; Richnow, H. H.; Häggblom, M. M.; Carbon isotopic fractionation during anaerobic biotransformation of methyl *tert*-butyl ether and *tert*-amyl methyl ether. *Environ. Sci. Technol.* 2005, 39, 103-109.

[82] Somsamak, P.; Richnow, H. H.; Häggblom, M. M.; Carbon isotope fractionation during anaerobic degradation of methyl tert-butyl ether under sulfate-reducing and methanogenic conditions. *Appl. Environ. Microb.* 2006, 72, 1157-1163.

[83] Barth, J. A. C.; Slater, G.; Schüth, C.; Bill, M.; Downey, A.; Larkin, M.; Kalin, R. M.; Carbon isotope fractionation during aerobic biodegradation of trichloroethene by Burkholderia cepacia G4: a tool to map degradation mechanisms. *Appl. Environ. Microb.* 2002, 68, 1728-1734.

[84] Bloom, Y.; Aravena, R.; Hunkeler, D.; Edwards, E.; Frape, S. K.; Carbon isotope fractionation during microbial dechlorination of trichloroethene, *cis*-1,2-dichloroethene, and vinyl chloride: Implications for assessment of natural attenuation. *Environ. Sci. Technol.* 2000, 34, 2768-2772.

[85] Brungard, K. L.; Munakata-Marr, J.; Johnson, C. A.; Mandernack, K. W.; Stable carbon isotope fractionation of trans-1,2-dichloroethylene during co-metabolic degradation by methanotrophic bacteria. *Chem Geol.* 2003, 195, 59-67.

[86] Chu, K. H.; Mahendra, S.; Song, D. L.; Conrad, M. E.; Alvarez-Cohen, L.; Stable carbon isotope fractionation during aerobic biodegradation of chlorinated ethenes. *Environ Sci Technol.* 2004, 38, 3126-3130.

[87] Griebler, C.; Adrian, L.; Meckenstock, R. U.; Richnow, H. H.; Stable carbon isotope fractionation during aerobic and anaerobic transformation of trichlorobenzene. *Fems Microbiol. Ecol.* 2004, 48, 313-321.

[88] Hunkeler, D.; Aravena, R.; Evidence of substantial carbon isotope fractionation among substrate, inorganic carbon, and biomass during aerobic mineralization of 1,2-dichloroethane by Xanthobacter autotrophicus. *Appl. Environ. Microb.* 2000, 66, 4870-4876.

[89] Kaschl, A.; Vogt, C.; Uhlig, S.; Nijenhuis, I.; Weiss, H.; Kästner, M.; Richnow, H. H.; Isotopic fractionation indicates anaerobic monochlorobenzene biodegradation. *Environ. Toxicol. Chem.* 2005, 24, 1315-1324.

[90] Nijenhuis, I.; Andert, J.; Beck, K.; Kastner, M.; Diekert, G.; Richnow, H. H.; Stable isotope fractionation of tetrachloroethene during reductive dechlorination by Sulfurospirillum multivorans and Desulfitobacterium sp strain PCE-S and abiotic reactions with cyanocobalamin. *Appl. Environ. Microb.* 2005, 71, 3413-3419.

[91] Nikolausz, M.; Nijenhuis, I.; Ziller, K.; Richnow, H. H.; Kastner, M.; Stable carbon isotope fractionation during degradation of dichloromethane by methylotrophic bacteria. *Environ. Microbiol.* 2006, 8, 156-164.

[92] Slater, G. F.; Sherwood Lollar, B.; Sleep, B. E.; Edwards, E. A.; Variability in carbon isotopic fractionation during biodegradation of chlorinated ethenes: Implications for field applications. *Environ. Sci. Technol.* 2001, 35, 901-907.

[93] Hunkeler, D.; Aravena, R.; Berry-Spark, K.; Cox, E.; Assessment of degradation pathways in an aquifer with mixed chlorinated hydrocarbon contamination using stable isotope analysis. *Environ. Sci. Technol.* 2005, 39, 5975-5981.

[94] Zwank, L.; Berg, M.; Elsner, M.; Schmidt, T. C.; Schwarzenbach, R. P.; Haderlein, S. B.; New evaluation scheme for two-dimensional isotope analysis to decipher biodegradation processes: Application to groundwater contamination by MTBE. *Environ. Sci. Technol.* 2005, 39, 1018-1029.

[95] Mancini, S. A.; Lacrampe-Couloume, G.; Jonker, H.; Van Breukelen, B. M.; Groen, J.; Volkering, F.; Sherwood Lollar, B.; Hydrogen isotopic enrichment: An indicator of biodegradation at a petroleum hydrocarbon contaminated field site. *Environ. Sci. Technol.* 2002, 36, 2464-2470.

[96] Griebler, C.; Safinowski, M.; Vieth, A.; Richnow, H. H.; Meckenstock, R. U.; Combined application of stable carbon isotope analysis and specific metabolites determination for assessing *in situ*-degradation of aromatic hydrocarbons in a tar oil-contaminated aquifer. *Environ. Sci. Technol.* 2004, 38, 617-631.

[97] Vieth, A.; Müller, J.; Strauch, G.; Kästner, M.; Gehre, M.; Meckenstock, R. U.; Richnow, H. H.; *In situ*-biodegradation of tetrachloroethene and trichloroethene in contaminated aquifers monitored by stable isotope fractionation. *Isot. Environ. Healt. S.* 2003, 39, 113-124.

[98] Mancini, S. A.; Hirschorn, S. K.; Elsner, M.; Lacrampe-Couloume, G.; Sleep, B. E.; Edwards, E. A.; Sherwood Lollar, B.; Effects of trace element concentration on enzyme controlled stable isotope fractionation during aerobic biodegradation of toluene. *Environ. Sci. Technol.* 2006, 40, 7675-7681.

[99] Morrill, P. L.; Lacrampe-Couloume, G.; Slater, G. F.; Sleep, B. E.; Edwards, E. A.; McMaster, M. L.; Major, D. W.; Sherwood Lollar, B.; Quantifying chlorinated ethene degradation during reductive dechlorination at Kelly AFB using stable carbon isotopes. *J. Contam. Hydrol.* 2005, 76, 279-293.
[100] Haderlein, S. B.; Schmidt, T. C.; Elsner, M.; Zwank, L.; Berg, M.; Schwarzenbach, R. P.; Response to comment on "New evaluation scheme for two-dimensional isotope analysis to decipher biodegradation processes: Application to groundwater contamination by MTBE". *Environ. Sci. Technol.* 2005, 39, 8543-8544.
[101] Kopinke, F. D.; Georgi, A.; Richnow, H. H.; Comment on "New evaluation scheme for two-dimensional isotope analysis to decipher biodegradation processes: Application to groundwater contamination by MTBE". *Environ. Sci. Technol.* 2005, 39, 8541-8542.
[102] Steinbach, A.; Seifert, R.; Annweiler, E.; Michaelis, W.; Hydrogen and carbon isotope fractionation during anaerobic biodegradation of aromatic hydrocarbons - A field study. *Environ. Sci. Technol.* 2004, 38, 609-616.
[103] Kopinke, F. D.; Georgi, A.; Voskamp, M.; Richnow, H. H.; Carbon isotope fractionation of organic contaminants due to retardation on humic substances: Implications for natural attenuation studies in aquifers. *Environ. Sci. Technol.* 2005, 39, 6052-6062.
[104] Hunkeler, D.; Chollet, N.; Pittet, X.; Aravena, R.; Cherry, J. A.; Parker, B. L.; Effect of source variability and transport processes on carbon isotope ratios of TCE and PCE in two sandy aquifers. *J. Contam. Hydrol.* 2004, 74, 265-282.
[105] Bronstein, I. N.; Semendjajev, K. A.; Handbook of Mathematics; Springer.: Berlin, 2004.
[106] Morrill, P. L.; Sleep, B. E.; Slater, G. F.; Edwards, E. A.; Sherwood Lollar, B.; Evaluation of isotopic enrichment factors for the biodegradation of chlorinated ethenes using a parameter estimation model: Toward an improved quantification of biodegradation. *Environ. Sci. Technol.* 2006, 40, 3886-3892.
[107] van Breukelen, B. M.; Extending the Rayleigh equation to allow competing isotope fractionating pathways to improve quantification of biodegradation. *Environ. Sci. Technol.* 2007, 41, 4004-4010.
[108] Huang, W. E.; Oswald, S. E.; Lerner, D. N.; Smith, C. C.; Zheng, C. M.; Dissolved oxygen imaging in a porous medium to investigate biodegradation in a plume with limited electron acceptor supply. *Environ. Sci. Technol.* 2003, 37, 1905-1911.
[109] Cirpka, O. A.; Frind, E. O.; Helmig, R.; Numerical simulation of biodegradation controlled by transverse mixing. *J. Contam. Hydrol.* 1999, 40, 159-182.
[110] Thornton, S. F.; Quigley, S.; Spence, M. J.; Banwart, S. A.; Bottrell, S.; Lerner, D. N.; Processes controlling the distribution and natural attenuation of dissolved phenolic compounds in a deep sandstone aquifer. *J. Contam. Hydrol.* 2001, 53, 233-267.
[111] Thullner, M.; Mauclaire, L.; Schroth, M. H.; Kinzelbach, W.; Zeyer, J.; Interaction between water flow and spatial distribution of microbial growth in a two-dimensional flow field in saturated porous media. *J. Contam. Hydrol.* 2002, 58, 169-189.
[112] Barczewski, B.; Marshall, P.; The influence of sampling methods on the results of groundwater quality measurements. In: Contaminant transport in groundwater; Kobus, H. E., Kinzelbach, W., Eds.; Balkema: Rotterdam, 1989; 33-39.
[113] Poulson, S. R.; Drever, J. I.; Stable isotope (C, Cl, and H) fractionation during vaporization of trichloroethylene. *Environ. Sci. Technol.* 1999, 33, 3689-3694.

[114] Wang, Y.; Huang, Y. S.; Hydrogen isotope fractionation of low molecular weight *n*-alkanes during progressive vaporization. *Org. Geochem.* 2001, 32, 991-998.

[115] Wang, Y.; Huang, Y. S.; Hydrogen isotopic fractionation of petroleum hydrocarbons during vaporization: implications for assessing artificial and natural remediation of petroleum contamination. *Appl. Geochem.* 2003, 18, 1641-1651.

[116] Smallwood, B. J.; Philp, R. P.; Allen, J. D.; Stable carbon isotopic composition of gasolines determined by isotope ratio monitoring gas chromatography mass spectrometry. *Org. Geochem.* 2002, 33, 149-159.

[117] Harrington, R. R.; Poulson, S. R.; Drever, J. I.; Colberg, P. J. S.; Kelly, E. F.; Carbon isotope systematics of monoaromatic hydrocarbons: vaporization and adsorption experiments. *Org. Geochem.* 1999, 30, 765-775.

[118] Huang, L.; Sturchio, N. C.; Abrajano, T.; Heraty, L. J.; Holt, B. D.; Carbon and chlorine isotope fractionation of chlorinated aliphatic hydrocarbons by evaporation. *Org. Geochem.* 1999, 30, 777-785.

[119] Lahvis, M. A.; Baehr, A. L.; Baker, R. J.; Evaluation of volatilization as a natural attenuation pathway for MTBE. *Ground Water.* 2004, 42, 258-267.

[120] Poulson, S. R.; Drever, J. I.; Colberg, P. J. S.; Estimation of K-oc values for deuterated benzene, toluene, and ethylbenzene, and application to ground water contamination studies. *Chemosphere.* 1997, 35, 2215-2224.

[121] Slater, G. F.; Ahad, J. M. E.; Sherwood Lollar, B.; Allen-King, R.; Sleep, B.; Carbon isotope effects resulting from equilibrium sorption of dissolved VOCs. *Anal. Chem.* 2000, 72, 5669-5672.

[122] Schüth, C.; Taubald, H.; Bolano, N.; Maciejczyk, K.; Carbon and hydrogen isotope effects during sorption of organic contaminants on carbonaceous materials. *J. Contam. Hydrol.* 2003, 64, 269-281.

[123] Caimi, R. J.; Brenna, J. T.; Quantitative evaluation of carbon isotopic fractionation during reversed-phase high-performance liquid chromatography. *J. Chromatogr. A.* 1997, 757, 307-310.

[124] Fischer, A.; Bauer, J.; Meckenstock, R.; Stichler, W.; Griebler, C.; Maloszewski, P.; Kästner, M.; Richnow, H. H.; A multitracer test proving the reliability of stable isotope fractionation analysis for assessing anaerobic degradation in a BTEX contaminated aquifer. *Environ. Sci. Technol.* 2006, 40, 4245-4252.

[125] Elsner, M.; Haderlein, S. B.; Kellerhals, T.; Luzi, S.; Zwank, L.; Angst, W.; Schwarzenbach, R. P.; Mechanisms and products of surface-mediated reductive dehalogenation of carbon tetrachloride by Fe(II) on goethite. *Environ. Sci. Technol.* 2004, 38, 2058-2066.

[126] Dayan, H.; Abrajano, T.; Sturchio, N. C.; Winsor, L.; Carbon isotopic fractionation during reductive dehalogenation of chlorinated ethenes by metallic iron. *Org. Geochem.* 1999, 30, 755-763.

[127] Schüth, C.; Bill, M.; Barth, J. A. C.; Slater, G. F.; Kalin, R. A.; Carbon isotope fractionation during reductive dechlorination of TCE in batch experiments with iron samples from reactive barriers. *J. Contam. Hydrol.* 2003, 66, 25-37.

[128] Elsner, M.; Cwiertny, D. M.; Roberts, A. L.; Sherwood Lollar, B.; 1,1,2,2-Tetrachloroethane reactions with OH-, Cr(II), granular iron, and a copper-iron bimetal: Insights from product formation and associated carbon isotope fractionation. *Environ. Sci. Technol.* 2007, 41, 4111-4117.

[129] Elsner, M.; McKelvie, J.; Lacrampe-Couloume, G.; Sherwood Lollar, B.; Insight into methyl tert-butyl ether (MTBE) stable isotope fractionation from abiotic reference experiments. *Environ. Sci. Technol.* 2007, 41, 5693-5700.
[130] Bear, J.; Dynamics of fluids in porous media; Elsevier: New York, 1972.
[131] Abe, Y.; Hunkeler, D.; Does the Rayleigh equation apply to evaluate field isotope data in contaminant hydrogeology? *Environ. Sci. Technol.* 2006, 40, 1588-1596.
[132] Aguilera, D. R.; Jourabchi, P.; Spiteri, C.; Regnier, P.; A knowledge-based reactive transport approach for the simulation of biogeochemical dynamics in Earth systems. *Geochem. Geophy. Geosy.* 2005, 6, -.
[133] Regnier, P.; O'Kane, J. P.; Steefel, C. I.; Vanderborght, P.; Modeling complex multi-component reactive-transport systems: towards a simulation environment based on the concept of a Knowledge Base. *Appl. Math. Model.* 2002, 26, 913-927.
[134] Gelhar, L. W.; Welty, C.; Rehfeldt, K. R.; A Critical-Review of Data on Field-Scale Dispersion in Aquifers. *Water Resour. Res.* 1992, 28, 1955-1974.
[135] Mak, K. S.; Griebler, C.; Meckenstock, R. U.; Liedl, R.; Peter, A.; Combined application of conservative transport modelling and compound-specific carbon isotope analyses to assess in situ attenuation of benzene, toluene, and o-xylene. *J. Contam. Hydrol.* 2006, 88, 306-320.

In: Environmental and Regional Air Pollution
Editors: D. Gallo, R. Mancini
ISBN: 978-1-60692-893-6

Chapter 3

A REVIEW ON DRY SPELLS STATISTICS IN CATALONIA (NE SPAIN)

X. Lana[1], A. Burgueño[2], M. D. Martínez[3] and C. Serra[1]
[1]Department of Physics and Nuclear Engineering,
Polytechnic University of Catalonia, Barcelona, Spain
[2]Department of Meteorology and Astronomy, Faculty of Physics,
University of Barcelona, Spain
[3]Department of Applied Physics,
Polytechnic University of Catalonia, Barcelona, Spain

1. INTRODUCTION

Meteorological drought results from a shortage of precipitation. Its definable characteristics include the magnitude of the precipitation deficit and how quickly it develops, duration, spatial extent, and timing (AMS Statement, 2004). Climatological analyses of droughts can be obtained from various indices applicable at different time scales, usually 1, 3, 6, 9, 12 and 24 months, such as the Palmer Drought Severity Index (PDSI) (Palmer, 1965; Soulé, 1992) and the Standard Precipitation Index (SPI) (McKee et al., 1993). The PDSI is mainly a hydrological index. It is not used internationally because its empirical nature restricts its broad applicability. The SPI is based on widely measured precipitation data and can probabilistically describe precipitation shortages at any desired time scale, which have permitted its application around the world (Keyantash and Dracup, 2002; Heim, 2002).

It has also been recently applied in the Iberian Peninsula (Lana et al., 2001; Vicente-Serrano et al., 2004; Vicente-Serrano, 2006a, b; Vicente-Serrano and López-Moreno, 2006; Vicente-Serrano and Cuadrat-Prats, 2007), and in Europe (Lloyd-Hughes and Saunders, 2002). An alternative methodology, based solely in daily rainfall measures, can be derived from the distribution of the dry spells in relation to different daily rain amount thresholds, as will be shown.

According to Kiktev et al. (2003), for middle latitudes (30°-60°N) and the period 1950-1995, most of Europe and East Asia are characterized by a positive trend in the annual maximum of consecutive days with daily precipitation below 1.0 mm. This general behavior becomes of larger relevancy for the Mediterranean area, where an increase of extreme daily rainfall has been observed during 1951-1995 in spite of a decrease in total amounts (Trigo et al., 2000; Alpert et al., 2002).

Thus, in the Mediterranean area, the prospect is for a larger frequency of drought periods, with associated impacts on agriculture, on the exploitation of water resources and also on socio-economic activities.

In the Iberian Peninsula, different trend analyses of daily rainfall point to an increase of light rainfall events at the cost of a decrease of more intense events (Goodess and Jones, 2002; Gallego et al., 2006; Rodrigo and Trigo, 2007). Different regional climate models, as also global models considering different greenhouse gases emission scenarios applied to the Mediterranean region, point to a lengthening of the long dry spells, due to a reduction in the rain amounts along the 21st century (Ghan and Shippert, 2006), particularly in the summer season (Voss et al., 2002; Gibelin and Déqué, 2003; Räisänen et al., 2004, Sánchez et al., 2004; Kundzewicz et al., 2006).

These observed shifts in the spatial and temporal rainfall distribution are connected to the dynamical effects of the enhanced greenhouse gases and sulphate aerosols in the atmosphere. Thus, on the extratropical cyclone activity, Geng and Sugi (2003) deduced that the total cyclone density will decrease significantly around the year 2050 in the middle latitudes of the Northern Hemisphere during December-January-February (DJF) and June-July-August (JJA) seasons. Moreover, while a reduction of weak and medium-strength cyclone density is foreseen, the density of strong cyclones will increase by more than 20% in JJA.

The reasons of these changes are linked to the decrease of baroclinicity in the lower troposphere, mainly caused by the decrease of meridional temperature gradient. Previous studies on cyclones over the North Atlantic by Knippertz et al. (2000) also pointed towards a reduction of the number of total cyclones and an increase of strong cyclones, with a poleward and eastward shift of their tracks. This shift of the storm tracks has also been found by other authors (Ulbrich and Christoph, 1999; Pinto et al., 2007).

This behaviour is consistent with the 1948-2002 arctic cyclone activity (60°-90°N), which shows that the number and intensity of cyclones entering the Arctic from midlatitudes have increased, thus suggesting a shift of storm tracks into the Arctic, particularly in summer (McCabe et al., 2001; Zhang et al., 2004a). The Mediterranean Basin would also suffer a decrease of the overall cyclone activity and an increase of deep cyclones in a doubled CO_2 climate scenario (Lionello et al., 2002).

As a second order consequence, the observed northward shift of storm tracks supports more stable circulation conditions over Europe, manifested in a higher persistence of the atmospheric circulation in last decades, with the consequent exacerbation of impacts on the occurrence and severity of temperature extremes (Kyselý and Domonkos, 2006; Kyselý, 2007).

The contents of this review paper cover the dry spell statistics developed in Catalonia along last 10 years, paying particular attention to their statistical distribution, time trends and some fractal properties linked to dry spell length predictability.

2. Study Area, Databases and Regional Climate

2.1. Study Area

Catalonia extends over a surface of 32.000 km^2 at the NE of the Iberian Peninsula; with nearly 400 km of Mediterranean coast (Figure 1). As main orographic characteristics, it is worth mentioning the Pyrenees and Pre-Pyrenees Ranges, which mitigate the effects of northern advections for the rest of the country; the Littoral and Pre-Littoral Ranges, sheltering the Central Basin against eastern advections and, at the same time, enhancing their effects on the Littoral and Pre-Littoral domains; and the Central Basin itself, subject to the effects of frontal passages and western advections, mitigated by the remoteness to the Atlantic coast. Within the Pyrenees Range, distinctions have to be made between the Eastern Pyrenees regime, strongly influenced by the vicinity to the Mediterranean Sea, and the north face of the Western Pyrenees, with clear signs of Atlantic influence.

The Mediterranean Sea stands out by its relatively warm sea-water temperatures, particularly in the summer season. A series of measurements of the sea-surface temperature made along the 1974-2005 period at 2.8 km from the coast (42° 38' 42"N 3° 11' 42" E) reaches the monthly maximum in August, 23 °C, and the monthly minimum in February, 12 °C. The linear trend of the weekly measurements is of 0.34 °C/decade (Vargas-Yañez et al., 2005; Salat and Pascual, 2006), which is close to the measured trend of inland air temperatures in this period.

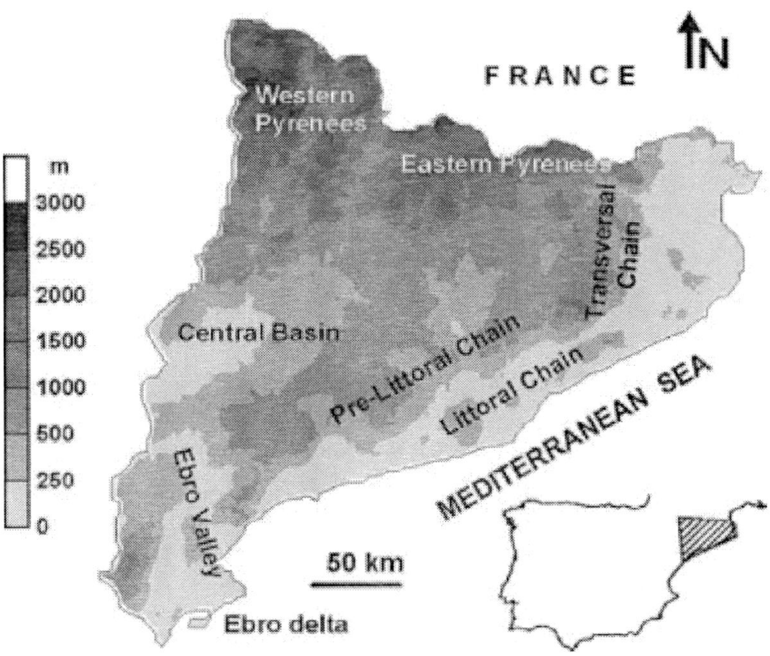

Figure 1. Main orographic patterns of Catalonia.

The Mediterranean regime is mainly governed by the influence of westerlies in winter and the subtropical anticyclone in summer. The vicinity of the sea and neighboring lands with high relief produce a great regional variety of weather and climate. In the Western

Mediterranean, a large majority of the depressions are originated in the same region. These are the Saharan and Genoa type depressions (Barry and Chorley, 2003; Jansà et al., 2001). As a difference with the central and south-west coast of the Iberian Peninsula, with winter precipitation mainly explained by the North Atlantic Oscillation, the eastern façade of the Iberian Peninsula keeps a negative correlation with the Western Mediterranean Oscillation (Martín-Vide and López-Bustins, 2006).

2.2. Databases

The daily rain amounts recorded in Catalonia have been obtained from the *Instituto Nacional de Meteorología* (Spanish Ministry of Environment). The length and size of the databases have changed along the years on account of the kind of studies. Thus, in Lana and Burgueño (1998a), data were obtained from 78 rain gauges covering an average of 34 years; in Serra et al. (2006), from 40 rain gauges for the 1950-2000; and in Lana et al. (2006), from 75 rain gauges (Figure 2) and the same recording period. For this last database, the homogeneity of every rain gauge record and a certain lack of data for some gauges have been extensively analyzed, quantified and discussed by Lana et al. (2004) and Burgueño et al. (2005), who have used the same database to investigate different patterns of the daily rainfall regime.

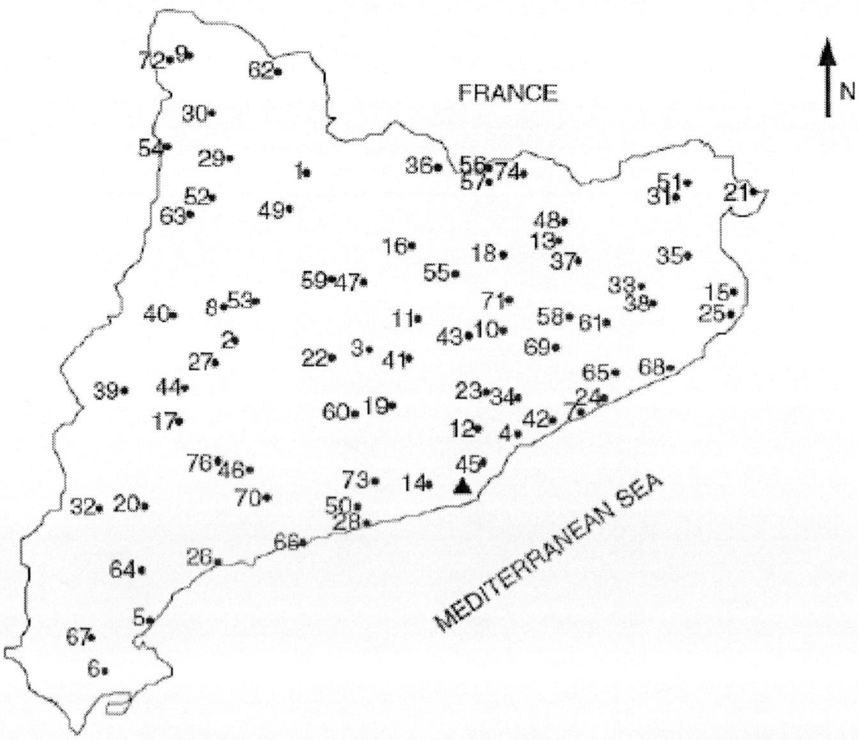

Figure 2. Rain gauge network of the *Instituto Nacional de Meteorologia*, Spain. Fabra Observatory is indicated by a solid triangle.

Missing data in some gauges can reach 10 years, usually not distributed in very short gaps but in continuous periods close to 1 year. Consequently, a dry spell interrupted by a missing period is discarded for any kind of statistical analysis and the next dry spell begins after the last recorded rainy day. Alternatively, missing daily data could have been estimated by considering statistical methods and data corresponding to neighboring gauges. Nevertheless, the strong spatial variability of the daily pluviometric regime of Catalonia (Lana et al., 2004) advises against this substitution of missing data. Even though the Western Pyrenees Range is poorly covered, the recording continuity of the remaining gauges is good. The database for the fractal analysis is rather different than those described before, because this analysis is not focused in Catalonia but it takes part of a larger scale study covering the whole Iberian Peninsula (Martínez et al., 2007). The dataset consists then of long daily pluviometric records from 43 rain gauges, four of them placed in Catalonia.

Dry spells have been derived in reference to four threshold levels of daily rainfall: 0.1, 1.0, 5.0 and 10.0 mm/day. In this way, the spell length is defined as the number of consecutive days with rain amounts below the threshold level considered. The first level is the resolution of the pluviometers; it distinguishes between wet and dry days and would be associated with episodes leading to drought. The second threshold is considered in many other dry spell analyses and comparisons with our results are then possible. The third and especially the fourth level could represent the beginning of a process of compensation for previous drought episodes. From this point of view, larger threshold levels, for instance, 20.0 and 30.0 mm/day would be better signs of the balance of a drought period. Unfortunately, on average, spells lengthen with the increase in threshold levels, thereby diminishing the number of spells detected. This number is then not high enough to establish reliable statistics for many of the rain gauges. Consequently, threshold levels higher than 10.0 mm/day have not been considered.

2.3. Regional Climate

The closeness to a warm sea and the distant Atlantic Ocean, together with a complex orography, produce a particular scenario for the spatial distribution of rainfall amounts, in which eastern and southern advections are the most significant contributors in spring and especially in autumn. They mainly affect the littoral and pre-littoral domains, but sometimes extend further inland.

The Föhn effect through the Iberian Peninsula under the western advections and through the Pyrenees under the western and northwestern circulations would explain the null or scarce rainfalls in some areas of Catalonia, except for the north face of Western Pyrenees. When a nuclei of low-pressure crosses Catalonia, the rain amounts are relevant. Moreover, these systems are sometimes reactivated when they reach the Mediterranean Sea and can generate eastern advections.

Long dry spells usually occur in the hot and cold seasons because of an anticyclonic persistence which often affects the whole territory. Along the hot season, the income of humid air from the Mediterranean together with cold air masses at high levels of the atmosphere can interrupt summer droughts in mountain ranges such as the Pyrenees and Pre-Pyrenees by convective phenomena (Lana et al., 2006).

3. Distribution of Dry Spell Lengths

3.1. Previous Studies

Previous studies on spatial patterns of dry spells have modeled annual extreme spells by means of the Gumbel and Jenkinson dsitributions (Lana and Burgueño, 1998a), the probability of transition from wet to dry spells by means of Markov chains (Lana and Burgueño, 1998b), as well as the Poisson distribution of repeated long dry spells (Lana and Burgueño, 1998c). The parameters of the different distributions have been estimated by moments or from probability papers. At a different time scale, with monthly instead of daily data, Lana et al. (2001) also analysed patterns of monthly shortage and excess of rain amounts by applying the SPI.

3.2. Annual Extreme And Partial Duration Dry Spell Distributions

a) Modeling

Annual extreme spells (AES) and partial duration series (PDS) of dry spells for 0.1, 1.0 and 5.0 mm/day thresholds are obtained from daily rainfall measures along the second half of the 20th century (Lana et al., 2006). The threshold of 10.0 mm/day is not considered because of the insufficient number of dry spells. AES and PDS are modeled by the generalized extreme value (GEV) and the generalized Pareto (GP) distributions and their parameters are estimated by L-moments (Hosking and Wallis, 1997). Spell lengths derived for different return periods from AES spells and from PDS series are compared. These series are generated from the original data set according to the mean-excess plot algorithm (Begueria, 2005).

Both statistical distributions can be written in terms of the reduced variable y

$$y = -\frac{1}{k}\ln\{1-\kappa(x-\xi)/\alpha\}, \quad \kappa \neq 0 \tag{1}$$

$$y = (x-\xi)/\alpha, \quad \kappa = 0 \tag{2}$$

where scale α , shape κ and location ξ are the distribution parameters. The GEV distribution is given by

$$F(x;\alpha,\kappa,\xi) = \exp\{-\exp(-y)\} \tag{3}$$

x representing the annual extremes to be analysed. The GP distribution is

$$F(x;\alpha,\kappa,\xi) = 1-\exp(-y) \tag{4}$$

where x now represents the values exceeding a selected percentile.

The L-skewness, τ_3, and the L-kurtosis, τ_4, depend solely on shape parameter κ , which permits writing the Euclidean distance D as

$$D = \left\{(t_3 - \tau_3(\kappa))^2 + (t_4 - \tau_4(\kappa))^2\right\}^{1/2} \quad (5)$$

where t_3 and t_4 are empirical estimations of L-skewness and L-kurtosis. Euclidean distance D is adopted as a measure of the fit between empirical data and the theoretical model. The lowest value of D will then indicate the best distribution function.

The computation of the value x_T linked to a return period T_r is straightforward for the GEV by remembering that:

$$x_T = \frac{\alpha}{\kappa}\{1 - g(T_r)\} + \xi \quad (6)$$

with

$$g(T_r) = \{-\ln[1 - 1/T_r]\}^k \quad (7)$$

Similarly, for GP:

$$x_T = \frac{\alpha}{\kappa}\left\{1 - T_r^{-k}\right\} + \xi \quad (8)$$

The application of Eqs. (6) and (8) is straightforward for the AES strategy, where there is only one extreme value per year. For the PDS sampling strategy, however, T_r must be replaced with $\beta \cdot T_r$, with β defined as the average number of spells per year (Vicente-Serrano and Beguería, 2003).

b) Results

Figure 3 compares the empirical L-skewness and L-kurtosis (triangles) of the AES series for each rain gauge with the theoretical GEV (solid line) and GP (dashed line) models for dry spell lengths. The open squares represent the Gumbel distribution, which is a particular case of the GEV model. Although, for most of gauges, the best fit is obtained for the GEV model, a non-negligible number of gauges are better modeled by the GP distribution. Figure 4 depicts the same comparison for the PDS series, generated considering spell lengths exceeding a threshold level corresponding to the 95th percentile of the empirical spells. A clear shift towards the GP distribution is observed for all the threshold levels and only a few gauges follow the GEV model.

Figure 5 illustrates the spatial distribution of the spell lengths for a threshold level of 5.0 mm/day for the PD strategy and return periods of 2, 5, 10 and 25 years. These maps have been drawn by choosing the best option for each rain gauge, among the GEV and GP models, according to the minimum L-skewness-kurtosis distance D. A common pattern is the presence of a clear north or northeast to south or southwest gradient, with a nucleus of maximum length in the southern corner of Catalonia for 0.1 and 1.0 mm/day, which drifts towards the Central Basin for 5.0 mm/day.

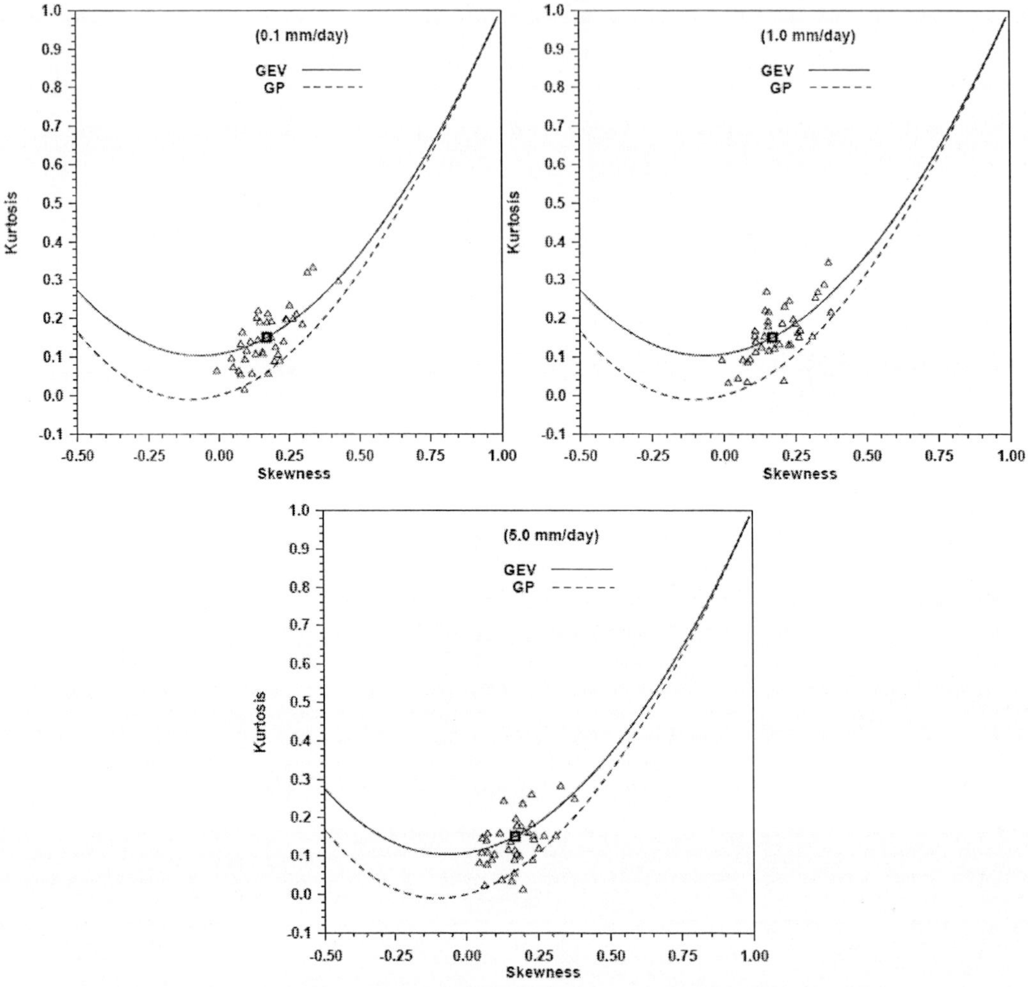

Figure 3. Empirical (open triangles) and theoretical L-skewness and L-kurtosis for thresholds rain amounts of 0.1, 1.0 and 5.0 mm/day according to the AE option. The solid line represents the GEV distribution, dashed line the GP distribution and the open square the Gumbel model. (*Source: Lana et al., 2006*).

Possible influences originated by the orography and vicinity to the Mediterranean Sea are not so evidently manifested here as in other kind of daily rainfall analyses (Lana et al., 2004). The Pre-Pyrenees and Pyrenees Ranges are characterized by the shortest dry spells associated with the 95th percentile used in the PDS strategy, whatever the daily rainfall threshold considered. The effects of the Littoral and Pre-Littoral Ranges are not detected, whereas the effects of the Transversal Range are only present for 5.0 mm/day. Additionally, east-west gradients, attributable to the influence of the Mediterranean Sea, are not observed.

It is also interesting to compare return period maps obtained here with those derived by Lana and Burgueño (1998c), who assumed a Poisson distribution to quantify the probabilities of repeated long dry episodes in Catalonia. Comparisons are possible for return periods of 5, 10 and 25 years and a threshold level of 0.1 mm/day. A north-south gradient and a nucleus of long spells close to the Ebro Delta were also detected.

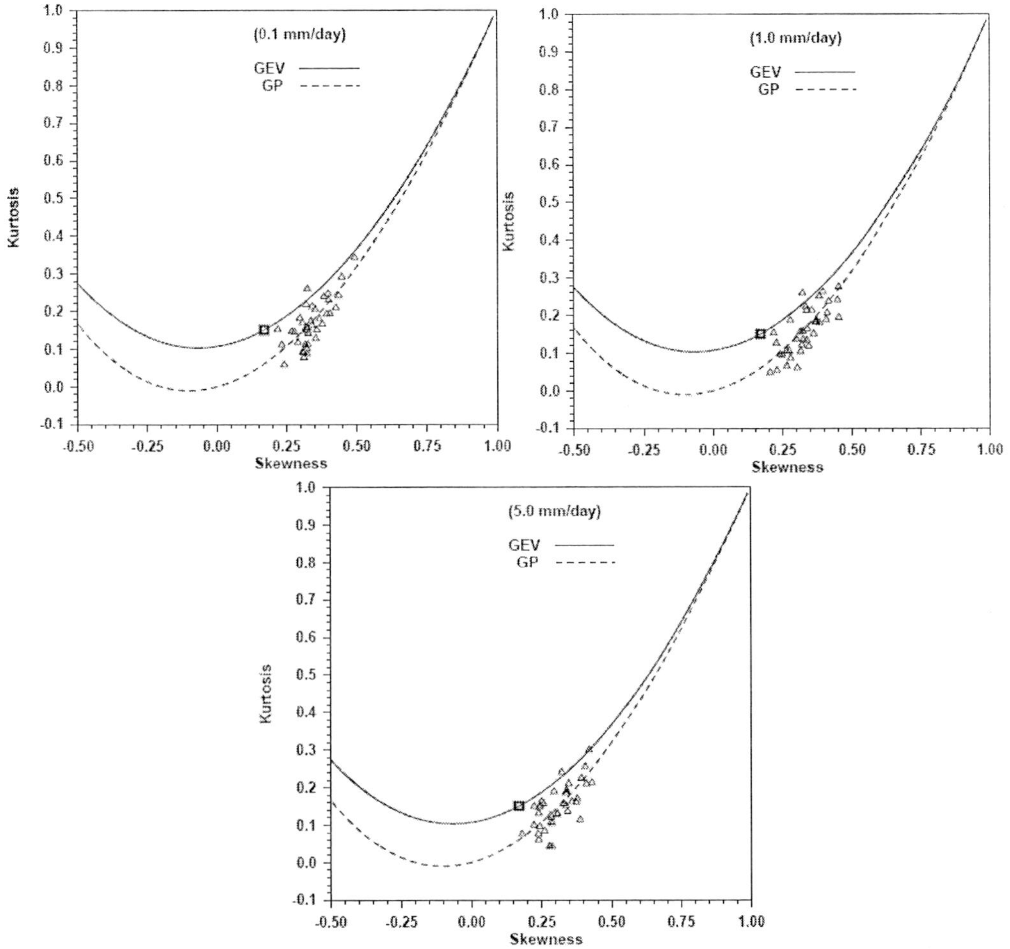

Figure 4. Empirical (open triangles) and theoretical L-skewness and L-kurtosis for thresholds rain amounts of 0.1, 1.0 and 5.0 mm/day according to the PD option. The solid line represents the GEV distribution, dashed line the GP distribution and the open square the Gumbel model. (*Source: Lana et al., 2006*)

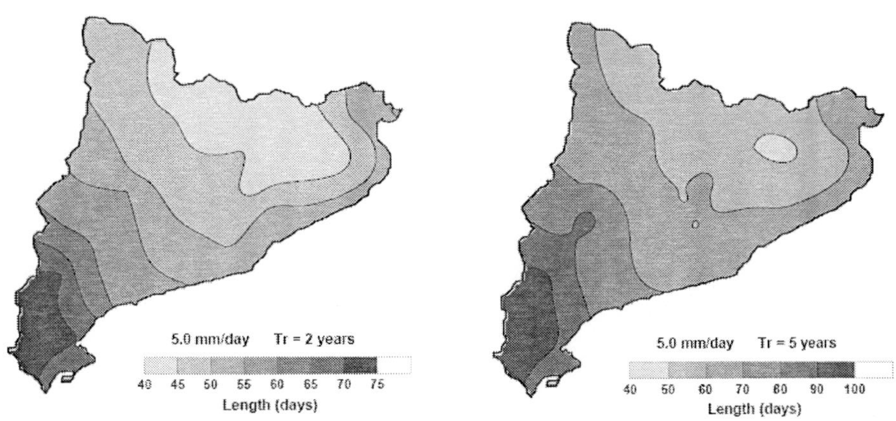

Figure 5. (Continued).

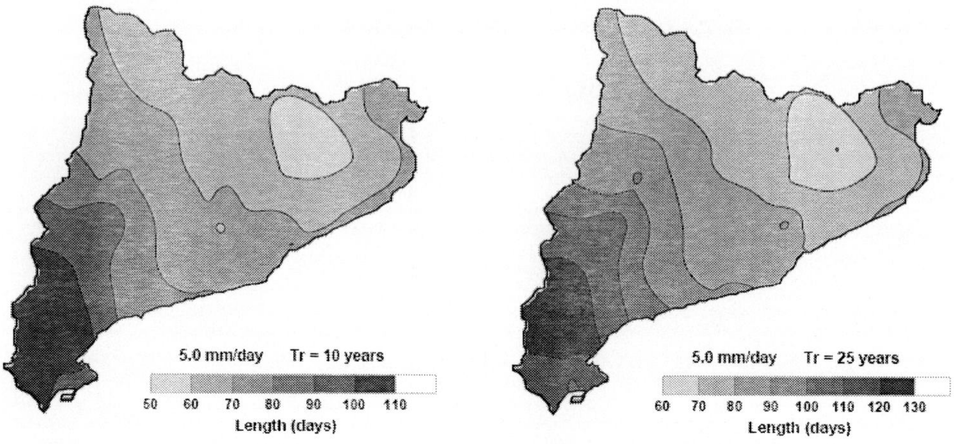

Figure 5. Dry spell lengths for return periods of 2, 5, 10 and 25 years and threshold levels of 5.0 mm/day, when the PD sampling strategy is considered. (*Source: Lana et al., 2006*).

Additionally, both the range and the spatial distribution of spells for every return period were quite similar to that obtained here. The delimitation of the western end of the Central Basin, observed in Lana and Burgueño (1998a), is not detected in Lana and Burgueño (1998c), in agreement with the present results. In short, improvements achieved with the GEV and GP distributions for both AES and PDS strategies agree better with the Poisson approach than with the Gumbel and Jenkinson models. Possibly, the use of the Gumbel model in Lana and Burgueño (1998a) biased their return period maps.

3.3. All Dry Spells Distribution

a) Modeling

In this section, the distribution of all available dry spell lengths is considered (Lana et al., 2007). Return period maps from 2 to 50 years obtained for threshold levels of 0.1, 1.0 and 5.0 mm/day are quite similar to those of Section 3.2. A new threshold of 10.0 mm/day can be investigated now because the number of empirical dry spells for this level is large enough when considering all dry spells, in contrast to what happened for AES or PDS strategies.

The spell length series for all gauges fit well the Weibull distribution, whatever the daily threshold. The L-skewness-kurtosis diagrams suggest that it should be the right option at present, and numerous examples in Climatology and other scientific fields support this possibility.

The cumulative expression of the Weibull distribution is (Benjamin and Cornell, 1970)

$$F(x;\kappa,u,\varepsilon) = P(x \le X) = 1 - \exp\left\{-\left(\frac{x-\varepsilon}{u-\varepsilon}\right)^{\kappa}\right\}; \quad x \ge \varepsilon; \quad u,\kappa > 0 \qquad (9)$$

and its parameters are the location ε, scale (u-ε) and shape κ. The location parameter ε represents the minimum possible value for the variable x, i.e., the minimum dry spell length, which in our analysis is 1 day. The parameters of the Weibull distribution are estimated by L-moments (Hosking and Wallis, 1997) due to the robustness of this procedure in comparison to other estimation methods, such as probability papers or moments. As mentioned earlier, the L-skewness, τ_3, and L-kurtosis, τ_4, depend solely on the shape parameter κ, which permits an easy and accurate validation of the closeness of the dry spell empirical distribution to the Weibull model in terms of the Euclidean distance D (Eq. 5).

For the Weibull distribution, the return period, T_r, can be defined as

$$T_r = \exp[(x_T - \varepsilon)/(u - \varepsilon)]^{\kappa} \tag{10}$$

where x_T is the corresponding dry length. In contrast to annual extreme analysis (AES), where there is one extreme sample per year and return periods are then given in years, the number of dry spells usually exceeds one per year and return periods would not be multiple of years. This shortcoming can be avoided, as when considering PDS strategy, introducing a slightly different definition of x_T (Vicente-Serrano and Beguería, 2003)

$$x_T = \varepsilon + (u - \varepsilon)[\ln(\beta \cdot T_r)]^{1/k} \tag{11}$$

where β represents the average number of dry spells per year. In this way, x_T is expressed in days and return period in years.

b) Results

A global overview of the fit between empirical data and the Weibull distribution is obtained from the distance D. The set of Figures 6 shows the empirical L-skewness-kurtosis diagrams for all the threshold levels considered and several theoretical distributions. Specifically, the generalized logistic (GLO), generalized Pareto (GPA), Pearson Type-3 (PE3), exponential (E) and Weibull L-skewness-kurtosis curves are plotted. GLO and E distributions must be discarded for all the thresholds. GPA model could only be acceptable for a few gauges. Weibull and PE3 distributions depict a quite similar evolution. After a detailed revision of distances D, it can be reasonable assumed that Weibull distribution fits empirical data slightly better than PE3.

Figure 7 depicts the spatial patterns of the dry spell lengths for the threshold level of 5.0 mm/day and return periods of 2, 5, 10, 25 and 50 years. A common pattern is a north or northeast to south or southwest gradient with minimum dry spell lengths in the Pyrenees domain and maximum lengths in the south or south-west of Catalonia. The same pattern is observed for the rest of thresholds. A second outstanding feature is the clockwise rotation of this gradient as threshold level increases. The longest dry spells shift from the southern coast towards western Catalonia, whatever the return period. In the Pyrenees, the shortest lengths shift from Western to Eastern Pyrenees as threshold level increases. This pattern is observed again for all return periods. Although average dry spells, parameters u of the Weibull distribution and return period maps depict a clear spatial variability, the test of regional homogeneity proposed by Hosking and Wallis (1997) suggests signs of spatial homogeneity for 5.0 and 10.0 mm/day thresholds. Conversely, the dry spell distributions would be spatially heterogeneous for 0.1 and 1.0 mm/day.

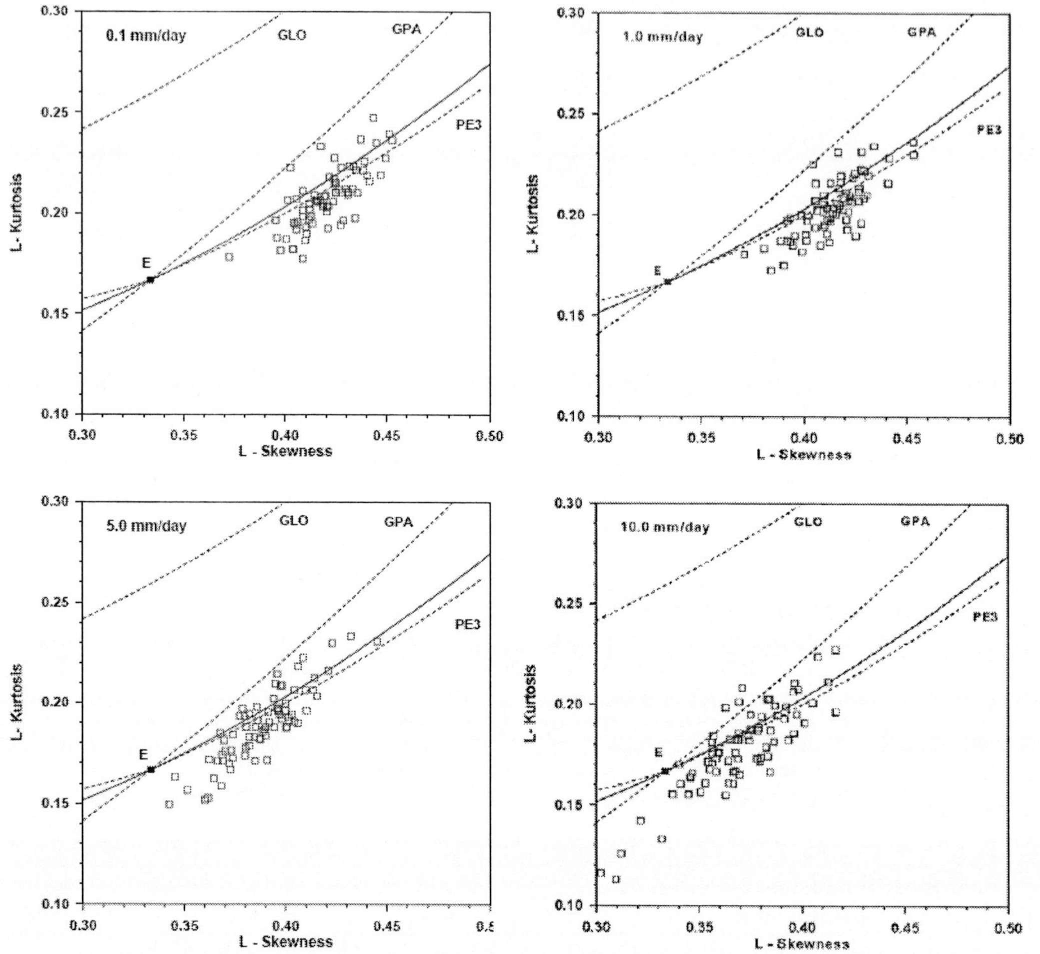

Figure 6. L-skewness-kurtosis diagrams for empirical data and Weibull distribution for 0.1, 1.0, 5.0 and 10.0 mm/day. Solid line represents the Weibull distribution. Dashed lines design GLO, GPA and PE3 distributions. Solid square represents E distribution. (*Source: Lana et al., 2007*).

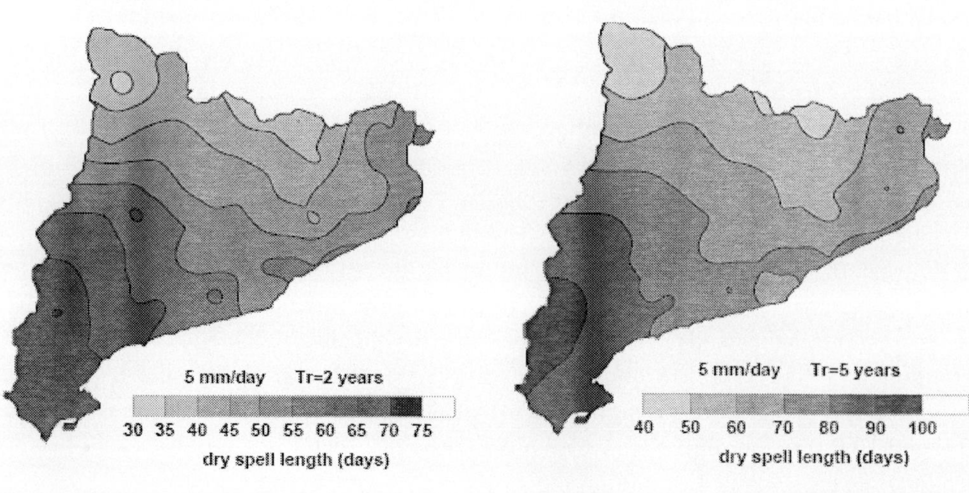

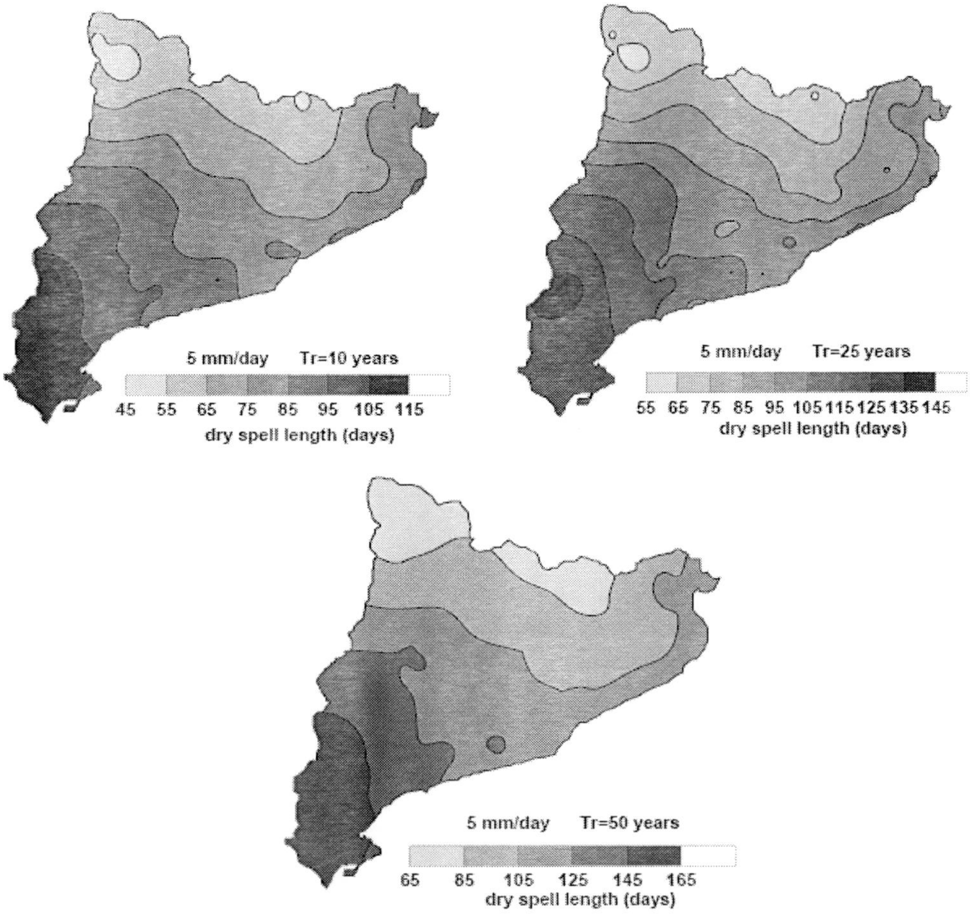

Figure 7. Return period maps of dry spell lengths for 2, 5, 10, 25 and 50 years and a threshold level of 5.0 mm/day, according to a Weibull distribution with parameters estimated by L-moments. (*Source: Lana et al., 2007*).

4. Time Trends of Dry Spells

4.1. Methodology

As indicated in Section 1, the present climatic change in the Mediterranean area should modify the distribution of dry spell lengths, thus suggesting the possibility of relevant time trends in several basic variables characterizing the dry spell behavior. The first variable considered is the number of dry spells, N; the second is their maximum recorded length, Lmax, and the third, their mean length, L. These three variables are analyzed at annual and winter and summer half-year scales, and four threshold levels (0.1, 1.0, 5.0 and 10.0 mm/day) are distinguished.

Local time trends at every rain gauge are evaluated by linear regression and their significance (95% confidence level) is verified according to the Mann-Kendall statistical test

(Sneyers, 1990). Field significant trends are also investigated by Monte Carlo techniques (Chu and Wang, 1997; Suppiah and Hennessy, 1998; Zhang et al., 2004b). These techniques are especially appropriate to assess if local trends occurred by chance or a field significant trend is very likely. Details concerning the steps of the Monte Carlo technique can be found in Serra et al. (2006).

4.2 Spatial Patterns of Mean Values of N, Lmax and L

The main spatial features of these three variables, <N>, <Lmax> and <L>, are summarized in Table 1, where the range of every variable for annual and summer and winter half-year scales is given. The mean number of dry spells in a year tends to decrease with the threshold level, which is in agreement with an increase of <Lmax> and <L>. Winter and summer-halves depict similar ranges except for <Lmax> that shows wider ranges for summer than for winter-half.

Table 1. Range of spatial variation of <N>, and <Lmax> and <L> in days

THRESHOLD	ANNUAL	WINTER-HALF	SUMMER-HALF
<N>			
0.1 mm/day	20-52	10-22	10-28
1.0 mm/day	21-45	10-20	10-26
5.0 mm/day	15-33	7-15	7-19
10.0 mm/day	8-26	3-11	3-15
<Lmax>			
0.1 mm/day	26-62	21-51	16-56
1.0 mm/day	30-62	28-52	20-56
5.0 mm/day	40-76	38-58	28-68
10.0 mm/day	45-105	44-74	32-74
<L>			
0.1 mm/day	5-15	4-16	4-16
1.0 mm/day	5-15	7-17	5-17
5.0 mm/day	9-21	11-21	7-23
10.0 mm/day	12-36	15-40	10-35

Source: Serra et al., 2006.

Figure 8 depicts the spatial distribution of <N>, <L> and <Lmax> for a threshold level of 1.0 mm/day. The values of <N> are characterized by a south-north gradient. Similar results are obtained for 0.1 mm/day. This spatial distribution indicates that the mean number of dry spells per year in the Pyrenees Range and neighboring areas is higher than in the rest of the country. Conversely, the distribution of <Lmax> for 1.0 mm/day depicts a north-south gradient. Whereas Pyrenees and surrounding areas show relatively low values of <Lmax> (30-38, 28-40 and 20-28 days for annual and winter and summer-halves respectively), the southern corner of Catalonia reaches values of 50-62 days for annual scale and 40-52 days for winter and 44-56 days for summer-half. The spatial distribution of <L> for 1.0 mm/day is also characterized by a north-south gradient, the maximum values in the southern end of Catalonia attaining 11-15 days for annual scale and 13-17 days for winter and summer-halves.

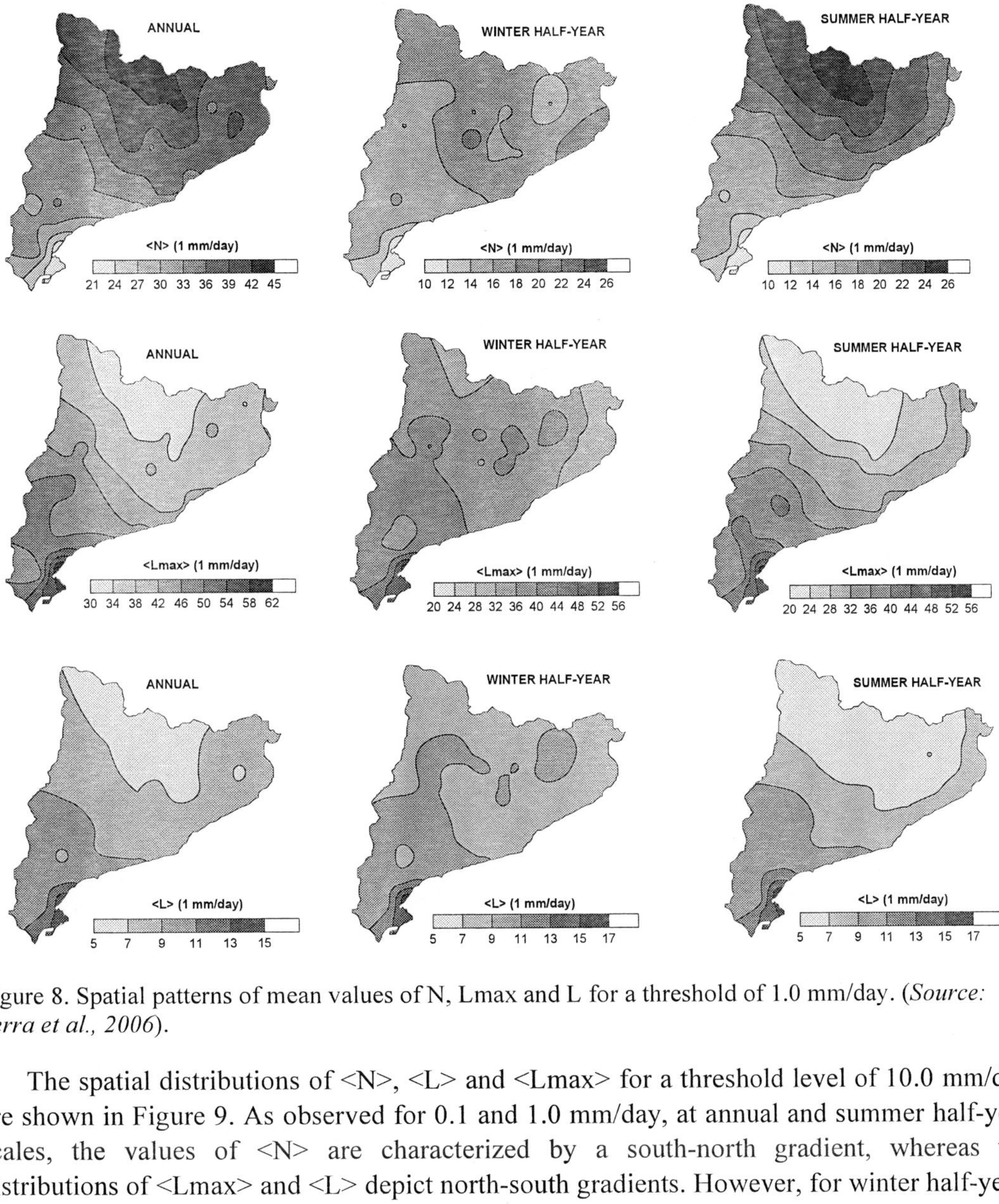

Figure 8. Spatial patterns of mean values of N, Lmax and L for a threshold of 1.0 mm/day. (*Source: Serra et al., 2006*).

The spatial distributions of <N>, <L> and <Lmax> for a threshold level of 10.0 mm/day are shown in Figure 9. As observed for 0.1 and 1.0 mm/day, at annual and summer half-year scales, the values of <N> are characterized by a south-north gradient, whereas the distributions of <Lmax> and <L> depict north-south gradients. However, for winter half-year, there are signs of southwest-northeast and northeast-southwest gradients respectively. The mean number of dry spells per year in the Pyrenees and surrounding areas reaches 17-26 for annual , 5-9 for winter and 9-15 days for summer-half respectively. The highest values of <Lmax> (85-105 and 62-74 days for annual and winter and summer-halves respectively) and <L> (28-36 and 35-40 for annual and winter and summer-halves respectively) tend to shift towards the southwest. These main features are also observed in the spatial distributions for 5.0 mm/day.

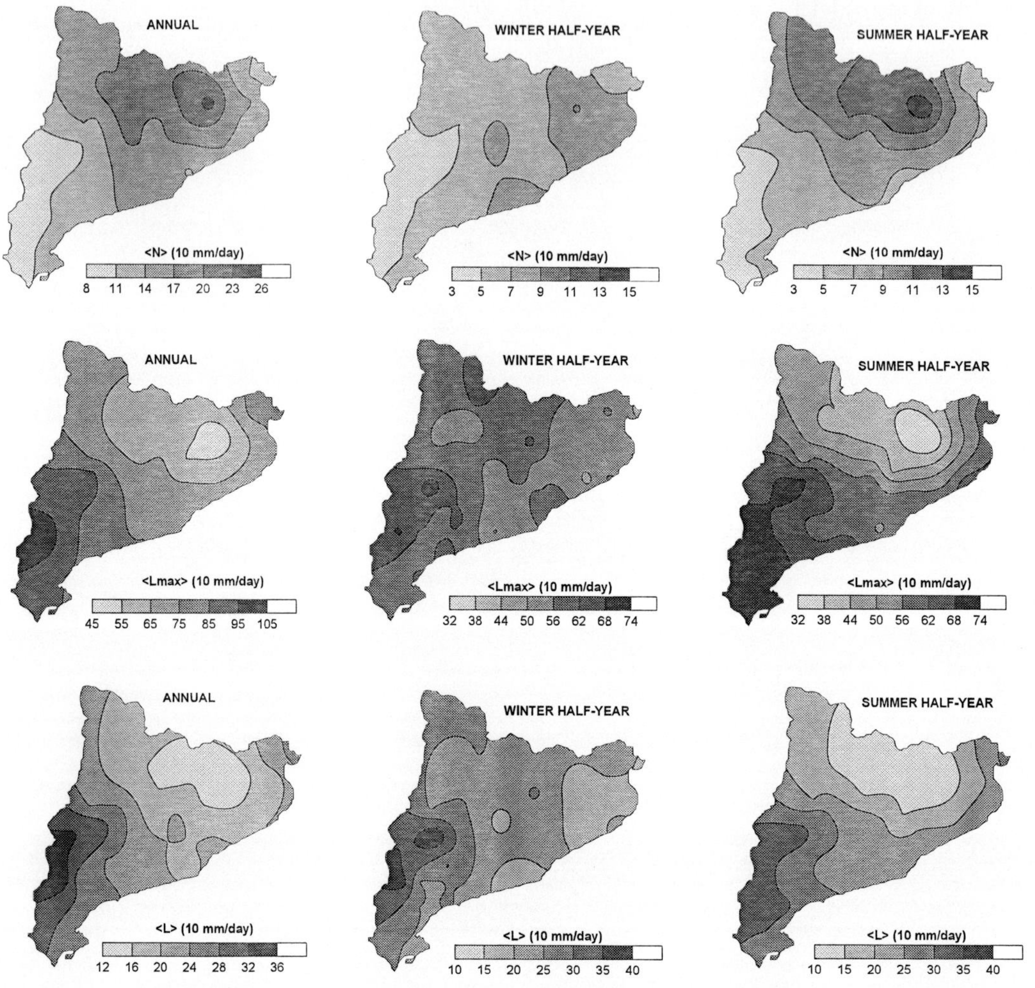

Figure 9. Spatial patterns of mean values of N, Lmax and L for a threshold of 10.0 mm/day. (*Source: Serra et al., 2006*)

As a summary, the spatial distributions of <N>, <Lmax> and <L> are rather heterogeneous due to the complex orography of the country, the atmospheric circulation patterns and the vicinity to the Mediterranean Sea (cyclogenesis in autumn) and remoteness to the Atlantic Ocean (poor contribution of Atlantic frontal passages to rain amounts except for the north face of Pyrenees). All these facts favor higher rain amounts in Pyrenees and Pre-Pyrenees Ranges in comparison with Central Basin and the Southern Mediterranean coast. The number of dry spell episodes is usually higher towards the north (Pyrenees and Pre-Pyrenees Ranges). Nevertheless, due to the pluviometric regime in this area, these dry spells are shorter, both in terms of <Lmax> and <L>, than in other places of Catalonia (Central Basin and southern Mediterranean coast), where extreme dry spell lengths could be a serious problem for agriculture, industry and water supplies.

4.3. Time trends

Time trends depict different spatial patterns for series of N, Lmax and L. For 0.1 mm/day, whereas significant local trends are scarce for Lmax, they are quite numerous for N and especially L. Threshold levels of 1.0, 5.0 and 10.0 mm/day are characterized by a relatively low number of significant trends except for the number of spells per year, N. Figure 10 illustrates three examples of spatial distribution of local trends corresponding to variable L for summer-half and 0.1 mm/day, and variable N for winter-half and thresholds of 5.0 and 10.0 mm/day. Additionally, Table 2 summarizes the number of positive and negative significant trends (95% confidence level) for N, Lmax and L at annual and summer and winter half-year scales.

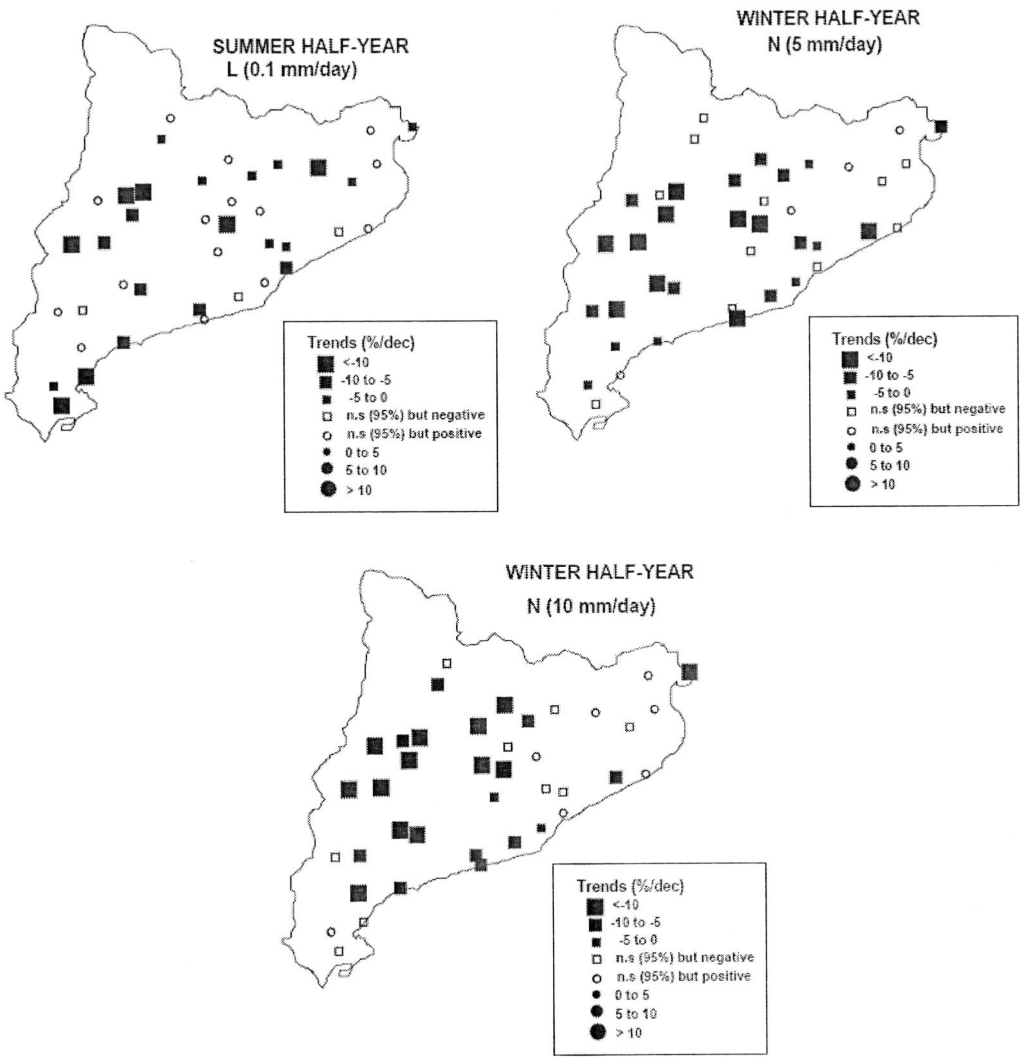

Figure 10. Three examples of significant field trends. Local trends are given in percentage of variation per decade. (*Source: Serra et al., 2006*).

Table 2. Number of positive and negative significant local trends for annual and summer and winter half-year series of N, Lmax and L. (*Source: Serra et al., 2006*)

	0.1 m/day		1.0 mm/day		5.0 mm/day		10.0 mm/day	
Trend Sign	+	–	+	–	+	–	+	–
Annual								
N	6	4	2	8	0	12	0	14
Lmax	2	1	3	1	9	0	5	0
L	0	23	0	8	1	1	2	1
Winter half-year								
N	2	11	1	14	0	25	0	24
Lmax	0	1	0	0	1	0	1	3
L	0	11	0	3	0	3	0	3
Summer half-year								
N	2	4	2	5	0	11	0	11
Lmax	2	2	4	1	7	0	6	0
L	0	22	1	6	0	2	2	1

Table 3. Statistical significance of field trends for annual and summer and winter half-years. (*Source: Serra et al., 2006*)

Threshold	0.1 mm/day	1.0 mm/day	5.0 mm/day	10.0 mm/day
Annual				
N	99.9%	99.9%	99.9%	99.9%
Lmax			99.9%	95.0%
L	99.9%	95.0%		
Winter half-year				
N	99.9%	99.9%	99.9%	99.9%
Lmax				
L	99.0%			
Summer half-year				
N			99.0%	
Lmax			99.0%	95.0%
L	99.9%			

A relevant question to consider is the statistical significance of field trends. Field trends are assumed as possible if their statistical significance, computed by the Monte Carlo algorithm mentioned before, equals or exceeds 95%. Table 3 details the results obtained. Whereas at annual scale the significant field trends are detected for the three variables at least for two of the four threshold levels, field significances are more scarce for summer and winter half-years, with the single exception of N for winter-half, which is field significant for all the thresholds.

4.4. Interpretation of Time Trends

As mentioned in the Introduction, the decreasing number of dry spells during the annual and winter half-year scales is in agreement with the overall decrease of the cyclonic activity in the Mediterranean, due to the north and eastward shift of cyclonic activity over Europe and the Northeast Atlantic. Nevertheless, the number of deep Mediterranean cyclones would increase (Knippertz et al., 2000; Lionello et al., 2002). These anomalies in the atmospheric circulation pattern could be consequence of a scenario of a doubling of the atmospheric CO_2 and it is in agreement with several General Circulation Models based on different greenhouse gas (GHG) forcing scenarios (Houghton et al., 2001). The reduction of the number of dry spells affects a larger number of gauges at 5.0 and 10.0 mm/day, particularly in the winter-half year. In spite of this, Lmax trends are positive for just a few gauges and L has a small number of trends of both signs. Therefore, there is not a simple link between the time behavior of the dry spell lengths (L and Lmax) and the circulation modifications generated by increasing GHG concentrations. Nevertheless, on the basis of hydrological studies, Voss et al. (2002) assume that expected 10-year return values of annual Lmax in 2060-89 would be 10 days longer than lengths found for Southern Europe from 1979 to 1999.

5. Predictability and Predictive Instability of Dry Spells

5.1. Fractal Concepts

The complexity of the daily pluviometric regime in Catalonia suggests immediately that the predictability of the dry spells (basically their length) will be difficult. In previous sections, some methods to obtain return period maps from different statistical assumptions and distribution functions have been revised. Now, the main objective is quantifying the difficulty to obtain a set of equations predicting a dry spell and its length.

The analysis is based on three concepts. First, the lacunarity, L (Maldenbrot, 1982), which provides a way of quantifying the gap size (dry spell length) distribution. It should be mentioned that several fractal sets with the same fractal dimension can be distinguished by their lacunarities (Lin and Yang, 1986; Turcotte, 1997). A detailed description of the lacunarity formulation for pluviometric and dry spell series can be found in Martínez et al. (2007). Second, the re-scaled analysis and the Hurst exponent, H (Feder, 1988; Goltz, 1997), would be a good descriptor of the predictability of the consecutive dry spells. H close to 0.5 implies randomness and the best prediction of a dry spell length is then the last empirical length. Values of H well above 0.5 indicate persistence in the series of spell lengths and the best prediction should be based then on the extrapolation of the time trend of the dry spell series. Finally, values of H well below 0.5 imply anti-persistence. Thus, the best prediction is an average of the previous spell lengths. The last concept is the Lyapunov exponents, which measure predictive instability, in other words, how small uncertainties, errors in the starting conditions (initial length) or erroneous equations describing the spell length evolution could lead to large errors in spell length predictions. The concept and computation of the Lyapunov exponents are based on the reconstruction theorem (Takens, 1981; Grassberger and Procaccia,

1983a, b). A detailed description of the computational algorithm of these exponents can be found in Eckmann et al. (1986), Stoop and Meier (1988) and Martínez et al. (2007).

The reconstruction theorem also permits the computation of the correlation dimension and the Kolmogorov entropy (Diks, 1999). The correlation dimension establishes the minimum number of equations describing the mechanism of the physical process analysed and the Kolmogorov entropy measures the "lost" of memory when the physical process passes through different states. In summary, whereas lacunarity and, especially, the Hurst exponent and the Lyapunov exponents are strongly related to the predictability and predictive instability of a physical process, the Kolmogorov entropy and the correlation dimension are more specifically linked to the number and type of required equations.

5.2. Results

a) Lacunarity

As mentioned before, the lacunarity, L, characterizes the dry spell length distribution. Thus, the highest L, the longest the dry spells. L is evaluated for different window lengths r from 1 to 100 days at steps of one day. For all the thresholds levels considered (0.1, 1.0, 5.0, 10.0 mm/day), the empirical L are quite well fitted by two power laws

$$L(r) = \alpha \cdot r^{\beta} \ ; r = 1, \ldots, r_c \tag{12}$$

$$L(r) = \delta \cdot r^{\varepsilon} \ ; r = r_c + 1, \ldots, r_{\max} \quad \text{(100 days)} \tag{13}$$

with negative exponents β and ε, always being $|\beta| > |\varepsilon|$. r_c represents the critical window length for which the power law changes from Eq. (12) to (13), its value depending on the threshold considered.

Figure 11 depicts the spatial distribution of the lacunarity throughout the Iberian Peninsula for a window length of 30 days and the four threshold levels considered (0.1, 1.0, 5.0 and 10.0 mm/day). It should be kept in mind that lacunarity characterizes the dry spell length distribution. Thus, the highest L, the longest the dry spells. A north-south gradient is observed in the Iberian Peninsula for all the thresholds. The north and especially the northeast of Spain are characterized by moderate lacunarities in comparison with those of the south and the southeast. For a window length of 1 day, the direction of the gradient changes to NW-SE (Martínez et al., 2007).

b) Predictability

Remembering that the predictability of the dry spell lengths is characterized by the Hurst exponent, the first relevant feature is the behavior close to randomness deduced for Catalonia, with small signs of persistence and anti-persistence, as H ranges from 0.4 to 0.6 for all the threshold levels. Only for 0.1 mm/day, a small area of Catalonia reaches values of H from 0.6 to 0.7. However, some other places of the Iberian Peninsula show clear signs of persistence (Martínez et al., 2007), with H attaining values of 0.7. Consequently, a certain difference exists between the predictability of spell lengths in Catalonia and in the rest of the Iberian

Peninsula. In places where H reaches values close to 0.7, time trends of dry spell lengths should be considered for a better prediction.

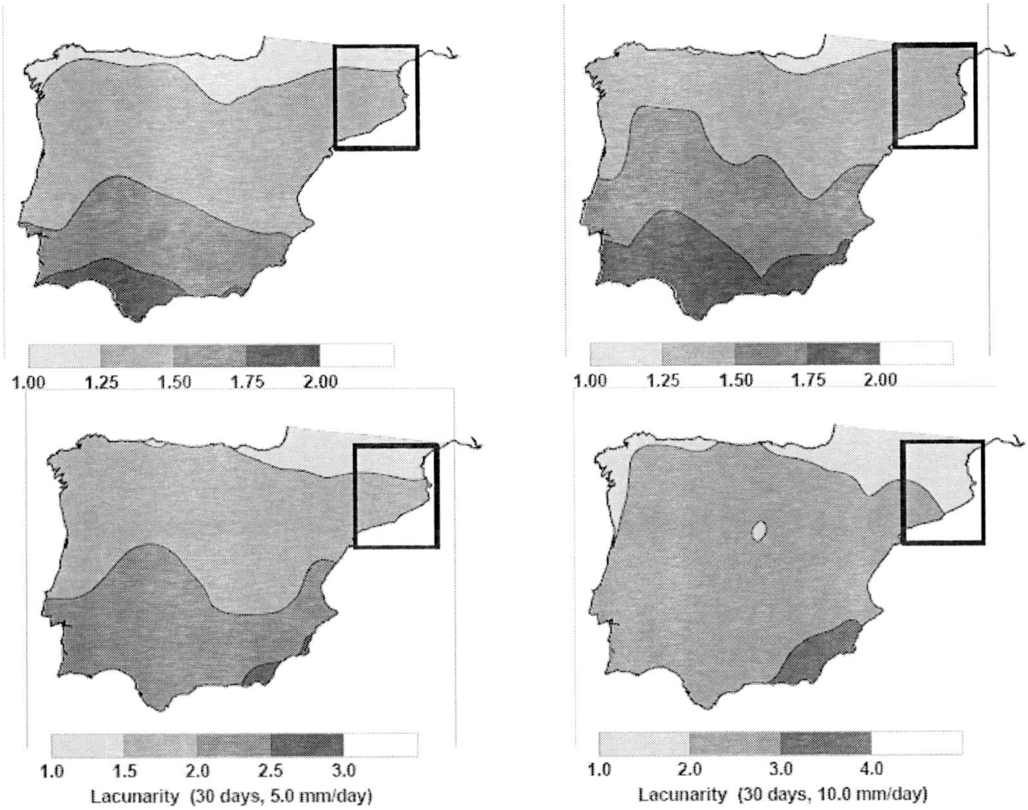

Figure 11. Spatial distribution of the lacunarity across the Iberian Peninsula for a window length of 30 days and the four threshold levels. Rectangles define approximately the limits of Catalonia. (*Source: Martínez et al., 2007*).

Figure 12 depicts an example of re-scaled analysis for the dry spell lengths derived from daily pluviometric records of the Fabra Observatory (FAB), close to Barcelona city, assuming a threshold level of 0.1 mm/day. The re-scaled analyses, R/S, has been computed by considering segments of spell length series ranging from 10 to 1000 elements, a good fit of the empirical data to a log-log representation being obtained, as re-scaled analysis demands. The square regression coefficient of 0.991 is very good and the uncertainty on H very small. This would be a clear example of random behavior of the spell lengths, as H is very close to 0.5.

c) Predictive Instability

Although the predictive instability of a physical system is usually characterized by the first two positive Lyapunov exponents, to be exact, it should be necessary to compute a number of exponents equal to the embedding dimension d_E, which is a concept related to the reconstruction theorem and the computation of the correlation dimension. The algorithm above mentioned computes a number of exponents equal to the embedding dimension d_E. Nevertheless, the first two positive exponents constitute a good measure of the instability.

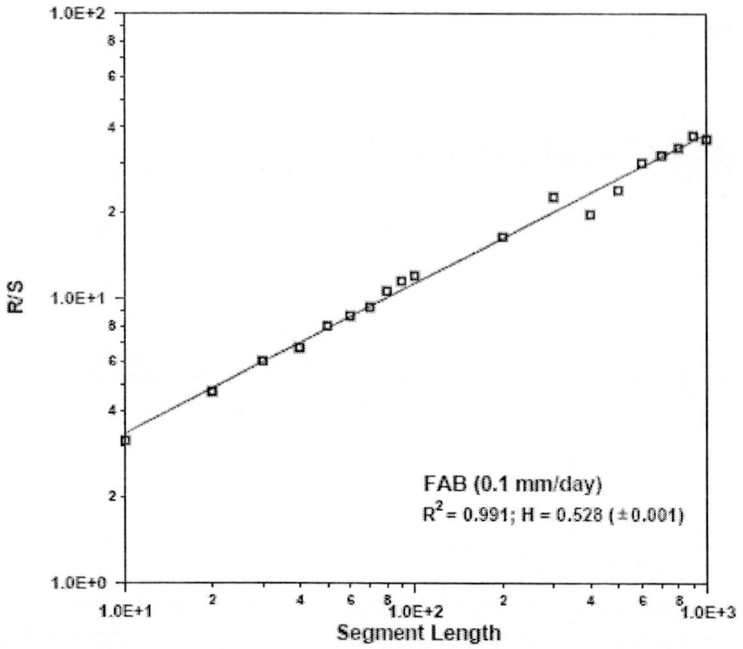

Figure 12. Re-scaled analysis for the Fabra Observatory and a threshold level of 0.1 mm/day.

It should be taken into consideration that a reasonable estimation of the Lyapunov exponents demands long series of data. Remembering that the number of dry spells diminishes with the increasing threshold level, only the level of 0.1 mm/day can be confidently used. Spell lengths in Catalonia (Figure 13) are characterized by the smallest values, both for the first and second Lyapunov exponents, in comparison with the rest of the Iberian Peninsula. Thus, instability exists when predicting dry spell lengths in Catalonia, but this predictive instability would be higher for the southern half of the Iberian Peninsula.

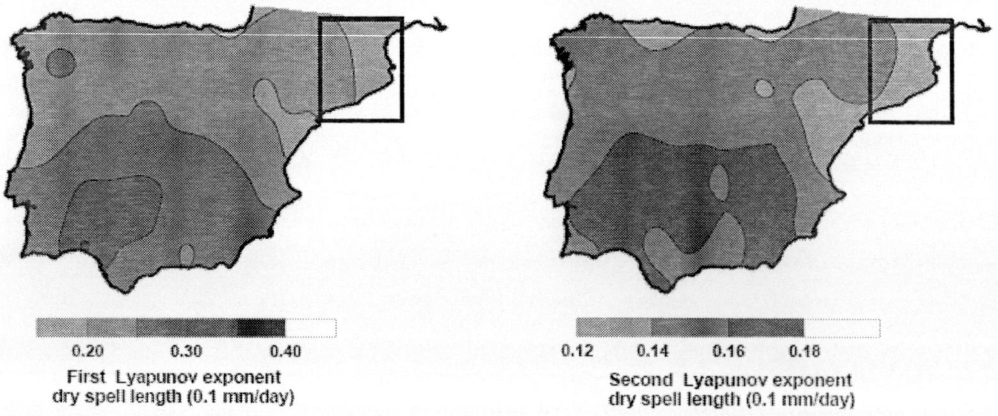

Figure 13. Spatial distribution of the first and second Lyapunov exponents throughout the Iberian Peninsula. Rectangles define approximately the limits of Catalonia. (*Source: Martínez et al., 2007*).

A question to be carefully considered is the embedding dimension d_E and the number of iterations required to an accurate computation of the Lyapunov exponents. Both questions are solved empirically. The algorithm is designed to compute, for many different embedding dimensions, a number of Lyapunov exponents equal to d_E, with a high number of iterations. In this way, the algorithm searches for stationary values of the exponents. An alternative way for the estimation of d_E is based on computing the correlation dimension (Diks, 1999). Figure 14 depicts an example for the series of dry spell lengths associated with a threshold of 0.1 mm/day obtained for the Fabra Observatory. Only the first two positive exponents are represented because they are enough to quantify the instability of a possible system of equations predicting successive spell lengths.

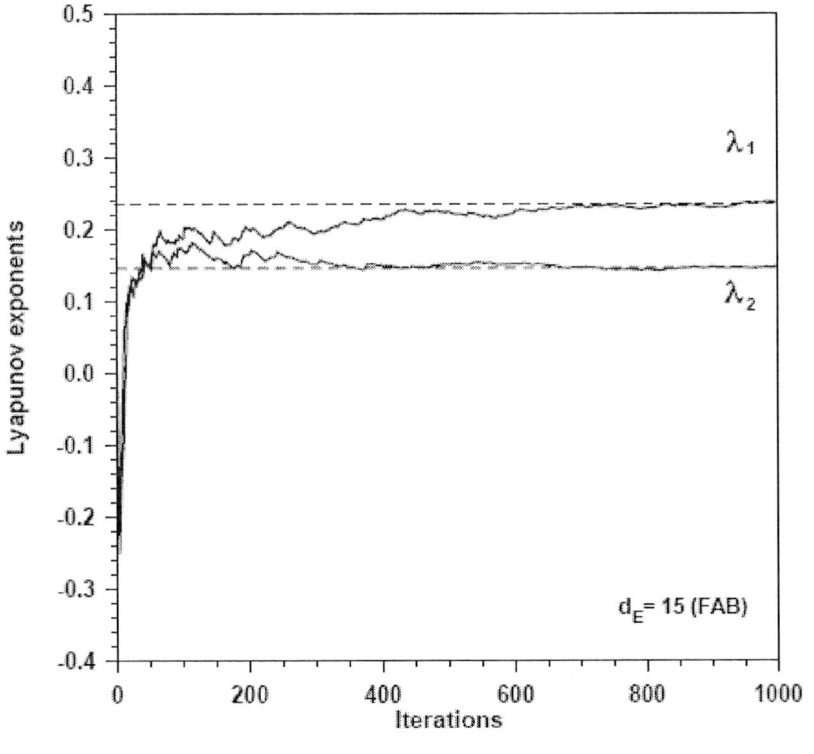

Figure 14. Evolution of the first two positive Lyapunov exponents for the Fabra Observatory after 1000 iterations of the algorithm.

It is also worthy of mention that a relatively high embedding dimension (15) has been necessary to achieve stationary values of the exponents. This is usually a sign of notable random component of the physical system governing the series of spell lengths and it is in agreement with the Hurst exponent derived for the Fabra Observatory. For the rest of rain gauges in Catalonia embedding dimensions are very similar, suggesting that the randomness of the spell lengths is quite common. One thousand iterations are sufficient for an accurate estimation of the first two Lyapunov exponents for all the rain gauges.

CONCLUSIONS

Several studies concerning exclusively Catalonia (NE Spain) or the whole Iberian Peninsula manifest the complex behavior of daily rainfall and dry spell lengths. With respect to the dry spell lengths in Catalonia, the orography and the dominant atmospheric circulation patterns are the main causes of their complex spatial distribution and temporal evolution. Rainfall amounts strongly depend on Mediterranean cyclogenesis in spring and, especially, in autumn and on convective phenomena in summer. Frontal passages from the Atlantic are not so important, except for the north face of the Pyrenees Range. Consequently, anomalies in these circulation patterns could imply relevant changes on the dry spell series. The different analyses are based on daily records of rain gauges belonging to the *Instituto Nacional de Meteorología* (Spanish Government) and series of dry spells are derived for threshold levels of daily rainfall equal to 0.1, 1.0, 5.0 and 10.0 mm/day.

Two attempts of statistical modeling of the dry spell lengths have been also introduced. The first is based on extreme analysis and the AES and PDS sampling strategies. Generalized extreme values and Generalized Pareto distributions fit quite well empirical data and return period maps depicting NE-SW or N-S gradients are obtained. The second is based on the use of series of all dry spells. In this case, the Weibull function satisfies the empirical distribution and the return period maps derived show very similar characteristics to those obtained from the AES and PDS strategies. In addition, to consider all dry spells permits to extent the analysis to a higher threshold (10.0 mm/day). Although average dry spells and return period maps suggest spatial variability, the homogeneity test shows signs of spatial coherency for threshold levels of 5.0 and 10.0 mm/day. Conversely, for 0.1 and 1.0 mm/day, the dry spells are statistically distributed with very heterogeneous spatial characteristics.

The climatic change effects are clearly manifested according to the significant field trends detected at annual and summer and winter half–year scales. The annual number of dry spells tends to diminish and some significant field trends indicate a tendency to an increase of Lmax at local scale in a few places. All these time trend patterns would be in agreement with a future reduction of the cyclonic activity in the Mediterranean but, at the same time, an increase of deep cyclones.

The spatial distribution of the three variables characterizing the dry spells, <N>, <L> and <Lmax>, is mainly governed by latitudinal gradients which is a common pattern for other rainfall variables analyzed for Catalonia. A characterization of the complex evolution and difficulty of prediction is manifested by means of fractal concepts as lacunarity and, especially, the Hurst and the Lyapunov exponents.

ACKNOWLEDGEMENTS

The authors are indebted to the *Instituto Nacional de Meteorología* (Spanish Ministry of the Environment) and the Fabra Observatory (*Reial Acadèmia de Ciències i Arts, Barcelona*) for providing gently the pluviometric data series.

References

Alpert P., Ben-Gai T., Baharad A., Benjamini Y., Yekutieli D., Colacino M., Diodato L., Ramis C., Homar V., Romero R., Michaelides S., Manes A. (2002): "The paradoxical increase of Mediterranean extreme daily rainfall in spite of decrease in total values". *Geophys. Res. Lett.*, 29(11), 10.1029/2001GL013554.

AMS Statement (2004): "AMS statement on meteorological drought". *Bulletin American Meteorological Society,* 85, 771-773.

Barry R.G., Chorley R.J. (2003): "*Atmosphere, Weather and Climate*", Eighth ed., Routledge, London, p.421.

Begueria, S. (2005). "Uncertainties in partial duration series modeling extremes related to the choice of the threshold value". *J. Hydrol.*, 303, 215-230.

Benjamin J.R., Cornell C.A. (1970): "*Probability, Statistics and Decision for Civil Engineers*", McGraw-Hill, New York, 685 pp.

Burgueño A., Martínez M.D., Lana X., Serra C. (2005): "Statistical distribution of the daily rainfall regime in Catalonia (NE Spain) for the years 1950-2000". *Int. J. Climatol.*, 25, 1381-1403.

Chu, P.S., Wang, J.B. (1997). Recent climate change in the tropical western Pacific and and Indian Ocean regions as detected by by outgoing long wave radiation. *J. Ckimate*, 10, 636-646.

Diks, C.: Non Linear Time Series Analysis. Methods and Applications. *Non Linear Time Series and Chaos*, 4, World Scientific, London, 1999.

Eckmann, J.P., Oliffson, S., Ruelle, D., Ciliberto, S. (1986). Lyapunov exponents from time series. *Phys. Rew*. A, 34(6), 4971-4979.

Fedder, J. Fractals. Edit. Plenum, New York, 1988.

Gallego M.C., García J.A., Vaquero J.M., Mateos V.L. (2006): "Changes in frequency and intensity of daily precipitation over the Iberian Peninsula". *J. Geophys. Res.*, 111, D24105, doi:10.1029/2006JD007280.

Geng Q., Sugi M. (2003): "Possible change of extratropical cyclone activity due to enhanced greenhouse gases and sulphate aerosols. Study with a high-resolution AGCM", *Journal of Climate,* 16, 2262-2274.

Ghan S.J., Shippert T. (2006): "Physically based global downscaling: climate change projections for a full century", *J. Climate*, 19, 1589-1604.

Gibelin A., Déqué M. (2003): "Anthropogenic climate change over the Mediterranean region simulated by a global resolution model", *Clim. Dynam.* 20, 327-339.

Goltz, C.: Fractal and chaotic properties of earthquakes. In: *Lecture Notes in Earth Sciences* 77, Springer, Berlin, 1997.

Goodess C.M., Jones P.D. (2002): "Links between circulation and changes in the characteristics of Iberian rainfall". *Int. J. Climatol.*, 22, 1593-1615.

Grassberger P., Procaccia I. (1983a). Characterization of strange attractors. *Phys. Rev. Lett.*, 50, 346-349.

Grassberger P., Procaccia I. (1983b). Estimation of the Kolmogorov entropy from a chaotic signal. *Phys. Rev*. A, 28, 2591.

Heim R. R. (2002): "A review of twentieth-century drought indices used in the United States". *Bulletin American Meteorological Society*, 83, 1149-1165.

Hosking J.R.M., Wallis J.R. (1997): "*Regional frequency analysis. An approach based on L-moments*". Cambridge University Press. Cambridge, 244 pp.

Houghton, J.T., Ding, Y., Griggs, D.J., Noguer, M., van der Linden, P.I., Dai, X., Maskel, K., Johson, C.A. (Edit.). 2001. *Climatic change 2001: the scientific basis: International Panel on Climatic Change*. Cambridge. Cambridge University Press, 881 pp.

Jansà A., Genoves A., Picornell M.A., Campins J., Riosalido R., Carretero O. (2001): "Western Mediterranean cyclones and heavy rain. Part 2: Statistical approach". *Meteorol. Appl.,* 8, 43-56.

Keyantash J., Dracup J.A. (2002): "The quantification of drought: an evaluation of drought indices", *Bulletin American Meteorological Society*, 83, 1167-1180.

Kiktev D., Sexton D.M.H., Alexander L., Folland C.K. (2003): "Comparison of modeled and observed trends in indices of daily climate extremes". *Journal of Climate*, 16, 3560-3571.

Knippertz P., Ulbrich U., Speth P. (2000): "Changing cyclones and surface wind speeds over the North Atlantic and Europe in a transient GHG experiment", *Climate Research*, 15, 109-122.

Kundzewicz Z.W., Radziejewski M., Pińskwar I. (2006): "Precipitation extremes in the changing climate of Europe", *Clim. Res*., 31, 51-58.

Kyselý J., Domonkos P. (2006): "Recent increase in persistence of atmospheric circulation over Europe: comparison with long-term variations since 1881", *Int. J. Climatol*., 26, 461-483.

Kyselý J. (2007): "Implications of enhanced persistence of atmospheric circulation for the occurrence and severity of temperature extremes", *Int. J. Climatol*., 27, 689-695.

Lana X., Burgueño A. (1998a): "Spatial and temporal characterization of annual extreme droughts in Catalunya (NE Spain)". *Int. J. Climatol*., 18, 93-110.

Lana X., Burgueño A. (1998b): "Daily dry-wet behaviour in Catalonia (NE Spain) from the viewpoint of first and second order Markov chains". *Int. J. Climatol*., 18, 793-815.

Lana X., Burgueño A. (1998c): "Probabilities of repeated long dry episodes based on the Poisson distribution: an example for Catalonia (NE Spain)". *Theor. Appl. Climatol*., 60, 111-120.

Lana X., Serra C., Burgueño A. (2001): "Patterns of monthly rainfall shortage and excess in terms of the Standard Precipitation Index for Catalonia (NE Spain)". *Int. J. Climatol*., 21, 1669-1691.

Lana X., Martínez M.D., Serra C., Burgueño A. (2004): "Spatial and temporal variability of the daily rainfall regime for Catalonia (NE Spain), 1950-2000", *Int. J. Climatol*., 24, 613-641.

Lana X., Burgueño A., Martínez M.D., Serra C. (2006): "Statistical distributions and sampling strategies for the analysis of extreme dry spells in Catalonia (NE Spain)", *Journal of Hydrology*, 324, 94-114.

Lana X., Martínez M.D., Burgueño A., Serra C. (2007). Return period maps of dry spells for Catalonia (NE Spain) based on the Weibull distribution. *Hydrological Sciences Journal* (In press).

Lin, B., Yan, Z.R. (1986). A suggested lacunarity expression for Sierpinski carpets. *J. Phys. A. Math. Gen.,* 19, L49-L52.

Lionello G.N., Dalan F., Elvini E. (2002): "Cyclones in the Mediterranean region: the present and the doubled CO_2 climate scenarios", *Climate Research*, 22, 147-159.

Lloyd-Hughes B., Saunders M.A. (2002): "A drought climatology for Europe". *International Journal of Climatology*, 22, 1571-1592.

Mandelbrot, B.B.: *The Fractal Geometry of Nature*. Freeman, San Francisco, 1982.

Martín-Vide J., López-Bustins J.A. (2006): "The western Mediterranean oscillation and rainfall in the Iberian Peninsula". *Int. J. Climatol.*, 26, 1455-1475.

Martínez, M.D., Lana, X., Burgueño, A., Serra, C. (2007). Lacunarity, predictability and predictive instability of the daily pluviometric regime in the Iberian Peninsula. *Nonli. Processes Geophys.*, 14, 109-121.

McCabe G.J., Clark M.P., Serreze M.C. (2001): "Trends in Northern Hemisphere surface cyclone frequency and intensity", *J. Climate*, 14, 2763-2768.

McKee T.B., Doesken N.J., Kleist J. (1993): "The relationship of drought frequency and duration to time scales". Preprints, Eighth Conf. on Applied Climatology, Anaheim, CA, Amer. *Meteor. Soc.*, Boston, MA; 179-184.

Palmer W.C. (1965): "Meteorological drought", *Research Paper* No. 45, US Weather Bureau, Washington, DC.

Pinto J.G., Ulbrich U., Leckebusch G.C., Spangehl T., Reyers M., Zacharias S. (2007): "Changes in storm track and cyclone activity in three SRES ensemble experiments with the ECHAM5/MPI-OM1 GCM". *Clim. Dyn.*, 29, 195-210.

Räisänen J., Hansson U., Ullerstig A., Doscher R., Graham L.P., Jones C., Meier H.E.M., Samuelsson P., Willén U. (2004): "European climate in the late twenty-first century: regional simulations with two driving global models and two forcing scenarios", *Clim. Dynam.*, 22, 13-31.

Rodrigo F.S., Trigo R.M. (2007): "Trends in daily rainfall in the Iberian Peninsula from 1951 to 2002". *Int. J. Climatol.*, 27, 513-529.

Salat J., Pascual J. (2006): "Principales tendencias climatológicas en el Mediterráneo nordoccidental a partir de más de 30 años de observaciones oceanográficas y meteorológicas en la costa catalana", Publicaciones de la Asociación Española de Climatología (AEC), Serie A, 5, 283-290.

Sánchez E., Gallardo C., Gaertner M.A., Arribas A., Castro M. (2004): "Future climate extreme events in the Mediterranean simulated by a regional climate model: a first approach", *Global and Planetary Change*, 44, 163-180.

Serra C., Burgueño A., Martínez M.D., Lana X. (2006): "Trends in dry spells across Catalonia (NE Spain) during the second half of the 20th century". *Theor. Appl. Climatol.*, 85, 165-183.

Sneyers, R. (1990). On the statisrical analysis of series of observation. *Technical Note* 415, WMO, Geneva, 192 pp.

Soulé P.T. (1992): "Spatial patterns of drought frequency and duration in the contiguous USA based on multiple drought event definitions", *Intern. J. Climatol.*, 12, 11-24.

Stoop, R., Meier, P.F. (1988). Evaluation of Lyapunov exponents and scaling functions from time series. *J. Opt. Soc. Am.* (B), 5, 1037-1045.

Suppiah, R., Henessy, K.J. (1998). Trends in total rainfall, heavy rain events and number of dry days in Australia, 1910-1990. *Int. J. Climatol.*, 10, 1141-1164.

Takens F. 1981. Detecting strange attractors in turbulence. In: *Lecture Note in Mathematics*. Rand D.A. and Young L.S. Editors. Springer, Berlin.

Trigo I.F., Davies T.D., Bigg G.R. (2000): "Decline in Mediterranean rainfall caused by weakening of Mediterranean cyclones". *Geophysical Research Letters*, 27(18), 2913-1916.

Turcotte, D.L.: *Fractal and Chaos on Geology and Geophysics* (2nd edition). Cambridge University press, 1997.

Ulbrich U., Christoph M. (1999): "A shift of the NAO and increasing storm track activity over Europe due to anthropogenic greenhouse gas forcing". *Climate Dynamics*, 15, 551-559.

Vargas-Yañez M., Salat J., Fernández de Puelles M.L., López-Jurado J.L., Pascual J., Ramírez T., Cortés D., Franco I. (2005): "Trenes and time variability in the northern continental shelf of the western Mediterranean", *J. Geophys. Res.*, 110, C10019, doi:10.1029/2004JC002799.

Vicente-Serrano S.M., Beguería S. (2003): "Estimating extreme dry-spell risk in the middle Ebro Valley (North-Eastern Spain): A comparative analysis of partial duration series with a Pareto distribution and annual maxima series with a Gumbel distribution". *Int. J. Climatol.,* 23, 1103-1118.

Vicente-Serrano S.M., González-Hidalgo J.C., de Luis M., Raventós J. (2004): "Drought patterns in the Mediterranean area: the Valencia region (eastern Spain)", *Clim. Res.*, 26, 5-15.

Vicente-Serrano S.M. (2006a): "Spatial and temporal analysis of droughts in the Iberian Peninsula (1910-2000)", *Hydrological Sciences Journal*, 51, 83-97.

Vicente-Serrano S.M. (2006b): "Differences in spatial patterns of drought on different time scales: an analysis of the Iberian Peninsula". *Water Resources Management*, 20, 37-60.

Vicente-Serrano S.M., López-Moreno J.I. (2006): "The influence of atmospheric circulation at different spatial scales on winter drought variability through a semi-arid climatic gradient in northeast Spain", *Int. J. Climatol.*, 26, 1427-1453.

Vicente-Serrano S.M., Cuadrat-Prats J.M. (2007): "Trends in drought intensity and variability in the middle Ebro valley (NE of the Iberian Peninsula) during the second half of the twentieth century", *Theor. Appl. Climatol.*, 88, 247-258.

Voss R., May W., Roeckner E. (2002): "Enhanced resolution modeling study on athropogenic climate change: changes in extremes of the hydrological cycle", *Intern. J. Climatol.*, 22, 755-777.

Zhang X., Walsh J.E., Zhang J., Bhatt U.S., Ikeda M. (2004a): "Climatology and Interannual variability of Arctic Cyclone Activity: 1948-2002", *J. Climate*, 17, 2300-2317.

Zhang X., Zwiers, F.W., Li, G. (2004b). Monte Carlo experiments on the detection of trends in extreme values. *J. Climate* 17, 1945-1952.

In: Environmental and Regional Air Pollution
Editors: D. Gallo, R. Mancini
ISBN: 978-1-60692-893-6

Chapter 4

DETERMINATION OF CHROMIUM IN SEAWATER BY MINICOLUMN FIELD SAMPLING TECHNIQUE

M. C. Yebra-Biurrun
Department of Analytical Chemistry, Nutrition and Bromatology,
Faculty of Chemistry, University of Santiago de Compostela,
15782-Santiago de Compostela, Spain

ABSTRACT

A simple field sampling/preconcentration strategy for determination and speciation of Cr(III) and Cr(VI) in seawater is described. Seawater was collected in-situ by pumping it through a minicolumn packed with a poly(aminophosphonic acid) chelating resin (PAPhA, 20 mesh) connected to a field flow preconcentration system (FFPS). This packing can isolate and retain the desired chromium species, Cr(III) directly and Cr(VI) previous reduction to Cr(III) by ascorbic acid as reducing reagent. Minicolumns loaded with the analytes are returned to the laboratory where are sequentially inserted into a flow injection system for the on-line chromium species elution with diluted hydrochloric acid and flame atomic absorption spectrometric detection (FAAS). Factorial designs have been used to optimize the FFPS and the flow injection elution process for chromium speciation. A detailed study of the construction of these minicolumns and the stability of the chromium species retained in the packing is presented. With regard to the possible redox changes of the chromium species which can proceed after sampling before their determination, the field sampling/preconcentration technique was found to be optimal because it enables the separation of the species directly at a sampling site and thus avoids the risk of the conversion of Cr species during sample pretreatment and transfer to the laboratory. This procedure has been successfully applied to the determination of Cr(III) and Cr(VI) in seawater samples from Galicia (Spain).

INTRODUCTION

Chemical speciation is currently one of the most interested area of research in the field of environmental science, toxicology, nutritional sciences, environmental and occupational medicine and analytical sciences, since the toxicity and reactivity depends on its chemical form. In this sense, the contamination of the environment by chromium, especially hexavalent chromium, has become a major area of concern because it is used on a large scale in many different industries, such as metallurgical, electroplating, production of paints and pigments, tanning, wood preservation, chemicals production, etc., resulting an anthropogenic contamination of chromium in coastal sediments and seawater [1]. Trivalent chromium is one of the essential trace elements in the human body. It appears to play a role in the metabolism of glucose, proteins and certain lipids, mainly cholesterol. Nevertheless, the biochemical toxicology of different compounds varies considerably with the chemical form and entrance route into the body. Thus, Cr(VI) compounds are approximately 100 times more toxic than Cr(III) salts, owing to their high oxidation potential and the easy with which they penetrate biological membranes [2]. Therefore, it is of great importance in environmental and toxicological studies to determine the level and oxidation state of chromium because excessive amounts of this element, particularly in the more toxic Cr(VI) form, are detrimental to health since this element may be involved in the pathogenesis of certain diseases such as lung and gastrointestinal cancers [3-4]. For this reason, the speciation of inorganic chromium in environmental samples is required for accurate assessment of pollution levels. Therefore, total chromium measurement cannot be used to determine environmental impact due to the considerable difference in toxicity of the two elemental forms.

The exact chemical forms of chromium in surface waters are not well defined, although thermodynamic calculations predict that the predominant oxidation state of chromium should be chromium(VI) in oxygenated seawater at the natural pH (about 8.0). Therefore, most of the soluble chromium in marine media may be present as Cr(VI) (chromate ion, CrO_4^{2-}), a small amount may be present as Cr(III) organic complexes, aquahydroxy species ($(Cr(OH)_2^+ (H_2O)_4)$), organically bound Cr(III) to natural organic substances such as amino acids, carboxylic acids and humic acids or colloidally sorbed Cr(III). Hexavalent chromium is the major stable form of chromium in seawater; however, Cr(VI) may be reduced to Cr(III) by organic matter present in water, and may eventually deposit in sediments. Although there are few oxidants capable of converting Cr(III) to Cr(VI), and the oxidation kinetics are normally very slow, it has been suggested that unstable (including dissolved and colloidal) forms of Cr(III) can be converted to Cr(VI) relatively quickly by strong oxidants such as H_2O_2 that are produced photochemically in aerobic surface waters [5-8].

Normally, the content of chromium in natural waters is very low, it is found at the low μg/L level. The average concentration range of total chromium in seawater is 0.2–50 μg/L and total chromium content in fresh waters is between 0.1–6 μg/L [9]. Thus, when chromium is determined in seawater samples, the chromium contents are close to or below the detection limits for the detection technique. Furthermore, the direct determination of chromium in complex samples such as seawater presents difficulties owing to matrix interferences. Therefore, analytical techniques such as atomic spectrometry, spectrophotometry, chemiluminescence, fluorometry and isotope dilution mass spectrometry have been applied for the quantitative determination of the chromium species after their separation and/or preconcentration [10]. For this reason, it is very important have a robust separation technique in order to ensure the accurate detection of both chromium species. Off-line pretreatment procedures involving solvent extraction, ion exchange, coprecipitation, solid-phase extraction, etc., are proposed as very effective methods for such purpose [11-14]. However, they are often time-consuming, tedious to operate and involve complex procedures with high contamination risk. Thus, these procedures are incompatible with the high efficiency of the detection techniques used. The application of non chromatographic continuous separation techniques by using flow injection analysis (FIA) methodologies coupled to atomic absorption spectrometry (AAS), inductively coupled plasma atomic emission spectrometry (ICP-AES), and inductively coupled plasma mass spectrometry (ICP-MS) (although the detection power of ICP-MS permits the direct determination of trace metals at sub-ng/mL levels, interferences from polyatomic species arising from salt in seawater prevent a direct analysis), offer a great number of advantages over classical off-line procedures for the speciation of Cr(III) and Cr(VI). These include automated sample handling, use of a closed system to process the sample, decreasing sample and reagent consumption, shorting analysis time and a high improvement concerning accuracy and precision [15-18].

In general, flow injection (FI) procedures developed to remove matrix interferences and preconcentrate the target analyte involve the use of column reactors packed with sorbent materials such as C_{18}, activated alumina, etc. (chromium species are sorbed on these materials either directly or after preliminary complexation with various chelating reagents like dithiocarbamates or diphenylcarbazide) [19-20]. Furthermore, chelating resins such as Chelex-100, Muromac A-1, silica-immobilized 8-hydroxyquinoline, etc., have been employed for separation purposes. Thus, many works have been reported on the development and/or the evaluation of the resins for the speciation of Cr(III) and Cr(VI) [21-23].

When is accomplished the chromium speciation, two different preconcentration strategies have been employed: 1) only one specie is preconcentrated (Cr(III) or Cr(VI)), while the other specie is often complexed or measured directly; 2) both species are preconcentrated, either in the same place or in different places. In the first alternative, in some cases, Cr(III) is oxidized to Cr(VI) and total chromium is measured, being the Cr(III) content calculated as the difference, or in other cases reduction of Cr(VI) to Cr(III) is carried out, total chromium is measured, and the Cr(VI) content is calculated from the difference.

The major problem to establish a reliable methodology for chromium speciation is related to the preservation of natural species present in the sample, since most of the procedures involve several chemical stages, increasing the possibility of analyte losses, sample contamination and chemical transformation of the chromium species [24]. The accepted method for preserving water samples for trace metals determination involves the addition of a concentrated acid (nitric or hydrochloric) to reduce the pH below 2. In this way, the losses of metal ions by adsorption onto the container surface and the removal of dissolved metal through hydrolysis were avoided. It was reported that the amount of trace metal loss was directly related to the length of storage time and inversely related to the concentration of the trace metal. Nevertheless, the conventional sample preservation method presents five main disadvantages: 1) transporting and manipulating a volume of concentrated acid to each sampling site, which implies safety hazards for the sampler and the environment; 2) contamination of the sample by the acid added; 3) prolonged exposure of the sample to the atmosphere with the possibility of contamination from dust-borne heavy metals; 4) return to the laboratory several containers containing the samples with a volume of 0.5- 1 L; 5) to have one place in the laboratory for sample storage. In order to avoid these drawbacks involving the sampling, storage and analyte determination, the minicolumn field sampling (MFS) technique was proposed. In this technique flow systems are used to process the water samples at the sampling site and trace elements of interest are immobilized on minicolumns. These minicolumns are returned to the laboratory where they are directly inserted into a FI system for on-line elution and quantitative analysis. In addition to the advantages of FIA methodology, MFS technique achieves the following: high efficiency, rapidity, improvement in data quality and ultratrace analysis, improvements in collection, transporting and storage of seawater samples (the samples occupy a small space for convenient storage in the laboratory). In this sense, for chromium determination in water, only there is one work available in the bibliography. Cox et al. [25] proposed the in situ preconcentration of Cr(III) and Cr(VI) in river waters. Water samples were immediately passed through microcolumns of activated alumina to isolate and retain the desired species. Microcolumns were then returned to the laboratory and inserted into a FI-AES system for elution/quantitation. With regard to the determination of other trace metals in water samples, at present, only a few methods involving MFS techniques are available in the literature. Jian et al. [26-27] developed a MFS technique to collect and preserve organomercury species using a microcolumn of sulphydryl cotton and detection by flow-injection atomic fluorescence spectrometry (FI-AFS). Gómez et al. [28] described a FI-ICP-MS system incorporating a microcolumn of sulphydryl cotton. With this microcolumn gold in natural waters can be preconcentrated in-situ. Fairman et al. [29] developed a field sampling strategy for the fast reactive aluminium fraction in natural waters, which reacts with 8-hydroxyquinoline, by minicolumns of Amberlite XAD-2 with subsequent detection by ICP-ES or ICP-MS. Nickson et al. [30] proposed a field sample unit for the selective in-situ preconcentration of trace elements from natural waters. The microcolumn contains an iminodiacetate resin and elution takes place into an ICP-ES system. Martínez-Blanco et al. [31] described a field sampling strategy for determination and speciation of mercury in river waters by ICP-MS. The minicolumn used is packed with C18 modified with sodium diethyldithiocarbamate. The investigation team of Professor Yebra have proposed the use of minicolumns packed with Amberlite XAD-4 impregnated with 1-(2-pyridylazo)-2-naphthol (PAN) and 4-(2-pyridylazo) resorcinol (PAR) and with the chelating resin Chelite P (aminomethylphosphonic groups) for the in-situ preconcentration of cadmium, copper and lead from seawater samples [32-38]. The device used for sampling/preconcentration (field flow preconcentration system, FFPS) facilitates several interesting on-site implementations by integrating sampling and sample preparation directly in the sampling site.

In this chapter, a field flow preconcentration system (FFPS) with minicolumns subsequently inserted to a flow-injection flame atomic spectrometric system (FI-FAAS) is described and optimized for Cr(III) and Cr(VI) determination in seawater samples. Minicolumns containing a chelating resin with aminophosphonic groups (PAPhA) were used for this purpose. Experimental designs were used to optimize the operational parameters of the FFPS and FI-FAAS systems. The procedure was applied to the determination of Cr(III) and Cr(VI) in seawater samples from Galicia (Spain).

EXPERIMENTAL

Instrumentation

The FFPS consists of a Gilson Minipuls-3 peristaltic pump, fitted with a poly(vinyl chloride) (PVC) tube, a 0.45 μm filter (Millipore) and a minicolumn manufactured from a A

Perkin Elmer 5000 atomic absorption spectrometer, furnished with a chromium hollow-cathode lamp, equipped with a deuterium lamp background corrector and an air-acetylene flame as the atomizer was used as the detector for Cr. The acetylene flow-rate was 4.5 L/min and an air flow-rate of 15.0 L/min was employed to obtain a clear yellow flame (reducing). The peak height was measured at 357.9 nm. The aspiration flow-rate of the nebulizer was adjusted to be the same as the flow-rate of the FI elution system (4.5 mL/min). The spectrometer output was connected to a Perkin Elmer 50 servograph recorder with a range of 5 mV. A Crison standard pH meter with electrode Ingold U 455 was used. The FI elution system comprised a Gilson Minipuls-3 peristaltic pump, fitted with poly(vinyl chloride) tubes and a Reodyne (model 5041) four-way valve.

Statistical analysis of the experimental designs was carried out by means of the STATGRAPHICS V.4.0 statistical package (Manugistic, Inc. Rockville, MD, USA).

Reagents and Solutions

All chemicals were of analytical-reagent grade. A standard solution of 1000 μg/mL Cr(III) was prepared from Cr(III) nitrate in 0.5 M nitric acid (BDH Chemicals, Poole, England). A solution of 1000 μg/mL Cr(VI) was prepared by dissolving 2.8290 g of $K_2Cr_2O_7$ (Merck) in 1 L of ultrapure water. From these solutions, other diluted standard solutions were prepared daily. Hydrochloric acid (Merck) diluted to 0.5 mol/L was used as eluent. Ascorbic acid (Merck) used as reducing reagent, 0.01 mol/L aqueous solution, prepared fresh daily. PAPhA chelating resin with poly(aminophosphonic acid) acid groups (20 mesh) was synthesized as described earlier [39]. Ultrapure water of 18.2 MΩ cm resistivity obtained from a Milli-Q water purification system (Millipore) was used for dilution and washing, respectively.

Minicolumns Preparation

The minicolumns were manufactured in the laboratory from PVC tubing (1.4 mm i.d.) and were packed with 0.05 g of 20 mesh PAPhA chelating resin (a small amount of resin beads was injected into the minicolumn with a syringe). The ends of the tube were fitted with glass wool to keep the beads in the tube.

Field Sample Preconcentration of Cr(III)

The field sampling/preconcentration was developed using the field flow preconcentration system (FFPS) depicted in Figure 1 (FFPS1). The seawater sample (400 mL) was pumped at 4.0 mL/min and is filtered on-line with a 0.45 μm filter. Then, the sample stream merged with a nitric acid 10^{-3} mol/L stream flowing at 0.5 mL/min, and the resulting mixture of these two streams was homogenized into a mixing coil. After, the homogenized stream goes to through the minicolumn. Thus, the Cr(III) was retained on the minicolumn, and the sample matrix was sent to waste.

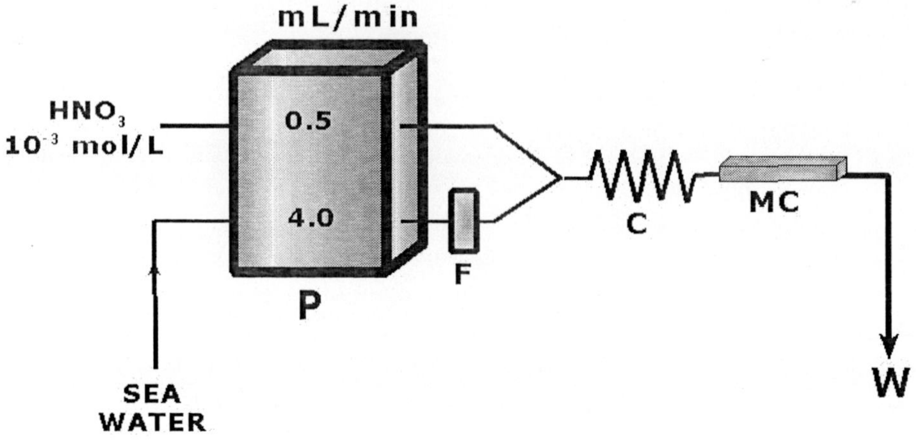

Figure 1. Field flow preconcentration system (FFPS1) for Cr(III) retention. P, peristaltic pump; C, mixing coil; F, filter; MC, minicolumn; W, waste.

Field Sample Preconcentration of Total Chromium

The field sampling/preconcentration was developed using the field flow preconcentration system (FFPS) depicted in Figure 2 (FFPS2). The seawater sample (400 mL) was pumped at 4.0 mL/min and is filtered on-line with a 0.45 μm filter. Then, the sample stream merged with a ascorbic acid 0.01 mol/L stream and the Cr(VI) is reduced to Cr(III) into a reaction coil. After, this stream goes to through the minicolumn. Thus, the Cr(III) formed was retained on the minicolumn, and the sample matrix was sent to waste.

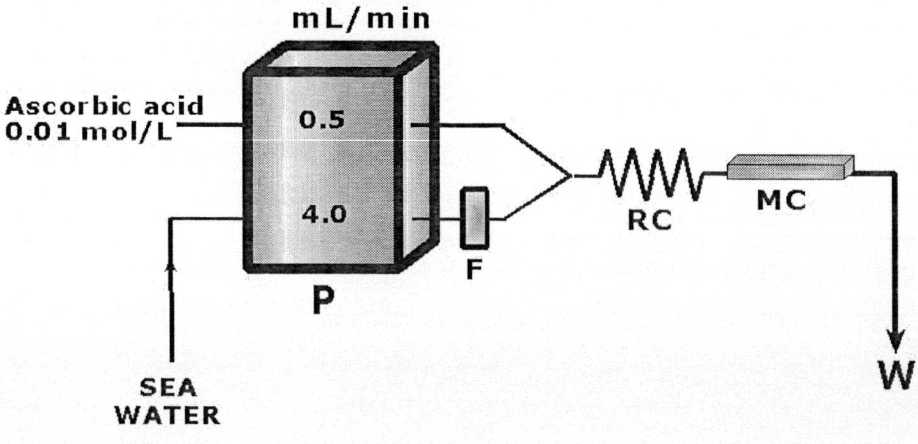

Figure 2. Field flow preconcentration system (FFPS2) for total Cr retention. P, peristaltic pump; RC, reaction coil; F, filter; MC, minicolumn; W, waste.

Minicolumns Washing

After loading with Cr(III) or total chromium (in the Cr(III) form), minicolumns were washed with ultrapure water and then, the residual internal fluid was drawn off with an air stream in order to remove the washing solution. The dry minicolumns were then disconnected from the FFPS, placed in a portable refrigerator and returned to the laboratory where they are stored in a refrigerator until the analysis is accomplished. Calibration and blanks columns were prepared in the laboratory using the FFPS by pumping the appropriate chromium standard solution or ultrapure water.

On-Line Elution and Flame Atomic Absorption Determination of Chromium

The chromium detection was performed by connecting sequentially each minicolumn to the FI manifold shown in Figure 3. Minicolumns were located immediately before the detector to obtain minimum eluent dispersion. Analysis procedure is based on injection of 150 μL of eluent (0.5 mol/L hydrochloric acid) into a carrier (ultrapure water pumped at a flow rate equal to the nebulizer aspiration flow, 4.5 mL/min). Thus, chromium was released directly into the nebulizer of the flame atomic absorption spectrometer. Blanks (ultrapure water and artificial seawater) were run in parallel, the absorbance of which ranged between 0.000 and 0.001.

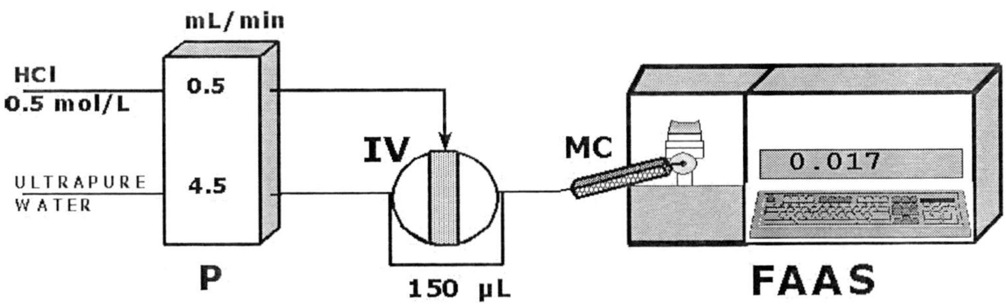

Figure 3. Flow injection manifold for elution-determination of chromium. P peristaltic pump; MC, minicolumn; IV injection valve; FAAS, flame atomic absorption spectrometer.

RESULTS AND DISCUSSION

The method described in this chapter, is based on the retention of Cr(III), but not Cr(VI) by the PAPhA chelating resin because this resin exhibits no affinity for the Cr(VI). Therefore, this chemical form of chromium is not retained on the chelating resin used. The optimization of the overall method here proposed was realized first for each FFPS (FFPS1 for Cr(III) retention (Figure 1) and FFPS2 for total chromium retention, in the Cr(III) form (Figure 2)), and finally, for the FI manifold used for on-line chromium elution (Figure 3).

Optimization of FFPS1: Cr(III) Retention

The variables affecting the FFPS1 were the sample pH, the sample flow rate and the mixing coil length. A mixed level factorial design 3*2^2 type V resolution allowing 5 degrees of freedom and involving 12 non-randomized runs plus one centerpoint was built for a screening study of the behavior of the main factors affecting the FFPS1. The upper and lower values given to each factor were selected from available data and experience gathered in preliminary experiments (Table 1). Sample pH was adjusted on-line including a nitric acid solution stream flowing at 0.5 mL/min. The concentration of this solution of nitric acid ranges between 0 mol/L to obtain a sample pH value of 8 and 10^{-2} mol/L to obtain a sample pH value of 3, since the pH of seawater approximately is equal to 8. The conclusions of this screening study were that the sample pH and the sample flow rate were influential factors in the ranges under study. Both influential factors with a negative influence, at 95% confidence, they overtake the limit of significance. The analysis of the results produced the standardized main effects Pareto chart shown in Figure 4. In this chart, bar lengths are proportional to the absolute value of the estimated effects helping in comparing the relative importance of effects. The mixing coil length was not statistically influential factor in the range under study. However, the results showed better recoveries with the maximum value tested and with the center point value (62.5 cm). Thus, this value was selected for subsequent experiments in order to guarantee a complete sample homogenization to achieve the desired sample pH.

Lower values for the sample pH (between 3-6) and sample flow rate (1-5 mL/min) were tested using a central composite design 2^2 + star design involving eight non-randomized runs plus two center points. Analyzing the results of this second experimental design, only an influence of the sample flow rate was observed. The sample pH was not a statistical influential factor within the range 3-6. As expected, the maximum retention of Cr(III) occurs in this pH range, at lower pH values Cr(III) is not chelated quantitatively and at pH values higher than 6 Cr(III) probably forms hydroxides.

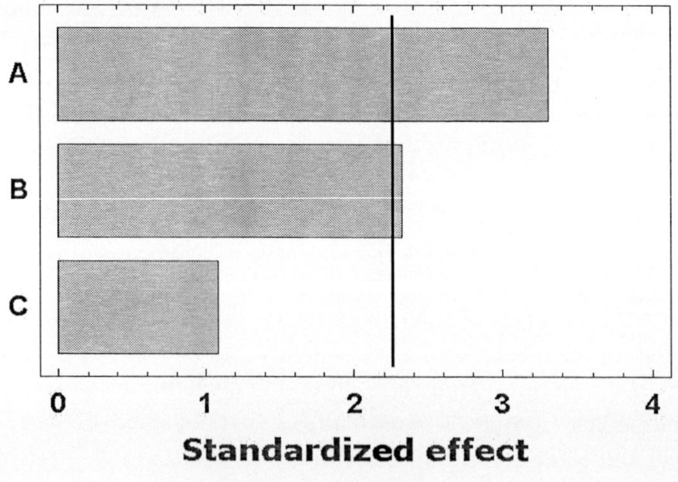

Figure 4. Pareto chart of main effects in mixed level factorial design 3*2^2 for optimization of FFPS1: Cr(III) retention.

Thus, in order to obtain a sample pH less than 6, a concentration of 10^{-3} mol/L for the solution of nitric acid was chosen (flow rate 0.5 mL/min).

With the aim of achieve the maximum sampling frequency, the influence of the sample flow rate was fine tuned in the univariate way by fixing the other variables at their optimal values. Sample flow rates between 3-5.5 mL/min were studied. The results obtained showed that the maximum chromium recovery was obtained between 3-4.0 mL/min; while a sample flow rate higher than 4.5 mL/min does not provide a quantitative recovery owing to very short residence times of the seawater sample into the minicolumn containing the chelating resin, which results in an incomplete Cr(III) retention. Therefore, a sample flow rate of 4.0 mL/min was selected for further experiments.

Table 1. Ranges assessed and optimal values for variables influencing the FFPS1

Variable	Key	Tested range	Optimum value
Sample pH	A	3-8	3-6
Sample flow-rate (mL/min)	B	2-6	4
Mixing coil length (cm)	C	25-100	62.5

Optimization of FFPS2: Total Cr Retention

This FFPS was used for sampling/preconcentration of total chromium in the Cr(III) form. Therefore, it must include a reducing stream, which transforms Cr(VI) to Cr(III). Ascorbic acid was chosen as reducing reagent because previous experiences demonstrated that with this reducing reagent a high reaction kinetic is obtained.

Taken into account that the optimum sample pH is between 3-6, and the sample flow rate should not be higher than 4 mL/min, the variables affecting the FFPS2 were the ascorbic acid solution concentration and the reaction coil length. These two variables were studied by using the univariate method. The ascorbic acid flow rate was fixed at 0.5 mL/min. Firstly, the effect of ascorbic acid concentration in the range 0.005-0.01 mol/L was studied by using solutions containing 15 μg/L Cr(VI). In the range tested, the pH ranges within the optimum pH interval (3-6). Results obtained indicated that the reduction was complete for all the concentration range studied, but in order to ensure reduction of larger concentrations of Cr(VI), the final choice was 0.01 mol/L ascorbic acid. The influence of the reaction coil was investigated between 20 and 100 cm. Short reactor coils, less than 50 cm, resulted in short residence times (too short for complete chromium reduction). A coil length of 50 cm was therefore chosen.

Optimization of the On-Line Chromium Elution

The variables affecting the on-line chromium elution were eluent concentration (hydrochloric acid), eluent volume and elution flow-rate. A mixed level factorial design 3*2^2 type V resolution allowing 5 degrees of freedom and involving 12 non-randomized runs plus one centerpoint was built for a screening study of the behavior of the main factors affecting the FI-FAAS determination of chromium. The upper and lower values given to each factor were selected from available data and experience gathered in preliminary experiments (Table 2). From the results of these analytical data can be concluded that the chromium elution process is affected by two statistically significant factors: eluent concentration and eluent volume. These two factors were affected by a positive sign (low eluent concentrations and eluent volumes are not enough for chromium release from the chelating resin). The analysis of the results produced the standardized main effects Pareto chart shown in Figure 5.

According to the results, a new experimental design would be desirable shifted in the direction of higher values of eluent volume and concentration (because higher concentrations could dissolve the metallic parts of the spectrometer nebulizer further experiments have not performed at more elevated values of this factor) and in the direction of high values of eluent volumes. However, due to good chromium recoveries achieved that made to think about nearness of system optimum and in accordance with previous, further experiments were performed in order to fine tune these variables without necessity of applying another factorial design. It is important to consider the main objective of this work was to optimize the system by reducing the number of experiments and the process time. In order to fine tune, three new experiments were conducted to fine tune eluent volume. The considered values were 100, 150, 200 and 300 μL. The experiments performed with 150, 200 and 300 μL gave similar quantitative recovery, while the recovery obtained with a volume of eluent of 100 μL was not quantitative. Therefore, in order to achieve a higher preconcentration factor a volume of 150 μL was chosen as optimum.

Finally, with the aim of decrease the eluent concentration, an experiment was carried out keeping all the other factor at optimum values and varying the eluent concentration between 0.5-3 mol/L. The results obtained demonstrated that in all the concentration range tested, a quantitative elution of chromium was achieved. Therefore, hydrochloric acid 0.5 mol/L is suitable to release the chromium retained on the minicolumn.

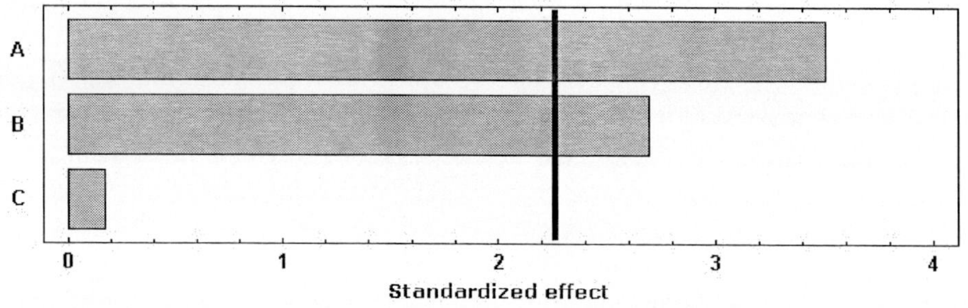

Figure 5. Pareto chart of main effects in mixed level factorial design 3*2^2 for optimization of the on-line chromium elution.

Table 2. Ranges assessed and optimal values for variables influencing the FI-FAAS chromium determination

Variable	Key	Tested range	Optimum value
Eluent concentration (HCl) (mol/L)	A	0.1-3	0.5
Eluent volume (μL)	B	90-250	150
Elution flow-rate (mL/min)	C	2.5-4.5	4.5

Since, eluent flow rate does not affect the chromium release in the range studied (according to factorial design) this variable has been fixed at a low value, but on the other hand, higher aspiration flow rates allow abundant chromium population in the flame (high sensitivity), previous experiments suggest that 4.5 mL/min is a good value for this purpose. The last column in Table 2 shows the optimal values of factors affecting the studied system for chromium determination.

Analytical Figures of Merit

Calibration graphs were run (n=7) under the optimal conditions. One, run by direct aspiration: absorbance = $8 \times 10^{-5} + 4.9 \times 10^{-5}$ C; C= μg/L Cr(III), and the other obtained with the FI-FAAS system: absorbance = $8 \times 10^{-4} + 0.130$ C; C= μg/L Cr(III) (sample volume 400 mL).

An experimental enrichment factor of 2653 was obtained for 400 mL of sample volume. This enrichment factor was calculated as the ratio between the slope of the calibration graph obtained for the sample volume used and the off-line direct aspiration of chromium. As the sample volume was 400 mL and the injected eluent volume was 150 μL, the theoretical preconcentration factor should be 2667, which demonstrates the good performance and the minimum dispersion of the FI system.

The calibration graph for total chromium (Cr(VI) reduced to Cr(III)) obtained under the optimal conditions was absorbance = $9 \times 10^{-4} + 0.130$ C; with C in μg/L. This equation was practically the same as that obtained with Cr(III) standard solutions. Therefore, one can make the calibration graph for total chromium determination with Cr(III) standards to avoid the reduction process with the standard solutions and so, speed up analysis time.

The detection limit was calculated as three times the standard deviation of the peak height for 30 determinations of the blank (ultrapure water) and was 0.01 μg/L, for a sample volume of 400 mL in both determinations (Cr(III) and total Cr). The quantitation limit, calculated as the blank signal plus ten times its standard deviation, was 0.03 μg/L, for a sample volume of 400 mL in both determinations (Cr(III) and total Cr). The preconcentration factor and detection limit can be further improved by increasing sample loading volume, without degradation in the efficiency due to the favorable kinetics of the present system. The high capacity of the minicolumn allowed large sample volumes to be preconcentrated without degradation in performance. The precision of the method obtained for standard solutions containing 0.1 (n = 11) and 1.5 (n =11) μg/L of chromium were 3.2 and 2.7%, respectively, expressed as relative standard deviation in both determinations (Cr(III) and total Cr).

Stability of the Minicolumns

Once loaded with the seawater sample stability studies were made, in order to establish how much time can be stored the minicolumns before the elution/detection step. Minicolumns were kept 1, 15, 20 and 30 days at room temperature and at 4°C (refrigerator). Results obtained indicated that these minicolumns containing the poly(aminophosphonic acid) chelating resin are stable during at least 30 days at both studied temperatures, which facilitates their storage.

Study of Interferences

A study of potential interferences and matrix ions of seawater in the determination of chromium was performed. A relative error of less than 5% was considered to be within the range of experimental error. Synthetic solutions containing 20000 mg/L Na, K as their chlorides and other important elements present in seawater such as NO_3^-, HPO_4^{2-}, IO_3^-, Br^-, Mg^{2+}, Ca^{2+}, Fe^{3+} and Li^+ (concentrations chosen according to the elemental composition of seawater [9]) were used as samples. The results obtained in these experiments demonstrate that the presence of large amounts of alkali and alkaline earth metals in the sample have no significant effect on the preconcentration of chromium (0.1 μg/L). The effects of other ions at studied concentrations are also negligible (did not cause an effect of more than ± 5% on the chromium absorbance signal).

Analysis of Seawater Samples

The proposed method, involving the use of FFPS and FI-FAAS was applied to the determination of Cr(III) and total chromium, which allowed calculation of the Cr(VI) concentration by difference in seawater samples of Galician coast (Spain). The results obtained are show in Table 3.

Table 3. Results obtained for chromium determination in seawater samples

Sample N°	Cr(III) μg/L	Total Cr μg/L	Cr(VI) μg/L
1	0.2	0.7	0.5
2	0.5	1.2	0.7
3	0.1	0.4	0.3
4	0.3	0.7	0.4
5	0.3	0.8	0.5
6	0.2	0.9	0.7
7	0.1	0.6	0.5

Conclusions

The field flow preconcentration technique involving a minicolumn containing a poly(aminophosphonic acid) chelating resin (PAPhA), and the subsequent elution/determination by a flow-injection-flame atomic absorption spectrometric system was successfully applied to the determination of Cr(III) and total chromium in seawater samples. Simplicity of the manifold is a great advantage of this procedure in comparison to many methods existing for chromium preconcentration. These advantages are the following: several samples can be preconcentrated at the same time and allows the use of the same minicolumn several operational cycles. Furthermore, the proposed FFPS is a powerful tool for the speciation of chromium in seawater samples because a low volume of eluent and the allowed high sample volumes lead to a high preconcentration factor, which was adequate for chromium detection by FAAS. This alternative sampling/preconcentration approach avoids the problems associated with sample transport and storage, minimizing problems of contamination, losses and changes in the oxidation state of the analyte. In addition, the stability of minicolumns containing the resin PAPhA is higher, which facilitates their storage.

In short, this chapter traces the simplification of sample preparation technologies and their integration with sampling or separation/quantification steps in the analytical processes, which are significant challenges to, and opportunities for, the contemporary analytical chemist. These developments will eventually enable attainment of a major goal of the analytical chemists to perform the sample pretreatment at the place where the sample is taken, rather than moving the sample to a laboratory, as is traditional. The on-site sample preconcentration approach reduces errors and the time associated with sample transport and storage, resulting in more accurate, precise, and faster analytical data in analytical processes involving environmental problems.

References

[1] A. M. Zayed, N. Terry, Chromium in the environment: factors affecting biological remediation, *Plant and Soil* (2003) 249: 139–156.

[2] M. Stoeppler, *Hazardous metals in the environment, techniques and instrumentation in analytical chemistry*, Elsevier, Amsterdam, 1992.

[3] A.B. Gwizala, S.K. Johnson, S. Mollah, R.S. Houk, Speciation of chromium (III) and chromium (VI) using pneumatically assisted electrospray mass spectrometry, , *J. Anal. At. Spectrom.* (1997) 12: 503-506.

[4] C. Barnowski, N. Jakubowski, D. Stuewer, J.A.C. Broekaert, Speciation of chromium by direct coupling of ion exchange chromatography with inductively coupled plasma mass spectrometry, *J. Anal. At. Spectrom.* (1997) 12: 1155-1161.

[5] E. Nakayama, T. Kuwamoto, S. Tsurubo, H. Tokoro, T. Fujinaga, Chemical speciation of chromium in sea water. Part 1. Effect of Naturally Occurring Organic Materials on the Complex Formation of chromium (III), *Anal. Chim. Acta* (1981) 130: 289-294.

[6] E. Nakayama, T. Kuwamoto, S. Tsurubo, T. Fujinaga, Chemical speciation of chromium in sea water. Part 2. Effects of manganese oxides and reducible organic materials on the redox processes of chromium, *Anal. Chim. Acta* (1981) 130: 401-404.

[7] F. Ahern, J.M. Eckert, N.C. Payne, K.L. Williams, Speciation of chromium in sea water, *Anal. Chim. Acta* (1985) 175: 147-151.

[8] T.S. West, *The determination of trace metals in natural waters*, Blackwell, Oxford, 1988.

[9] H.J.M. Bowen, *Environmental chemistry of the elements*, Academic Press, London, 1979.

[10] M.J. Marqués, A. Salvador, A. Morales-Rubio, M. de la Guardia, Chromium speciation in liquid matrices: a survey of the literature, *Fresenius J. Anal. Chem.* (2000) 367:601-613.

[11] E. Nakayama, T. Kuwamoto, H. Tokoro, T. Fujinaga, Chemical speciation of chromium in sea water. Part 3. The determination of chromium species, *Anal. Chim. Acta* (1981) 131:247-254.

[12] H.F. Zhang, J. Holzbecher, D.E. Ryan Preconcentration and speciation of chromium in waters by coprecipitation with lead salts and neutron activation-γ-spectrometry, *Anal. Chim. Acta* (1983) 149:385-389.

[13] E. Andrés García; D. Blanco Gomis, Speciation analysis of chromium using cryptand ethers. *Analyst* (1997) 122: 899-902.

[14] A. Xue, S. Qian, G. Huang, L. Chen, Separation and preconcentration of chromium speciation on chitosan and determination by graphite furnace atomic absorption spectrometry, *J. Anal. At. Spectrom.* (2000) 15: 1513-1515.

[15] F. Lázaro Boza, D. Luque de Castro, M. Valcárcel Cases, Flow injection environmental analysis: a review, Analusis (1985) 13:147-159.

[16] M. Trojanowicz, R. L. Benson, P.J. Worsfold, Recent developments in water quality monitoring by flow injection analysis, *Trends Anal. Chem.* (1991) 10: 11-17.

[17] J. Atienza, M.H. Herrero, A. Maquieira, R. Puchades, Flow injection analysis of seawater. Part II. Cationic species, *Crit. Rev. Anal. Chem.* (1992) 23: 1-14.

[18] Z. Fang, Trends of flow injection sample pretreatment approaching the new millennium, *Anal. Chim. Acta* (1999) 400:233-247.

[19] T. Prasada Rao, S. Karthikeyan, B. Vijayalekshmy, C.S.P. Iyer, Speciative determination of chromium (VI) and chromium (III) using flow-injection on-line preconcentration and flame atomic-absorption spectrometric detection, *Anal. Chim. Acta* (1998) 369:69-77.

[20] M.J. Marqués, A. Morales-Rubio, A. Salvador, M. de la Guardia, Chromium speciation using activated alumina microcolumns and sequential injection analysis-flame atomic absorption spectrometry, *Talanta* (2001) 53:1229-1239.

[21] P.A. Sule, J.D. Ingle, Determination of the speciation of chromium with an automated two-column ion-exchange system, *Anal. Chim. Acta* (1996) 326:85-93.

[22] R.M. Cespón-Romero, M.C. Yebra-Biurrun, M.P. Bermejo-Barrera, Preconcentration and speciation of chromium by the determination of total chromium and chromium (III) in natural waters by flame atomic absorption spectrometry with a chelating ion-exchange flow injection system, *Anal. Chim. Acta* (1996) 327:37-45.

[23] S. Hirata, K. Honda, O. Shikino, N. Maekawa, M. Aihara, Determination of chromium (HI) and total chromium in seawater by on-line column preconcentration inductively coupled plasma mass spectrometry, *Spectrochim. Acta* (2000) 55B: 1089-1099.

[24] W. Sirinawin, S. Westerlund, Analysis and storage of samples for chromium determination in seawater, *Anal. Chim. Acta* (1997) 356:35-40.

[25] A.G. Cox, C.W. McLeod, Field sampling technique for speciation of inorganic chromium in rivers, *Mikrochim. Acta* (1992) 109:161-164.

[26] W. Jian, C.W. McLeod, Field sampling technique for mercury speciation, *Anal. Proc.* (1991) 28: 293-294.

[27] W. Jian, M.L. Mena, C.W. McLeod, Microcolumn field sampling and flow injection-atomic fluorescence spectrometry for speciation of mercury in surface waters, *Intern. J. Environ. Anal. Chem.* (1994) 57: 99-106.

[28] M.M. Gómez Gómez, C.W. McLeod, Trace enrichment and determination of gold by flow injection inductively coupled plasma spectrometry. Part II. Inductively coupled plasma mass spectrometry, *J. Anal. At. Spectrom.* (1995) 10: 89-91.

[29] B. Fairman, A. Sanz-Medel, P. Jones, *J. Anal. At. Spectrom.* (1995) 10: 281-285.

[30] R.A. Nickson, S.J. Hill, P.J. Worsfold, Field preconcentration of trace metals from seawater and brines coupled with laboratory analysis using flow injection and ICP-AES detection, *Intern. J. Environ. Anal. Chem.* (1999) 75: 57-69.

[31] R. Martínez Blanco, M. Tagle Villanueva, J.E. Sánchez Uría, A. Sanz-Medel, Field sampling, preconcentration and determination of mercury species in river waters, *Anal. Chim. Acta* (2000) 419:137-144.

[32] M.C. Yebra, N. Carro, M.F. Enriquez, A. Moreno-Cid, A. Garcia, Field sample preconcentration of copper in sea water using chelating minicolumns subsequently incorporated on a flow-injection-flame atomic absorption spectrometry system, *Analyst* (2001) 126:933-937.

[33] M.C. Yebra, N. Carro, A. Moreno-Cid, L. Puig, Minicolumn field sampling and flow-injection-atomic absorption spectrometric determination of lead in seawater, *Field Anal. Chem. Technol.* (2001) 5: 239-243.

[34] M.C. Yebra, N. Carro, A. Moreno-Cid, Optimization of a field flow pre-concentration system by experimental design for the determination of copper in sea water by flow-injection-atomic absorption spectrometry, Spectrochim. Acta (2002), 57B: 85-93.

[35] M.C. Yebra, A. García, N. Carro, A. Moreno-Cid, L. Puig, Design of a field flow preconcentration system for cadmium determination in seawater by flow-injection-atomic absorption spectrometry, *Talanta* (2002) 56: 777-785.

[36] M.C. Yebra, J. Salgado, L. Puig, A. Moreno-Cid, Field preconcentration of cadmium from seawater by a minicolumn packed with Amberlite XAD-4/4-(2-pyridylazo) resorcinol and its flow-injection-flame atomic absorption spectrometric determination at the ng L^{-1} Level, *Anal. Bioanal. Chem.* (2002) 374: 530-534.

[37] M.C. Yebra, L. . Rodríguez, L. Puig, A. Moreno-Cid, Application of a Field Flow Preconcentration System with a Minicolumn Packed with Amberlite XAD-4/1-(2-pyridylazo)-2-naphthol and a Flow Injection-Flame Atomic Absorption Spectrometric System for Lead Determination in Sea Water, *Mikrochim. Acta* (2002) 140: 219-225.

[38] M.C. Yebra-Biurrun, A. Moreno-Cid, L. Puig, Minicolumn field preconcentration and flow injection flame atomic absorption spectrometric determination of cadmium in seawater, *Anal. Chim. Acta* (2004) in press.

[39] M.C. Yebra-Biurrun, A. Bermejo-Barrera y M.P. Bemejo-Barrera, Synthesis and characterization of a poly(aminophosphonic acid) chelating resin, *Anal. Chim. Acta* (1992) 264: 53-58.

In: Environmental and Regional Air Pollution
Editors: D. Gallo, R. Mancini
ISBN: 978-1-60692-893-6

Chapter 5

SPATIOTEMPORAL DISTRIBUTION OF SOIL NEMATODE COMMUNITIES ALONG A CHRONOSEQUENCE OF VEGETATION RESTORATION IN HORQIN SANDY LAND

Xiaoke Zhang and Wenju Liang*
Institute of Applied Ecology, Chinese Academy of Sciences,
Shenyang 110016, China

ABSTRACT

Spatiotemporal distribution of soil nematode communities was investigated along a chronosequence of vegetation restoration in Horqin Sandy Land, Northeast China. Soil nematode communities in different seasons (spring, summer and fall) and soil depths (0-10 cm, 10-20 cm and 20-30 cm) were compared in an age sequence of 0, 5, 10 and 22-year-old *Caragana microphylla* plantations (treatments). The results showed that the total abundance of soil nematodes increased with increasing age of *C. microphylla* plantation, but decreased with depth in spring and fall. The relative abundances of plant parasites decreased, while those of omnivore-predators and fungivores increased with increasing age of *C. microphylla* plantation. Thirty-four genera were observed and *Acrobeles* was dominant genus in all treatments. Significant differences in the numbers of total nematodes, bacterivores, fungivores, plant parasites, omnivores-predators, *Acrobeles* and *Eudorylaimus* were observed among seasons, treatments, and depths. The evenness (J′) and generic diversity (H′) all increased with increasing age of *C. microphylla* plantation and varied significantly among the different seasons and treatments ($P < 0.05$). The dominance (λ) was significantly higher in 22-year treatment than in 0-year and 5-year treatments. The distribution of soil nematode communities in spring had a similar trend with that in fall, but different with that in summer. The soil nematode community structure and its ecological indices along different chronosequence of vegetation restoration reflected the different restoration stage and degree of the sandy land

* Correspondence should be addressed: Prof. Wenju Liang, Institute of Applied Ecology Chinese Academy of Sciences, P. O. Box 417, Shenyang 110016. China. Tel.: +86-24-83970359; Fax: +86-24-83970300; Email: liangwj@iae.ac.cn

ecosystem. Vegetation rehabilitation played key roles for restoring the Horqin desertified ecosystem.

Keywords: *Caragana microphylla*; nematode communities; plantation chronosequence; spatiotemporal distribution; vegetation restoration; desertified ecosystem.

INTRODUCTION

Specific ecosystem disturbances in diverse regions of the world had led to the degradation of soil and vegetation. Dregne and Chou (1992) estimated that 3600 million ha were degraded in dryland zones alone, and 1000 million ha with degradation of both soil and vegetation. Khoshoo (1992) reported 175 million ha prone to severe erosion loss in India. Now desertification is one of the major environment problems which Chinese people faced and it brings threat to production and the living of people in Horqin Steppe, Northeast China (Jiang et al., 2003). The spatiotemporal heterogeneity is the main characteristics of the soil in sandy land. The sandy land is not favorable for the invasion, recruitment and survival of most plant species. Strategies for their rehabilitation/restoration, such as shrub establishment, played an important role in this region (Su et al., 2003). In the semi-arid Horqin Sandy Land, *Caragana microphylla*, a leguminous shrub, was the dominant sand-fixation plant species, and works on vegetation restoration had been made to stabilize mobile sand gradually.

Nematodes are the most abundant metazoan phylum and widely distribute in various habitats. The degradation or restoration of soil and vegetation would have negative or positive effect on the activities of soil nematodes which are largely controlled by soil biological and physical conditions (Jiang et al., 2007). Certain parameters of soil nematodes such as density, biomass and generic diversity tended to change locally on a spatiotemporal scale along with changes in vegetation types (Pen-Mouratov et al., 2003). The study of De Goede et al. (1994) on the relation of nematode community structure with soil and vegetation characteristics proved that nematode community structure varied as plant succession proceeded. Goralczyk (1998) investigated the bioindication of nematodes in a coastal dune and showed that vegetation succession was accompanied by a parallel succession of the nematode fauna. Soil nematodes had strong relationships with fundamental ecosystem processes such as decomposition and nutrient cycling, and interactions with the microbial community, plant growth and pedogenesis (Wall et al., 2002; Yeates, 2003; Wardle et al., 2004). Some studies argued that nematodes can be used as functional indicators of ecosystems to reflect the effect of fertilization (Hua et al., 2006), irrigation (Zhi et al., 2008), elevated CO_2 (Li et al., 2007), land use change (Ou et al., 2005; Hua et al., 2008), soil degradation and restoration in sand dune (Zhang et al., 2007b), and heavy metal stress (Zhang et al., 2006; Zhang et al, 2007a) on Chinese soil ecosystems. Study on the bioindication of soil nematode communities to the vegetation restoration is necessary to understand the mechanisms of recovery and enhance the sustainability in severely degraded soil ecosystem of the Horqin Sandy Land.

The objectives of this study were to describe the spatiotemporal distribution of soil nematode communities along a chronosequence of vegetation restoration, and to utilize the bioindication of soil nematodes to reflect the condition of plantation rehabilitation in Horqin Sandy Land.

METHODS

Study Site

This study was conducted at the Wulanaodu Experimental Station of Desertification (43°02'N, 119°39'E), Chinese Academy of Sciences, western Horqin Sandy Land, Northeast China. The region is typical of a continental temperate semi-arid zone, with mean annual precipitation of approximately 340.5 mm. The annual mean temperature is 6.2°C, the non-frost period is 130 days, and evaporation is 2500 mm. The predominant native plants are *Bassia dasyphlla, Caragana micophylla* and *Pennisetum flaecidum* (Cao et al., 2004).

Experimental Design

Three vegetation restoration types were selected for study with various lengths of time (5, 10 and 22 years). These sites are of similar topography and soil type, and have not been grazed since enclosure. Caragana microphylla has been gradually planted on desertified sandy land in the experimental plots. Planting was arranged in belts (row spacing 1 x 1 m) and the orientation of the belts was perpendicular to the prevailing wind direction. Within each plantation (5-, 10-, 22-yr), four sites were selected as sampling (four replicates). Four non-vegetated dunes in the vicinity of the planted sites were also chosen as control (0-yr) for the chronosequence. Sixteen plots, 20 x 20 m each, were established in this study (Jiang et al., 2007) .

Soil Sampling and Nematode Examination

Samples from three soil depths (0–10 cm, 10–20 cm and 20–30 cm) were collected using a 5-cm diameter corer, under the canopy of *C. microphylla* in an age series of 0-, 5-, 10- and 22-yr plantations (four treatments) in spring (April 15), summer (August 12) and fall (October 21) of 2005. Composite samples were made up from 9 cores. All samples were stored in individual plastic bags, and immediately transferred to a 4 °C cold room.

Nematode was extracted by elutriation and sugar-centrifugation, and the nematode populations were expressed as individuals per 100 g dry soil (Liang et al., 2005a). Nematodes were identified to genus level using an inverted microscope. The classification of trophic groups was assigned to bacterivores, fungivores, plant parasites and omnivores-predators based on known feeding habitats or stoma and esophageal morphology (Yeates et al., 1993; Bongers and Bongers, 1998; Pen-Mouratov et al., 2003; Liang et al., 2005b).

Data Analysis

The data obtained were analyzed by the following approaches: (1) absolute abundance of individuals 100 g^{-1} dry soil (TNE); (2) trophic groups, which includes bacterivores (BF), fungivores (FF), plant parasites (PP) and omnivores-predators (OP); (3) Shannon-Wiener index (H′), a species diversity measure, which gives more weight to rare species, $H' = -\sum P_i(\ln P_i)$, where P_i is the proportion of each taxon in the total population; (4) evenness $J' = H'/\ln S$, S is the total number of genera; (5) Simpson index (λ), where $\lambda = \sum P_i$; (6) maturity index (MI), which measures disturbances, with a lower value indicating a more disturbed environment (Bongers, 1990; Yeates and Bongers, 1999; Ekschmitt et al., 2001; Liang et al., 2005b).

All statistical analyses were performed by SPSS software package. Difference at $P < 0.05$ level was considered as statistically significant.

Results

Spatiotemporal Variation of Total Nematodes

The total number of nematodes in different soil depths, plantation years and seasons ranged from 2 to 200 individuals per 100 g dry soils. In spring and fall, the total number of nematodes all increased significantly with increasing age of C. microphylla plantation ($P < 0.05$) (Figure 1). In the different seasons, the numbers of total nematodes were significantly higher in spring than in summer and fall ($P < 0.05$). The highest value appeared at 0-10 cm of the 22-year treatment in spring. Significant differences in the total nematodes were observed among different soil depths, plantation years and seasons ($P < 0.01$) (Table 2).

Spatiotemporal Variation of Trophic Groups

Bacterivores were found to be the most abundant trophic group in spring and fall (Table 1). In the summer, omnivores-predators at depth of 0-10 cm were dominant trophic group in 5-year and 10-year treatments. Plant parasites had higher abundance in summer than in spring and fall. With the increasing age of C. microphylla plantation, the relative abundance of plant parasite decreased, but omnivore-predators and fungivores increased. Except in summer, the responses of the four trophic groups in different depths were almost consistent. Significant differences in the numbers of bacterivores, fungivores, plant parasites and omnivores-predators were found among seasons, treatments and depths ($P < 0.01$) (Table 2). Significant interaction effects of season, treatment and depth were also found in the four trophic groups.

Table 1. Spatiotemporal distribution of trophic groups along the chronosequence of vegetation restoration (Individuals per 100 g dry soil)

Season	Year	Depth	BF	FF	PP	OP
Spring	0-year	0-10cm	3.50	0.75	1.00	3.00
		10-20cm	4.50	0.75	1.25	2.00
		20-30cm	2.25	0.00	1.50	2.25
	5-year	0-10cm	67.50	5.75	0.25	8.75
		10-20cm	41.75	4.75	4.25	17.00

Season	Year	Depth	BF	FF	PP	OP
		20-30cm	11.00	0.50	1.50	3.75
	10-year	0-10cm	93.50	7.25	3.50	55.50
		10-20cm	63.00	8.00	10.25	35.00
		20-30cm	30.50	2.50	12.75	25.00
	22-year	0-10cm	89.25	12.00	6.00	93.00
		10-20cm	58.50	5.50	6.25	70.75
		20-30cm	62.50	12.00	16.50	19.25
Summer	0-year	0-10cm	1.75	0.50	0.50	1.75
		10-20cm	2.50	0.00	2.50	2.50
		20-30cm	1.75	0.25	0.75	1.75
	5-year	0-10cm	9.50	1.00	17.25	22.50
		10-20cm	12.25	0.25	55.50	8.75
		20-30cm	5.50	0.25	12.75	6.25
	10-year	0-10cm	48.75	2.50	9.50	62.25
		10-20cm	30.50	4.75	21.50	22.50
		20-30cm	30.75	2.25	18.00	21.00
	22-year	0-10cm	40.50	4.00	5.00	20.00
		10-20cm	32.75	9.25	9.25	22.25
		20-30cm	15.25	3.50	8.25	13.25
Fall	0-year	0-10cm	4.50	0.25	1.50	1.00
		10-20cm	0.75	0.00	1.75	0.00
		20-30cm	0.25	0.00	0.25	0.50
	5-year	0-10cm	15.00	1.00	0.00	5.25
		10-20cm	38.25	0.75	0.50	5.00
		20-30cm	31.50	0.00	0.75	17.00
	10-year	0-10cm	10.50	1.00	1.25	3.25
		10-20cm	51.50	7.00	9.00	23.25
		20-30cm	51.75	2.00	11.25	19.00
	22-year	0-10cm	97.25	2.50	0.50	4.00
		10-20cm	61.00	22.00	15.75	21.00
		20-30cm	48.50	7.50	9.75	31.50

Table 2. Analysis of variance (F-values) for soil nematodes with season, depth and plantation chronosequence effect (*, $P < 0.05$; **, $P < 0.01$)

F-value	Season (S)	Treatment (T)	Depth (D)	S × T	S × D	T × D	S × T × D
TNE	37.27**	177.33**	47.83**	15.54**	9.54**	7.77**	3.29**
BF	31.28**	81.27**	28.25**	5.34**	4.31**	4.60**	3.03**
FF	4.61**	22.36**	7.62**	0.94**	1.82**	2.10**	1.49**

Table 2. (Continued)

F-value	Season (S)	Treatment (T)	Depth (D)	S × T	S × D	T × D	S × T × D
PP	21.09**	16.38**	7.48**	14.39**	3.75**	2.68**	2.13**
OP	33.71**	98.53**	26.87**	17.53**	15.82**	6.30**	6.96**
J′	11.21**	18.14**	7.36**	1.27	1.01	2.02*	0.87
λ	1.34	16.58**	1.85	1.68	2.86**	1.42	0.96
H′	3.73*	42.71**	1.34	2.14	1.66	1.44	0.73
MI	4.26*	2.43	4.10*	1.65	5.18**	1.63	1.55
Acrobeles	14.08**	9.71**	4.31**	9.21**	1.52	0.59	0.67
Chiloplacus	1.35	6.68**	2.87*	1.81	1.72	1.46	1.72
Eudorylaimus	21.63**	5.28**	2.82*	0.78	3.04**	2.57**	2.15**

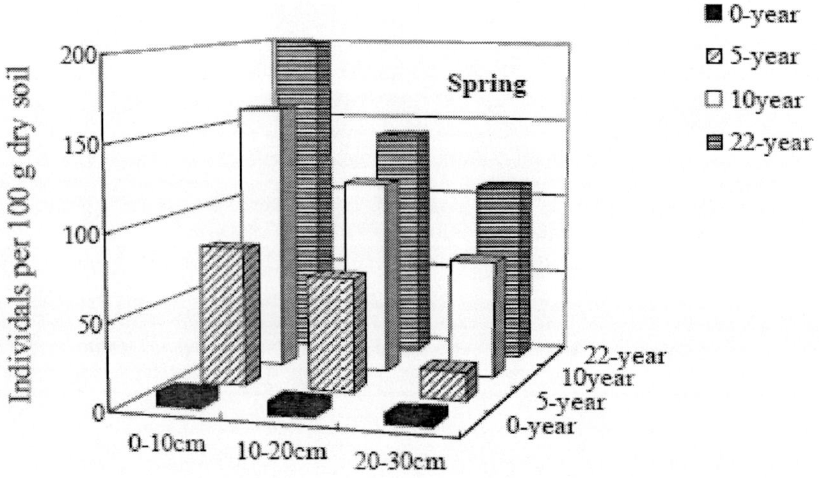

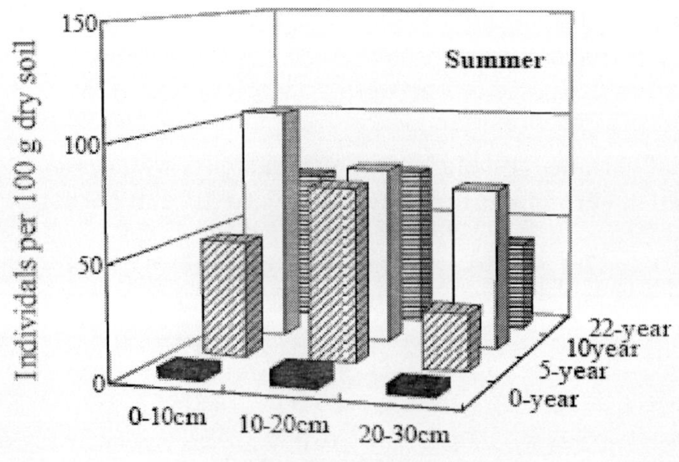

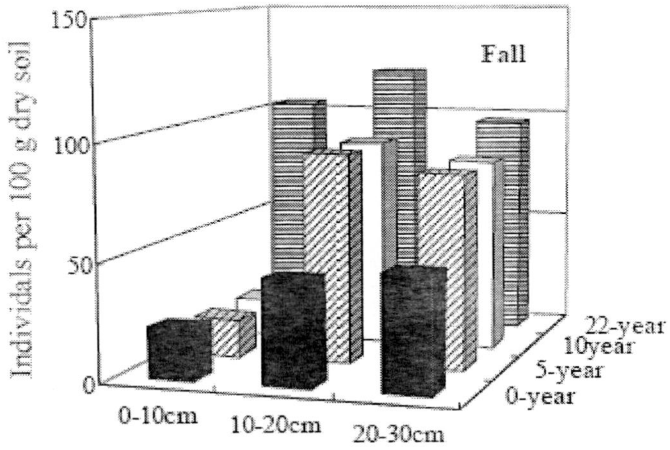

Figure 1. Spatiotemporal distribution of the total nematodes along the chronosequence of vegetation restoration.

Nematode Taxa

Thirty-four genera were found in samples from different chronosequence of vegetation restoration (Table 3). The mean number of nematode genera increased from 20 in the nonvegetation soil of mobile sand dunes to 31 in the 22-year treatment. *Acrobeles* was the dominant genera in all treatments. In spring and summer, the dominance of *Eudorylaimus* was all more than 10% in the control, 5-year and 10-year treatments. Significant differences in the numbers of *Acrobeles* and *Eudorylaimus* were observed among seasons, treatments and depths ($P < 0.05$) (Table 2). More *Scutylenchus* existed in non-vegetated control. *Heterodera* became the predominant genus of the 5-year treatment in summer, with 48.8% of the community. In the control and 5-year treatment, *Chiloplactus* was dominant genus in spring, accounting for nearly 17% and 10%, respectively. Significant differences in the number of *Chiloplacus* were observed among treatments ($P < 0.01$) and depths ($P < 0.05$). *Heterocephalobus* only appeared in the sandy land with vegetation restoration. *Discolaimus* and *Paraxonchium* belonged to the omnivore-predator disappeared in summer.

Table 3. Mean relative abundance (%) of soil nematodes along the chronosequence of vegetation restoration

Genera	Spring				Summer				Fall			
	0yr	5yr	10yr	22yr	0yr	5yr	10yr	22yr	0yr	5yr	10yr	22yr
Bitylenchus	2.0	0.7	1.6	5.1	6.0	0.6	8.9	1.6	8.1	0.3	0.7	1.8
Criconemoides	5.7	1.4	0.8	1.7	3.4	0.9	1.4	3.5	9.0	0.3	1.2	1.3
Crossonem	0.0	0.0	0.0	0.0	0.0	0.0	0.0	0.2	0.0	0.0	0.0	0.0
Filenchus	2.4	0.6	0.0	0.1	0.0	1.1	0.6	1.4	0.0	0.0	0.0	0.0
Hemicycliophora	0.7	0.0	2.3	0.0	0.0	0.0	0.0	0.0	0.7	0.0	0.1	0.0
Heterodera	0.4	4.5	0.0	0.4	0.0	48.8	4.6	4.0	0.0	0.5	2.2	0.0
Paratrichodocus	3.3	1.2	7.2	0.0	0.0	0.0	5.3	0.5	0.0	0.1	6.9	0.0

Table 3. (Continued)

Genera	Spring				Summer				Fall			
	0yr	5yr	10yr	22yr	0yr	5yr	10yr	22yr	0yr	5yr	10yr	22yr
Pratylenchus	0.0	0.0	0.0	0.0	0.0	0.0	0.2	0.1	0.8	0.1	0.0	0.5
Scutylenchus	5.9	0.0	3.4	1.6	12.1	2.2	0.5	7.3	22.4	0.2	1.5	6.2
Xphinema	0.4	0.0	0.0	0.3	1.3	0.0	0.0	0.6	1.3	0.2	0.0	0.2
Achromadora	0.0	0.1	0.9	0.8	0.0	0.0	0.4	1.8	0.0	0.5	0.9	0.3
Acrobeles	17.1	41.5	40.0	35.3	30.7	10.3	30.7	26.9	19.3	53.8	46.0	38.4
Acrobeloides	2.9	7.1	0.3	1.9	0.0	1.7	0.5	0.5	2.4	2.6	1.6	5.4
Alaimus	0.7	0.2	1.5	1.4	0.0	0.0	1.6	1.1	0.4	0.2	0.6	1.0
Anaplectus	0.0	0.0	1.5	2.5	3.5	0.2	1.8	4.2	1.4	1.3	5.6	0.7
Aulolaimus	0.0	1.7	0.0	0.0	0.0	0.0	0.0	0.0	0.0	0.0	0.1	0.0
Chiloplachus	17.3	10.0	2.1	5.1	7.2	4.7	3.9	7.4	13.0	4.0	5.0	8.5
Heterocephalobus	0.0	1.9	1.4	1.6	0.0	0.1	0.2	0.3	0.0	1.4	0.2	0.7
Mesorhabditis	0.0	0.0	0.2	0.0	0.0	3.6	0.3	0.4	0.7	0.0	0.0	0.0
Panagrolaimus	2.5	0.0	0.0	0.0	0.0	0.1	0.0	0.0	0.0	0.1	0.0	0.1
Plectus	0.0	0.0	0.0	0.1	0.0	0.0	0.0	0.0	0.0	0.0	0.0	0.0
Prismatolaimus	0.0	0.0	0.0	0.1	0.0	0.0	0.0	0.0	0.0	0.0	0.0	0.0
Rhabdolaimus	2.2	0.1	0.0	0.1	0.0	0.0	0.0	0.0	0.0	0.0	0.0	0.0
Wilsonema	0.0	0.0	0.5	1.8	0.0	0.0	1.4	1.0	0.0	0.4	0.7	0.6
Elaphonematidae	0.0	0.0	0.0	0.0	0.0	0.0	0.0	0.0	0.7	0.0	0.0	0.0
Aphelenchus	1.0	0.9	0.4	0.1	1.3	0.2	0.4	0.3	0.0	0.5	0.4	0.8
Ditylenchus	1.6	1.4	0.0	0.3	0.0	0.5	0.3	0.8	0.0	0.0	0.5	0.0
Dorylaimoides	0.0	0.0	0.1	0.5	1.3	0.3	0.0	0.3	1.6	1.0	0.1	0.8
Paraphelenchus	2.1	1.3	0.2	0.3	0.0	0.0	0.9	2.2	0.0	0.2	0.0	0.0
Tylencholaimus	0.8	1.4	2.8	6.4	0.8	0.0	1.4	4.4	2.2	0.1	2.7	5.7
Aporcelaimellus	0.0	0.0	1.3	2.0	0.0	2.3	5.1	5.5	0.6	4.5	2.7	1.1
Carcharolaimus	0.5	0.3	0.5	0.3	0.4	0.4	0.5	0.4	1.7	1.1	0.9	1.1
Clarkus	1.6	0.0	0.8	0.4	1.7	0.0	0.1	0.2	1.3	0.6	1.5	0.2
Discolaimium	2.3	3.0	1.2	1.6	3.9	0.6	0.9	2.5	1.0	5.8	3.1	4.2
Discolaimus	0.0	0.0	0.0	6.9	0.0	0.0	0.0	0.0	0.0	0.1	1.7	2.7
Eudorylaimus	18.9	14.1	17.7	7.2	14.8	11.5	14.0	7.5	5.4	5.0	1.9	1.0
Mesodorylaimus	0.0	0.0	0.0	0.0	0.0	0.0	0.1	0.0	0.0	0.0	0.0	0.2
Microdorylaimus	2.8	4.9	9.7	9.8	11.9	6.9	9.9	8.3	0.6	2.8	1.5	1.2
Nygolaimus	0.0	0.8	1.3	3.2	0.0	3.1	3.9	4.6	5.5	11.9	9.5	15.0
Oxydirus	0.0	0.2	0.0	0.0	0.0	0.0	0.0	0.0	0.0	0.0	0.0	0.0
Paraxonchium	0.0	0.0	0.0	0.6	0.0	0.0	0.0	0.0	0.0	0.1	0.1	0.5
Prionchulus	0.0	0.0	0.1	0.1	0.0	0.0	0.0	0.0	0.0	0.1	0.0	0.0
Pungentus	0.0	0.0	0.0	0.0	0.0	0.0	0.0	0.3	0.0	0.4	0.0	0.0
Sectonema	0.0	0.0	0.0	0.0	0.0	0.0	0.1	0.0	0.0	0.0	0.0	0.0
Tobrilus	0.0	0.0	0.0	0.5	0.0	0.0	0.0	0.0	0.0	0.0	0.0	0.0
Torumanawa	0.0	1.0	0.0	0.0	0.0	0.0	0.0	0.0	0.0	0.0	0.0	0.0

Ecological Indices

The generic diversity of soil nematode (H′) increased from the nonvegetation of mobile sand dunes to 22-year treatment (Table 4). The values of H′ were significantly different among different seasons ($P < 0.05$) and treatments ($P < 0.01$) (Table 2). The reverse trend appeared in the value of the Simpson index (λ). The mean value of λ decreased with the increasing age of *C. microphylla* plantation. There was a significant difference in the value of λ among the different treatments ($P < 0.01$). Significant differences in the values of MI were found among different seasons and soil depths, where 0-10 cm < 10-20 cm < 20-30 cm except in summer and fall. But no significant treatment effect was observed in the values of MI. The values of J′ were significantly affected by different treatments and soil depths ($P < 0.01$). The values of evenness (J′) were significantly higher in 5-year, 10-year and 22-year treatments than in control. Significant differences in the values of J′ were also found among different seasons ($P < 0.01$), where spring > fall > summer.

Table 4. Variation of ecological indices along the chronosequence of vegetation restoration

Indices	Season	Depth	0-year	5-year	10-year	22-year
J′	Spring	0-10cm	0.92	0.76	0.64	0.68
		10-20cm	0.95	0.78	0.75	0.78
		20-30cm	0.94	0.84	0.78	0.77
	Summer	0-10cm	0.93	0.79	0.72	0.75
		10-20cm	0.80	0.65	0.57	0.68
		20-30cm	0.65	0.51	0.71	0.66
	Fall	0-10cm	0.84	0.70	0.71	0.77
		10-20cm	0.76	0.62	0.66	0.70
		20-30cm	0.83	0.74	0.71	0.54
λ	Spring	0-10cm	0.31	0.42	0.36	0.23
		10-20cm	0.29	0.30	0.24	0.21
		20-30cm	0.64	0.34	0.22	0.20
	Summer	0-10cm	0.32	0.26	0.25	0.23
		10-20cm	0.37	0.53	0.19	0.18
		20-30cm	0.31	0.31	0.17	0.15
	Fall	0-10cm	0.33	0.37	0.34	0.48
		10-20cm	0.53	0.67	0.30	0.34
		20-30cm	0.34	0.36	0.29	0.22
H′	Spring	0-10cm	1.34	1.33	1.39	1.75
		10-20cm	1.29	1.55	1.90	1.91
		20-30cm	0.77	1.38	1.92	2.05
	Summer	0-10cm	1.05	1.37	1.51	1.85

Table 4. (Continued)

Indices	Season	Depth	0-year	5-year	10-year	22-year
		10-20cm	0.87	1.02	1.79	1.76
		20-30cm	1.09	1.36	1.78	1.93
	Fall	0-10cm	1.45	1.58	1.62	1.16
		10-20cm	0.92	0.76	1.80	1.66
		20-30cm	1.21	1.51	1.75	1.88
MI	Spring	0-10cm	2.86	2.33	2.89	3.18
		10-20cm	2.52	2.74	2.89	3.24
		20-30cm	3.25	2.66	3.01	2.88
	Summer	0-10cm	3.20	3.42	3.33	2.90
		10-20cm	3.03	2.83	2.95	3.04
		20-30cm	3.08	3.16	3.03	3.16
	Fall	0-10cm	2.40	2.58	2.62	2.12
		10-20cm	2.56	2.29	2.91	2.99
		20-30cm	3.41	3.01	2.74	3.18

Discussion

Temporal Variation of Soil Nematodes

In our study, soil nematodes exhibited different distributions in different seasons. In spring and fall, there was a significant chronological change in the total nematodes, and the longer the age of *C. microphylla* plantation was, the larger the number of total nematodes was. The results were similar with that of Panesar et al. (2001), who proved that the abundance of soil nematodes was lowest in the regeneration stands and highest in old-growth stands. This trend was interpreted by Panesar et al. (2001) as indicating that nematode population levels in the earlier succession stages would be restored to old-growth levels as they attain old-growth features with time.

In summer, the lowest numbers of total nematodes were found, and no clear chronological variation. From July to August in 2005, there was high temperature (34.4-36.8□) and much precipitation (80.1-117.9 mm) in the Horqin Sandy Land (Liu, 2006). So the temperature and precipitation may become the major limiting factors to affect the reproduction of soil nematodes in summer (Li et al., 2002).

The more plant parasites existed in summer can be explained that plantations with flourishing growth in summer provided the abundant food resources to plant parasites (Bongers and Ferris, 1999; Yeates and Bongers, 1999).

Spatial Variation of Soil Nematodes

In addition to temporal changes, nematode communities were sensitive to soil depths. Most of soil nematodes distributed in the upper layer of 0-10 cm in spring and the 10-20 cm in summer and fall, and smaller ones in 20-30 cm. This kind of superficial distribution also appeared in the study of Li et al. (2002), who argued that soil nematodes preferred to live in the soil environment with steady temperature and humid. Panesar et al. (2001) found that more soil nematodes were found in the upper organic horizons of the soil than in the lower mineral horizons. Spatial distribution effect was more significant in spring than that in summer and fall.

Changes in Nematode Ecological Indices

The Shannon diversity index (H′) was an important measure for the evaluation of sandy land ecosystem (Yeates and Bongers, 1999). The value of H′ increased with increasing plantation year, which indicated the gradual restoration process of biodiversity in sandy land. The evenness (J′) displayed the similar trend with that of H′. The dominance of soil nematodes can be reflected by Simpson index (Yeates and Bongers, 1999). The decreasing trend in Simpson index (λ) with increasing plantation year showed that the dominant species tend to be single in shorter plantation year and diverse in longer plantation. The MI was mainly influenced by seasons and soil depths but not by plantation chronosequence. From the analysis of the ecological indices, we can conclude that the main characteristics of stable soil ecosystem should be with high diversity, high evenness and low dominance. The ecological indices in spatial and temporal scale were less sensitive to chronological variation than the abundance of total nematodes and trophic groups.

Bioindication of Soil Nematodes

We had observed a vegetation restoration process through the spatiotemporal distribution of soil nematode community under the different chronological vegetation, and gained some understanding of the mechanisms of soil and vegetation restoration. The process of vegetation restoration on desertified land can be viewed as succession (Zhang et al., 2005) and we also regarded the variation of soil nematodes with chronosequence as the succession process of soil animal. We can conclude that the succession of vegetation was companied with the succession of soil nematode communities. The restoration process of soil and vegetation can be reflected by the distribution of soil nematodes community. The important reason which soil nematode can be regarded as bioindicator was that they potentially provided holistic measure of the biotic and functional status of soils (Ritz and Trudgill, 1999). Plantation rehabilitation became the driver of the restoration of sandy land ecosystem and played key roles for improving soil environment and restoring the desertification ecosystems in Horqin Sandy Land.

Acknowledgements

This study was financially supported by the Natural Science Foundation of China (No. 30700109). The authors appreciate the technical assistance of staffs at the Wulanaodu Experimental Station of Desertification, Institute of Applied Ecology, Chinese Academy of Sciences.

References

Bongers, T. (1990). The maturity index: An ecological measure of environmental disturbance based on nematode species composition. *Oecologia*, 83, 14–19.

Bongers, T., and Bongers, M. (1998). Functional diversity of nematodes. *Appl. Soil. Ecol.*, 10, 239–251.

Bongers, T., and Ferris, F. (1999). Nematode community structure as a bioindicator in environmental monitoring. *Trends Ecol. Evol.,* 14, 224–228.

Cao, C. Y., Jiang, D. M., Luo, Y. M., Kou, Z. W. (2004). Stability of *Caragana microphylla* plantation for wind protection and sand fixation. *Acta Ecol. Sin.,* 24, 1178–1186.

De Goede, R. G. M., and Bongers, T. (1994). Nematode community structure in relation to soil and vegetation characteristics. *Appl. Soil Ecol.*, 1, 29–44.

Dregne, H. E., and Chou, N. T. (1992). Global desertification dimensions and costs. In H. E. Dregne (Ed.), In *Degradation and Restoration of Arid Lands* (pp 249–282). Texas Tech University, Lubbock, TX, USA.

Ekschmitt, K., Bakonyi, G., Bongers, M., Bongers, T., Bosröm, S., Dogan, H., Harrison, A., Nagy, P., O'Donnell, A. G., Papatheodorou, E. M., Sohlenius, B., Stamou, G. P., Wolters, V. (2001). Nematode community structure as indicator of soil functioning in European grassland soils. *Eur. J. Soil. Biol.*, 37, 263–268.

Goralczyk, K. (1998). Nematodes in a coastal dune succession: Indicators of soil properties? *Appl. Soil. Ecol.*, 9, 465–469.

Hua, J. F., Jiang, Y., and Liang, W. J. (2006). Response of nematodes in a Hapli-Udic Argosol to urea amended with urease and nitrification inhibitors. *Pedosphere,* 16, 428–434.

Hua, J. F., Jiang, Y., and Liang, W. J. (2008). Effects of vegetation coverage on the spatial distribution of soil nematode trophic groups. *Front Biol. China*, 3, 63–67.

Jiang, D. M., Liu, Z. M., Cao, C. Y., Kou, Z. W., Wang, R.Y. (2003). Desertification and ecological restoration of Heerqin Sandy Land. China Environmental Science Press, Beijing, China.

Jiang, D. M., Li, Q., Liu, F. M., Jiang, Y., Liang, W. J. (2007). Vertical distribution of soil nematode in an age sequence of *Caragana microphylla* plantations in the Horqin Sandy Land, Northeast China. *Ecol. Res.*, 22, 49–56.

Khoshoo, T. N. (1992). Degraded lands for agroecosystems. In *Ecosystem Rehabilitation: Preamble to Sustainable Development*. Ed. M K Wali. Vol. 2, pp 3–17. SPB Academic Publishing, The Hague, The Netherlands.

Li, H. X., Liu, M. Q., Hu, F., Chen, X.Y., He, Y. Q. (2002). Nematode abundance under different vegetation restored on degraded red soil. *Acta Ecol. Sin.*, 22, 1882–1889.

Li, Q., Liang, W. J. Jiang, Y., Shi, Y., Zhu, J. G., Neher, D. A. (2007). Effect of elevated CO_2 and N fertilisation on soil nematode abundance and diversity in a wheat field. *Appl. Soil. Ecol.*, 36, 63–69.

Liu, F. M. (2006). Soil nematode communities under *Caragana Microphylla* plantations. Ph. D. *dissertation. Institute of Applied Ecology*, Chinese Academy of Sciences, Shenyang, China.

Liang, W. J., Jiang, Y., Li, Q., Liu, Y. J., Wen, D. Z. (2005a). Spatial distribution of bacterivorous nematodes in a Chinese Ecosystem Research Network (CERN) site. *Ecol. Res.,* 20, 481–486.

Liang, W. J., Lavian, I., and Steinberger, Y. (2005b). Diversity and dynamics of soil free-living nematode populations in a Mediterranean agroecosystem. *Pedosphere*, 15, 204–215.

Ou, W., Liang, W. J., Jiang, Y., Li, Q., Steinberger, Y. (2005). Vertical distribution of soil nematodes under different land use types in an aquic brown soil. *Pedobiologia*, 49, 139–148.

Panesar, T. S., Marshall1, V. G. and Barclay, H. J. (2001). Abundance and diversity of soil nematodes in chronosequences of coastal Douglas-fir forests on Vancouver Island, British Columbia. *Pedobiologia*, 45, 193–212.

Pen-Mouratov, S., Rakhimbaev, M., and Steinberger, Y. (2003). Seasonal and spatial variation in nematode communities in a Negev desert ecosystem. *J. Nematol*., 35: 157–166.

Ritz, K., and Trudgill, D. L. (1999). Utility of nematode community analysis as an integrated measure of the functional state of soils: perspectives and challenges. *Plant Soil*, 212, 1–11.

Su, Y. Z., and Zhao, H. L. (2003). Soil properties and plant species in an age sequence of *Caragana microphylla* plantations in the Horqin Sandy Land, north China. *Ecol. Eng.*, 20, 223–235.

Wall, J. W., Skene, K. R., and Neilson, R. (2002). Nematode community and trophic structure along a sand dune succession. *Biol. Fertil Soils*, 35, 293–301.

Wardle, D. A., Bardgett, R. D., Klironomos, J. N., Setälä, H., van der Putten, W. H., Wall, D. H. (2004). Ecological linkages between aboveground and belowground biota. *Science*, 304, 1629–1633.

Yeates, G. W. (2003). Nematodes as soil bioindicators: Functional and biodiversity aspects. *Biol. Fertil Soils*, 37, 199–210.

Yeates, G. W., and Bongers, Y. (1999). Nematode diversity in agroecosystems. *Agric. Ecosyst Environ.*, 74, 113–135.

Yeates, G. W., Bongers, T., de Goede, R. G. M., Freckman, D. W., Georgieva, S. S. (1993). Feeding habits in soil nematode families and genera: an outline for ecologists. *J. Nematol.*, 25, 315–331.

Zhang, J., Zhao, H., Zhang, T., Zhao, X., Drake, S. (2005). Community succession along a chronosequence of vegetation restoration on sand dunes in Horqin Sandy Land. *J. Arid. Environ.*, 62, 555–566.

Zhang, W. D., Wang, X. F., Li, Q., Jiang, Y., Liang, W. J. (2007a). Soil nematode responses to heavy mental stress. *Helminthologia*, 44, 87–91.

Zhang, X. K., Liang, W. J., Jiang, D. M., Liu, Z. M., Jiang, S. W. (2007b). Soil nematode community structure in a Chinese sand dune system. *Helminthologia*, 44, 204–209.

Zhang, X. K., Li, Q., Wang, S. B., Jiang, Y., Liang, W. J. (2006). Effect of zinc addition to soil on nematode community structure. *Bull. Environ. Contam. Toxicol.*, 76, 589–594.

Zhi, D. J., Li, H. Y., and Nan, W. B. (2008). Nematode communities in the artificially vegetated belt with or without irrigation in the Tengger Desert, China. *Eur. J. Soil Biol.*, 44, 238–246.

In: Environmental and Regional Air Pollution
Editors: D. Gallo, R. Mancini
ISBN: 978-1-60692-893-6

Chapter 6

PHYTOREMEDIATION OF MERCURY-CONTAMINATED MINE WASTES

Fabio N. Moreno[1]*, Joel B. Sígolo[1], Chris W. N. Anderson[2], Robert B. Stewart[2], John A. Meech[3] and Brett H. Robinson[4]

[1]Institute of Geosciences, Department of Environmental and Sedimentary Geology, University of São Paulo, SP, Brazil

[2] Institute of Natural Resources, Soil and Earth Sciences, Massey University, Palmerston North, New Zealand

[3] Environmental Research on Minerals, Metals and Materials (CERM3), University of British Columbia, Vancouver, BC, Canada

[4] Institute of Terrestrial Ecosystems, Swiss Federal Institute of Technology, Zurich, Switzerland

ABSTRACT

Although mercury (Hg) is one of the most toxic elements to living organisms, emission of Hg from natural sources worldwide has only a limited impact on life. However, over the past 150 years, anthropogenic Hg emissions have more than doubled the global Hg input from natural Hg sources (Hylander and Meili, 2003).

There have been sporadic episodes of human Hg poisoning from discharge of Hg-containing wastes to soil and water systems. Nevertheless, human activity still causes Hg releases into the environment. The main Hg pollution sources are atmospheric Hg discharges from coal-burning power plants, mining operations and incineration of urban and medical wastes (Morel et al., 1998).

The effect of Hg discharge into soils and aquatic systems is exacerbated by the transformation of inorganic Hg forms into methylmercury, which accumulates in living cells and tissues. It constitutes the greatest fraction of total mercury in living organisms and is quickly biomagnified in the aquatic and terrestrial food chains. Humans and wildlife can be exposed to methylmercury through ingestion of contaminated fish, which tend to accumulate high Hg concentrations in contaminated areas. Therefore,

* Corresponding author: Fabio Netto Moreno, fabionmoreno@terra.com.br), Affiliation: Institute of Geosciences, University of São Paulo, R. do Lago, 562, Cidade Universitária, CEP 05508-080, São Paulo, Brazil.

communities dependent on Hg-laden fish as a main food source frequently have elevated Hg levels in blood (Veiga and Hinton, 2002). Methylmercury is considered the sixth most toxic compound of a total of six million such toxins known to humankind (Malm, 2001).

The concept of using plants for environmental remediation has been intensively researched over the last 15 years. Phytoremediation embraces an array of low-cost plant-based technologies that could be potentially advantageous for remediation of Hg-polluted soils, especially in developing countries, where artisanal and small-scale mining for gold extraction has left a legacy of Hg pollution. Potentially, phytoremediation could remove Hg from the soil by growing successive crops of Hg-accumulating plants (phytoextraction). The Hg in the biomass could then be recovered, or stored in an area where it does not pose a risk to the environment. Alternatively, plants could be used to reduce the Hg flux in contaminated sites (phytostabilisation). This role relies on plants stabilising the soil surface, fixing Hg in the rootzone and returning water to the atmosphere via evapotranspiration. The reduction of Hg flux requires that the plants do not translocate large amounts of Hg into the shoots, where it may be consumed by herbivores.

Mercury in most soils is unavailable for uptake by plant roots. However, soil amendments can be used to increase the bioavailablility of Hg to plants. In this chapter, we examine the conditions by which the Hg concentration in the soil solution can be enhanced for plant uptake for three metal-contaminated mine sites where Hg pollution caused by anthropogenic or ASM activities are of concern.

In this chapter, we review the most important sources of Hg into the environment, the behaviour of Hg in soils, and describe plant Hg uptake experiments, with a view to elucidating the most viable strategies for the phytoremediation of Hg-contaminated sites worldwide. We describe laboratory and greenhouse experiments where the effect of total Hg concentrations, plant species, and humic acids was investigated on induced Hg solubilisation and plant-Hg accumulation. The results of a field-scale phytoextraction experiment in situ is also presented. Based on the soil Hg chemistry, thiol-containing ligands were tested for their potential to enhance Hg solubility for subsequent plant uptake. Research was conducted with mine wastes collected from Hg-contaminated sites in New Zealand, Brazil, and China.

The possibility of Hg (0) emissions to the atmosphere and/or Hg leaching to groundwater during the course of phytoextraction represents a threat to the health of humans and ecosystems and should not be overlooked. Therefore, we evaluated the generation of Hg-containing leachates and Hg vapour emissions from plant pots enclosed in gas-tight volatilisation chambers. These experiments provided a picture of Hg fate in the plant-soil-atmosphere continuum.

Finally, we discuss the role of phytostabilisation to reduce the flux of Hg in the environment.

Introduction

Artisanal and Small-Scale Mining (ASM)

Artisanal and small-scale mining (ASM) is any mining activity that does not follow the conventional technical approach adopted by organised mining companies. This means that the cycle of geological exploration, drilling, reserve proving and engineering studies are not part of ASM. Around 13 million miners are currently involved in ASM operations in approximately 55 countries. Between 80–100 million people worldwide directly or indirectly depend on this activity as a source of income (Veiga et al., 2006).

Environmental degradation and deplorable social conditions normally accompany ASM (Figure 1). The most profound impact of ASM is the pollution of the environment with metallic Hg. Mercury is a cheap, simple, and effective reagent for extracting alluvial gold, but its use results in environmental, social and human health problems (Veiga and Hinton, 2002).

There are two main processes for extracting gold from secondary gold ores (alluvial, colluvial or illuvial material): amalgamation and cyanidation. Most miners use amalgamation, which relies on the capacity of the two metals to interpenetrate each other in a deep sorption process (Veiga et al., 2006). Once formed, the dense Hg-Au amalgam grain can be separated from the rest of the sediment by gravitation. Gold extraction using Hg consists of four main steps: amalgamation, separation of the mineral portion, amalgam decomposition, and gold smelting.

Excess Hg, which is removed from the amalgam by squeezing the liquid Hg through a piece of fabric by hand, is re-bottled and used again. Amalgam decomposition frequently involves heating the amalgam, which usually contains 60% of gold, in open-air pans or shovels using a blowtorch. Gold-doré is the result of amalgam heating and contains about 20g of Hg per Kg of gold. The residual Hg is released from the gold-doré by smelting in gold shops. Environmental contamination due to ASM gold mining occurs mostly from atmospheric deposition of emitted Hg (0) as a result of amalgam decomposition and gold smelting. A further substantial fraction of Hg can also contaminate water and soils after discharge of the amalgamation tailings, which can contain between 200–500 mg/kg of residual Hg (Veiga et al., 2006). This practice has been implicated in the generation of Hg hot spots in adjacent soils and water bodies (Veiga and Hinton, 2002).

Figure 1. Artisanal gold mine site in the Brazilian Amazon.

The Hg problem due to ASM might be growing due to the gradual increase in gold price since 2003 (Veiga et al., 2006). As a result, Hg emissions from ASM may be increasing. It is estimated that between 80–100 tonnes are annually released to the environment due to ASM worldwide. Mercury emissions from China, the main source, comprise between 200–250 tonnes/annum. Indonesia releases 100–150 tonnes/annum, while Brazil, Bolivia, Colombia,

Peru, Philippines, Zimbabwe and Venezuela each releases from 10 to 30 tonnes of Hg annually (Veiga et al., 2006). The number of artisanal mine sites worldwide is ca. 160,000, with some 2,000 abandoned artisanal gold mine sites in the Brazilian Amazon (Veiga and Hinton, 2002).

Mercury in Soils

Most Hg that is released into soil is adsorbed onto the solid-phase of organic matter and onto soil minerals, such as sulphides and oxy-hydroxides of iron and aluminium (Evans, 1987). A substantial fraction, however, undergoes physical transformation (leaching, erosion, and volatilisation) and biochemical transformation (methylation, photochemical and biological reduction) (Morel et al., 1998). Mobilisation of Hg can occur through exchange reactions with sulphur-containing ligands and chloride ions, leading to enhanced Hg solubility in soil solution (Schuster, 1991). In weathered tropical soils, Hg bound to iron and aluminium oxy-hydroxides can be mobilised from the surface horizon through the erosion of deforested soils (Roulet et al., 2000). The removal of plant cover allows increased oblique runoff on slopes and subsequent depletion of iron oxy-hydroxides and Hg in the upper centimetres of the soil (Roulet et al., 1999). The mobilised Hg eventually forms complexes with dissolved organic constituents and reaches aquatic systems, where it can be exported to areas away from the pollution source (Oliveira et al., 2001; Veiga, 2004).

The strong affinity of Hg for organic matter influences Hg solid phase speciation more than inorganic mineral colloids. The abundance of functional groups in HS provides many mechanisms for Hg binding including chelation, ionic exchange, complex formation, adsorption, and precipitation. Because of these interactions, Hg adsorption onto the solid phase of clay minerals, silicates, and metal oxides/hydroxides will not profoundly affect Hg speciation in soils as it would do for metals such as Zn, Al, Cd and Ni (Wallschalger, 1998).

Mercury Remediation

Ex-situ remediation technologies such as excavation, physical separation, and hydrometallurgical treatments are expensive, particularly if Hg contamination is spread over a large area or extends below the water table. Thermal treatment (soil heating combined with soil vapour extraction) can be effective for Hg removal from solid media but is technically complicated and costly. Furthermore, soil heating releases Hg as a vapour into the environment and can have deleterious effects on the physical, chemical and biological properties of soils (Meagher et al., 2000; Hinton and Veiga, 2001). Chemical stabilisation, electrolytic extraction and chemical leaching, although expensive, have proved effective for the clean up of Hg-contaminated media (Villas Boas, 1997; Hinton et al., 2003). Other methods, such as the use of impermeable plastic lines and the burying of contaminants will merely postpone clean up for a later date (Meagher et al., 2000).

Phytoremediation has been proposed as an environmentally friendly technology for the clean up of heavy metal polluted soils. It has attracted attention in recent years due to the low cost of implementation and environmental benefits (Robinson et al., 1998). These features make phytoremediation particularly welcome in developing countries, where there is a lack of financial incentive to remediate and rehabilitate heavy-metal polluted areas.

PHYTOREMEDIATION

Phytoremediation refers to any plant-based system that is used to improve contaminated media (Pilon-Smiths, 2005). Phytoremediation uses the sun's energy to cleanse contaminated soil and is therefore inherently cheaper than other remediation technologies. Plants contain, sequester, remove, or degrade contaminants as part of their normal physiological processes such as water uptake, inorganic ion uptake, evapotranspiration, root exudation and turnover, and the photosynthetic production of phyto-chemicals. Rhizosphere microbes stimulated by plant roots can biodegrade organic contaminants (Tsao, 2003). The application of these plant-rhizosphere processes has been used to treat contaminated soils, sediments, surface and ground waters. Treatable contaminants include hydrocarbons, chlorinated compounds, and nitroaromatics in the organic contaminant class, as well as heavy-metals, metalloids, radionuclide, and salts in the inorganic contaminant class.

PHYTOEXTRACTION

Inorganic soil contaminants can be phytoextracted from the soil subsurface through repeated cropping and safe storage of the harvested plant biomass following accumulation of metals in the plant aerial tissues. Incineration of the biomass can reduce up to 95% the material to be disposed of in landfills (Robinson et al., 1998) (Figure 2).

Phytoextraction can be applied to the soil removal either of those elements that are essential (e.g., B, Cl, Co, Cu, Fe, Mn, Mo, Ni, Zn) or non-essential (e.g., Cd, Pb, Cr, Hg and As) to plants (McGrath, 1998). Plant species suitable for phytoextraction can be divided in two classes according to metal accumulation capacity and biomass production. The first class is a group of plants known as metal hyperaccumulators. By definition, hyperaccumulator plant species are those that can accumulate metals to a concentration at least 100 times higher than "normal" plants growing in the same environment (Robinson et al., 1998).

Plants that fall in this category can accumulate metals up to greater than 1% of the total plant dry weight (e.g the Zn hyperaccumulator *Thlaspi caerulensces*) but often do not produce high annual biomass (Robinson et al., 1998). So far, the metal hyperaccumulation trait has been observed for As, Ni, Cd, Co, Mn, Se, Tl, and Zn (Robinson et al., 1998; Ma et al., 2001) in more than 400 plant species. However, there are no reports of plants that can naturally hyperaccumulate Pb, Hg, Au and Ag.

A clear advantage of phytoextraction is the generation of metal-rich biomass that can be recyclable. This approach has been successfully used to recover Ni from the biomass of the South-African Ni-hyperaccumulator *Berkheya coddii* grown in a Ni-contaminated soil (Moreno and Sígolo, 2007). Incineration of the metal-rich biomass could reduce substantially the volume and mass of contaminant that would need to go to landfills.

Conventional approaches to the reclamation of metal contaminated sites (removal, isolation, and incineration) are expensive and vary as a function of the soil density, regional transportation and landfill costs. It is estimated that the overall cost for removing one meter of soil from a one-hectare site ranges between US$ 1.5 to 6.2 million. In comparison, the costs for site preparation, planting, and harvesting of plant material for the same hectare site drop to between US$ 5000 to US$ 125000 (McIntyre, 2003).

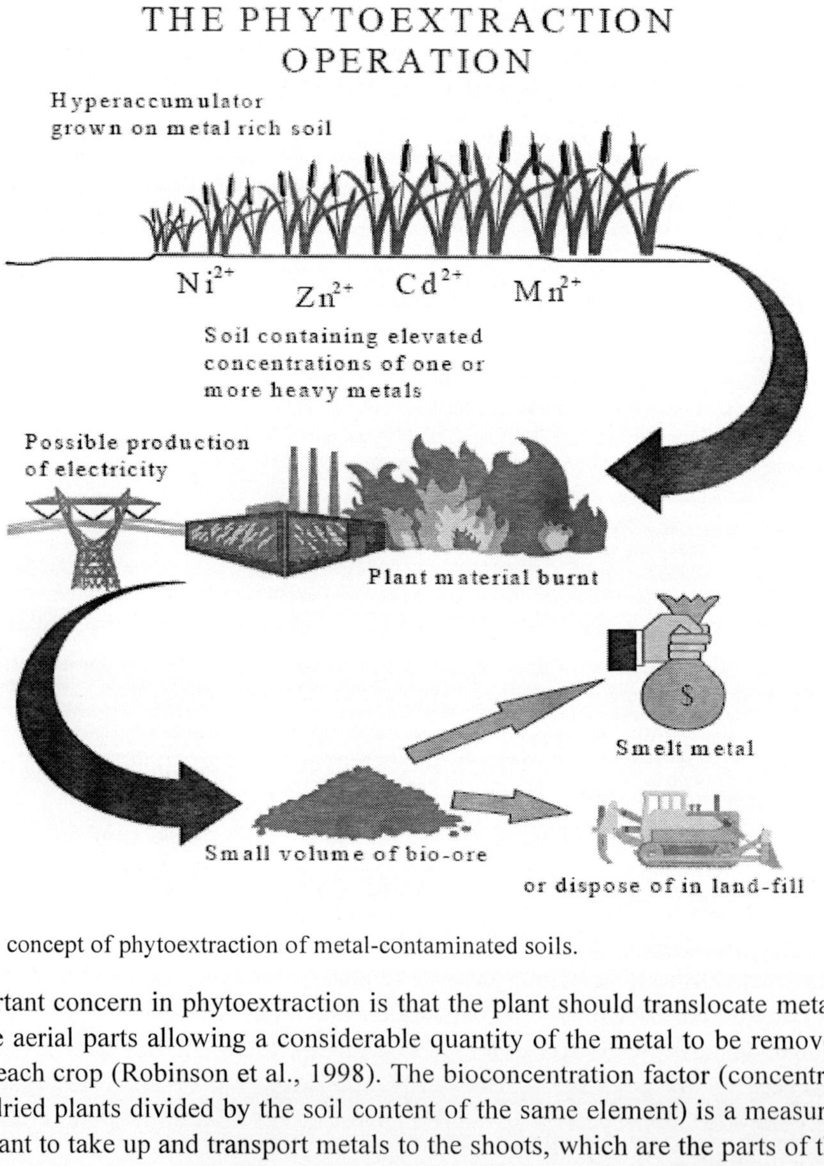

Figure 2. Basic concept of phytoextraction of metal-contaminated soils.

An important concern in phytoextraction is that the plant should translocate metals from the soil to the aerial parts allowing a considerable quantity of the metal to be removed from the soil with each crop (Robinson et al., 1998). The bioconcentration factor (concentration of the metal in dried plants divided by the soil content of the same element) is a measure of the ability of a plant to take up and transport metals to the shoots, which are the parts of the plant that can be harvested. Most plants have bioconcentration factors for metals of less than 1. Modeling has shown that for these plants, the number of crops required to reduce the concentration of the metal in the top soil by half is in excess of 100 (McGrath and Zhao, 2003). Phytoextraction is a realistic approach for plants with a bioconcentration factor above 5 and biomass production of at least 10 t/ha.

The limitations for phytoextraction relates thus to the long period required for site clean up, the restricted number of target metals that can be extracted, the lower depth of rooting systems, the low biomass production of most metal hyperaccumulator plant species, the phytotoxicity levels of metals in soils and insufficient knowledge about agronomic and management practices (Van Netel et al., 2007).

Ligand-Induced Phytoextraction

Some metals (e.g., Pb and Hg) are so strongly adsorbed to the solid phase of the soil that only trace concentrations will be available for plant uptake. Ligand-induced phytoextraction has been developed to deal with metal insolubility or with the low plant biomass problems of phytoextraction. In this case, chelates such as EDTA (ethylenediaminetetraacetic acid) are amended to soil in order to induce solubility and subsequent plant uptake of insoluble metals (Nowack et al., 2006). By using this approach, Pb has been shown to accumulate in shoot tissues of high biomass plant species (e.g., *Zea mays* and *B. juncea*) by factors of over 250 (Huang et al., 1997; Blaylock et al., 1997). The potential of ligand-induced phytoextraction for soil decontamination has been demonstrated in laboratory, greenhouse and field studies for Pb (Blaylock et al., 2000), Au (Anderson et al., 2005) and As (Tassi et al., 2001).

However, there is a concern about the leaching of metal-ligand complexes to groundwater. EDTA, for instance, is persistent in the environment being a common pollutant in almost all natural waters (Nowack et al., 2006). The implications of ligand-enhanced phytoextraction for groundwater pollution has been examined by Lombi et al. (2001), who found that EDTA complexes of Cu, Ni, Zn, Cd, and Pb metals were persistent in soil several weeks after EDTA treatment. Model studies have demonstrated that up to 90% of a chelant that is applied to soil is not taken up by plants and thus, can be leached down to groundwaters (Nowack et al., 2006). Biodegradable chelants such as EDDS (ethylenediaminedisuccinic acid) have been proposed as substitutes for EDTA in ligand-induced phytoextraction as metal-EDDS complexes have been shown to be easily biodegradable in soils (Tandy et al., 2006).

Substrate Types

Mercury-contaminated mine wastes from three different locations were investigated: the Tui base-metal mine, in the North Island of New Zealand; the Gold Mountain mine (GM), in North-Central China; and the Serra Pelada artisanal mine site (SP), in Northern Brazil. Selected geochemical characteristics of these substrates are presented in Table 1.

The Tui Base Metal Mine

The Tui base-metal mine is located on the NW flank on Mount Te Aroha, approximately 3 km north east of the township of Te Aroha, North Island of New Zealand (Figure 3). The cessation of activities in 1974 left a tailings dam containing 100,000 m^3 of sulphide-rich tailings. Metal-bearing minerals present in the tailings include sphalerite (ZnS), galena (PbS), chalcopyrite ($CuFeS_2$) and pyrite (FeS_2) with minor amounts of Cd, Ag and Au. The tailings have been have been left in contact with the air since closure of the mine. Oxidation of pyrite and other sulphide-bearing minerals has depressed the pH of surface tailings and increased metal bioavailability. Organic matter levels in this substrate are virtually absent. The nearby Tui stream and the local ground water supply have been severely contaminated by acid mine drainage (AMD) produced on site (Morrell et al., 1996).

(A)

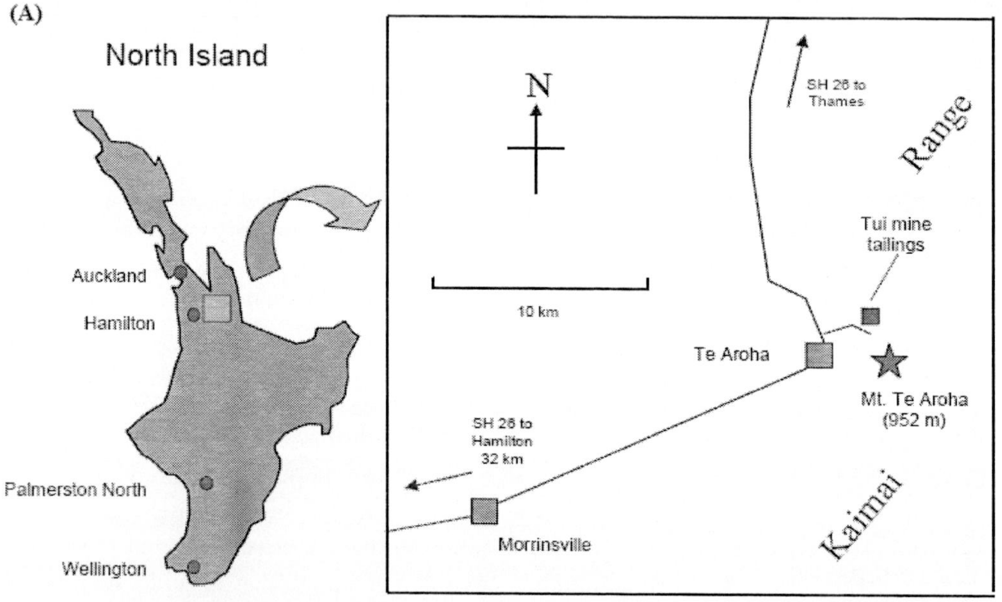

(B)

Figure 3. The Tui mine tailings. The tailings dam is located on the NW flank on Mount Te Aroha, approximately 3 km of the township of Te Aroha, North Island of New Zealand (A). Revegetation trials ongoing on the Tui mine tailings (B).

Table 1. Selected geochemical characteristics of Gold Mountain, Tui and Serra Pelada mine tailings used in this chapter

	Substrate Type		
	Gold Mountain[1]	Tui[2]	Serra Pelada[3]
Total Concentration			
Hg (mg/kg)[4]	100	2.82 ± 0.31	0.27 ± 0.06
Au (mg/kg)	1.58	0.5	0.09
Cu (mg/kg)	9356	na	1338
Mn (mg/kg)	na	na	188.4
Cd (mg/kg)	na	26	na
Zn (mg/kg)	na	5.400	na
Pb (%)	na	1.15	na
Fe (%)	4.48	Na	1.30
Other Characteristics			
PH[5]	9.4	3.9	5.4
Eh (mV)[5]	-137	400	93
Total Carbon (%)[6]	0.7	na	0.1
Total Nitrogen (%)[6]	0.07	na	0.02
OM (%)[6]	1.3	na	0.3
C:N[6]	10	na	5
Soil Volume (g/ml)[6]	1.42	1.5	1.30

[1]Analysis of total Au, Cu, Mn and Fe for the original mine tailings were carried out by ACME Labs, Vancouver, BC, Canada.

[2]Source: Moreno et al., 2004.

[3]Source: Cabral et al., 2002.

[4]Total Hg concentrations in the samples were determined through *aqua-regia* digestions; values are the mean of three replicates.

[5]The pH and Eh values are the mean of three measurements.

[6]Analyses of these soil parameters were carried out by Fertilizer and Lime Research Centre, Palmerston North, NZ.

Na= not analysed.

The Gold Mountain Mine

The Gold Mountain mine is a small-scale mine located in North Central China that extracts Au from ore using the amalgamation method. The exact location of the mine has been omitted so as to protect the local mining community. An aqueous slurry of the Au ore is ground with elemental Hg in locally manufactured mills. Gold liberated during rotation of iron wheels contacts Hg and forms an amalgam that may contain a 1:1 Au/Hg ratio. The Gold Mountain mine processes about 10 to 15 tonnes of ore daily and is responsible for discharge of an estimated 70 tonnes per year of Hg into the regional environment (Gunson and Veiga, 2004).

The Serra Pelada Artisanal Mine

The Serra Pelada mine is located within the Carajás mineral province, approximately 100 km southwest of Marabá in the State of Pará, Brazil (Figure 4).

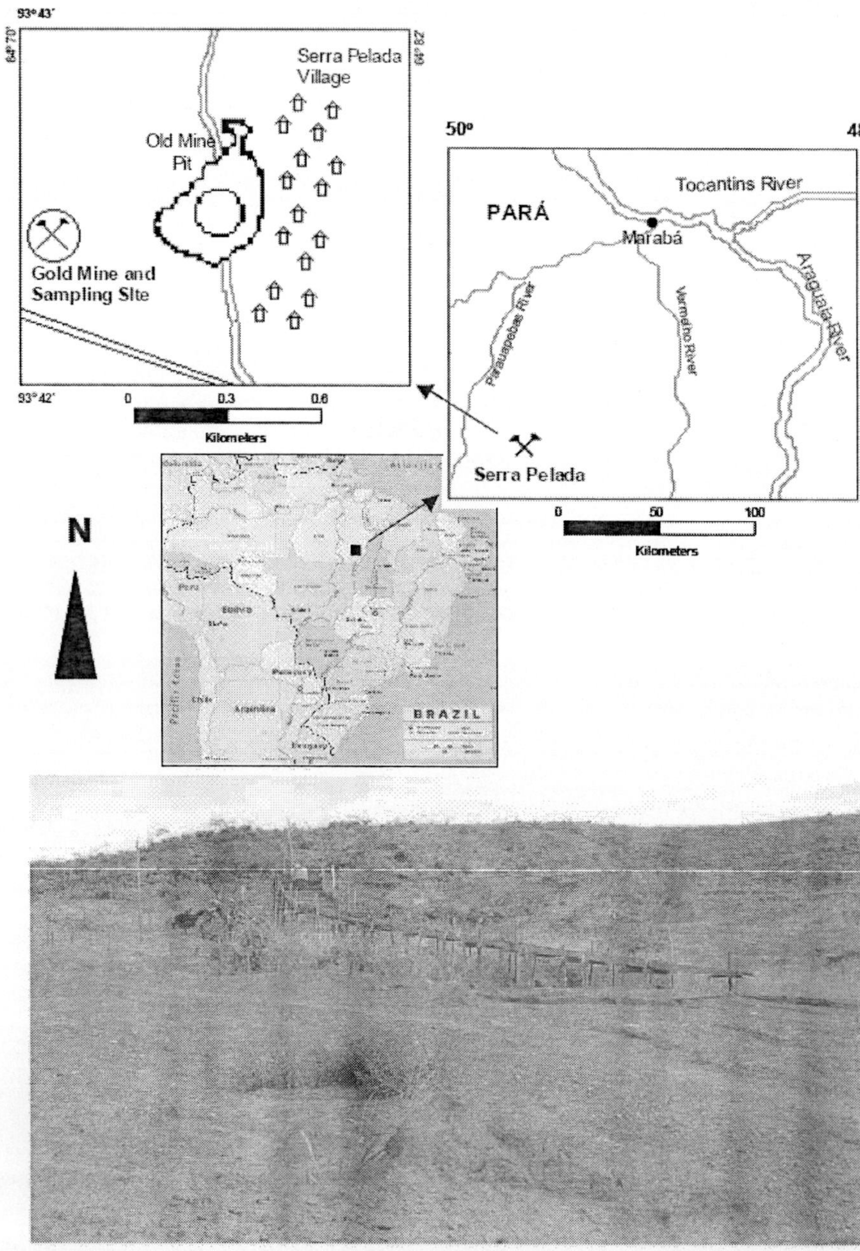

A

B

Figure 4. Location of the sampling site at the Serra Pelada artisanal gold mine, State of Pará, Brazil (A). View of a tailings area within the artisanal gold mine site of Serra Pelada (B). A sluice riffle box used for amalgamation of gold-rich gravels is visible in the background.

The mine attracted 20,000 miners within only a few months after gold was discovered, in January 1980. By 1990, the main pit was flooded and most of the gold miners had moved on. The Serra Pelada mine was responsible for officially producing 43 tons of gold in around 10 years. As the amalgamation method was the preferred method employed by local miners, there are many Hg-polluted spots in the 835ha of the Serra Pelada area. Deforestation is also prominent, as many hectares of native forest were destroyed in order to find exploitable ores. Loss of plant cover increased erosion and this has lead to a reduction in soil fertility. As the mineral resources in Serra Pelada were depleted, the community became increasingly dependant on agricultural activities for subsistence, and additional deforestation and soil erosion have occurred.

MERCURY SPECIATION IN SUBSTRATES

The Tui Base Metal Mine

The high sulphide content (>10%) and the presence of the sulphide-bearing minerals galena (PbS), sphalerite (ZnS), and chalcopyrite ($CuFeS_2$) in the Tui ore indicate that Hg in the tailings might be present as sulphidic Hg forms such as cinnabar, metacinnabar, Hg-polysulphides or Hg associated with Fe-sulphides. Trace quantities of cinnabar have been shown to be present in the Tui ore (Morrel, 1996). These Hg forms are insoluble, not easily altered, and seldom found as detrital material (Kabata-Pendias, 2000). Because of their insolubility, sulphidic forms of Hg are said to be unavailable and/or immobile in the environment.

The Gold Mountain Mine

Given that the original tailings samples exhibited mild reducing conditions (-137 mV), high pH (9.45) (Table 1), the speciation of solid phase Hg in the tailings substrate may be in the elemental Hg (0) form. In this oxidation state, Hg is volatile and thus some Hg may have been lost from the system due to volatilisation. However, it is possible that dissolved Hg species in the tailings were adsorbed to iron oxy-hydroxides, as it has been demonstrated for soils with neutral to alkaline conditions (Andersson, 1979). The fact that the mine tailings has an alkaline pH and contains 17 % Fe provides support to this assumption.

The Serra Pelada Artisanal Mine

Table 1 shows that SP mine tailings exhibit moderately oxidising conditions and pH 5.4. Therefore, Hg dispersed in this substrate during artisanal gold mining was likely to be stable in the elemental Hg (0) form. As the vapour pressure of Hg in the elemental state is high, some Hg fraction might have escaped from the system in the gaseous form. The SP tailings contain 1.15 % of Fe and 188 mg/kg of Mn, probably in the form of goethite (α-FeOOH) and manganese oxides, as shown in Table 1. Therefore, it is likely that another Hg fraction was bound to the solid phase of these Fe and Mn oxide minerals.

Enhancement of Hg Availability in Mine Wastes

The success of phytoextraction is dependent on the availability of the target metal in soil for plant uptake (Blaylock et al., 1997). Mercury has limited solubility in soils, and thus low availability for plant uptake. In general, only trace concentrations of Hg are found in soil solution, mostly as uncharged complexes (Schuster, 1991). Plant availability and uptake of Hg will, therefore, be dependent on the ability to control the processes that enhance the concentration of this element in the soil solution (Moreno et al., 2004).

In this section we describe the results for induced Hg solubility assays with the tested tailings in the presence of different chemical ligands. The effect of humic acid-amended substrates on induced Hg solubility was investigated only for the modified GM mine tailings. This substrate was prepared through dilution of the original GM mine tailings (100 mg/kg) with a 1:1 mixture of coarse and fine silica sand to give final Hg concentrations of 0, 1.25, 2.5 and 5 mg/kg. The humic acid (HA) powder was mixed with the substrate to give (w/w) HA concentrations of 0, 0.125, and 1.25 g/kg of substrate.

Effect of Ligand Type

The coordination chemistry of Hg indicates that this element will be present mostly as a complex in soil solution. Therefore, the partitioning of Hg from the solid phase into soil solution will occur as a consequence of coordinative reactions where Hg ions are exchanged with water molecules for some preferred ligands (Moreno et al., 2004).

Mercury is described as a "soft metal" and has a tendency to form complexes with soft ligands such as NH_3^+, CN^-, Cl^- and S-donors. This affinity can be more pronounced when SCN^- and $S_2O_3^-$ anions are present in the soil solution. The solution geochemistry of metal-thiocyanate complexes favours stable complex formation under moderately acidic and oxidizing conditions (Bowell et al., 1993). Conversely, stable Hg-thiosulphate complexes are likely to form in neutral to alkaline pH conditions (Wilkinson et al., 1987).

The geochemistry of GM and SP mine tailings (Table 1) indicates that Hg may form stable complexes with SCN^- and $S_2O_3^-$ anions present in the soil solution. The GM tailings is alkaline (pH 9.4) whereas the SP tailings exhibits more acidic (pH 5.4) and oxidizing conditions (Eh 93), thus supporting the formation of stable Hg-thiosulphate and Hg-thiocyanate complexes, respectively. Figure 5 A shows that the increase in the soluble Hg concentration relative to water in these substrates was possibly due to complex formation in the presence of the respective ammonium salts. For the Tui mine tailings, the soluble Hg concentrations were superior in the presence of the ammonium salts of thiosulphate ($[NH_4]_2S_2O_3$) and thiocyanate (NH_4SCN) plus hydrogen peroxide (Figure 5 B). The increased Hg solubility from the Tui mine tailings was likely due to two processes. The first occurred in the presence of $(NH_4)_2S_2O_3$ and likely involved extraction of Hg bound to the solid phase minerals (mostly sulphidic) possibly through the formation of a thiosulphate-Hg complex. For the second process, hydrogen peroxide may have oxidised sulphidic Hg forms, allowing formation of a stable Hg-SCN complex.

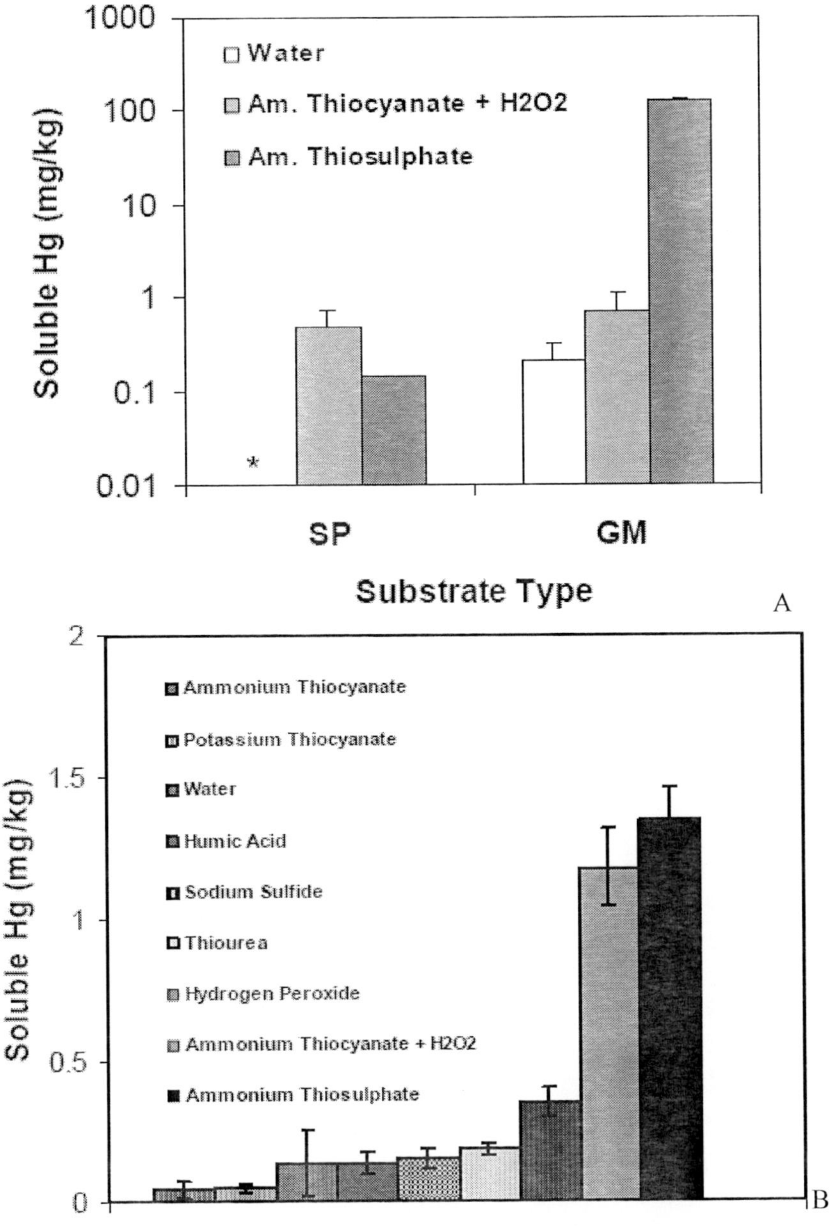

Figure 5. Effect of thioligands (applied at 2 g/L concentration) on soluble Hg concentrations of Serra Pelada (SP) and original Gold Mountain (GM) (A) and Tui (B) substrates. Bars denote ± 1 standard deviation from the mean of three replicates. The symbol (*) indicates Hg below detection limits (5ng/mL in the extract solutions). Ammonium thiocyanate was supplemented with hydrogen peroxide (H_2O_2) at 0.3%.

The kinetics of the formation/decomposition of Hg thiosulphate complexes is unclear. However, Bowell et al. (1993) modelled the stability fields for the gold thiosulphate system in a pH-Eh diagram. These authors demonstrated that the formation of a stable Au-thiosulphate complex would be expected in the pH range of 3 to 12. As both Hg and Au exhibits similar geochemistry, then is possible that the acidic conditions that prevail in the Tui mine tailings were not restrictive to the formation of a stable Hg-thiosulphate complex.

Effect of Humic Acids

Humic substances (HS) represent 50% of the natural organic matter in soils and contain a high proportion of sulphur-containing functional groups (Wallchläger et al., 1998 a). The soluble fraction of HS comprises fulvic and humic acids (HA), which are known Hg ligands. Since Hg complexes with humic substances are stable over the entire pH range from 1-14 (Wallchläger, 1996), these complexes can be abundant in the solution of organic and mineral rich horizon soils (Evans, 1986). Therefore, Hg-HA complexes are mobile in soils (Wallchläger et al., 1998 b) and HA have been demonstrated to enhance both the Hg availability in soils and Hg uptake by organisms (Veiga and Hinton, 2002).

The effect of Ha on the solubility of Hg in the GM mine tailings is shown in Figure 6. The Hg concentration in the extracts was almost two fold higher in the presence of $(NH_4)_2S_2O_3$ relative to water and averaged over 1 mg/kg. The Hg concentrations increased significantly with increasing Hg concentration in the water-soluble fraction of the HA-amended substrates. Conversely, humic acids had no significant effect in the Hg solubility in the presence of the thiosulphate ion in the soil solution.

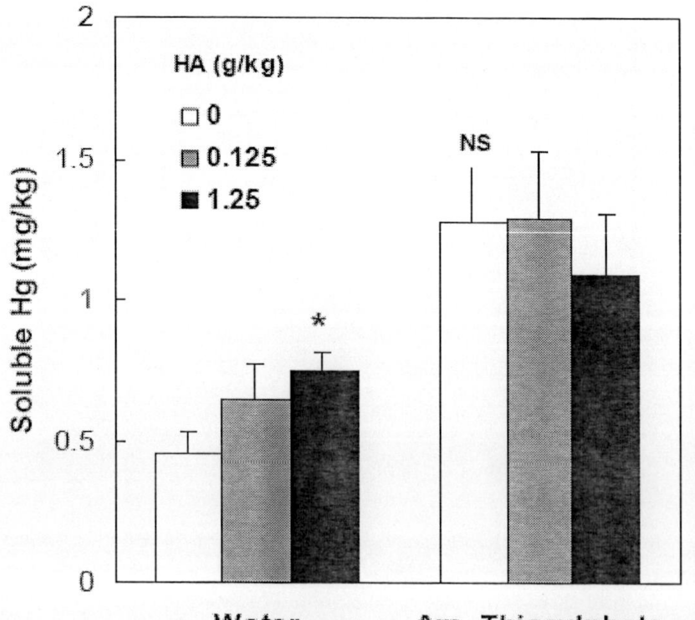

Figure 6. Effect of humic acids (HA) on soluble Hg concentrations for modified GM substrates containing 2.5 mg/kg of Hg. The substrate was treated with 2 g/L of ammonium thiosulphate. Bars denote ± 1 standard deviation from the mean of three replicates. The symbol (*) indicates statistically significant ($p < 0.05$). NS = not significant ($p > 0.05$). Water = control.

The enhanced Hg solubility in the water extracts for HA-amended substrates can only be explained through the formation of Hg-HA complexes. The fact that the pH of the modified GM substrates was around 8 and that humic acid is soluble at alkaline conditions (Wallschläger et al., 1996) provides support for this statement. Considering that soluble Hg-HA complexes were present in both water and $(NH_4)_2S_2O_3$ soluble fractions (Figure 6), we hypothesise that the total soluble Hg fraction of HA-amended substrates comprises a mixture of both Hg-HA and Hg-thiosulphate complexes. This result will have strong implications for the uptake of Hg by plants during the course of phytoextraction.

INDUCED PLANT-HG ACCUMULATION

Metal transport from the soil solution to plant roots can occur by diffusion as well as by convection with the transpiration water stream. Therefore, increasing the dissolved metal concentration by adding ligands or chelants to soils can enhance the diffusive and convective flux toward roots. Metals can be then delivered to the plant shoots via the xylem through the apoplastic pathway, an interconnected system that includes all cell walls in the root cortex (Nowack et al., 2006). The formation of Hg-complexes in the soil solution is thus, a prerequisite to induce the accumulation of Hg in aerial plant tissues. However, there is still a debate whether metal complexes are taken into the plant or not. It is believed that metal-complexes disrupt cells in the root cortex allowing the free passage of metal complexes into the xylem vessels, therefore facilitating shoot metal accumulation (Nowack et al., 2006).

In this section, we show the role of plant species and sulphur-containing ligands, substrate Hg concentration, substrate type and humic acids in the induced plan-Hg accumulation process. For plant growth, all substrates were supplemented with Osmocote (slow release fertiliser) at 5 g/kg. The acidic pH of Tui mine tailings was adjusted to 5.5 by addition of lime. Due to the phytotoxic Hg concentration (100 mg/kg) of the original GM mine tailings, the plant experiments on this substrate were carried out on a diluted substrate containing Hg at 1.25, 2.5 and 5 mg/kg, as explained before. Sulphur-containing ligands were applied at 2 g/kg rate in the form of solutions. All experiments were run for five days under greenhouse conditions.

Effect of Sulphur-Containing Ligands and Plant Species

Average background levels of mercury in plants are usually not superior to 0.1 mg/kg dry weight (Kabata-Pendias, 2000). By adding $(NH_4)_2$ S_2O_3 to Tui mine tailings, *B. juncea* was able to accumulate over 400 times this value in the aerial tissues, as shown in Figure 7 A. Recorded values in aerial tissues of *B. juncea* averaged 43 mg/kg, thus reaching a bioconcentration factor around 15 (Plant Hg 43/ Soil Hg 2.8, values in mg/kg). Induced plant-Hg accumulation for plant species grown in the modified GM mine tailings (Hg at 2.5 mg/kg) was also effective in the presence of thiosulphate salts (ammonium and sodium) (Figure 7 B). However, shoot Hg values were not as pronounced as those observed for *B. juncea* grown in the Tui substrate.

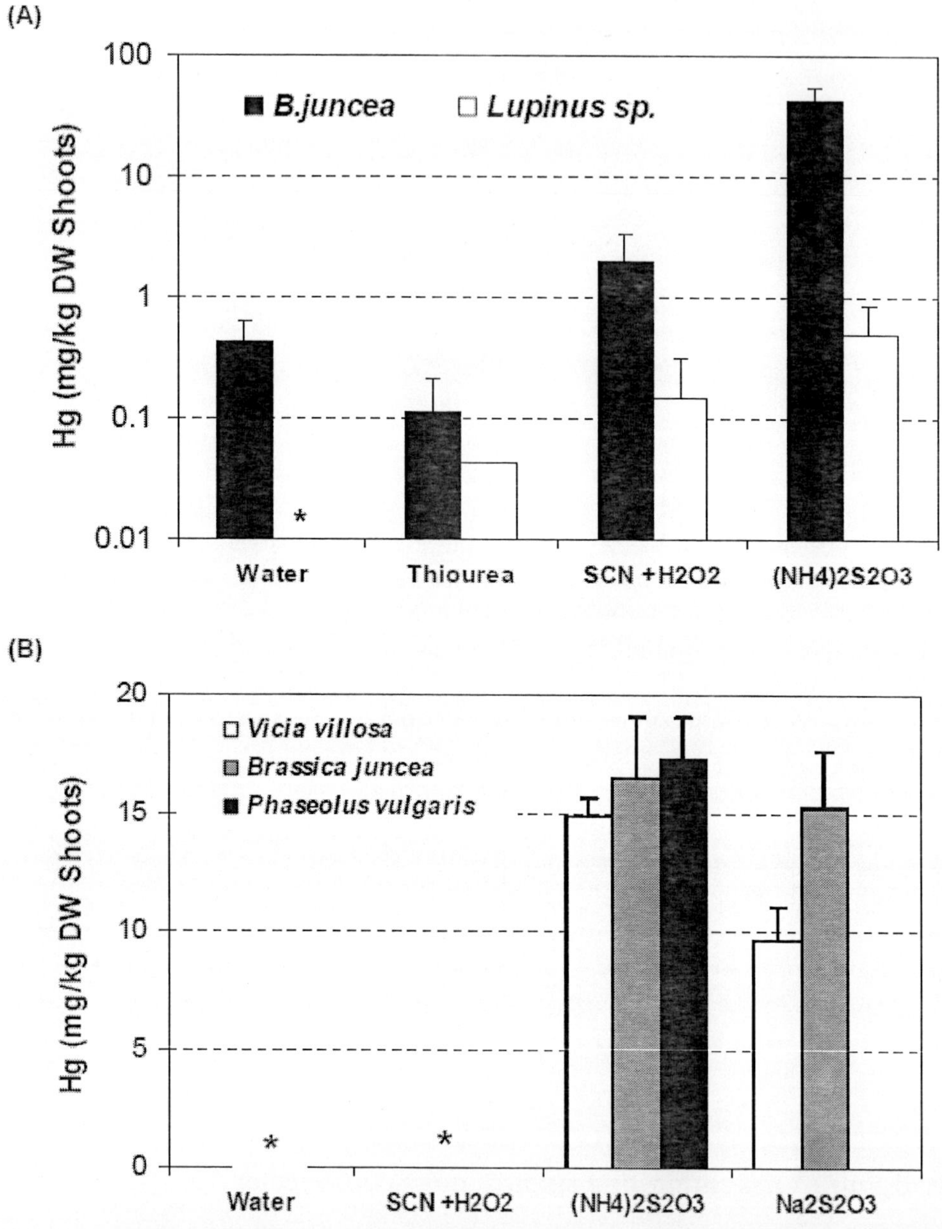

Figure 7. Shoot Hg concentrations in different plant species after application of sulphur-containing ligands at 2g/kg to Tui (a) modified GM mine tailings (b). Note that Water = control, SCN + H_2O_2 = ammonium thiocyanate + hydrogen peroxide (at 0.27%); $(NH_4)_2S_2O_3$ = ammonium thiosulphate. Bars denote ± 1 standard deviation from the mean of 5 replicates. The symbol (*) indicates Hg below detection levels (5 ng/mL in the plant digests). DW = dry weight.

Overall, plants species grown in the modified GM substrate accumulated over 15 mg/kg in shoot tissues in the presence of the ammonium salt, hence yielding a bioconcentration factor around 6. The thiosulphate-induced plant-Hg accumulation process was, therefore, effective both under the moderately acidic (pH 5.5) and the slightly alkaline (pH 8.2) conditions existing in the Tui and modified GM substrates, respectively.

Application of other sulphur-containing ligands, such as thiourea or thiocyanate, to the Tui or GM substrates did not induce Hg levels to the same extent. For the modified GM tailings, Hg shoot accumulation was below detection levels in the presence of other ligands (Figure 7 A and B).

The translocation of soluble Hg species was not only a function of ligand type but also of plant species. For example, application of thiosulphate to *Lupinus* sp. and *Vicia villosa* grown respectively in Tui and modified GM substrates did not enhance Hg accumulation in aerial tissues as it did for *B.juncea* and *Phaseolus vulgaris* under similar conditions. This discrepancy may result from intrinsic plant species characteristics such as root morphology, root surface area, and plant transpiration rates.

Effect of Substrate Type

In general, Hg has a low solubility regardless of the natural geochemical conditions that prevail in substrates. The application of thiosulphate substantially increased Hg solubility and subsequent shoot accumulation for *B. juncea* grown in Tui and Gold Mountain substrates, as shown in Figure 8. Thiocyanate was also effective for inducing Hg solubility in SP substrates (Figure 5 A) and, to some extent, shoot-Hg accumulation. Although shoot tissues accumulated around 2.3 times more Hg than controls, this difference was not statistically significant. This is possibly due to the low levels of total and hence soluble Hg in the SP substrate (Table 1), which limited the effect of NH_4SCN on induced plant-Hg accumulation. Therefore, it is plausible that thioligand-induced plant-Hg accumulation will work only above a threshold value for total Hg in substrates.

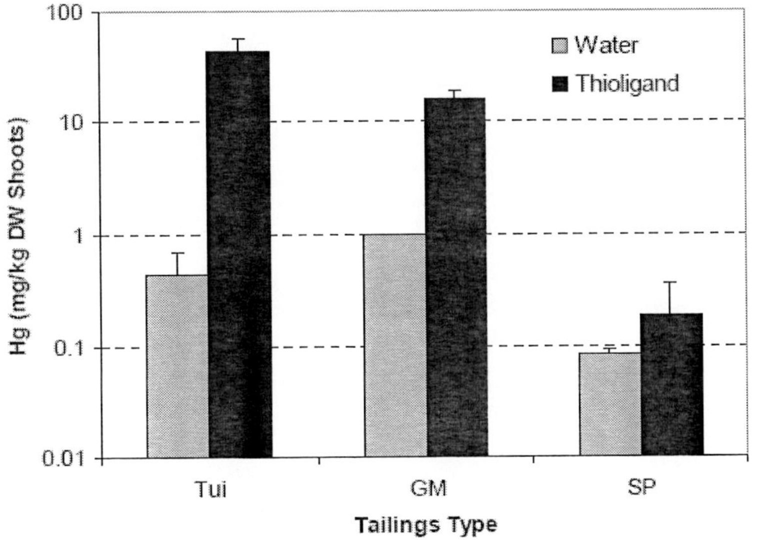

Figure 8. Shoot Hg concentrations for *B. juncea* plants after application of sulphur-containing ligands to mine tailings substrates. Tui and modified Gold Mountain (GM) mine tailings were treated with ammonium thiosulphate at 2 g/kg. Serra Pelada substrates (SP) was treated with ammonium thiocyanate at 2 g/kg supplemented with hydrogen peroxide at 0.3 %. Bars denote ± 1 standard deviation from the mean of at least 3 replicates.

Effect of Substrate Hg Concentration

There is still no evidence of a physiological role for Hg in living organisms. Therefore, Hg is a non-essential element and can be toxic to higher plants. Reduced levels of biomass and plant transpiration rates are among the primary effects of Hg toxicity to aquatic and terrestrial plants (Moreno et al., 2008). As metal removal by plants relies on biomass production and metal-accumulating capacity, these toxic effects of Hg can limit the effectiveness of phytoextraction. Conversely, if soil Hg concentrations are not a limiting factor for plant growth, then increasing soil Hg concentration can increase the Hg content of plants (Figure 9 A).

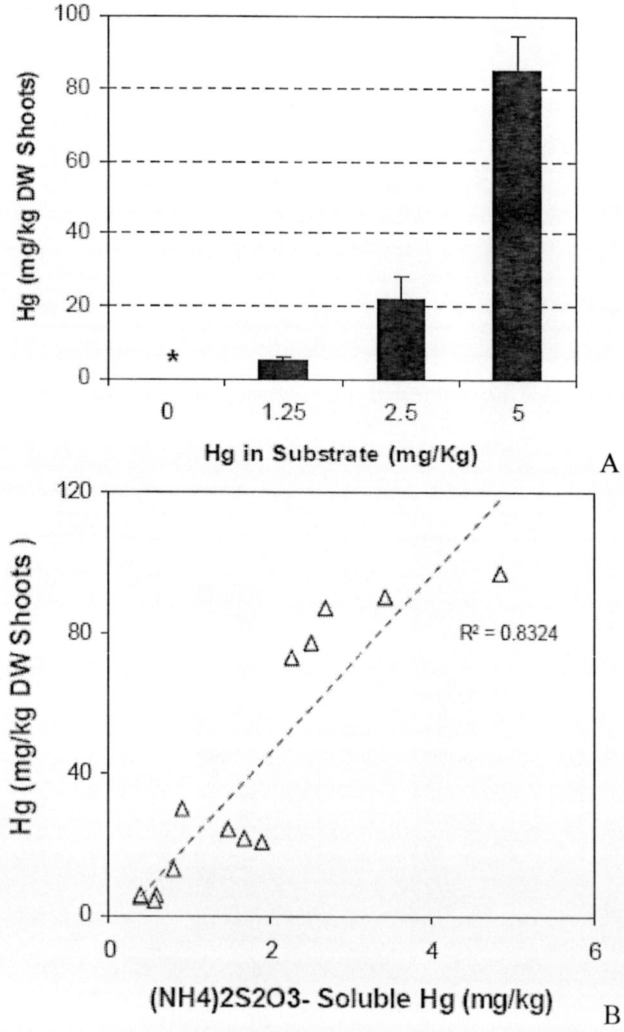

Figure 9. Accumulation of Hg in shoots of $NH_4S_2O_3$-treated *B.juncea* plants as a function of total (A) and $(NH_4)_2S_2O_3$-soluble (B) concentrations in modified GM substrates containing 2.5 mg/kg of Hg. Bars denote ± 1 standard deviation from the mean of five replicates. DW = dry weight, $(NH_4)_2S_2O_3$ = ammonium thiosulphate. An asterisk indicates Hg below detection levels (5 ng/mL in the plant digests).

For example, Figure 9 (A) shows that Hg accumulation in the shoots of $(NH_4)_2S_2O_3$-treated *B. juncea* plants was significantly enhanced as a function of increasing Hg concentrations in the modified GM substrates. At the highest substrate level (5 mg/kg), Hg was accumulated in shoots at an average of 85 mg/kg, hence yielding a concentration factor of 17. The plot of the shoot Hg concentration versus the $(NH_4)_2S_2O_3$-extractable Hg concentration (Figure 9 B) confirms this trend as it shows a positive and linear response of Hg in shoots to the concentration of available Hg in the substrates.

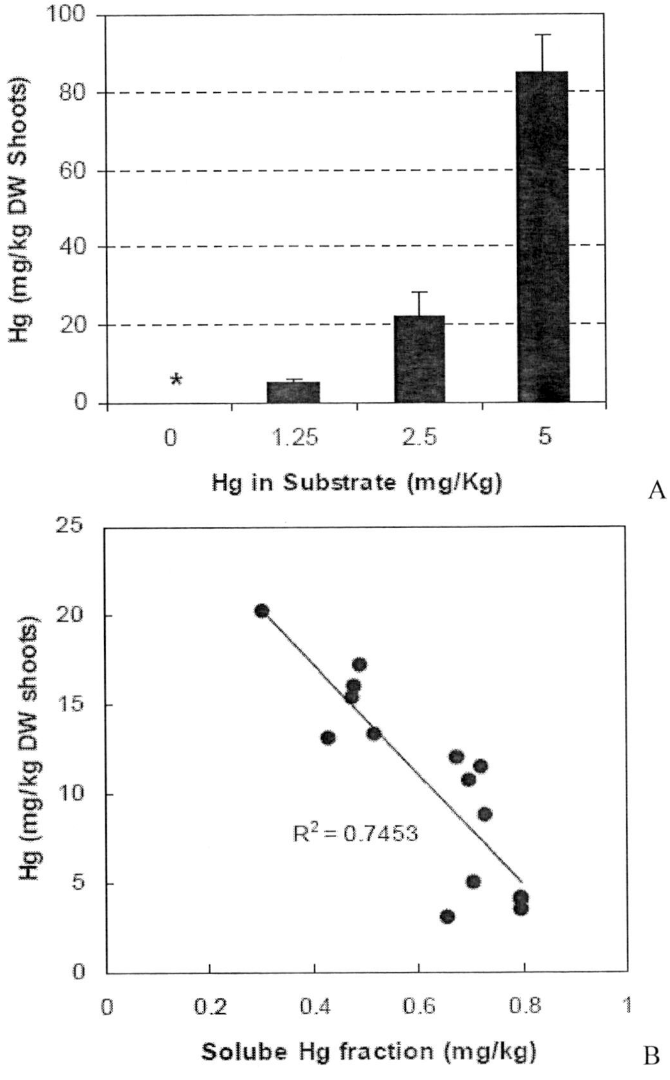

Figure 10. Accumulation of Hg in shoots of $(NH_4)_2S_2O_3$-treated *B.juncea* plants as a function of the humic acid content (g/kg) (A) and the $(NH_4)_2S_2O_3$-soluble Hg fraction (B) for modified GM substrates containing 2.5 mg/kg of Hg. Bars denote ± 1 standard deviation from the mean of five replicates. DW = dry weight, $(NH_4)_2 S_2O_3$ = ammonium thiosulphate.

Effect of Humic Acid

Some studies have shown that organic matter limits the dissociation of Hg bound to soil organic components. This effect translates into low Hg availability in soils and, therefore, into restricted Hg plant uptake as demonstrated by soil and hydroponic studies with *Vulpia myuros* L. (Zorro fescue), *Lemna minor* (Duckweed), *Brassica chinesis* and *Lactuca sativa* (Heeraman *et al.*, 2001, Mo *et al.*, 1989, Wang *et al.*, 1997). The reason for the decline in the plant Hg concentrations was attributed to a decrease in the soluble Hg fraction due to Hg complexation with either organic matter or humic acid. Figure 10 (A) shows that the shoot Hg concentration of *B. juncea* was significantly decreased in the presence of HA thus corroborating with previous studies. Root to shoot transport, therefore, appears to be inhibited by humic acid. These results were confirmed by plotting the plant Hg concentration in *B. juncea* as a function of the soluble Hg concentration of HA-amended substrates, as shown in Figure 10 (B). It is possible that that Hg-S_2O_3 complexes were preferentially selected for shoot transport over other dissolved Hg species present in the substrate such as HA-Hg complexes, which were probably adsorbed by root tissues.

Ligand-Induced Phytoextraction Field Trials

Three field plots with dimensions of 5 x 5 m were established on the tailings dam of the abandoned Tui base-metal mine. The plots were fertilised with NPK fertiliser at 75 g/m^2 and had their pH adjusted to 5.5 by addition of lime. Organic matter (mushroom compost) was added at a rate of 3.2 L/m^2 in order to supplement plant nutritional needs. Around 75 grams of *B. juncea* seeds were planted in two rows. After 6 weeks of plant growth, $Na_2S_2O_3$ was applied to the field plots at a concentration of 5 g/kg of substrate. Two weeks after the treatment, biomass from the plots was harvested and processed for Hg analyses. Results for ligand-induced phytoextraction on the Tui tailings are shown in Figure 11 and Table 2. The addition of $Na_2S_2O_3$ to substrates enhanced Hg solubility, leading to increased Hg accumulation in the shoots of *B. juncea.* The extractable Hg concentration rose proportionally to increasing concentrations of $Na_2S_2O_3$ added to the substrates.

Table 2. Phytoextraction results for the Tui mine tailings after application of sodium thiosulphate ($Na_2S_2O_3$) at 5 g/kg (unless otherwise stated) to Tui tailings field plots[*]

Tui Plot	Shoot Hg (mg/kg DW)	Harvested Biomass (kg, DW)	Equivalent Biomass (t/ ha)[a]	Hg Extraction Yield (g/ ha)[b]
1[c]	2.99 ± 0.5	0.19	0.38	1.14 ± 0.2
2	9.77 ± 1.2	1.24	2.49	24.39 ± 3.1
3	2.93 ± 0.7	1.16	2.33	6.84 ± 1.7

[*]Values are the mean ± 1 standard deviation from 5 replicates (1 replicate = 0.1 g of ground roots and shoots digested in 15 ml HNO_3,).

[a]Calculated on a basis of 5 m^2 area for each plot

[b] Hg values for extraction yields (g /ha) are the product of shoot Hg concentrations (mg/ kg DW) and equivalent plant biomass production (t/ ha).

[c]$Na_2S_2O_3$ applied at 2.5 g/kg.

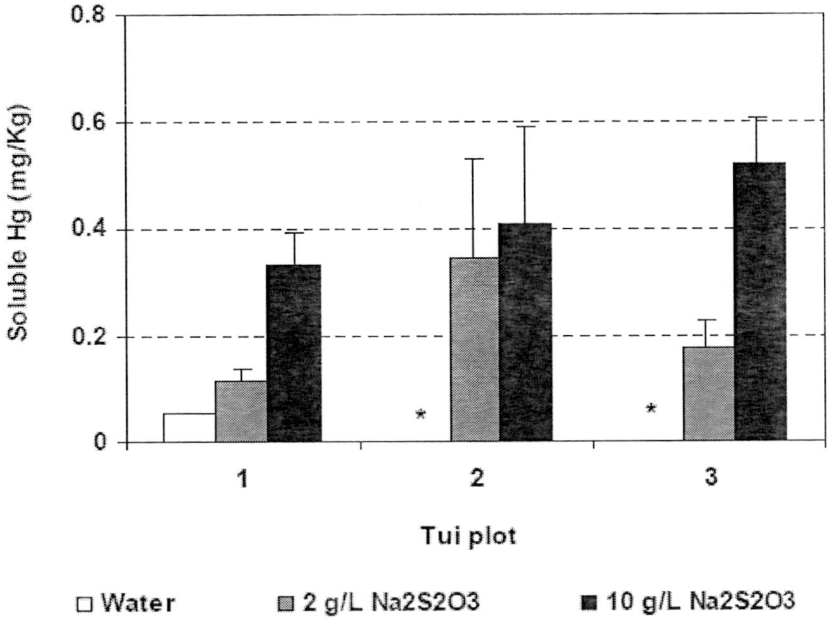

Figure 11. Extractable Hg concentrations as a result of sodium thiosulphate ($Na_2S_2O_3$) application to Tui mine tailings at concentrations of 2 and 10 g/L. Bars denote ± 1 standard deviation from the mean of 3 replicates. An asterisk indicates Hg below detection levels (5 ng/mL in the extract solutions).

The thiosulphate treatment induced a significant increase in the shoot Hg concentration relative to the control treatments, which had Hg values below detection levels (data not shown). Mercury concentrations in shoots of *B. juncea* were higher in plot number 2, where the assayed value averaged around 10 mg/kg (Table 2), thus reaching a bioconcentration factor above 3 (plant Hg 9.77 /soil Hg 2.82, values in mg/kg).

The phytotoxic conditions that prevail in the substrates of the Tui base metal mine did not prevent plant growth, as average plant height on site was around 25 cm (Figure 12). One week after treatment with sodium thisosulphate solutions, however, phytotoxic symptoms (e.g., wilting and chlorosis) were evident. Estimates of biomass production indicated a maximum dry matter yield of around 2.5 tonnes per hectare. Considering that plants on site had a relatively short growing season (6 weeks) and that the sulphide-rich tailings of the Tui mine exhibit high levels of the toxic metals Zn, Cu, Mn, Pb and Ag (Morrell et al., 1996), the low levels of plant biomass production on this site are no surprising. As a result, Hg-extraction yields were also low. Due to its highest biomass production and shoot Hg concentration, the maximum Hg-extraction yield was achieved at plot number 2 with an average of around 25 g Hg/ha (Table 2). The addition of organic matter to Tui substrates might also have played a negative effect on shoot Hg translocation, as highlighted by the shoot Hg concentrations for $(NH_4)_2S_2O_3$-treated *B. juncea* plants grown in the humic acid amended substrates.

However, one should be bear in mind that the maximum Hg-extraction yield obtained in this experiment was over 30 times the maximum yield obtained for *Hordeum vulgare* (barley), which extracted 0.71 g Hg/ha from a Hg-contaminated site in Almadén, Spain (Rodriguez et al., 2003) without the application of synthetic chelates. Overall, ligand-induced phytoextraction increased the Hg removal efficiency by plants grown *in situ* thus encouraging more detailed studies under field conditions.

Figure 12. Experimental field plot at the Tui mine tailings (North Island, NZ) before application of sodium thiosulphate ($Na_2S_2O_3$) to the substrate (March/2004). Biomass for *B. juncea* plants grown at the toxic tailings yielded a maximum of 2.5 tonnes/ha after 6 weeks of planting.

LEACHING

The application of chelants to soils to induce a higher metal concentration in plants raises environmental concerns due to the potential contamination of groundwater via the leaching of mobilised metals. Leaching can occur when the rates of solubilising agents applied to the soil are above the levels of metals that the plants can remove from soil solution (Moreno et al., 2004). This excess metal may leach below the root zone into groundwater, a process exacerbated by macropore flow (Bundt et al., 2000). Modelling calculations for the application of EDTA revealed that 90% of the applied chelant is not taken up by plants during the course of ligand-induced phytoextraction (Nowack et al., 2006). This explains the enhanced generation of Hg-containing leachates after the addition of $(NH_4)_2S_2O_3$ to substrates, as shown in Table 3. The generation of leachates also appeared to be affected by a plant-chemical factor. However, the experimental protocol utilised in the experimental method did not allow for explanation of discrepancies resulting from plant species variability.

Given the potential for leachate generation, ligand-induced phytoextraction may pose an unacceptable environmental risk unless the link to receiving waters can be broken. In this case, the operation could be carried out ex situ. Geo-textiles and subirrigation drainage systems would intercept leachates and pump it back to the irrigation supply, thus preventing leaching and the potential for ground water contamination. Mercury-leachate generation also could be managed by using plant species with high transpiration rates, such as willows and poplars or by intensive cropping practices and careful management of the irrigation system (Moreno et al., 2004). Additional research should be carried out to examine the potential for chemical and biological degradation of Hg-thiosulphate complexes in revegetated mine substrates.

Table 3. Mercury concentrations in leachates (μg/L) collected from plant pots after application of ammonium thiosulphate ($[NH_4]_2S_2O_3$) at 2 g/kg to Tui and modified GM mine tailings[*]

			Treatment	
Substrate	Plant Species	Rep.	Water[a]	$(NH_4)_2S_2O_3$[b]
Tui[c]	*B. Juncea*	2	9.1 ± 11.2	953 ± 246
	Lupinus sp.	3	20.8 ± 6.6	3963 ± 1063
	B. Juncea	3	BDL	2762 ± 506
GM[d]	*P. vulgaris*	3	1.3 ± 0.5	1068 ± 332
	V. villosa	3	BDL	1868 ± 492
GM + HA[e]	*B. juncea*	3	3 ± 3.4	2592 ± 132

[*]Values are the mean ± 1 standard deviation from 3 replicates unless otherwise stated.
[a]Water = control;
[b]$(NH_4)_2S_2O_3$ = ammonium thiosulphate;
[c]Tui mine tailings, 2.82 mg Hg/kg of substrate, Table 1;
[d]GM, modified Gold Mountain mine tailings, 2.5 mg Hg/kg of substrate;
[e]GM + HA, modified GM substrates amended with humic acid at 1.25 g/kg.
BDL = below detection levels (5 ng/mL in the aqueous solutions)

VOLATILISATION

Mass balance calculations for greenhouse experiments investigating induced plant-Hg accumulation revealed a substantial Hg fraction in the substrates that could not be accounted for (Moreno et al., 2004). This fraction suggested that Hg (0) was volatilised from the substrate as a result of chemical and biological transformations. We, therefore, carried out a volatilisation study for ligand-induced plant-Hg accumulation using *B. juncea* plants grown in the Hg-contaminated mine tailings. Each plant pot was enclosed in a gastight acrylic chamber and these were subsequently placed inside a plant controlled light and temperature growht chamber (Figure 13). Volatile Hg was driven by air into two-successive trap solutions containing 5% $KMnO_4$ dissolved in 2N H_2SO_4 The use of this experimental apparatus has allowed Hg recoveries of around 90% for *B. juncea* plants cultured in Hg-spiked solutions (Moreno et al., 2008).

Figure 13. Gastight volatilisation chambers used for trapping volatile Hg released from the plant-substrate system. *B. juncea* plants grown on Tui, modified Gold Mountain (GM) and Serra Pelada (SP) substrates were treated either with $(NH_4)_2S_2O_3$ or NH_4SCN and immediately enclosed within the chambers.

Figure 14 shows the daily Hg volatilisation rates for thioligand and water-treated *B. juncea* plants and control pots (without plants) on a mass basis. Mercury volatilisation was ubiquitous regardless of the substrate type, existence of vegetation, or application of ligands. Overall, the presence of plants considerably enhanced the Hg volatilisation rates over unplanted (barren) mine tailings.

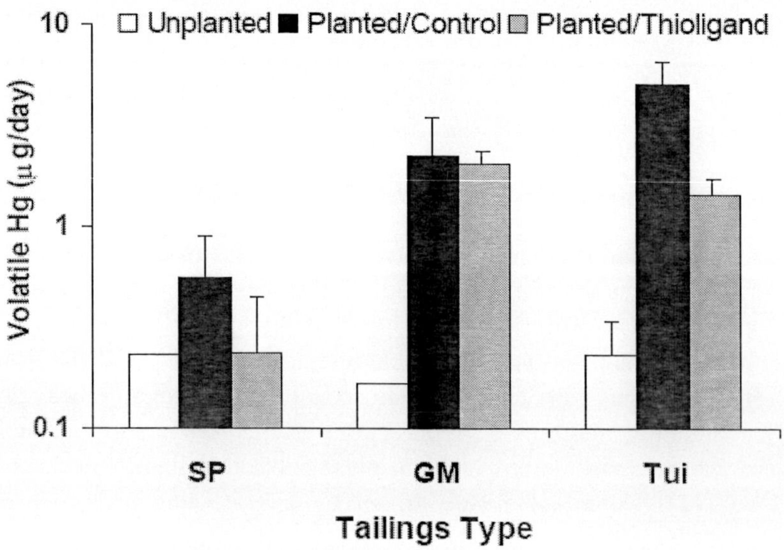

Figure 14. Volatilisation rates from *B. juncea* plants after application of sulphur-containing ligands to mine tailings. Tui and modified Gold mountain (GM) mine tailings were treated with ammonium thiosulphate at 2 g/kg. Serra Pelada substrates (SP) were treated with ammonium thiocyanate at 2 g/kg supplemented with hydrogen peroxide at 0.3%. Water was used as a comparison to thioligand treatments (planted/control). Pots without plants (barren pots) were used as controls. Bars denote ± 1 standard deviation from the mean of 3 replicates.

For Tui and SP mine tailings, Hg volatilisation was more pronounced for water-treated plants. For these substrates, the Hg volatilisation rate exceeded the Hg volatilisation rates from thioligand-treated plants at least by a factor of 2. Volatilisation rates between water and $(NH_4)_2S_2O_3$-treated plants grown in the modified GM substrate, on the other hand, were not statistically different. Application of thioligands did not cause any increase in the Hg volatilisation from substrates and in fact, for two substrates, there was a trend for a reduced Hg flux. A final point to note is that Hg volatilisation from barren substrates indicates that barren mine tailings appear to be inherently prone to biological or chemical transformations.

The quantitative capture of volatile Hg in the potassium permanganate solution under acid conditions can be written by the following equation:

$$2KMnO_{4(aq)} + 3Hg\ (0)_{(g)} + 4H_2SO_{4(aq)} \rightarrow 2MnO_{2(prec.)} + 3HgSO_{4(prec.)} + K_2SO_{4(aq)} + 4H_2O \quad [1]$$

Since elemental Hg (0) is oxidized by potassium permanganate, then we would presume that the predominant Hg form released from barren and planted substrates was the inorganic vapour Hg (0). The Hg volatilisation from tested mine tailings, therefore, may be the result of on site biological and sunlight-mediated reduction of Hg (see Morel et al., 1998 for a detailed explanation on Hg photoreduction).

The role of bacteria on Hg volatilisation from Hg-contaminated environments has been well documented (Barkay et al., 2003). The reduction of Hg (II) to Hg (0) as well as the degradation of organomercury is carried out by Hg (II) resistant bacteria (both gram positive and negative). Bacteria resistant to Hg (II) produce a flavin-containing disulphide oxireductase known as mercuric reductase (MR) that catalyses the reaction:

$$Hg(SR)_2 + NADPH + H^+ \rightarrow Hg\ (0) + NADP^+ + 2RSH \quad [2]$$

The MR removes Hg from stable thiol salts $[Hg(SR)_2]$ by eletrochemical reduction in a NADPH-coupled redox reaction (equation 2). The resulting Hg (0) is rather volatile [Henry's constant (H) = 0.3] and escapes from the organism (Barkay et al., 2003). This strategy has been shown to be an effective solution to bacterial Hg-exposure, as Hg (0) is less toxic than methylmercury compounds (Meagher et al., 2000).

The fact that planted substrates volatilised more Hg than the barren pots (Figure 14) indicates that there is a plant-mediated factor in the volatilisation of Hg from substrates. The enhanced Hg (0) emissions from planted substrates could be the result of biological Hg reduction carried out by Hg (II) resistant bacteria living in the rhizosphere or inside the roots. Experiments with *B. juncea* plants grown in Hg (II)-containing solutions have shown that around 95% of the total Hg mass volatilised from the plant originated from the root compartment, thus implicating the role of rhizosphere processes in Hg volatilisation (Moreno et al., 2008). Alternatively, increased soil moisture due to air-H_2O emissions from plants inside the gas chamber could have enhanced the photoreduction of Hg (II) from the substrate surface (Gustin et al., 2004). Further experiments should be carried out to assess the contribution of biological and photochemical transformations to the volatilisation of Hg from planted substrates.

All tested substrates showed some degree of Hg (0) volatilisation with and without the presence of plants suggesting that this process is likely to happen spontaneously from any barren, revegetated, or phytoremediated Hg-contaminated mine site. Although important from an environmental point of view, the scaling up of the volatilisation data to a field-scale scenario would be unrealistic as Hg (0) emissions may be related to biotic/abiotic factors that could not be controlled under the conditions utilised in the experimental method. The significance of Hg (0) emissions from a phytoremediation operation would depend, therefore, on the elucidation of the biological/chemical pathways by which Hg (II) is reduced in substrates and how plant roots affect this process.

Bacterial resistance to inorganic and organic mercury compounds is found in a wide diversity of Gram-positive and Gram-negative bacteria (Barkay et al., 2003). This fact has motivated the University of Georgia (USA) to conduct research on the development of transgenic Hg-tolerant plants able to volatilise Hg (0) to the atmosphere. At the present time, the microbial *mer*A and the *mer*B genes, expressing mercuric ion reductase and organomercury lyase enzymes respectively, have been inserted into herbaceous (Brassica sp. and Tobacco), trees (Poplar trees), and wetland (Rice) species (Meagher and Heaton., 2005). These Hg-resistant plants are, therefore, capable of phytoremediating soils and sediments polluted with inorganic and organic Hg species. Two issues that may retard the large-scale application of transgenic plant-based systems is negative public perception of the release of genetically modified plants into the environment and concerns regarding the release of plant-Hg (0) vapours to the atmosphere.

PHYTOSTABILISATION

The ability of plants to precipitate or immobilise metals in the organic fraction of soils, onto root surfaces, or within root tissues, can be harnessed to prevent the migration of contaminants off site or into ground water systems. This latter application is referred to as phytostabilisation (Tsao, 2003). Its primary purpose is to reduce metal mobility and bioavailability hence minimising the associated ecological (food chain entrance) and environmental (metal distribution to other compartments) risks. Phytostabilisation can occur through sorption, precipitation, complexation, or metal valence reduction. For instance, the enzymatic reduction of toxic and more soluble Cr (VI) to inert and less soluble Cr (III) has been demonstrated to occur in the fine lateral roots of *Eichornia crassipes* (Lytle et al., 1998).

Phytotoxicity levels of metals in substrates may limit plant growth development thus requiring the application of chemical-immobilising agents. Besides improving soil physical properties, water holding capacity and water infiltration, the application of peat, biosolids and other organic amendments can increase cation exchange capacity, provide a source of carbon and protons and stimulate microbial activity. Reduction of metal bioavailability is achieved by adsorption, complexation and redox reactions thus allowing development of plant covers (Adriano et al., 2004; Moreno and Sígolo, 2007).

The main advantages of phytostabilisation are the following: (1) it eliminates the necessity of toxic material/biomass disposal; (2) it is an effective system to the natural dispersal of wind blow dust or water borne wastes. Additionally, soil erosion can be prevented as root systems aggregate soil into large clumps. Organic carbon in barren soils can be built up as a result of root exudation and the deposition of organic material that sloughs off from root tissues, thus enhancing soil biodiversity. The limitations of phytostabilisation include: contaminant remaining in the site, application of extensive fertiliser or chemical amendments, mandatory monitoring is required and the immobilisation of contaminants may be largely due to chemical amendments (Adriano et al., 2004).

Mercury Retention by the Root System

Successful phytostabilisation of Hg-contaminated substrates demands the use of plants with efficient root-Hg accumulation capacity and low Hg translocation to the shoot (Wang, 2005). Our studies show that untreated plants grown in mine tailings retained Hg preferentially in root tissues with minimal translocation to shoots, regardless of the substrate type and tested plant species. In fact, the analytical equipment did not detect Hg accumulation in shoots for most tested species (Table 4). In addition, the accumulation of Hg by roots was enhanced in the presence of both low and high levels of humic acids, as shown for *B. juncea* plants grown in the humic-acid amended modified GM substrate. Further experiments showed that the root-Hg accumulation pattern was positive and linearly correlated to the concentration of Hg in the water-soluble phase of humic–acid amended substrates (Figure 15). These results, consequently, indicate the strong affinity of roots for Hg, whose accumulated fraction may contain Hg adsorbed onto and/or taken up (absorbed) into root cells.

Table 4. Plant-Hg concentrations for untreated plant species grown in Tui, modified Gold Mountain (GM) and Serra Pelada (SP) mine tailings for 5 weeks. Values are mean of 5 replicates (unless otherwise stated) ± 1 standard deviation.

			Plant Hg (mg/kg)	
Substrate	Plant Species	Rep.	Root	Shoot
Tui[c]	*B. Juncea*	5	18.9 ± 15	0.4 ± 0.2
	Lupinus sp.	5	1.05 ± 0.7	BDL
	B. Juncea	5	9.8 ± 1.7	BDL
GM[d]	*P. vulgaris*	5	4.2 ± 0.3	BDL
	V. villosa	5	5.5 ± 0.8	BDL
GM + HA 0.125[a]	*B. juncea*	5	6.5 ± 0.8	BDL
GM + HA 1.25[b]	*B. juncea*	5	13.7 ± 1.1	BDL

[a]Modified GM mine tailings amended with humic acid at 0.125 g/kg;
[b]Modified GM mine tailings amended with humic acid at 1.25 g/kg;
BDL = below detection leveles (5 ng/mL in the plant digests).

The results for plant species grown in the contaminated mine tailings corroborate with previous experiments with *B.juncea* plants exposed to Hg at concentrations ranging from 0.05 to 10 mg/L in hydroponics solutions, where almost all Hg is bioavailable (Moreno et al., 2008). For all solution concentrations, roots accounted for 98–99 % of the total plant Hg, thus demonstrating the ability of plant roots to retain Hg and control its transport to shoots. Cavalinni et al. (1999) demonstrated that most Hg taken up by roots of *Triticum durum* cv.

Creso after 25 days exposure to ^{203}Hg-spiked solutions was found on cell walls at the external layers of the cortical cylinder. The binding of Hg to specific sites on the cell wall can restrict transport to sensitive sites in shoots and was suggested, therefore, as an important factor to Hg tolerance in plants (Cavallini et al., 1999).

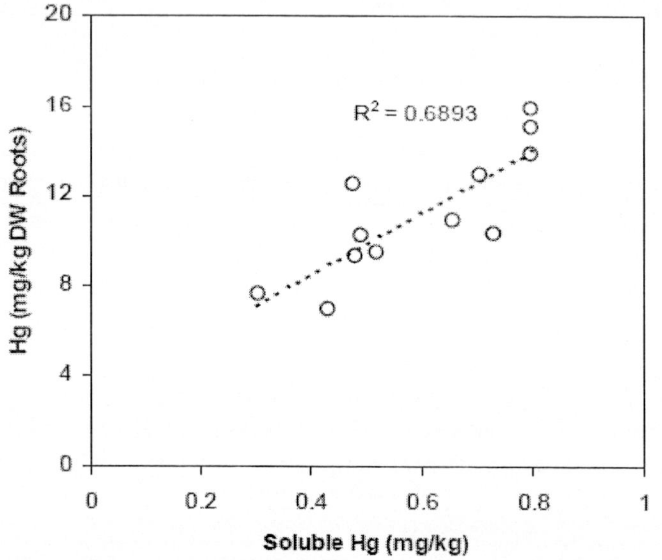

Figure 15. Accumulation of Hg in roots of untreated *B. juncea* plants as a function of the water-soluble phase in Gold Mountain substrate amended with humic acid at 1.25 g/kg containing 2.5 mg/kg of Hg. DW = dry weight.

Although preliminary, the results of this investigation provide conspicuous evidence that the root-Hg accumulation process may be used to phytostabilise Hg in contaminated mine tailings where phytoextraction is not possible or feasible. Further research should be carried out to investigate the role of plant roots and organic amendments on the reduction of the Hg bioavailable and mobile fractions in the soil profile.

CONCLUSION

Laboratory, greenhouse and field results from this research revealed that plants can remove Hg from mine tailings through ligand-induced phytoextraction. Providing that metal levels in substrates are non-phytotoxic and support biomass growth, this simple technique would allow for off-site transport and safe disposal of this contaminant. Although greenhouse studies indicated that single *B. juncea* crops could be induced to extract between 500 to 850 g of Hg per hectare (10 tonnes/ha x 50–85 mg of Hg/kg of shoots DW), the maximum yield for this species in the Tui field was around 25 g Hg/ha. To be economically feasible, phytoextraction yields at least 20 times higher would be needed to reduce considerably soil Hg concentrations. In order to maximise plant-Hg uptake, attention should be given to the ligand-irrigation system and plant water demands. The possibility of leaching soluble Hg species after ligand application requires such an operation to be performed ex situ where measures (e.g., geomembrane lining) can be implemented to avoid groundwater contamination. Risk analysis should be performed to quantify the risks of shoot-accumulated Hg contaminating the food chain.

Phytoremediation offers advantages for the management of Hg-polluted mining areas. The establishment of a vegetative cover is a better option than inaction, as Hg (0) is spontaneously emitted from any barren or naturally-revegetated mine site, and AMD can severely contaminate local surface and groundwater supplies. Phytostabilisation could be used in situ to prevent Hg flow to other compartments. Plant roots would stabilise the substrate, reduce the Hg bioavailable fraction and Hg leaching. Ligand-induced phytoextraction could be used ex situ to remove Hg from substrates and to generate a low volume disposal residue. After reaching acceptable soil Hg levels, the substrate would allow new land uses. Additionally, a plant-based remedial system could be coupled to profit-making operations (forestry, carbon credits generation or biomass energy programs).

In spite of its apparent limitations, there are some examples of successful commercial applications of phytoextraction for Pb- and As-contaminated soils (Edenspace, 2008). The scaling up of ligand-induced Hg phytoextraction or phytostabilisation to a commercial scale, however, will require time and multidisciplinary research efforts. This would include the contribution of expertise from a wide array of disciplines, such as soil science, geology, genetics, biochemistry, microbiology, botany, environmental engineering, and human health. It is imperative that scientific knowledge be disseminated to the public in comprehensible form, as phytoremediation is strongly dependant upon their support.

Acknowledgements

We gratefully acknowledge the National Council for Scientific and Technological Development (CNPq/Brazil) for providing a post-doctoral scholarship to the first author. The authors are also in debt to lab and field assistance provided by the technical staff of Massey University, especially to Mr. Bob Toes and to Mr. Ian Fuckert.

References

Adriano, D.C., Wenzel, W.W., Vangronsveld, J. and Bolan, N.S. (2004). Role of assisted natural remediation in environmental clean up. *Geoderma* , *122,* 121-142.

Anderson, C., Moreno, F. and Meech, J. (2005). A field demonstration of gold phytoextraction technology. *Minerals Engineering, 18*, 385-392.

Andersson, A. (1979). Mercury in soils. In: J.O. Nriagu (Ed.), *The Biogechemistry of Mercury in the Environment* (pp. 79-106).Amsterdan, The Netherlands: Elsevier Biomedical Press.

Barkay, T., Miller, S.M. and Summers, A. O. (2003). Bacterial mercury resistance from atoms to ecosystems. *FESM Microbiology Reviews, 27*, 355-384.

Blaylock, M.J., Salt, D.E., Dushenkov, S., Zakharova, O., Gussman, C., Kapulnik, Y., Ensley, B. and Raskin, I. (1997). Enhanced accumulation of Pb in indian mustard by soil-applied chelating agents. *Environmental Science and Technology, 31*, 860-865.

Blaylock, M.J. (2000). Field demonstrations of phytoremediation of lead-contaminated soils. In: N. Terry, and G. Bañuelos (Eds.) *Phytoremediation of Contaminated Soil and Water* (pp. 1-12). Florida : Lewis Publishers.

Bowell, R.J., Foster, R.P. and Gize, A.P. (1993). The mobility of gold in tropical rain forest soils. *Economic Geology, 88*, 999-1016.

Bundt, M., Albrecht, A., Froidevaux, P., Blaser, P. and Fluhler, H. 2000. Impact of preferential flow on radionuclide distribution in soil. *Environmental Science and Technology, 34*, 3895–3899.

Cabral, A.R., Lehmann, B., Kwitko, R. and Cravo Costa, C.H. (2002). The Serra Pelada Au-Pd-Pt deposit, Carajás mineral province, Northern Brazil: reconnaissance, mineralogy and chemistry of very high grade palladian gold mineralization. *Economic geology, 97*, 1127-1138.

Cavallini, A., Natali, L., Durante, M. and Maserti, B. (1999). Mercury uptake, distribution and DNA affinity in durum wheat (*Triticum durum* Desf.) plants. *Science of the Total Environment, 233/234*, 119-127.

Evans, L.J. (1989). Chemistry of metal retention by soils. *Environmental Science and Technology, 23(9),* 1046-1056.

Heeraman, D.A., Claassen, V.P. and Zasoski, R.J. (2001). Interaction of lime, organic matter and fertilizer on growth and uptake of arsenic and mercury by Zorro fescue (*Vulpia myuros* L.). *Plant and Soil, 234*, 215- 231.

Hinton, J.J. and Veiga, M.M. (2001). Mercury contaminated sites: a review of remedial solutions. In: *Proceedings of the National Institute of Minamata Disease Forum (NIMD)* (pp 73-84), Minamata City, Japan.

Hylander, L.D. and Meili, M. (2003). 500 years of mercury production: global annual inventory by region until 2000 and associated emissions. *Science of the Total Environment, 304,*13-27.

Gunson, A.J. and Veiga, M..M. (2004). Mercury and artisanal mining in China. *Environmental Practice, 6(2)*, 109-120.

Gustin, M. S., Ericksen, J. A., Schorran, D. E., Johnson, D. W., Lindberg, S. E. and Coleman, J. S. (2004). Application of controlled mesocosms for understanding mercury air-soil-plant exchange. *Environmental Science and Technology, 38*, 6044-6050.

Kabata-Pendias, A. (2000). *Trace Elements in Soils and Plants* (3rd Edition). Florida: CRC Press.

Lytle, C.M., Lytle, F.W., Yang, N., Qian, J-H., Hansen, D., Zayed, A. and Terry, N. (1998). Reduction of Cr(VI) to Cr(III) by wetland plants: potential for in situ heavy metal detoxification. *Environmental Science and Technology, 32,* 3087-3093.

Ma, L.Q., Komar, K.M. and Tu, C. (2001). A fern that accumulates arsenic. *Nature, 409*, 579.

McGrath, S.P. (1998). Phytoextraction for soil remediation. In: R.R Brooks (Ed.) *Plants that Hyperaccumulate Heavy Metals* (pp 261-287). Cambridge, United Kingdom: CAB International.

McGrath, S.P. and Zhao, F.J. (2003). Phytoextraction of metals and metalloids from contaminated soils. *Current Opinion in Biotechnology, 14*, 277-282.

McIntyre, T. (2003). Phytoremediation of heavy metals from soils. In: D.T. Tsao (Ed.) *Phytoremediation* (pp. 97-124). Berlin,Germany:Springer.

Malm, O. (2001). Autismo é associado ao uso excessivo de vacinas. *Ciência Hoje, 30(175),*17-18.

Meagher, R.B., Rugh, C.L. Kandasamy, M.K., Gragson, G. and Wang, N.J. (2000). Engineering phytoremediation of mercury pollution in soil and water using bacterial genes. In N. Terry, and G. Bañuelos (Eds.), *Phytoremediation of Contaminated Soil and Water* (pp 201-219). Florida: Lewis Publisher.

Meagher, R.P. and Heaton, A.C.P. (2005). Strategies for the engineered Phytoremediation of toxic element pollution: mercury and arsenic. *Journal of Industrial Microbiology and Biotechnology,32,* 502-513.

Mo, S.C., Choi, D.S. and Robinson, J.W. (1989). Uptake of mercury from aqueous solutions by duckweed: the effects of pH, copper, and humic acid. *Journal of Environmental Science and Health, A24*, 135-146.

Morel, F.M.M., Kraepiel, A.M.L. and Amyot, M. (1998). The chemical cycle and bioaccumulation of mercury. *Annual Reviews in Ecological Systems, 29,*543-566.

Moreno, F. N., Anderson, C. W. N., Stewart, R. B. and Robinson, B. H. (2004). Phytoremediation of mercury-contaminated mine tailings by induced plant-Hg accumulation. *Environmental Practice, 6(2)*, 165-175.

Moreno, F.N. and Sígolo, J.B. (2007). Fitoestabilização controlada: proposta de processo de revitalização para passivos de areias de fundição. In: E. N. Moeri, D. Rodrigues, and A. Nieter (Eds.) *Áreas Contaminadas : Remediação e Revitalização* (pp.81-99). São Paulo, São Paulo: Signus.

Moreno, F. N., Anderson, C. W. N., Stewart, R. and Robinson, B. H. (2008). Phytofiltration of mercury-contaminated water: plant accumulation and volatilisation aspects. *Environmental and Experimental Botany, 62*, 78-85.

Morrell, W.J., Stewart, R.B., Gregg, P.E.H., Bolan, N.S. and Horne, D. (1996). An assessment of sulfide oxidation in abandoned base-metal tailings, Te Aroha, New Zealand. *Environmental Pollution,* 94(2), 2176-225.

Nowack, B., Schulin R. and Robinson, B. H. (2006). Critical assessment of chelant-enhanced metal phytoextraction. *Environmental Science and Technology, 40 (17)*, 5225-5232.

Oliveira, S.M.B., Melfi, A.J., Fostier, A.H., Forti, M.C., Favaro, D.I.T. and Boulet, R. (2001). Soil as an important sink for mercury in Amazonia. *Water, Air, and Soil Pollution, 126*, 321-337.

Pilon-Smits, E. (2005). Phytoremediation. *Annual Review on Plant Biology, 56*, 15-39.

Robinson, B.H., Leblanc, M., Petit, D., Brooks, R.R., Kirkman, J.H. and Gregg, P.E.H. (1998). The potential of *Thlaspi caerulencens* for phytoremediation of contaminated soils. *Plant and Soil,* 203, 47-56.

Rodriguez, L., Lopez-Bellido, F.J., Carnicer, A. and Alcade-Morano, V. (2003). Phytoremediation of Hg-polluted soils using crop plants. *Fresenius Environmental Bulletin, 12(9)*, 967-971.

Roulet, M., Lucotte, M., Farella, N., Serique, G., Coelho, H., Souza Passos, C.J., De Jesus S. E., Scavone, A. P., Mergler, J. R. D. and Amorin, M. (1999). Effects of recent human colonization on the presence of mercury in Amazonian ecosystems. *Water, Air, and Soil Pollution, 112(3-4)*, 297-313.

Roulet, M., Lucotte, M., Canvel, R., Farella, N., Courcelles, M., Guimarães, D. Mergler, J.R.D. and Amorin, M. (2000). Increase in mercury contamination recorded in lacustrian sediments following deforestation in Central Amazon. *Chemical Ecology, 165(3-4)*, 243-266.

Schuster, E. (1991). The behaviour of mercury in the soil with special emphasis on the complexation and adsorption processes- a review of the literature. *Water, Air, and Soil Pollution, 56*, 667-680.

Tandy, S., Schulin, R. and Nowack, B. (2006). Biodegradation and speciation of residual SS-ethylenediaminedisuccinic acid (EDDS) in soil solution left after soil washing. *Environmental Pollution, 142*, 191-199.

Tsao, D.T. (2003). *Phytoremediation*. Berlin, Germany: Springer.

Tassi, E. and Barbafieri, M. (2004). Phytodec: um projeto europeu para uso da vegetação no gerenciamento de áreas contaminadas. In: E. Moeri, R. Coelho, and A. Marker (Eds.) *Remediação e Revitalização de Áreas Contaminadas* (pp. 159-168). São Paulo, São Paulo: Signus.

Wallschläger, D., Desai, V.M.M. and Wilken, R. (1996). The role of humic substances in the aqueous mobilization of mercury from contaminated floodplain soils. *Water, Air, and Soil Pollution, 90*, 507-520.

Wallschläger, D., Desai, V.M.M., Spengler, M. and Wilken, R. (1998 a). Mercury speciation in floodplain soils and sediments along a contaminated river transect. *Journal of Environmental Quality,* 27,1034-1044.

Wallschläger, D., Desai, V.M.M., Spengler, M., Windmöller, C.C. and Wilken R. (1998 b). How humic substances dominate mercury geochemistry in contaminated floodplain soils and sediments. *Journal of Environmental Quality, 27*, 1044-1057.

Wang, D.Y., Qing, C.L., Guo, T.Y. and Guo, Y.J. (1997). Effects of humic acid on transport and transformation of mercury in soil-plant systems. *Water, Air, and Soil Pollution, 95*, 35-43.

Wang, Y., Stauffer, C., Keller, C. and Greger, M. (2005). Changes in Hg fractionation in soil induced by willow. *Plant and Soil, 275,* 67-75.

Wilkinson, G., Gillard, R. D. and McCleverty, J. A. (1987). *Comprehensive Coordination Chemistry: the synthesis, reactions, properties and applications of coordination compounds* (Volume 2). London,United Kingdom: Pergamon Press.

Van Netel, L., Mertens, J., Oorts, K. and Verheyen, K. (2007). Phytoextraction: how far from practice. *Environmental Pollution, 150*, 34-40.

Veiga, M.M. and Hinton, J.J. (2002). Abandoned artisanal gold mines in the Brazilian Amazon: a legacy of mercury pollution. *Natural Resources Forum, 26,*15-26.

Veiga, M.M. (2004). Mercury pollution: revealing sources and suggesting solutions. *Environmental Practice,* 6(2), 97-98.

Veiga, M.M., Maxson, P.A. and Hylander, L.D. (2006). Origin and consumption of mercury in small-scale gold mining. *Journal of Cleaner Production, 14*, 436-447.

In: Environmental and Regional Air Pollution
Editors: D. Gallo, R. Mancini
ISBN: 978-1-60692-893-6

Chapter 7

THE USE OF ANIMALS AS INDICATORS OF HUMAN MERCURY EXPOSURE

Marta López-Alonso*
Universidade de Santiago de Compostela,
Departamento de Patoloxía Animal,
Facultade de Veterinaria, 27002 Lugo, Spain

ABSTRACT

Mercury is a highly mobile environmental pollutant, and small environmental releases of mercury can result in exposure of animals and human foodstuffs large enough to result in toxic effects. Point discharges, particularly coal burning power stations, but also other industrial plants and emissions from waste incinerators and crematoria, are major sources of environmental mercury and can result in marked spatial variation in mercury deposition and subsequent uptake by biota. Although chemical analysis of air, soil or water provides information on mercury environmental contamination, these analyses alone can be inadequate to assess bioavailability or predict subsequent toxicity to animals and humans. Animals in these environments, however, integrate exposure spatially, temporally and across media, and consequently can be expected to exhibit a wide amplitude of exposure and biological response. In this chapter, the use of different animal species as bioindicators of human mercury exposure are evaluated.

INTRODUCTION

Mercury (Hg) is a highly mobile environmental pollutant (WHO, 1990, 1991), small environmental releases of which can result in exposures of animals and human foodstuffs large enough to result in toxic effects (Sell et al., 1975; Simpson et al., 1997). Methylmercury (MeHg), the common and more toxic environmental form, can biomagnify at successive trophic levels in the aquatic food chain (Fortin et al., 2001). MeHg irreversibly destroys the

---

* Tel: + 34 982 28 59 00; Fax: + 34 982 28 59 40; e-mail: marta.lopez.alonso@usc.es

neurons of the central nervous system, and toxicity in mammals is primarily manifested as central nervous system damage, including sensory and motor deficits and behavioural impairment (Strom, 2008). Initial symptoms of MeHg exposure include anorexia and lethargy, eventually progressing to muscle ataxia, motor control deficits, visual impairment, and, finally, convulsions preceding death. Mammalian MeHg exposure can result in reproductive and developmental disorders, as MeHg is readily transferred across the placenta and concentrates selectively in the foetal brain (Wolfe et al., 1998). Other reproductive effects of MeHg exposure range from developmental alterations in the foetus, which can produce physical and behavioural deficits after birth, and eventually foetal death (Strom, 2008).

Current levels of mercury in the environment are about threefold greater than pre-industrial values (Lamborg et al., 2002; Lindberg et al., 2007); approximately two-thirds of the atmospheric Hg is of human origin, and the remaining amount is from natural sources (Mason et al., 1994; Morel et al., 1998). Anthropogenic Hg in the environment is commonly derived from a variety of sources. The most important man-made sources are point discharges associated with the combustion of fossil fuels, particularly coal-burning power plants (Hg concentrations in coal are 10–10,000 fold higher than in fuel oil and natural gas; Chu and Porcella, 1995; Flores and Martins, 1997; EWG, 1999) and other industrial plants, and emissions from waste incinerators and crematoria (Nriagu and Pacyna, 1988; Mukherjee et al., 2000; Pirrone et al., 2001), chlor-alkali plants (Parks and Hamilton, 1987; Lodenius, 1998) and mining operations (Bonzongo et al., 1996; Lacerda, 1997). Although more general combustion of fossil fuels, such as from domestic sources, results in diffuse inputs of Hg to the atmosphere that subsequently cause widespread low-level contamination (Hoffman et al., 2001), the major contribution of point sources to total anthropogenic emissions can lead to marked spatial variation in Hg deposition (Carpi, 1997; Nadim et al., 2001). Natural sources of Hg input to the atmosphere include forest fires, dust particles, volcanic eruptions, and degassing from water surfaces (Rasmussen, 1994).

Although industrial use of Hg has declined markedly since the 1990s and earlier, global consumption of coal for energy production is predicted to increase 74% between 2004 and 2030 (Energy Information Administration, 2007). In this context, it is increasingly necessary to develop biomonitoring tools that can be used to assess the health of ecosystems and their component parts and to provide early warning of potential harm to human health (Ratcliffe et al., 1996).

A monitoring programme can be defined as a repeated measurement of a variable so that a trend is defined and can be related to a goal or standard, whether or not it is mandatory (O'Brien et al., 1993). Once a programme of monitoring is determined, decisions have to be made about what to measure, where and when to measure, by what method and to what purpose. A monitoring programme can be made by the direct analysis of the pollutant or toxin in soil, air or water samples, or alternatively in animals from these environments. Although chemical analysis of soil, air, and water provide information on contaminant concentrations, these analyses alone can be inadequate to assess bioavailability or predict subsequent toxicity to animals and humans. Animals in these environments, however, integrate exposure spatially, temporally, and across media, and consequently can be expected to exhibit a wide amplitude of exposure and biological response (Talmage and Walton, 1991).

The suitability of many animal species as biomonitors of pollutants in the terrestrial environment has been widely investigated (MARC, 1990; O'Brien et al., 1993) and some selection criteria have been proposed. Where human health is the primary concern, biomonitors should ideally be (1) widely distributed to allow cross-habitat comparisons; (2) territorial and non-migratory so that measured concentrations can be linked to source areas; (3) numerous so that it is feasible to collect sufficient samples to ensure that resultant data are representative of the whole population; (4) sensitive enough to be of use in predicting route and severity of human exposure; (5) near the top of a food chain so that any biomagnification is detected; (6) broadly similar to Man in their physiological and toxicological responses to contaminants; (7) sufficiently large so that adequate tissue samples can be collected for chemical and histopathological study; (8) relatively long-lived so that the effects of long-term chronic exposure can be evaluated (Holden, 1973; Landres at al., 1988; O'Brien et al., 1993; Esselink et al., 1995).

Many animal groups have been hailed as valuable sentinels of Hg exposure in different environments. While wild species (mainly birds and mammals) have been largely used as indicators in their normal habitats, less information is available for livestock in farmland and other rural areas or pets in the domestic environment.

Wildlife Species

Certain wild animals represent excellent models, or sentinels, to address issues related to environmental pollution, as they can provide integrative data on both exposure (i.e., information on the type, amount, and availability of contaminants) and effect (i.e., information on sublethal and clinical health responses) (Scheuhammer et al., 2007).

Terrestrial mammalian wildlife, in particular, possesses multiple characteristics that are favourable to their inclusion in studies concerning environmental and human health. Mammalian wildlife have physiological systems that are similar to those of humans and rodents in mediating the uptake, distribution, metabolism, and elimination of toxicants. Furthermore, humans and many species of wildlife inhabit similar ecosystems and are exposed to common climates, food sources and pollutants. Free-ranging wildlife can integrate ecological factors and real-world complexities (i.e., stresses associated with disease, human disturbance, or temperature) that are not attainable in controlled laboratory bioassays. As such, the toxicological information derived from mammalian wildlife may have more applicability to humans than data collected from non-mammalian models, such as fish, reptiles and birds.

Of the wild mammalian species, wild mink (*Mustela vison*) and river otter (*Lutra canadensis*) have been the most extensively used as biomonitors of environmental Hg. This can be attributed to their widespread distribution, availability of tissues through trappers and top position of the food chain (Fortin et al., 2001). Both species are mostly piscivorous, have a fairly small home range, live close to watersheds, and prefer to catch prey in rivers, streams and ponds (Gilbert and Nancekivel., 1982). A large number of studies have reported Hg concentrations in the tissues of wild mink and river otter showing geographic variations between areas based on degree of contamination and bedrock geology (Bank et al., 2006; Basu et al., 2006; Strom, 2008). Hg concentrations in mink and otter fluctuated in parallel to Hg concentrations in fish, crayfish and bottom sediments of rivers, suggesting that the two wild furbearer species could be useful indicators of Hg environmental contamination (Fortin et al., 2001). Hg accumulates at higher concentrations in fur followed by liver, kidney, muscle and brain, and highly significant relationships were found to exist between Hg concentrations in fur and internal organs (Strom, 2008). Numerous cases of Hg toxicity in wild mink and river otter in the proximity of Hg polluted rivers (as downstream of a pulp and paper mill, chlor-alkali plants and other Hg related industries) appear described in the scientific literature (for review see Mason and Wren, 2001). Poisoned animals behave erratically and are easily catch by hand or in traps. Post-mortem exam of these animals reveal Hg tissue concentrations comparable to or higher than Hg concentrations in experimentally poisoned animals.

MeHg is the most relevant form of Hg to which mink and otter are exposed because more than 95% of the Hg in fish in the organic form. The margin by which MeHg induces toxicity in these furbearing species is quite narrow, and animals ingesting 0.5 μg/g Hg were clinically normal, whereas 1 μg/g Hg caused mortality in adults (Basu et al., 2006). Furthermore, death is imminent once the initial symptoms of Hg intoxication are observed, and efforts to reverse the toxicity with chelating agents have proven unsuccessful. With one week, incoordination and anorexia can rapidly progress to ataxia, tremors, convulsions, and high pitched vocalizations. The progression of Hg neurotoxicity in mink and otter is quite similar to that reported in humans and rodents (Watanabe and Satoh, 1996). At ecologically relevant exposure levels, Hg can affect the reproductive system. Litter size, kit weight, and hepatic lipid concentrations were all lower in animals exposed to 0.22 μg/g dietary Hg compared to non exposed controls (Halbrook et al., 1997)

The potential for birds as monitors of Hg pollution in the terrestrial environment is evident from the variety of studies which reveal that tissue Hg concentrations reflect the degree of environmental contamination (for review see Samiullah, 1990). The atmospheric deposition of Hg is an important contributor to declining local and global bird populations (Evers et al., 2008; Scheuhammer and Sandheinrich, 2008). Waterborne Hg from legacy sources such as mining, chlor-alkali plants, and manufacturing facilities have also been shown to negatively impact breeding.

Feathers are an excellent, non-invasive tissue to sample for chemicals that are largely endogenously derived, such as Hg (Burger, 1993; Hollamby et al., 2006). Feathers are the major excretory pathway for Hg (Furness et al., 1986) from 70 to 93% of the body burden of mercury is in the plumage (Burger, 1993). Lewis and Furness (1991) showed that 49% of a single dose of MeHg was sequestered in feathers. Whereas blood Hg reflects recent dietary intake, feathers indicate cumulative exposure; Hg can only enter feathers while they are growing and connected to the blood supply (Cristol et al., 2008). Thus, feathers can be used as a biomonitoring tool for exposure assessment. In addition, feathers can be collected non invasively, which allows their use with threatened or endangered species, and feathers can be collected repeatedly from the same individuals over many years (Burger and Gochfeld, 1997). Routine measurement of Hg levels in unhatched eggs can provide useful information both on the exposure of the adult birds and on possible impact on hatchability and survival, as well as to identify spatial patterns of contamination as well as temporal trends (Burger, 1993). Several laboratory egg-injection and controlled-feeding studies have demonstrated that environmentally relevant Hg exposure reduces egg hatchability and chick survival (Heinz et al., 1996; Heinz and Hoffman, 1998).

Long-lived obligate piscivores such as the common loon (*Gavia immer*) or bold eagle (*Haliaeetus leucocephalus*) are considered important bioindicators of environmental Hg loads (Fournier et L., 2002; Evers, 2006; Wolfe et al., 2007; Scheuhammer et al., 1998, 2008) because of the ability of MeHg to biomagnify through the food-chain (Burgess and Hobson, 2006) and bioacumulate over an individual's lifetime (Evers et al., 2008). The U.S.

Environmental Protection Agency has identified bold eagles and common loons as species with high risk for Hg contamination (USEAP, 1998) and has selected those species as primary indicators for monitoring environmental Hg in North America. Numerous studies of Hg concentrations in these species blood, feathers (Meyer et al., 1998; Evers et al., 1998; Burgess et al. 2005; Champoux et al. 2006) and eggs (Scheuhammer et al. 2001; Evers et al. 2003) provide a robust geographic Hg exposure profile. These and other studies indicate loons breeding in northeastern United States have the greatest risk of adverse effects from Hg due to elevated Hg exposure associated with relatively high rural atmospheric deposition (up to 32 ug/m^2 per year of wet and dry deposition; Miller et al., 2005) and enhanced Hg methylation in sensitive environments (Driscoll et al., 2007). Biological Hg hotspots have been identified in some areas where concentrations are high enough to result in population-level impacts to breeding loons (Evers et al., 2007).

Substantial evidence exists for adverse neurological, physiological, and reproductive effects to these piscivorous species associated with environmental Hg levels. Neurotoxic effects of MeHg in free-living loons and bold eagles include reduced back-riding by chicks (Nocera and Taylor, 1998) and lowered chick-feeding rates by adults (Counard, 2000). Recently, changes in neurotransmitter receptor concentration and other neurochemical effects have been correlated with brain Hg concentrations in wild loons (Scheuhammer et al., 2008). Physiological responses to elevated MeHg levels include reduced diving frequency of free-living birds (because Hg inhibits heme production; Olsen et al. 2000) and the production of smaller eggs (Evers et al., 2003). Sublethal Hg dosing studies in laboratory conditions indicate that histological, immunological, and biochemical biomarkers are effective for detecting adverse effects on juvenile piscivorous birds (Spalding et al., 2000; Henny et al., 2002; Hoffman et al., 2005), including loons (Kenow et al., 2003; 2007). Negative impacts of MeHg exposure on reproductive success and survival of adult loons are of ultimate concern. There are inherent difficulties in relating Hg exposure to reproductive success that relate to both intrinsic (e.g., density dependence, species longevity) and extrinsic (e.g., weather, habitat quality) factors as well as anthropogenic stressors (e.g., recreational disturbances and other contaminants). Still, there are studies that have identified Hg as a cause of reduced reproductive success in loons and bold eagle. Barr (1986) found whole body Hg levels in prey fish of 0.30 μg/g wt.w. resulted in reproductive impairments on breeding, and no loon reproduction when Hg in fish exceeded 0.40 μg/g wt.w. More recently, Burgess and Meyer (2008) found a significant negative correlation between Hg in fish and loon productivity with 50% fewer fledged young in breeding loons foraging on fish averaging 0.21 μg/g wt.w. and no reproductive success if prey fish exceeded 0.41 μg/g wt.w.

Although most studies on the effects of Hg have focused on piscivorous wildlife, there is growing evidence that insectivorous birds may also be at risk in some environments. Hg accumulation in tree swallows (*Tachycineta bicolor*) has been documented in several studies in Canada and the northeastern United States (Custer et al., 2006, 2007; Longcore et al., 2007a,b; Brasso and Cristol, 2008; Hothem et al., 2008), Hg concentrations in tree swallow tissues being reflective of sediment Hg concentrations over small geographic and temporal timescales. Brasso and Cristol (2008) recently reported very high Hg concentrations in adult swallows, being comparable with other piscivorous species as belted kingfishers (*Ceryle alcyon*); across 2 years of study only 11% of adult swallows sampled in a Hg contaminated area had blood mercury under 1.00 μg/L and only 3% were below 0.50 μg/L. The amount of MeHg in feathers was associated with a lower linear growth rate of weight in tree swallow during early age (2–10 days) although asymptotic mean weight during days 11-16 was not different (Longcore et al., 2007a). Hg concentrations were 15-40 times higher in the eggs and livers of tree swallows that nested along a gold mining Hg contaminated area than in the same samples above the mining-impacted areas; only 70-74% of the eggs hatched, which is significantly less than the average for these species (>85% of eggs) and might indicate an reproductive impact of Hg in birds in this area (Custer et al., 2007). Appreciable Hg residues were found in tree swallows from the Acadia National Park, and area remote from point mercury discharge (Longcore et al., 2007b); some sampled egg exceeded the threshold (i.e. 800-1000 ng/g wt.w.) of embryotoxicity established for Hg; on average total Hg biomagnified 2 to 4 fold from food to eggs and 9 to 18-fold from food to feathers.

LIVESTOCK SPECIES

Animals in livestock production largely fulfill the general criteria of selection for biomonitoring species if they are reared on grass or locally-grown fodder (Ronneau et al., 1983; Ronneau and Cara, 1984; Morcombe et al., 1994; Petersson Grawé et al., 1997). In addition, domestic animals have the advantage that they are an important source of food for humans and so provide a direct measure of pollutant transfer to humans (O'Brien et al., 1993).

In rural or agricultural areas the use of livestock as biomonitors of pollution has specific advantages over direct measurement of soil samples (López-Alonso et al., 2002). Variation in tissue residues between animals is a measure of the variation in bioavailable, rather than total, toxic levels in the environment and, arguably, is a more useful measure of contamination. Residues are also an integrated measure of contamination over the foraging area of the calf and so may be less prone than soil cores to large inter-sample variations caused by small-scale heterogeneity in contaminant distribution. Additionally, because farmers are required to keep good animal records and to take calves to slaughterhouses for butchering, well-documented samples are naturally collected at a relatively small number of sites, making sampling of environmental contamination across wide spatial scales easy and cheap compared with field sampling of soil cores.

Biomonitoring programmes have been carried out in farm species (mainly cattle, pig and poultry) in many countries (Korsrud et al., 1985; Vos et al., 1986; Jorhem et al., 1991; Niemi et al., 1991; Salisbury et al., 1991; Kluge-Berge et al., 1992; Falandysz, 1993a,b; Raszyk et al., 1996; Ulrich et al., 2001; López Alonso et al., 2003a, 2007a) with the purpose of avoiding the distribution of foodstuffs that could pose a risk to human health if consumed. These studies have allowed identifying the main sources of Hg in feed for livestock animals. In agricultural regions organomercurial seed dressings have traditionally been the major source of mercury to farm animals (Janicki et al., 1987); these fungicidal treatments have been restricted or banned in many countries in recent years and so their use has drastically reduced. Currently, fish meal added to feed is the main source of mercury for livestock (Salisbury el at., 1991; Zarski et al., 1997). As indicated, mercury accumulates in the food chain, particularly in fish, and in recognition of this the maximum limits for fishmeal is higher (0.5 mg/kg) than in other feed materials (EFSA, 2008). Hair and feathers also accumulate a large amount of MeHg, and the practice of recycling poultry (chicken, turkey) feathers as feather meal (under the form protein hydrolysate) to feed back to farmed animals could represent an additional source of MeHg contamination (EFSA, 2008).

In industrialised and urban regions, farm animals are additionally exposed to mercury emitted from various anthropogenic sources, the most important of which is the combustion of fossil fuels which accounts for up to half of all man-made emissions (Nriagu and Pacyna, 1988; Mukherjee et al., 2000). Results of a biomonitoring study of Hg in NW Spain (López-Alonso et al., 2003b) demonstrated a strong relationship between renal Hg residues in cattle and the proximity of animals to coal-fired power plants. In this region most of the total Hg atmospheric emissions are the result of combustion processes in energy production and manufacturing industry. The two coal-fired power plants in Galicia burn lignite, which has a high mercury content (Peña, 1993), and account for 47% of the total Hg emissions from all the coal-fired power plants in the country (Ecologistas en acción, 2000). This study demonstrated that these power stations are significantly associated with elevated Hg accumulation by calves, and, presumably, by other grazing livestock and wild mammals. This association was apparent for calves from throughout the whole of Galicia and, overall, differences between calves in their proximity to the power plants accounted for approximately 10% of the total variation in their renal Hg concentrations. The power plants appeared to be a major Hg source for calves directly downwind, distance from the source alone explaining some two-thirds of the variation between animals in their renal Hg levels. Cattle also proved to be a good bioindicator of Hg contamination for more diffuse Hg emissions in an industrialized area where there are no dominant point sources of Hg emission, as coal power plants, but manufacturing industries (particularly iron-steel plants and other industries involved in metal smelting) are the most important sources of atmospheric Hg emissions. Hg emissions from the industrial zone appeared to affect Hg accumulation by downwind calves, renal mercury concentrations being significantly correlated with distance to the industrial zone.. In both regions, the effects of atmospheric Hg emissions have a detectable or significant effect some 140–200 km downwind, suggesting that cattle can be used as bioindicators to map variation in the extent of contamination with Hg that is bioavailable for accumulation by biota.

PETS OR COMPANION ANIMALS

Pets or companion animals fulfill most of the general selection criteria for biomonitoring species, and have proved to be particularly valuable tools for domestic human exposure because they are exposed to many of the same environments and may eat the same food as people (NRC, 1991). Pets may be even more directly exposed than their owners to Hg contained in soil or house dust and airborne particles. Dogs often behave like young children, eating or chewing everything they can find, whereas cats are more representative of adult humans. Cats and dogs respond to many toxic insults, including Hg, in similar ways to humans, and they can develop similar environmentally-induced diseases through the same pathogenic mechanisms. When humans are exposed to Hg in complex environments (e.g., in the home or in the work place), it may be difficult to estimate the degree of exposure with conventional procedures for measuring ambient concentrations and to calculate their intake from contaminated media. Studies involving domestic animals are free from some of the confounding effects (such as lifestyle and occupational risk factors) that can make the results of human studies difficult to interpret. Blood and other tissues can be easily collected from companion animals during routine clinical exams, surgical procedures or upon death of the animal (pets are generally not sacrificed for study). Finally, as in humans, endpoints measured in biological samples from animals represent an integrated measure of exposure to ambient environmental contamination.

One of the earliest examples of animals as bioindicators of Hg contamination correspond to the "dancing cats" that consumed MeHg-contaminated fish from Minemata bay in Japan in the 1950s, before a similar disease appeared in fisherman and their families, becoming excellent sentinels of Hg exposure to humans. Pet cats were also proposed as sentinels and used to study the clinical and pathological effects of a MeHg poisoning in Ontario (Takeuchi et al., 1977). The identification of two cats that showed clinical signs, mercury levels and histopathological effects similar to the domestic cats poisoned in Minemata after eating fish from an MeHg-contaminated river allowed inferring of the gravity of the possible health consequences for the humans living near and eating fish from the contaminated river. The susceptibility of cats to MeHg toxicity has been known for several decades. Even if toxicological signs were not present, cats fed tuna fish showed a modified behaviour: they were less active, vocalized less, and spent more time on the floor and more time eating than cats fed commercial beef cat food (Houpt et al., 1988). Exposure to MeHg in fish nowadays constitutes a public health problem in most regions of the world because atmospheric Hg is deposited in waterways and in the sea, where it is taken up by aquatic organisms. On its way up the food chain it becomes concentrated and can reach high levels in large predatory fish, such as tuna, shark and swordfish. MeHg contamination in fish is of such importance that the main health organizations have issued warnings about the risk of eating large amounts of fish. The limit for consumption of three piscivorous fish (shark, swordfish, and tuna) in the general adult population is one meal per week, and for pregnant women, women of child-bearing age and young children it is no more than one meal per month. Hg is of particular danger to the developing foetus, and can cause permanent damage when the nervous system is developing. In the domestic environment, pet cats can act as sentinels of excessive MeHg exposure if they eat the same food as their owners. This could be especially valuable in population groups at high risk of MeHg exposure, i.e., people who consume large amounts of fish, pregnant women and children.

Although less susceptible to cats, dogs have also proven to be good indicators or sentinels of Hg exposure in humans. Smith and Armstrong (1975) examined Hg concentrations in various food items for a group of native Inuit in Northern Canada. Though the region is far from industrial sources of Hg contamination, humans had above average blood Hg levels, which were thought to be due to eating contaminated seal meat and liver. While the humans' diet was presumed to be sufficiently varied to avoid intoxication, it was noted that sled dogs owned by the Inuit and fed a nearly exclusive diet of seal had much higher Hg levels, and it was suggested that these dogs could act as sentinels to predict what human levels might approach if seasonal population constraints of their other primary food items (caribou and char) forced the Inuit to consume seal as a greater proportion of their diet. More recently, López-Alonso et al. (2007b), in a biomonitoring study using dogs as indicators of toxic metal exposure in NW Spain, found that dogs from urban areas showed kidney mercury levels three times higher than dogs from rural areas. The region in which that study was performed is relatively unpolluted, and there are no major point sources of atmospheric mercury pollution (e.g., major industries or coal-fired power plants); such atmospheric mercury pollution as it exists is probably related to more diffuse emission sources mostly located in urban habitats, such as combustion of fossil fuels in small industries, heaters or vehicles (Hoffman et al., 2001). Since pets closely share the same environment as their owners, and are therefore supposed to be exposed at least in part to the same pollutants, pets could be suggested as surrogate indicators of human metal exposure.

REFERENCES

Bank, M.S.; Burgess, J.R.; Evers, D.C. and Loftin, C.S. (2006). Mercury Contamination of Biota from Acadia National Park, Maine: A Review. *Environmental Monitoring and Assessment*, 126, 105–115.

Barr, J.F. (1986). Population dynamics of the common loon (*Gavia immer*) associated with mercury-contaminated waters in northwestern Ontario. Occasional Paper Number 56, *Canadian Wildlife Service*, Ottawa, Ontario, Canada.

Basu, N.; Scheuhammer, A.M.; Bursian, S.J.; Elliott, J.; Rouvinen-Watt, K.and Chan, H.M. (2006). Mink as a sentinel species in environmental health. *Environmental Research*, 103, 130–44.

Bonzongo, J.; Heim, K.; Warwick, J.and Lyons, W. (1996). Mercury levels in surface waters of the Carson River – Lahontan Reservoir System, Nevada: Influence of historic mining activities. *Environmental Pollution*, *92*, 193–201.

Brasso, R.L. and Cristol, D.A. (2008). Effects of mercury exposure on reproductive success of tree swallows (*Tachycineta bicolor*). *Ecotoxicology*, 17, 133–141.

Burger, J. (1993). Metals in avian feathers: bioindicators of environmental pollution. *Reviews in Environmental Toxicology*, 5, 203–311.

Burger, J.; Gochfeld, M. (1997). Risk, mercury levels, and birds: Relating adverse laboratory effects to field monitoring. *Environmental Research,* 75, 160–172.

Burgess, N.M. and Hobson, K.A. (2006). Bioaccumulation of mercury in yellow perch (Perca flavescens) and common loons (*Gavia immer*) in relation to lake chemistry in Atlantic Canada. *Hydrobiologia*, 567, 275–282.

Burgess, N.M. and Meyer, M.W. (2008) Methylmercury exposure associated with reduced productivity in common loons. *Ecotoxicology*, 17, 81-93.

Burgess, N.M., Evers, D.C. and Kaplan, J.D. (2005). Mercury and other contaminants in common loons breeding in Atlantic Canada. *Ecotoxicology*, 14, 241–252.

Carpi, A. (1997). Mercury from combustion sources: A review of the chemical species emitted from their transport in the atmosphere. *Water, Air and Soil Pollution*, 98, 241–254.

Champoux, L.; Masse, D.C.; Evers, D.C.; Lane, O.; Plante, M. and Timmerman, S.T.A. (2006). Assessment of mercury exposure and potential effects on Common Loons (*Gavia immer*) in Quebec. *Hydrobiologia*, 567, 263–274.

Chu, P. and Porcella, D.B. (1995). Mercury stack emissions from U.S. electric utility power plants. *Water, Air and Soil Pollution*, 80, 135–144.

Counard, C.J. (2000). Mercury exposure and effects on Common Loon (*Gavia immer*) behavior in the Upper Midwestern United States. Unpublished MS thesis, University of Minnesota, St. Paul, MN.

Cristol, D.A.; Brasso, R.L.; Condon, A.M.; Fovargue, R.E.; Friedman, S.L.; Hallinger, K.K.; Monroe, A.P. and White, A.E. (2008). The movement of aquatic mercury through terrestrial food webs. *Science*, 320, 335.

Custer, C.M.; Custer, T.W. and Hill, E.F. (2007). Mercury exposure and effects on cavity-nesting birds from the Carson River, Nevada. *Archives of Environmental Contamination and Toxicology*, 52, 129–136.

Custer, C.M.; Custer, T.W.; Warburton, D.; Hoffman, D.J.; Bickham, J.W. and Matson' C.W. (2006). Trace element concentrations and bioindicator responses in tree swallows from northwestern Minnesota. *Environmental Monitoring and Assessment*, 118, 247–266.

Driscoll, C.T.; Han, Y.J.; Chen, C.Y.; Evers,,D.C.; Lambert, K.F.; Holsen; T.M.; Kamman, N.C. and Munson, R. (2007). Mercury contamination in remote forest and aquatic ecosystems in the northeastern U.S.: Sources, transformations and management options. *BioScience*, 57, 17–28.

Ecologistas en acción (2000). Mercurio estado Español. http://www.ecologistasenaccion.org/.

EFSA (2008). Opinion of the Scientific Panel on Contaminants in the Food chain on a request from the European Commission on mercury as undesirable substance in feed. *The EFSA Journal*, 654, 1–74.

Energy Information Administration (2007). International energy outlook 2007. Office of integrated analysis and forecasting. US Department of Energy, Washington, DC. http://www.eia.doe.gov/oiaf/ieo/index.html.

Esselink, H.; van der Geld, F.M.; Jager, L.P.; Posthuma-Trumpie, G.A.; Zoun, P.E.F. and Baars, A.J. (1995). Biomonitoring heavy metals using the Barn Owl (*Tyto Alba Guttata*): Sources of variation especially relating to body condition. *Archives of Environmental Contamination and Toxicology*, 28, 471–486.

Evers, D.C. (2006). Loons as biosentinels of aquatic integrity. *Environmental Bioindicators*, 1, 18–21.

Evers, D.C.; Han, Y.J.; Driscoll, C.T.; Kamman, N.C.; Goodale, M.W.; Lambert, K.F.; Holsen, T.M.; Chen, C.Y.; Clair, T.A. and Butler, T. (2007). Identification and evaluation of biological hotspots of mercury in the Northeastern U.S. and Eastern Canada. *BioScience*, 57, 29–43.

Evers, D.C.; Kaplan, J.D.; Meyer, M.W.; Reaman, P.S.; Braselton, W.E.; Major, A.; Burgess, N. and Scheuhammer, A.M. (1998). A geographic trend in mercury measured in common loon feather and blood. *Environmental Toxicology and Chemistry*, 17, 173–183

Evers, D.C.; Savoy, L.J.; DeSorbo, C.R.; Yate, D.E.; Hanson, W.; Taylor. K.M.; Siegel, L.S.; Cooley, J.H.; Bank, M.S.; Major, A.; Munney, K.; Mower, B.F.; Vogel, H.S.; Scholh, N.; Pokras, M.; Goodale, M.W. and Fair, J. (2008). Adverse effects from environmental mercury loads on breeding common loons. *Ecotoxicology*, 17, 69–81.

Evers, D.C.; Taylor, K.M.; Major, A.; Taylor, R.J.; Poppenga, R.H. and Scheuhammer, A.H. (2003). Common loon eggs as indicators of methylmercury availability in North America. *Ecotoxicology*, 12, 69–81.

EWG (Environmental Working Group) (1999). Mercury Falling: An analysis of mercury pollution from coal-burning power plants. http://www.clearair.net.

Falandysz, J. (1993a). Some toxic and essential trace metals in swine from Northern Poland. *The Science of the Total Environment*, 136, 193–204.

Falandysz, J. (1993b). Some toxic and essential trace metals in cattle from the northern part of Poland. *The Science of the Total Environment*, 136, 177–191.

Flores, R.M.D.M. and Martins, A.F. (1997). Distribution of trace elements in egg samples collected near coal power plants. *Journal of Environmental Quality*, 26, 744–748.

Fortin, C.; Beauchamp, G.; Dansereau, N.; Lariviere, N. and Belanger, D. (2001). Spatial Variation in Mercury Concentrations in Wild Mink and River Otter Carcasses from the James Bay Territory, Québec, Canada. *Archives of Environmental Contamination and Toxicology*, 40, 121–127.

Fournier, F.; Karasov, W.H.; Kenow, K.P.; Meyer, M.W. and Hines, R.K. (2002). The oral bioavailability and toxicokinetics of methylmercury in Common Loon (*Gavia immer*) chicks. *Comparative Biochemistry and Physiology. Part A*, 133, 703–714.

Furness, R.W.; Muirhead, S.J. and Woodburn, D.M. (1986). Using bird feathers to measure mercury in the environment: Relationship between mercury content and moult. *Marine Pollution Bulletin,* 17, 27–30.

Gilbert, F.F. and Nancekivell, E.G. (1982). Food habilts in mink (*Mustela vison*) and otter (*Lutra canadensis*) in northeastern Alberta. *Canadian Journal of Zoology*, 60, 1282–1288.

Halbrook, L.A.; Lewis; R.J. Aulerich and S.J. Bursian, (1997). Mercury accumulation in mink fed fish collected from streams on the Oak Ridge Reservation, *Archives of Environmental Contamination and Toxicology* 33, 312–316.

Heinz, G. H. and Hoffman, D. J. (1998). Methylmercury chloride and selenomethionine interactions on health and reproduction in mallards". *Environmental Toxicology and Chemistry,* 17, 139–145.

Heinz, G.H. (1996). Mercury poisoning in wildlife. In: A. Fairbrother, L.N. Locke and Hoff GL (Eds) *Noninfectious diseases of wildlife*, 2nd ed. Iowa State University Press, Ames, IA, pp 118–127.

Henny, C.J.; Hill, E.F.; Hoffman, D.J.; Spalding, M.G. and Grove, R.A. (2002). Nineteenth century mercury: hazard to wading birds and cormorants of the Carson River, Nevada. *Ecotoxicology*, 11, 213–231.

Hoffman, D.J.; Rattner, B.A.; Scheunert, I. and Korte; F. (2001). Environmental contaminants. In: R.F. Shore, R.F. and B.A. Rattner (Eds) *The ecotoxicology of wild mammals*. John Wiley and Sons, London, pp. 1–48.

Hoffman, D.J.; Spalding, M.G. and Frederick, P.C. (2005). Subchronic effects of methylmercury on plasma and organ biochemistries in great egret nestlings. *Environmental Toxicology and Chemistry*, 24, 3078–3084.

Holden, A.V. (1973). International cooperative study of organochlorine and mercury in wildlife. *Pesticides Monitoring Journal*, 7, 37–52.

Hollamby, S; Afema-Azikuru, J; Waigo, S.M.; Cameron, K.; Gandolf, A.; Norris, A. and Sikarskie, J.G. (2006). Suggested guidelines for use of avian species as biomonitors. *Environmental Monitoring And Assessment*, 118, 13–20.

Hothem, RL; Trejo, BS; Bauer, ML, John J. Crayon (2008). Cliff swallows Petrochelidon pyrrhonota as bioindicators of environmental mercury, Cache Creek watershed, California *Archives Of Environmental Contamination And Toxicology*, 55, 111–121.

Houpt, K.A.; Essick, L.A.; Shaw, E.B.; Alo, D.K.; Gilmartin, J.E.; Gutenmann, W.H.; Littman, C.B. and Lisk, D.J. (1988). A tuna fish diet influences cat behaviour. *Journal of Toxicology and Environmental Health*, 24, 161-72.

Janicki, K,; Dobrowolski, J. and Krasnicki, K. (1987). Correlation between contamination of the rural environment with mercury and occurrence of leukaemia in men and cattle. *Chemosphere*, 16, 253–257.

Jorhem, L.; Slorach, S.; Sundstrom, B. and Ohlin B. (1991). Lead, cadmium, arsenic and mercury in meat, liver and kidney of Swedish pigs and cattle in 1984-88. *Food Additives and Contaminants*, 8, 201–212.

Kenow, K,P,; Grasman, K.A.; Hinesm R,K,; Meyer, M.W.; Gendron-Fitzpatrick, A.; Spalding, M.G. and Gray, B.R. (2007). Effects of methylmercury exposure on the immune function of juvenile common loons (*Gavia immer*). *Environmental Toxicology and Chemistry*, 26, 1460–1469.

Kenow, K.P.; Gutreuter, S.; Hines, R.K.; Meyer, M.W.; Fournier, F. and Karasov, W.H. (2003). Effects of methyl mercury exposure on the growth of juvenile common loons. *Ecotoxicology*, 12,171–182.

Kluge-Berge S; Skjerve E; Sivertsen T; Godal A. 1992. Lead, cadmium, mercury and arsenic in Norwegian cattle and pigs. Proceedings of the 3rd World Congress Foodborne Infections and Intoxications. Berlin, pp.745–748.

Korsrud, G.O.; Meldrum, J.B.; Salisbury, C.D.; Houlahan, B.J.; Saschenbrecker, P.W. and Tittiger, F. (1985). Trace element levels in liver and kidney from cattle, swine and poultry slaughtered in Canada. *Canadian Journal of Comparative Medicine*, 49, 159–163.

Lacerda, L. (1997). Global mercury emissions from gold and silver mining. *Water Air and Soil Pollution*, 97, 209–221.

Lamborg, C.H.; Fitzgerald, W.F.; Damman, W.H.; Benoit, J.M.; Balcom, P.H. and Engstrom, D.R. (2002). Modern and historic atmospheric mercury fluxes in both hemispheres: global and regional mercury cycling implications. *Global Biogeochemistry Cycles*, 16, 204.

Landres, P.B.; Verner, J. and Thomas, J.W. (1988). Ecological uses of vertebrate indicator species: A critique. *Conservation Biology*, 284,316–328.

Lewis, S.A. and Furness, R,W. (1991). Mercury accumulation and excretion in laboratory reared black-headed gull *Larus ridibundus* chicks. *Archives of Environmental Contamination and Toxicoolgy*, 21, 316–320.

Lindberg, S.; Bullock, R.; Ebinghaus, R.; Engstrom, D.; Feng, X.M.; Fitzgerald, W.; Pirrone, N.; Prestbo, E. and Seigneur, C. (2007). A synthesis of progress and uncertainties in attributing the sources of mercury in deposition. *Ambio*, 36, 19–32.

Lodenius, M. (1998). Dry and wet deposition of mercury near a chlor-alkali plant. *The Science of the Total Environment*, 213, 53–56.

Longcore, J.R.; Dineli, R. and Haines, T.A. (2007a). Mercury and growth of tree swallows at Acadia National Park, and at Orono, Maine, USA. *Environmental Monitoring and Assessment*, 126, 117–127.

Longcore, J.R.; Haines, T.A.; Halteman, W.A. (2007b). Mercury in tree swallow food, eggs, bodies, and feathers at Acadia National Park, Maine, and an EPA Superfund site, Ayer, Massachusetts. *Environmental Monitoring and Assessment*, 126, 129–143.

López-Alonso, M.; Benedito, J.L.; Miranda, M.; Castillo, C.; Hernández, J.; Shore, R. (2002). Cattle as biomonitors of soil arsenic, copper and zinc concentrations in Galicia (NW Spain). *Archives of Environmental Contamination and Toxicology*. 43, 103-108.

López-Alonso, M.; Benedito, J.L.; Miranda, M.; Castillo, C.; Hernández, J. and Shore, R.F. (2003a). Mercury concentrations in cattle from NW Spain. *The Science of the Total Environment*, 302, 93–100.

López-Alonso, M.; Miranda, M.; Castillo, C.; Hernández, J.; García, M. and Benedito, J.L. (2007a). Toxic and essential metals in liver, kidney and muscle of pigs at slaughter in Galicia (NW Spain). *Food Additives and Contaminants*, 24, 943–954.

López-Alonso, M.; Benedito, J.L.; Miranda, M.; Fernández, J.A.; Castillo, C.; Hernández , J.; Shore, R.F. (2003b). Large-scale spatial variation in mercury concentrations in cattle in NW Spain. *Environmental Pollution*, 125, 173–181.

López-Alonso, M.; Miranda, M.; García-Partida, P.; Cantero, F.; Hernandez, J.; Benedito, J.L. (2007b). Use od dogs as indicators of metal exposure in rural and urban areas in NW Spain. *The Science of the Total Environment*, 372, 668–675.

MARC. (1990). *Biological Monitoring: Animals*. Monitoring and Assessment Research Centre (MARC) Report Number 37, 767 pp.

Mason, C.F. and; Wren, C.D. (2001). Carnivora. In: R.F. Shore and B.A. Rattner. *Ecotoxicology of wild animals*, Jonh Willey and Sons Ltd. Chichester.

Mason, R.; Fitzgerald, W. and Morel, F. (1994). The biogeochemical cycle of elemental mercury: Anthropogenic influences. *Geochimica et Cosmochimica Acta*, *58*, 3191–3198.

Meyer, M.W.; Evers, D.C.; Hartigan, J.J. and Rasmussen, P.S. (1998). Patterns of common loon (*Gavia immer*) mercury exposure, reproduction, and survival in Wisconsin, USA. *Environmental Toxicology and Chemistry*, 17, 184–190.

Miller, E.K.; Vanarsdale, A.; Keeler, G.J.; Chalmers, A.; Poissant, L.; Kamman, N.C. and Brulotte, R. (2005). Estimation and mapping of wet and dry mercury deposition across northeastern North America. *Ecotoxicology*, 14, 53–70.

Morcombe, P.W.; Petterson, D.S.; Masters, H.G.; Ross, P.J. and Edwards, J.R. (1994). Cadmium concentrations in kidneys of sheep and cattle in Western Australia. I. Regional Distribution. *Australian Journal of Agriculture Research*, 45, 851–862.

Morel, F., Kraepiel, A. and Amyot, M. (1998). The chemical cycle and bioaccumulation of mercury. *Annual Review of Ecology and Systematics*, 29, 543–566.

Mukherjee, A.B.; Melanen, M.; Ekqvist, M. and Verta, M. (2000). Assessment of atmospheric mercury emissions in Finland. *The Science of the Total Environmen*, 259, 73–83.

Nadim, F.; Perkins, C.; Liu, S.; Carley, R.J. and Hoag, G.E. (2001). Long-term investigation of atmospheric contamination in Connenticut. *Chemosphere*, 45, 1033–1043.

National Research Council, Committee on Animals as Monitors of Environmental Health Hazards (1991). *Animals as sentinels of environmental health hazards*. Washington, DC: National Academy Press.

Niemi, A.; Venäläinen, E.R.; Hirvi, T.; Hirn, J. and Karppanen, E. (1991). The lead, cadmium and mercury concentrations in muscle, liver and kidney from Finnish pigs and cattle during 1987-1988. *Zeitschrift für Lebensmittel Untersuchung und Forschung*, 192, 427–429.

Nocera, J. and Taylor, P. (1998). In situ behavioral response of common loons associated with elevated mercury exposure. *Conservation Ecology* 2, 10. Available from the Internet. URL: http://www.consecol.org/vol2/iss2/art10/.

Nriagu, J.O. and Pacyna, J.M. (1988). Quantitative assessment of worldwide contamination of air, water and soils by trace elements. *Nature* 333, 134–139.

O'Brien, D. J.; Kaneene, J. B. and Poppenga, R. H. (1993). The use of mammals as sentinels for human exposure to toxic contaminants in the environment. *Environmental Health Perspectives* 99, 351–368.

Olsen, B.; Evers, D.C. and DeSorbo, C. (2000). Effect of methylated mercury on the diving frequency of the Common Loon. *Journal of Ecology Research*, 2, 67–72.

Parks, J. and Hamilton, A. (1987). Accelerating recovery of the mercury-contaminated Wabigoon-English River system. *Hydrobiologia*, 149, 159–188.

Peña, R.M. (1993). Balance de elementos en el ciclo de combustión de una central térmica de carbón. Master. University of Santiago de Compostela (Spain).

Petersson Grawé, K.P.; Thierfelder, T.; Jorhem, L. and Oskarsson, A. (1997). Cadmium levels in kidneys from Swedish pigs in relation to environmental factors - temporal and spatial trends. *The Science of the Total Environment*, 208, 111–122.

Pirrone, N.; Costa, P.; Pacían, J.M.; Ferrara, R. (2001). Mercury emissions to the atmosphere from natural and anthropogenic sources in the Medierranean region. *Atmospheric Environment*, 35, 2997–3006.

Rasmussen, P.E. (1994). Current Methods of Estimating Atmospheric Mercury Fluxes in Remote Areas. *Environmental Science and Technology*, 28, 2233–2241.

Raszyk, J.; Gajduskva, V.; Ulrich, R.; Nezveda, K.; Jarosova, A.; Sabatova, V.; Docekalova, H.; Salava, J.; Palac, J. and Scjöndorf, J. (1996). Evaluation of the presence of harmful pollutants in fattened pigs. *Veterinarni Medicina*, 9, 261–266.

Ratcliffe, H.E.; Swanson, G.M. and Fischer, L.J. (1996). Human exposure to mercury: a critical assessment of the evidence of adverse health effects. *Journal of Toxicology and Environmental Health*, 49, 221–270.

Ronneau, C. and Cara, J. (1984). Correlations of element deposition on pastures with analysis of cows' hair. *The Science of the Total Environment*, 39, 135–142.

Ronneau, C.; Detry, M.; Hallet, J.P.H. and Lardinois, P. (1983). Concentration of some elements in the hair of cattle as an indicator of contamination by air pollutant deposition on grass. *Agriculture, Ecosystems and Environment*, 10, 285–298.

Salisbury, C.D.C.; Chan, W. and Saschenbrecker, P. (1991). Multielement concentrations in liver and kidney tissues from five species of Canadian slaughter animals. *Journal of the Association of Official Analytical Chemists*, 74, 587–591.

Samiullah, Y. (1990). *Biological Monitoring of Environmental Contaminants: Animals*. Global Environment Monitoring System. A Technical Report. pp. 72–74.

Scheuhammer, A.M. and Sandheinrich, M.B. (2008). Recent advances in the toxicology of methylmercury in wildlife. *Ecotoxicology*, 17, 67–68.

Scheuhammer, A.M.; Atchison, C.M.; Wong, A.H.K. and Evers, D.C. (1998). Mercury exposure in breeding common loons (*Gavia immer*) in central Ontario, Canada. *Environmental Toxicology and Chemistry*, 17, 191–196.

Scheuhammer, A.M.; Basu, N.; Burgess, N.M.; Elliot, J.E.; Campbell, G.D.; Wayland, M.; Champoux, L. and Rodrigue, J. (2008) Relationships among mercury, selenium, and neurochemical parameters in common loons (*Gavia immer*) and bald eagles (*Haliaeetus leucocephalus*). *Ecotoxicology*, 17, 93–101.

Scheuhammer, A.M.; Meyer, M.W.; Sandheinrich, M.B. and Murray, M.W. (2007). Effects of environmental methylmercury on the health of wild birds, mammals, and fish. *AMBIO: A Journal of the Human Environment*, 36, 12–19.

Scheuhammer, A.M.; Perrault, J.A. and Bond, D.E. (2001). Mercury, methylmercury, and selenium concentrations in eggs of Common Loons (*Gavia immer*) from Canada. *Environmental Monitoring and Assessment*, 72, 79–94.

Sell, J.L.; Deitz, F.D. and Buchanan, M.L. (1975). Concentration of mercury in animal products an soils of Norht Dakota. *Archives of Environmental Contamination and Toxicology*, 3, 278–288.

Simpson, V.R.; Stuart, N.C.; Munro, R.; Hunt, A. and Livesey, C.T. (1997). Poisoning of dairy heifers by mercurous chloride. *Veterinary Record*, 140, 549–552.

Smith, T.G. and Armstrong, F.A.J. (1975). Mercury in seals, terrestrial carnivores, and principal food items of the Inuit, from Holman, N.W.T. *Journal of the Fisheries Research Board of Canada*, 32, 795–801.

Spalding, M.G.; Frederick, P.C.; McGill, H.C.; Bouton, S.N.; Richey, L.J.; Schumacher, M.; Blackmore, C.G. and Harrison, J. (2000). Histologic, neurologic, and immunologic effects of methylmercury in captive great egrets. *Journal of Wildlife Disease*, 36, 423–435.

Strom, S.M. (2008). Total Mercury and Methylmercury Residues in River Otters (*Lutra canadensis*) from Wisconsin. *Archives of Environmental Contamination and Toxicology*, 54, 546–554.

Takeuchi, T.; D'Itri, F.M.; Fischer, P.V.; Annett, C.S. and Okabe, M. (1977). The outbreak of Minamata disease (methyl mercury poisoning) in cats on Northwestern Ontario reserves. *Environmental Research*, 13, 215–28.

Talmage, S.S. and Walton, B.T. (1991). Small mammals as monitors of environmental contaminants. *Reviews of Environmental Contamination and Toxicology*, 119, 48–99.

Ulrich, R.; Raszyk, J. and Napravnik, A. (2001). Variations in contamination by mercury; cadmium and lead on swine farms in the district of Hodonin in 1994 to 1999. *Veterinarni Medicina*, 46, 132–139.

United States Environmental Protection Agency (USEPA) (1998). *Mercury in solids and solutions by thermal decomposition, amalgamation, and atomic absorption spectrophotometry.* EPA Method 7473 Report, January 1998. 15 p.

Vos, G.; Teeuwen, J.J.M.H. and Van Delft, W. (1986). Arsenic, cadmium, lead and mercury in meat, livers and kidneys of swine slaughtered in the Netherlands during the period 1980–1985. *Zeitschrift für Lebensmittel Untersuchung und Forschung*, 183, 397–401.

Watanabe, C. and Satoh, H. (1996). Evolution of our understanding of methylmercury as a health threat, *Environmental Health Perspectives*, 104 (Suppl. 2), 367–379.

WHO, (1990). *Environmental Health Criteria 101: Methylmercury.* World Health Organisation, Geneva.

WHO, (1991). *Environmental Health Criteria 118: Inorganic mercury.* World Health Organisation, Geneva

Wolfe, M.; Schwarzbach, F.S. and Sulaiman, R.A. (1998). Effects of mercury on wildlife: a comprehensive review. *Environmental Toxicology and Chemistry*, 17, 146–160.

Wolfe, M.F.; Atkeson, T.; Bowerman, W.; Burger, K.; Evers, D.C.; Murray, M.W. and Zillioux, E. (2007). Wildlife indicators. In: R. Harris; D.P. Krabbenhoft; R. Mason; M.W. Murray; R. Reash and T. Saltman (Eds) *Ecosystem response to mercury contamination: indicators of change*. CRC Press, SETAC, Webster, NY, pp 123–189.

Zarski, T.P.; Zarska, H.; Samek, M. and Siluk, I. (1997). Mercury contamination of the tissues of cattle from various regions of Poland. *Annals of Warsaw Agricultural University / Veterinary Medicine*, 20, 107–111.

In: Environmental and Regional Air Pollution
Editors: D. Gallo, R. Mancini
ISBN: 978-1-60692-893-6

Chapter 8

ENVIRONMENTALLY FRIENDLY REFRIGERANTS: HFC REFRIGERANTS' CONDENSATION INSIDE SMOOTH HORIZONTAL TUBES

Ahlèm Boussetta[a], Kamel Halouani[a], Abdelmoneim Ayadi[b], Belkacem Zeghmati[c] and Riad Benelmir[d]
[a]METS-IESG, IPEIS, Route Menzel Chaker, B.P 805, 3000 Sfax, Tunisie
[b]LRAE, ENIS, B.P W, Sfax, Tunisie
[c]MEPS, 52, avenue de Villeneuve, Perpignan, France
[d]LEMTA (umr 7563 CNRS-INPL-UHP), ESSTIN 2 Rue Jean Lamour, Vandoeuvre, France

ABSTRACT

A theoretical analysis of heat transfer and pressures drop during the condensation of pure HFC refrigerants (and HCFC22) inside a horizontal smooth tube is developed. The tube wall is subject to a constant heat flux density. Heat and mass transfer through the liquid are described by classical equations of continuity, momentum and energy. An overall good agreement was obtained with the experimental data of local heat transfer coefficient and pressure drop in the literature.

Keywords: Condensation, pure refrigerants, horizontal tube, local heat transfer, pressure drop, modeling.

NOMENCLATURE

A cross section area [m^2]
c_p specific heat at constant pressure [$Jkg^{-1}K$]
d diameter [m]

f friction factor

Fr^* modified Froude number, $Fr^* = \dfrac{G_v}{\left[gd\rho_v\left(\rho_l - \rho_v\right)\right]^{1/2}}$

F_g component of the gravity [Nm^{-2}]
G mass velocity [kgm^{-2}s^{-1}]
g gravitational acceleration [ms^{-2}]
h heat transfer coefficient [Wm^{-2}K^{-1}]
h_{fg} latent heat of condensation [Jkg^{-1}]
J superficial velocity [ms^{-1}]
L length of the tube [m]
$\dot{m}$ masse flux rate [kgm^{-2}]
P total pressure [Pa]
$\overline{q}$ heat flux density [Wm^{-2}]
T temperature [°C]
S perimeter of the tube [m]
x tangential coordinate for condensate flow film [m]
y radial coordinate for condensate flow film [m]
z axial coordinate for condensate flow film [m]
u, v, w are velocity components in x, y and z directions respectively [ms^{-1}]
ΔP difference of pressure by unit of length [Pa.m^{-1}]

Greek Symbols

α vapor void fraction
δ film thickness [m]
θ peripherial angle [rd]
θ_m stratified angle [rd]
λ thermal conductivity [Wm^{-1}K^{-1}]
μ dynamic viscosity [kgm^{-1}s^{-1}]
ν cinematic viscosity [m^{2}s^{-1}]
ρ mass density [kgm^{-3}]
τ viscous shear [Nm^{-2}]
x vapor title

Subscripts

a acceleration
c condensate
f frictional
h hydraulic
g gravity

gliss glide
i interfacial
l liquide
v vapor
s saturation
p wall size in the entry of the tube

INTRODUCTION

The knowledge of condensation heat transfer of refrigerants inside tubes is required for the design and sizing of heat exchangers in refrigeration and air conditioning applications. Condensation studies of pure refrigerants inside horizontal smooth tubes were the subject of several works [1-21]. Torikoshi [5] carried out an experimental study of the condensation inside an horizontal smooth tube of pure fluids such as HCFC22, HFC134a and HFC32 for inlet mass velocity ranging between 40 to 280 kg/m²s. For all these refrigerants, the heat transfer coefficients increase with their inlet mass velocity. For HFC134a and HFC32 refrigerants, the transfers are respectively about 10% and 65% superior to those obtained with HCFC22 because the latent heat of HFC32 and the Prandtl number of HFC134a are higher than those of HCFC22. The results show that the vapor density affects considerably the heat transfer coefficient and the pressure drop during condensation. An experimental analysis of the condensation of HCFC22 and HFC125 [7] shows that the pressure drop and the heat transfer coefficient increase with the mass velocity of these refrigerants for inlet mass velocity values ranging between 125 to 375 kg/m^{2}s. The condensation of HCFC22 and HFC134a inside horizontal smooth tubes was An studied experimentally by Barrada et al. [14] who showed that, for important inlet vapor mass velocity, the heat transfer coefficient increases quickly with the vapor fraction in the annular flow region. For small inlet mass velocity, corresponding to stratified flow regime the heat transfer coefficient evolves slightly with the vapor fraction until a value of 50%. A comparison between pressure drop values deduced from Lockart-Martinelli correlation [22] and those measured showed an overall average deviation of 30% for high values of inlet mass velocity. Therefore, the pressure drop and the heat transfer coefficient during condensation are strongly influenced by the thermo-physical properties and the inlet mass vapor velocity of the refrigerant.

Dobson and Chato [17] have developed a detailed analysis of condensation in smooth horizontal tubes. They proposed a new correlation for the condensation in annular and wavy flow regimes, based on two broad flow regime categories: the gravity and the shear flow regimes. The correlation was evaluated with various experimental results obtained for HCFC22, HFC134a, CFC12, with mean deviations varying between 4.4 and 13.7%. This correlation has been developed for smooth tubes with small diameters (3 and 7 mm), for mass flux ranging from 25 to 800 kg/m²s, heat flux density varying between 5 and 15 kW/m² and vapor saturation temperatures ranging between 35°C and 45°C, and vapor fraction varying between 90 and 10%. Several earlier correlations such as proposed by Akers and Rosson [1], Altman et al. [2] have been reported but were not suited for new refrigerants. As indicated by Tichy et al. [3] and Thome [8], the range of test conditions for each correlation is very important when comparing data. It was, therefore, decided to consider correlations that suited the experimental test conditions. A discussion of these papers is given in Boisseux [18].

Most of the local correlations were deduced from empirical or semi-empirical models of the condensation and could be applied to their specific flow regimes. Indeed, it is difficult to develop a theoretical model due to the complexity of heat transfer between the two phases flow during condensation. Indeed, Chitti et al. [13] have developed a theoretical model, validated with experimental data for HCFC22 and refrigerant mixtures, but the local heat transfer was always limited to partial condensation over a limited mass flux range between 250 and 350 kg/m²s.

In this work, a theoretical analysis is proposed for pure fluids condensation in stratified flow inside horizontal smooth tube based on a mathematical formulation of two phases flow equations. An analytical model is developed to study the local heat transfer around the tube perimeter and the pressure drop along the tube cross sections.

MODEL MATHEMATICAL FORMULATION

Figure 1 shows the proposed physical model constituted of a horizontal smooth tube with an inside diameter *d*, and a length *L*. The wall of the tube is subject to a constant heat flux density $\overline{q}$. We associated a Cartesian coordinate system where *x* is the tangential direction around the periphery, *y* is the radial direction, *z* is the axial direction and θ is the inclination angle, which varies with the distance *x* around the periphery. *u*, *v* and *w* are respectively the velocities in *x*, *y* and *z* directions.

At the entry of the horizontal tube (*z=0*), a saturated vapor at a temperature T_{s_0} and a mass flux velocity G_{v_0} , flows in the *z*-direction and condenses around the internal wall surface. Along the tube, the vapor temperature is supposed to be equal to the local saturation temperature (T_s = $T_s(z)$). The liquid flows down tangentially in the *x*-direction, around the tube perimeter under the interaction of the interfacial liquid-vapor shear stress $\tau_i(z)$ and the tangent component of the gravity F_{gx}. At the bottom of the tube, the accumulated liquid flows axially, in the *z*-direction under the interaction of the interfacial liquid-vapor shear stress. The thickness of this layer is characterized by the stratification angle θ_m.

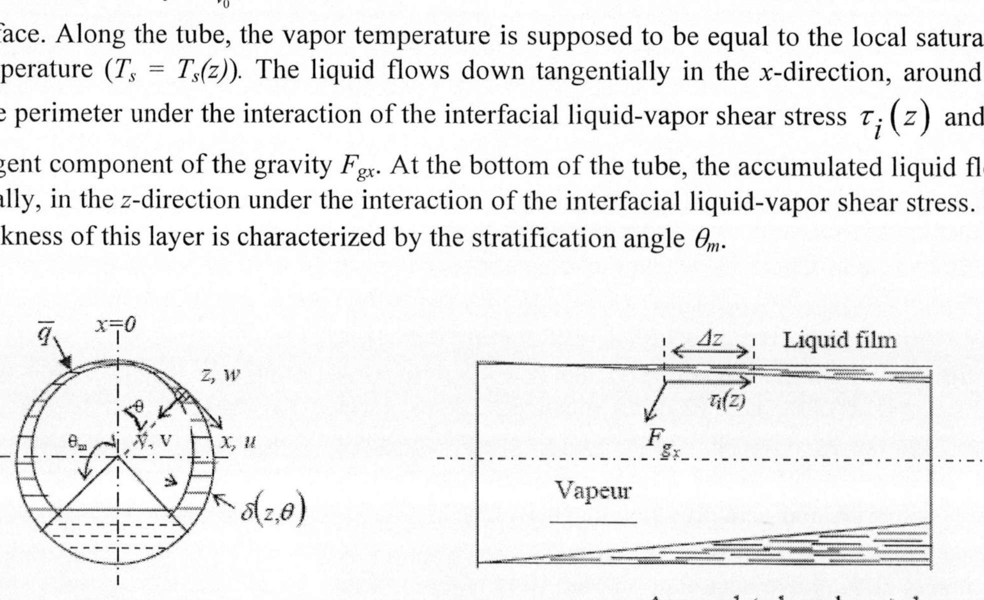

Figure 1. Physical model of stratified flow condensation inside horizontal smooth tube.

The heat transfer in the liquid film is governed by the field equations [23-25]. The analysis of heat and mass transfers in the liquid film can be simplified by the following assumptions:

a) the liquid and the vapor phases behavior is a Newtonian,
b) the thermo-physical properties of fluids are calculated at the local saturation temperature and their variations are neglected in a tube cross section,
c) the vapor-condensate interface is smooth (not wavy),
d) the flow regime is stratified along the length of the tube (the modified Froude number is less than 2.06 [26]),
e) the liquid film flow is laminar and stationary (the film thickness is very small compared to the tube diameter).

Taking into account the above assumptions [27-28], the mass, momentum and energy balances in the liquid film yield to the following governing equations written in the *(O,x,y,z)* coordinate system:

- Continuity equation

$$\frac{\partial u}{\partial x}+\frac{\partial v}{\partial y}+\frac{\partial w}{\partial z}=0 \qquad (1)$$

- x-component of momentum equation

$$\frac{\partial^2 u}{\partial y^2}=-\frac{(\rho_l-\rho_v)g\sin\theta}{\mu_l}=-\frac{F_{g_x}}{\delta\mu_l} \qquad (2)$$

where: $F_{g_x}=\delta(\rho_l-\rho_v)g\sin\theta$

- z-component of momentum equation

$$\frac{\partial^2 w}{\partial y^2}=\frac{1}{\mu_l}\frac{\partial P}{\partial z} \qquad (3)$$

- Energy equation

$$u\frac{\partial T}{\partial x}+v\frac{\partial T}{\partial y}=\frac{\lambda}{\rho_l c_p}\frac{\partial^2 T}{\partial y^2} \qquad (4)$$

To close the above equations system, the following boundary conditions and interfacial balance equations are considered:

- Boundary conditions for film velocity and temperature:

At $y = 0$ (tube wall) $u = w = 0$ (5)

At $y = \delta$ (interface liquid-vapor) $T = T_s(z)$ (Thermal equilibrium) (6)

- x-tangential component of interfacial momentum balance (no slide interface)

$$\left.\frac{\partial u_l}{\partial y}\right)_{y=\delta} = \left.\frac{\partial u_v}{\partial y}\right)_{y=\delta} = 0 \tag{7}$$

- z-tangential component of interfacial momentum balance

$$\mu_l \left.\frac{\partial w_l}{\partial y}\right)_{y=\delta} = \mu_v \left.\frac{\partial w_v}{\partial y}\right)_{y=\delta} = \tau_i(z) \tag{8}$$

- Interfacial mass balance

$$\dot{m}_{l_i} = \rho_l \left(v - u\frac{\partial \delta}{\partial x} - w\frac{\partial \delta}{\partial z} \right)_{y=\delta} \tag{9}$$

Determination of the Liquid Film Local Thickness

The integration of equations (2) and (3) across the film thickness, associated to the boundaries conditions (5) and (7) gives:

$$u(y) = \frac{(\rho_l - \rho_v) g \sin\theta}{\mu_l}\left(\delta y - 0.5y^2\right) = \frac{F_{gx}}{\mu_l}\left(y - 0.5\frac{y^2}{\delta} \right) \tag{10}$$

$$w_l(y) = \frac{B_1}{2} y^2 + \left(\frac{\tau_i}{\mu_l} - B_1 \delta \right) y \tag{11}$$

where: $B_1 = \frac{1}{\mu_l}\left(\frac{dP}{dz}\right)$

The equation of the local film thickness is obtained by integration of the continuity equation (1) with respect to y direction:

$$\frac{\partial}{\partial x}\int_0^{\delta} u dy + \frac{\partial}{\partial z}\int_0^{\delta} w dy = \left(u\frac{\partial \delta}{\partial x} + w\frac{\partial \delta}{\partial z} - v\right)_{y=\delta} = \left(\frac{\overline{q}}{\rho_l h_{fg}}\right) \tag{12}$$

The combination of the equations (10), (11) and (12) gives a partial differential equation of the local film thickness $\delta(\theta, z)$ distribution:

$$\frac{\partial}{\partial x}\left(\frac{F_g}{3\mu_l}\delta^2\right) + \frac{\partial}{\partial z}\left(\frac{B_1\delta^3}{6} + \left(\frac{\tau_i}{\mu_l} - B_1\delta\right)\frac{\delta^2}{2}\right) = \frac{\overline{q}}{\rho_l h_{fg}} \tag{13}$$

This equation can be expressed as:

$$C_l\delta^2 \sin\theta\left(\frac{\partial \delta}{\partial \theta}\right) + \frac{C_l\delta^3}{3}\cos\theta - \frac{\rho_l}{\mu_l}\left(\delta^2\frac{dP}{dz}\right)_f - \tau_i\delta\Bigg)\left(\frac{\partial \delta}{\partial z}\right) + \left(\frac{\rho_l\delta^2}{2\mu_l}\right)\left(\frac{d\tau_i}{dz}\right)$$
$$-\left(\frac{\rho_l\delta^3}{3\mu_l}\right)\frac{d}{dz}\left(\frac{dP}{dz}\right)_f - \left(\frac{\overline{q}}{h_{fg}}\right) = 0$$

(14)

Where: $C_l = \dfrac{2g\rho_l(\rho_l - \rho_v)}{\mu_l d}$

The film thickness is determined by the resolution of the equation (14) by using Matlab Runge-Kutta function (ODE45). The local frictional pressure drop $\left.\frac{dP}{dz}\right)_f(z)$ and the liquid-vapor shear stress $\tau_i\,(z)$ are determined by the finite difference method.

Analysis of Pressure Drop

During the condensation inside horizontal smooth tubes, the pressure drop is generated by the liquid streaming and the vapor flow along the tube under the interaction of the pressure strengths and the liquid-vapor shear stress. For the liquid and vapor phases, the local pressure drop can be yield as the following relations from one dimensional momentum balance equations of liquid and vapor phase respectively:

$$A(1-\alpha)\frac{dP_l}{dz} = -\tau_i S_i - \tau_{pl} S_{pl} - A(1-\alpha)\rho_l w_l \frac{dw_l}{dz} + A(1-\alpha)\rho_l g_z - \dot{m}_i S_i (w_i - w_l) \tag{15}$$

$$A\alpha\frac{dP_v}{dz} = -\tau_i S_i - \tau_{pv} S_{pv} - A\alpha\rho_v w_v \frac{dw_v}{dz} + A\alpha\rho_v g_z - \dot{m}_i S_i (w_i - w_v) \tag{16}$$

The frictional pressure drop is calculated from the momentum vapor equation (16) and can be written as:

$$\left.\frac{dP}{dz}\right)_f = \frac{-\tau_{pv} S_{pv} - \tau_{iv} S_i}{A_v} \tag{17}$$

where: $\tau_{pv} = \frac{1}{2} f_v \rho_v w_v^2$; $\tau_i = \frac{1}{2} f_i \rho_v \left(w_v^2 - w_l^2\right)$

f_i is the interfacial friction factor given by Dorokhov expression [29]: $f_i = f_v \frac{(1-0,25b)^2}{(1+0,25b)^{0,2}}$

with: $b = -2\tilde{J}/f_v$; $\tilde{J} = \bar{q}/h_{fg}\rho_v w_v$

The vapor friction factor f_v can be expressed in the form proposed by Lockart and Martinelli [22], defined by the generalized Blasius correlation: $f_v = c_v \,\mathrm{Re}_v^{-m}$, where:

$c_v = 16$ $m = 1$	for laminar flow
$c_v = 0.046$ $m = 0.2$	for turbulent flow

In addition, it may be assumed that $w_v^2 >> w_l^2$ and that the interfacial shear stress is equal to the vapor shear stress at the wall ($\tau_i = \tau_{pv}$). The frictional pressure drop can be also written as:

$$\left.\frac{dP}{dz}\right)_f = -\tau_i\left(\frac{S_{pv} + S_i}{A_v}\right) = -\frac{4\tau_i}{d_{hv}} \tag{18}$$

The interfacial shear stress coefficient τ_i is evaluated by using the two phase flow model [29]: $\tau_i = \frac{1}{2} f_i \rho_v w_v^2$.

The overall average pressure drop can be calculated by the integration of the local pressure drop along the tube:

$$\Delta P = \frac{1}{L}\int_0^L \left.\frac{dP}{dz}\right)_t (z)dz \tag{19}$$

Local and Average Local Heat Transfer Coefficient

In a stratified flow (figure 1), heat transfer through the liquid film is mainly governed by conduction. Then, at a wall position (z,θ), the local heat transfer coefficient by conduction is calculated by:

$$h(z,\theta) = \frac{\overline{q}}{\left[T_s(z) - T_p(z,\theta)\right]} \tag{20}$$

where:

$T_s(z)$: vapor saturation wall temperature

$T_p(z,\theta)$: inner wall temperature

$\overline{q}$: average heat flux density imposed at the tube wall

At a cross section located at a distance z from the inlet of the horizontal tube, the local average heat transfer coefficient around the perimeter of the tube is defined as:

$$h_c(z) = \frac{1}{\theta_m}\int_0^{\theta_m} h(z,\theta)d\theta \tag{21}$$

The average local heat transfer coefficient $h(z)$ around the tube perimeter is obtained by Butterworth equation [30] by combining the solution of the equation (21) and the Meyer prediction [31] for the accumulated liquid at the bottom of the horizontal tube:

$$h(z) = \frac{\theta_m}{\pi} h_c(z) + \frac{\pi - \theta_m}{\pi} h_l(z) \tag{22}$$

Results and Discussion

In this paragraph we propose to use the developed model to simulate the condensation inside horizontal smooth tubes, at stratified two phases flow, of four pure refrigerants HCFC22, HFC134a, HFC32 and HFC125. To validate the model, simulation conditions are chosen on the base of the experimental data available in literature [5,20]. The analysis is developed here by means of local and average local heat transfer coefficient around the tube perimeter and the average pressure drop evolutions along the tube.Figures 2a to 5a show the evolution of the local liquid film thickness $\delta(z,\theta)$ around the perimeter of the tube, at a cross section z=0,04 m from the tube inlet [20], for different values of inlet mass velocity G_{vo} and heat flux density $\overline{q}$ corresponding to stratified flow conditions [20]. The liquid film is thinner at the top of the tube, its thickness increases with the inclination angle θ because of the film falling. For high inlet vapor mass velocities the film thickness decreases because the liquid film is streaming by the vapor flow. The same figures show also that for constant inlet mass velocity G_{vo} and saturation temperature $T_{so,}$ the liquid film thickness increases with the heat flux density. Indeed, the heat transfer through the tube wall allows vapor condensation yields and then the increase of liquid film thickness.

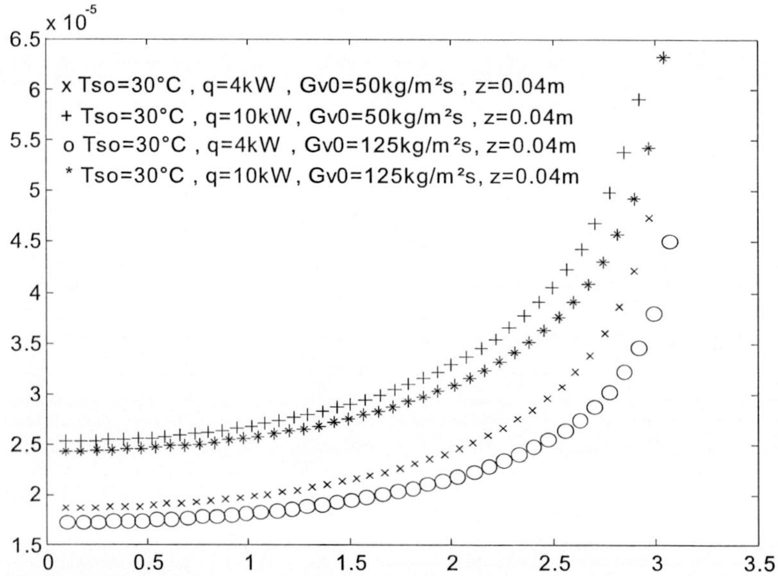

Figure 2a. HCFC22 liquid film thickness evolution around the tube perimeter.

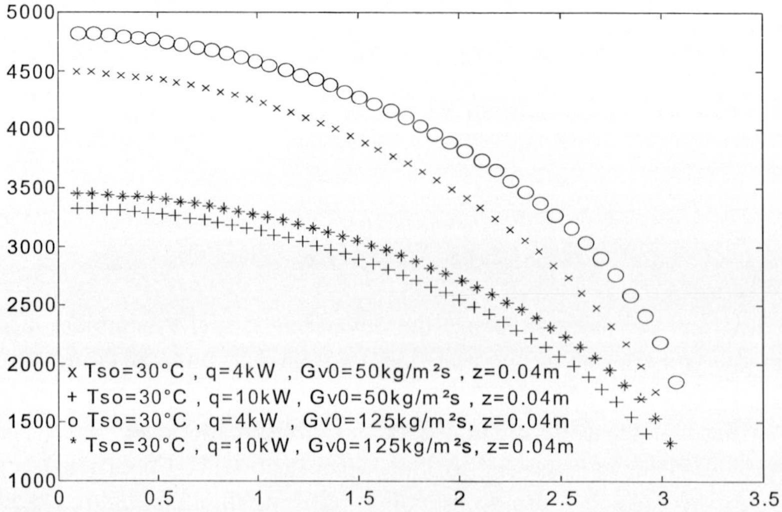

Figure 2b. HCFC22 local heat transfer coefficient evolution around the tube perimeter.

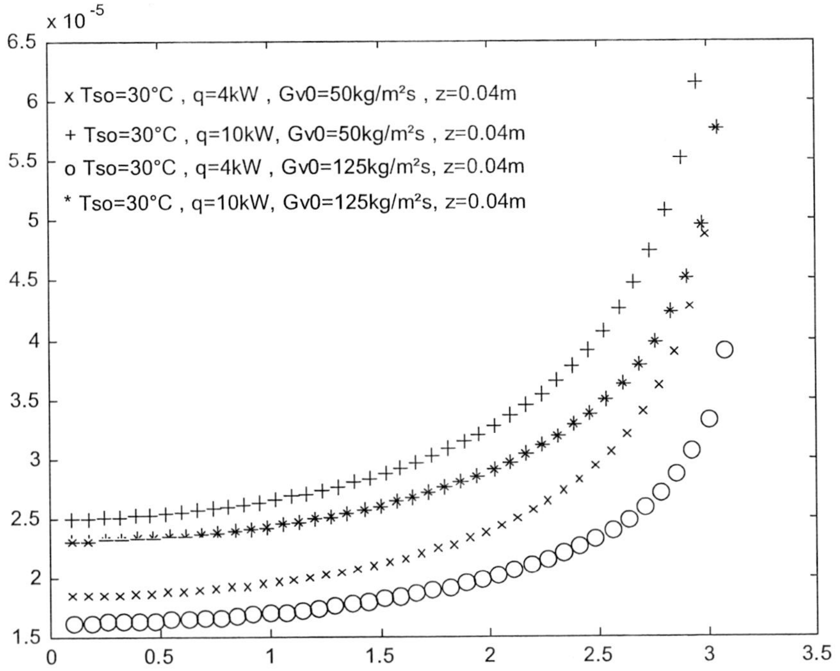

Figure 3a. HFC134a liquid film thickness evolution around the tube perimeter.

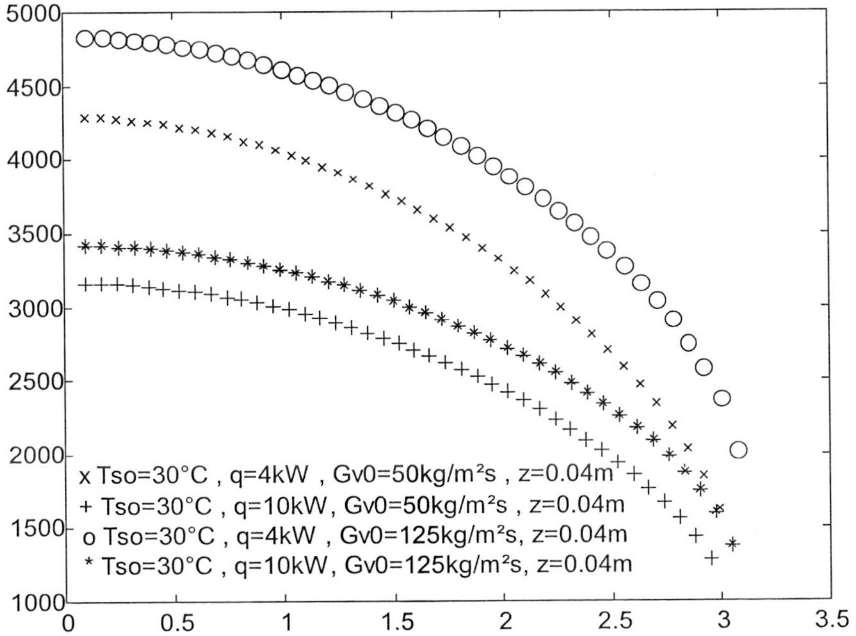

Figure 3b. HFC134a local heat transfer coefficient evolution around the tube perimeter.

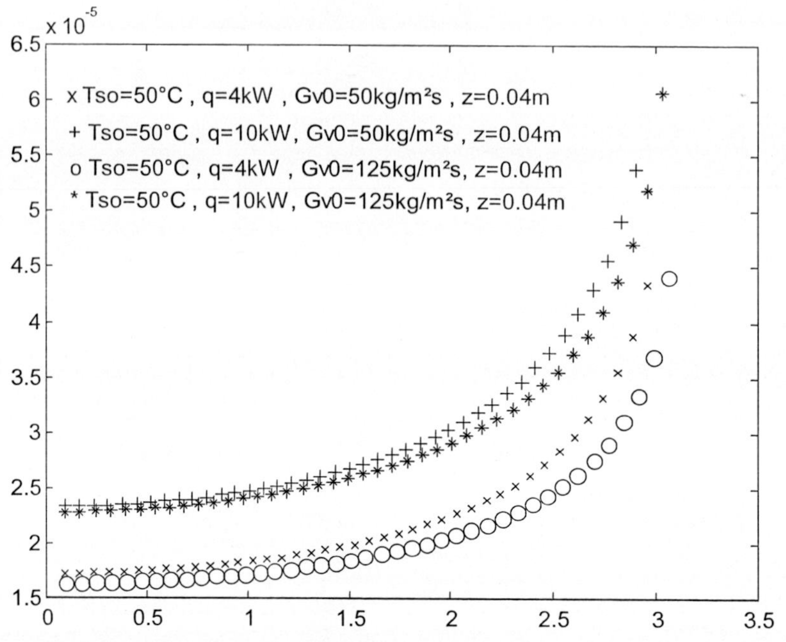

Figure 4a. HFC32 liquid film thickness around the tube perimeter.

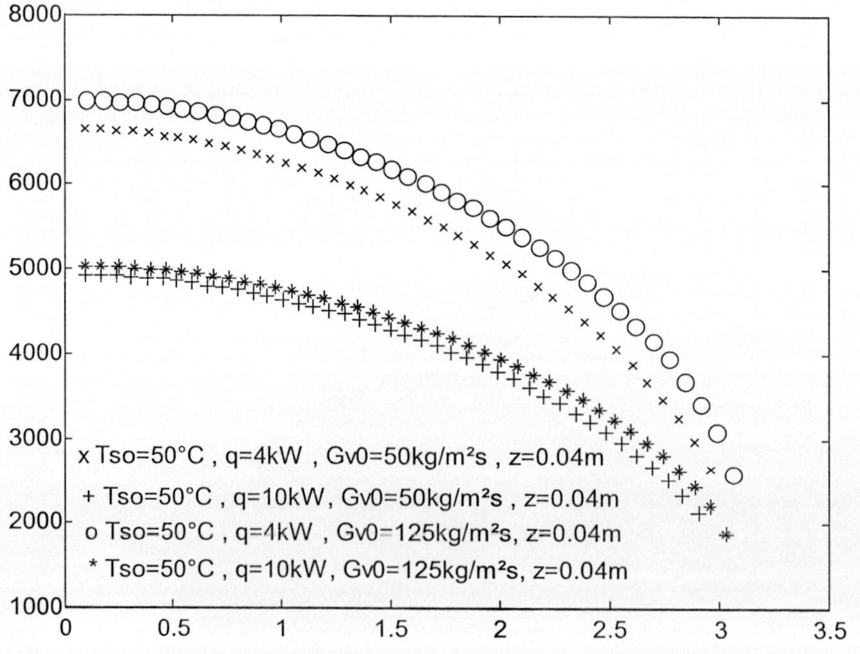

Figure 4b. HFC32 local heat transfer coefficient evolution around the tube perimeter.

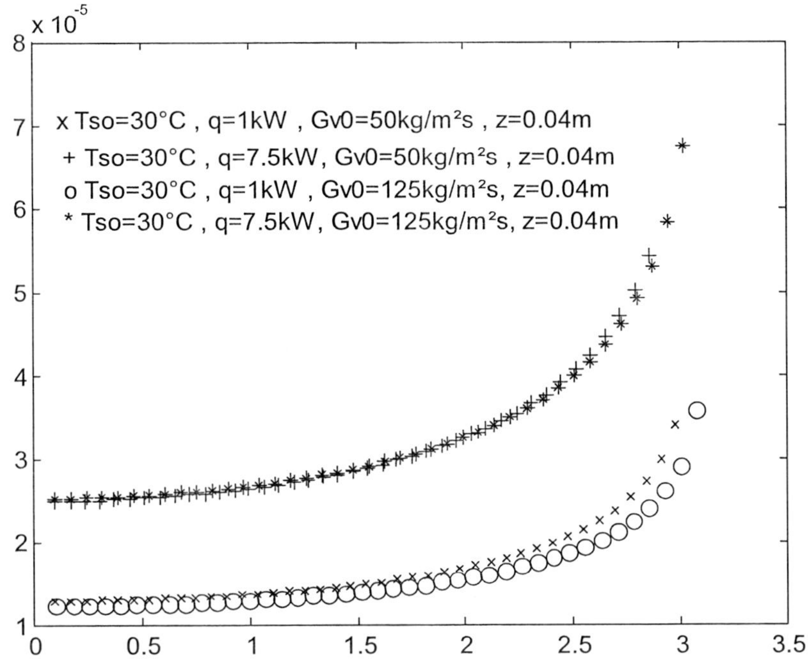

Figure 5a. HFC125 liquid film thickness around the tube perimeter.

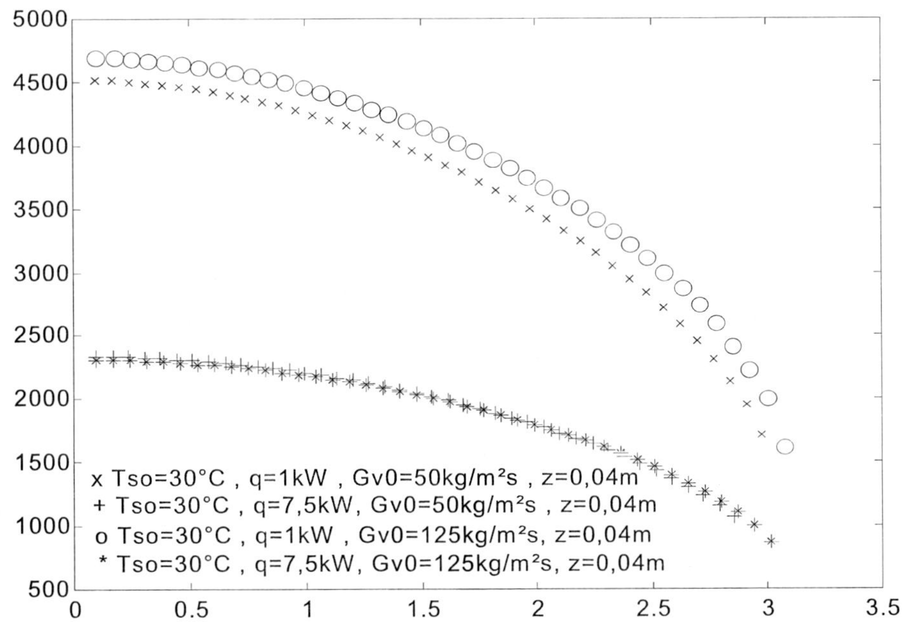

Figure 5b. HFC125 local heat transfer coefficient evolution around the tube perimeter.

Figures 2b to 5b show the evolution of the local heat transfer coefficient $h(z, \theta)$ around the perimeter of the tube, at a cross section z=0,04 m from the tube inlet [20], for different values of inlet mass velocity G_{vo} and heat flux density $\overline{q}$ corresponding to stratified flow conditions [20]. The local heat transfer coefficient, which is inversely proportional to the film thickness decreases continuously around the periphery of the tube. In the stratified flow zone, the local heat transfer coefficient decreases with the angular position because the film thickness increases (figures 2a to 5a). The local heat transfer coefficient is very sensible to the increase of the heat flux, and decreases quickly around the circumference of the test cross section.

Figure 6 represents the evolution of the average local heat transfer coefficient $h(z)$ with respect to the inlet vapor mass velocity G_{vo} and the vapor saturation temperature T_{so} for heat flux densities $\overline{q}$ ranging from 11 to 62 kW/m² corresponding to stratified flow regime [20]. In theses conditions, the heat transfer coefficient varies slightly with the flow vapor fraction until a value of 0.7. Beyond this zone, the heat transfer coefficient increases very quickly with the vapor fraction. The slope evolution of the heat transfer coefficient with respect to the mass velocity (figure 6) can be explained by the change of the vapor phase during the flow which is governed by the condition of Lockart and Martinelli [22]. A comparison is made with experimental results of Cavallini et al. [20] shows an average deviation of about 2.2% for HCFC22, 5.7% for HFC134a, 37.7% for HFC32 and 5.3% for HFC125. The important deviation for R32 is due to its high density that effects the liquid thermal conductivity which has an important effect on the heat transfer coefficient.

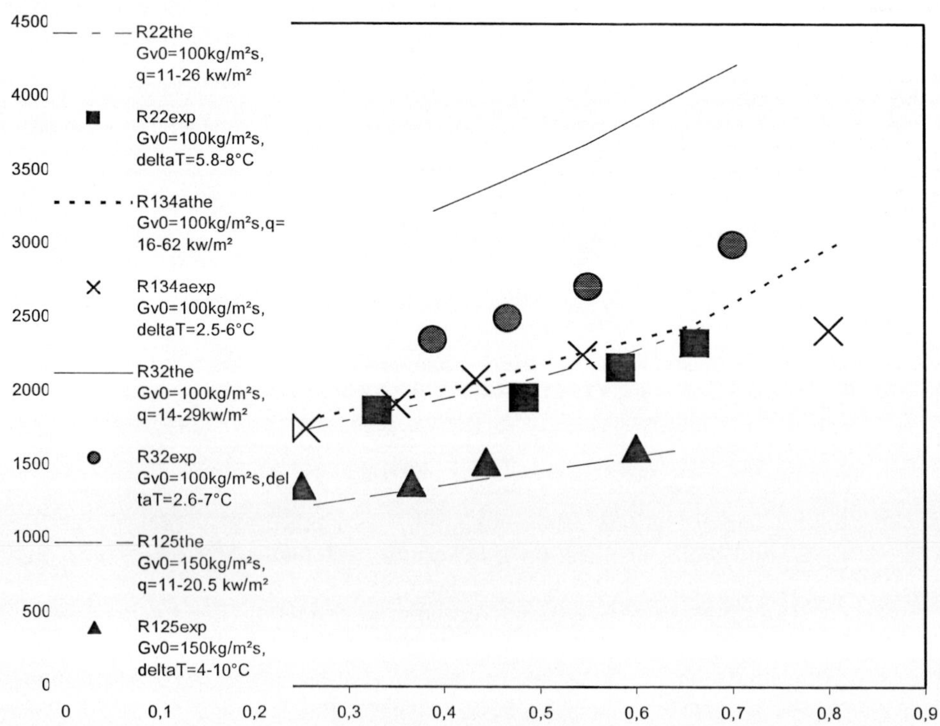

Figure 6. Comparison of the evolution of the average local heat transfer coefficient with experimental data [20].

As shown in figure 7, the absolute value of the average pressure drop is an increasing function of the inlet vapor mass velocity. Indeed, the frictional pressure drop generated by the stratified liquid layer at the bottom of the tube increases with the inlet vapor mass velocity. This phenomenon is more and more amplified along the tube and accompanied by an increase of the average pressure drop. A comparison between our results and those of Torikoshi [5] gives an overall average deviation of 1% for the three refrigerants.

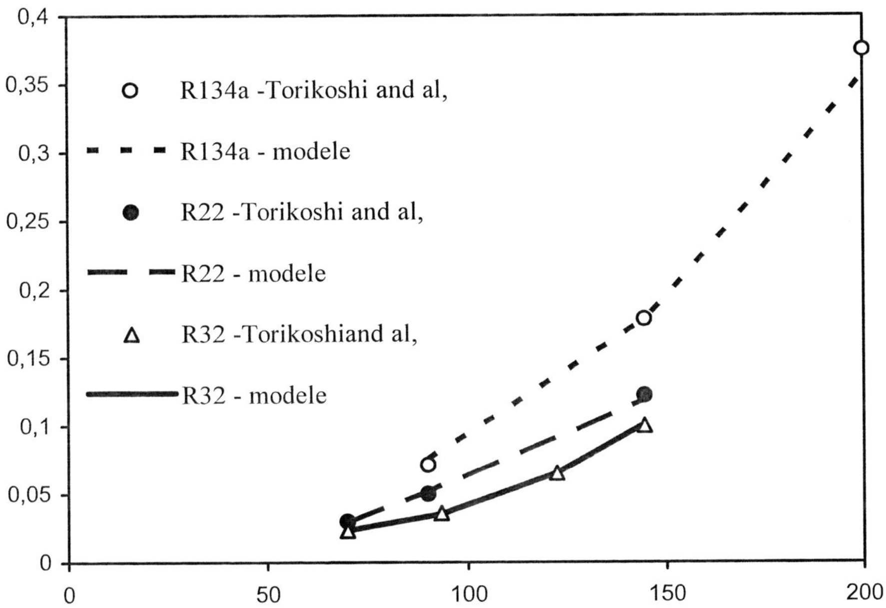

Figure 7. Comparison of pressure drop evolution with experimental data [5].

Conclusion

An analytical model of two phases flow equations for the condensation of pure vapor fluid inside a horizontal smooth tube at stratified flow regime have been developed and used to study the effects of the inlet mass vapor velocity, the vapor saturation temperature and the heat flux density on the local heat transfer and the average pressure drop during the condensation at the stratified regime of the four refrigerants HCFC22, HFC134a, HFC32 and HFC125. An overall good agreement between the obtained results and the experimental data available in the literature. The developed model will be extended, in a next work, to study the condensation of non-azeotropic refrigerants mixture.

References

[1] WW. Akers, HF Rosson, Condensation inside a horizontal tube, *Chem. Eng. Prog. Symp.* Ser, 56, 145-9, 1960.

[2] M. Altman, F.W. Staub and R.H. Norris, Local and average heat transfer and pressure drop for refrigerants evaporating in horizontal tubes, *Journal of Heat Transfer, Transactions of ASME,* August, pp.18, 1960.

[3] JA. Tichy, NA Macken, WMB. Duval, An experimental investigation of heat transfer in forced convection condensation of oil-refrigerant mixtures, *ASHRAE Transactions,* 91 (Part 1a), 297-308, 1985.

[4] S. Koyama, L. Gao et T. Fujii, Condensation du CFC-12 et son substitut le HFC-134a à l'intérieur d'un tube horizontal lisse, Actes 18e Congr. Int. Froid, Montréal/Proc., pp.509-513, tome2, 1991.

[5] K. Torikoshi, T. Ebisu, Heat transfer and pressure drop characteristics of R-134a, R-32 and mixture of R-32/R134a inside horizontal tube, *ASHRAE Transactions: Research,* pp. 90-96, 1993.

[6] M.K. Dobson, J.C. Chato, D.K. Hinde, S.P. Wang, Experimental evaluation of internal condensation of refrigerants R12 and R134a, ASHRAE Trans, US, vol.100, N°1, 1994.

[7] T.M. Doerr, S.J. Eckels and M.B. Pate. In-tube condensation heat transfer of refrigerant mixtures. *ASHRAE Transactions,* 1994.

[8] JR. Thome, Two phase heat transfer to new refrigerants, Proceedings of the Tenth International Heat Transfer Conference, Brighton, UK, 19-41, 1994.

[9] S. Koyama, L. Gao et T. Fujii, Condensation of pure vapor refrigerants inside horizontal smooth tube, Eurotherm Semin., Paris/Adem/GRETh, pp.132-138, 1995.

[10] D.W. Shao, E. Granryd, An experimental comparison of R502 and R404a in a condensing tube, Proc. Eurotherm Seminar, *Heat Transfer in Condensation,* N°47, pp. 125 – 131, 1995.

[11] Shao, D.W., Granryd, E., Heat transfer and pressure drop of HFC134a-oil mixtures in horizontal condensing tube, International journal of refrigeration, vol.18, N°8, pp.524–533, 1995.

[12] TN. Tondon, HK Varma, CP Gupta, Heat transfer during forced convection condensation inside horizontal tube, *Int. J. Refrig,* 18 (3), 210-4, 1995.

[13] M.S. Chitti and N.K. Anand, A heat transfer correlation for condensation inside horizontal smooth tubes using the population balance approach, *Int. J. Heat Mass transfer,* 39 (14), 2974-56, 1996.

[14] N. Barrada, Ch. Marvillet, A. Bontemps et Daoudi. Transferts de chaleur lors de la condensation du mélange HFC22/HFC134a à l'intérieur d'un tube lisse horizontal. *Int. J. of Refrigeration*, Vol. 19, n°7, pp. 463-472, 1996.

[15] X. Liu, Condensing end evaporating heat transfer and pressure drop characteristics of HFC-134a and HCFC-22, ASEM Trans, vol.119, February 1997.

[16] D.W. Shao, E. Granryd, Experimental and theoretical study on flow condensation with non-azeotropic refrigerant mixtures of R32/R134a, *International journal of refrigeration,* vol.21, N°3, pp.230–246, 1998.

[17] M.K. Dobson, J.C. Chato, Condensation in smooth horizontal tubes, Journal of Heat Transfer, *transactions of ASME,* 120, 193-213, 1998.

[18] X. Boissieux, Two phase local heat transfer correlations for non-ozone depleting refrigerant-oil mixtures, Ph.D. thesis, University of Brighton, UK, 1998.

[19] J. Eckels, J. Unruh, Local heat transfer coefficients during condensation of R22 and R32/R125 mixtures, *HVAC and Research,* vol.5, N°1, pp.59-76, 1999.

[20] A. Cavallini, G. Censi, D. Del Col, L. Doretti, G.A. Longo, L. Rossetto, Experimental investigation on condensation heat transfer and pressure drop of new HFC refrigerants (R134a, R125, R32, R410a, R236a) in a horizontal smooth tube, *Int. J. of Refrigeration,* 24, pp. 73-87, 2001.

[21] PK Panday, Condensation en film, In: Ecole d'été du groupement universitaire de thermique (GUT) "Transfert de chaleur et de la matière avec changement de phase", 1990.

[22] R.W. Lockhart et R.C. Martinelli, Proposed correlation of data for isothermal tow-phase, two-component flow in pipes, *chem. Eng. Pro., vol. 45*, p 39-48, 1949.

[23] K. Halouani et A. Ayadi, Analyse locale des mécanismes de la condensation en régime stratifié à l'intérieur d'un tube horizontal, *Entropie,* N°207, pp.53-60, 1997.

[24] K. Halouani, A. Ayadi, F. Halouani et M. Feidt, Etude hydrodynamique et thermique de la condensation en film à l'intérieur d'un tube horizontal, *Rev. Gé. de Thermique,* Vol. 36, n°9, pp.655-665, 1997.

[25] A. Boussetta, K. Halouani, A. Ayadi et B. Zeghmati, Modélisation du transfert de chaleur et de la perte de charge lors de la condensation d'un réfrigérant pur à l'intérieur d'un tube lisse horizontal, *Entropie, N°243* pp. 32-39, 2002.

[26] R.G. Sardesai, R.G. Owen and D.J. Pulling. Flow regimes for condensation of a vapor inside a horizontal tube. *Chem. Eng. Science, Vol. 36,* n°9, pp. 1173-1180, 1981.

[27] K. Halouani, Contribution à la modélisation de la condensation en régime stratifié à l'intérieur d'un tube horizontal - Application à la condensation de la vapeur d'eau. Thèse de 3ème cycle, Ecole Nationale d'Ingénieurs de Sfax, 1996.

[28] I.Y. Chen, Heat transfer analysis of falling-film horizontal tube evaporator, Doctoral Thesis, University of Wisconsin-Milwaukee, Milwaukee, Wisconsin, 1984.

[29] A.R. Dorokhov, Effect of condensing vapor velocity on heat transfer. In *the Problems of Hydrodynamics and Heat Transfer,* pp.54-58. Izd. SO AN SSSR, Novosibirsk, 1972.

[30] Butterworth D. et Hewitt G. F. *Two-phase flow and heat transfer,* Oxford University Press, England, (1977).

[31] J. A. Myers, *Condensation inside a horizontal pipe, studied with a heat meter,* Ph.D. Thesis, Univ. of Kansas, Lawrence, Kansas, 1964.

In: Environmental and Regional Air Pollution
Editors: D. Gallo, R. Mancini

ISBN: 978-1-60692-893-6

Chapter 9

PROFILING THE IMPLEMENTING OF PRODUCT STEWARDSHIP: A CASE STUDY OF SOME CHEMICAL INDUSTRIES IN GERMANY

Dieudonne Alemagi*

Environemntal and Resource Management,
Department of Industrial Sustainability,
Brandenburg University of Technology,
Cottbus, Germany

ABSTRACT

This paper had the fundamental objective of assessing product stewardship initiatives designed by a plethora of chemical industries in Germany. The paper justifies the need for product stewardship in Germany's chemical industry's and presents the results of a survey of some chemical industries in Germany in June 2003. A total of 122 questionnaires were administered to environmental managers in the selected industries using systematic random sampling. The return rate was 51% corresponding to 66 industries. Survey results revealed that despite the inadequacy in the implementation of some product stewardship tools like emission reduction (particularly solid waste emission), implementation of product stewardship tools by the sampled industries was advanced. It was concluded that most of the attempts made thus far by industries to address product stewardship have been major successes. However, it was emphasized that further improvement could be achieved if incentives for product stewardship implementation are promoted and the merits of strategic research in the area of reduction in solid waste emission is further developed and encouraged.

Keywords: Germany, Product Stewardship, Product Stewardship Tools, Chemical Industries.

---

* Correspondence to the author can be made by post, email or telephone at: Dieudonne Alemagi, Department of Industrial Sustainability, Brandenburg University of Technology, Cottbus; Postfach 101344, 03013 Cottbus, Germany. Email: dalemagi@yahoo.co.uk , Telephone: (0049) 0355 694 653, Fax :(0049)0355 694 700.

1. Introduction

The global production of chemicals increased from 1 million tonnes in 1930 to 400 million tonnes in 2001. The world production of chemicals in 2001 was estimated at 1,632 billion Euros. The European Union chemical industry accounted for 32% of this total amount, which generated a trade surplus of 41 Billion Euros (CEFIC 2002). The Federal Republic of Germany is the largest chemical producer in Europe and most of Germany's largest chemical industries are located in North – Rhine Westphalia State of (Figure 1).

Figure 1. Location of major chemical industries in the North – Rhine Westphalia State of Germany (Source: Lemke, K.W, 2002).

The chemical industry is one of the key branches of the German economy and the growth engine for innovations and technological advance. The production and industrial uses of chemicals by Germany's chemical industry makes a vital contribution to the economic and social well being of citizens particularly in terms of trade and employment.

Chemicals, their production and uses bring about benefits on which modern society is entirely dependent (table 1).

This, notwithstanding, chemicals, their production and uses by chemical industries have caused serious damage to human health resulting in suffering and premature death as well as damage to the environment in general. Well-known examples include asbestos that is known to cause lungs cancer and benzene which leads to leukemia. Abundant use of DDT resulted to reproductive disorder in birds (Commission of the European Communities, 2001). The incidence of some disease, e.g. testicular cancer in young men and allergies, has increased significantly over the last decades. While the underlying reason for this has not been identified, there is justified concern that certain chemicals play a causative role for allergies. Corporate initiatives have been formulated by chemical industries in an attempt to mitigating the adverse impacts of chemicals on the environment. In Germany, the chemical industry has made key strides in the arena of Product Stewardship as a rigorous strategy to attain excellence in their pursuit for environmental protection.

Table 1. Industrial uses of some chemicals (Source: Clarkson R.J, 1998)

Chemical	Uses
Aluminium	Cooking foil and food containers. In alloys for making aircraft, Electric cables
Ammonia	Manufacture of nitric acid and ammonium salts
Chlorine	Batteries
Nitrate	Fertilizers/Explosives
Sulphate	Fertilizers
Chlorine	As a bleach, for cotton and paper, in making PVC
Copper	Electric cables
Silver Compounds	Photography
Zinc	Galvanizing iron, making brass
Sulphuric Acid	Phosphate and ammonium fertilizers, detergent pains
Sodium Hydroxide	For making soap
Polymers (in general)	Electrical insulators
Steel (Stainless)	Large- scale chemical reaction vessels and Cutlery production
Steel (Mild)	Car bodies
Calcium Carbonate	Manufacturer of iron and cement

The European Chemical Industry Council (CEFIC) defines product stewardship or Extended Product Responsibility (EPR) as the responsible and ethical management of the safety, health and environmental aspects of a product throughout its total life cycle. In other words, products must be managed and used safely every step of the way, through development, manufacture, packaging, distribution, use and ultimate disposal (CEFIC, 2002). Product stewardship is an essential part of the Responsible Care Program within the chemical industry and as Cramer *et al* (1999) opine, this branch of the industry is setting a trend for other branches. However, effluents emanating from chemical industries are still a nuisance to safety, health and the environment.

The purpose of this paper is therefore to evaluate initiatives taken by a plethora of chemical industries in Germany towards the implementation of product stewardship and to prescribe robust recommendations for improvement.

The paper is organized as follows. In the following section I begin by providing a background to the paper. This sector of the paper presents the importance of Germany's Chemical industries from a socio-economic standpoint. In the penultimate part of this section, tool for product stewardship are outlined.

This section terminates with a justification for product stewardship in Germany's chemical industry. The research methodology is outlined in the next section of the paper. In the section that follows, the research results are presented and reflected upon. Finally, the paper is concluded with a way forward with regards to product stewardship enhancement in Germany's chemical industry.

2. BACKGROUND

2.1. The Socio- Economic Importance of Germany´s Chemical Industry

The German chemical industry does not only consist of large international companies. In the year 2000, the share of small and medium sized enterprises (< 500 employees) was quite considerate as well. Small and Medium Size companies (SMEs) provided 35 percent of all jobs in Germany's chemical industry and realize 36% of total chemical sales in 2001. The German chemical industry is the third biggest producer of chemical products worldwide. In terms of employment, this industry employs more people than any other European country or Japan (Table 2). As detailed in table 2, Germany was the second largest exporter of chemicals in 2001 (largest exporter: USA), with a world market share of nearly 12.2 percent (VCI, 2002).

Table 2. Indicators of chemical industry in the world main chemical producer nations

2001	USA	Japan	Germany	France	Italy	UK
Sales (in bill. Euro)	476.0	237.9	133.9	85.0	72.9	56.9
Employees (in 1000)	1033	366	467	239	207	235
Exports (in bill. Euro)	87.6	37.6	81.4	50.2	25.5	46.0
Imports (in bill. Euro)	86.2	25.8	54.8	40.2	33.7	39.0

Source: VCI, 2002.

From a European perspective, Germany is Europe's largest producer of chemicals. It realized 26% of the total European Union (EU) sales in 2001 (Table 3). Indeed, the chemical industry in Germany offers almost every third job within the European chemical industry.

Table 3. Germany's Chemical Industry and its contribution to the EU economy (

Major Contribution	Amount (in billion Euro)	Share of total EU	Rank within the EU
Sales	133.9	26.0%	1
Investment	7.0	27.2%	1
Imports of chemical products	54.8	18.7%	1
intra EU imports	42.2	-	1
Exports of chemical products	81.4	22.2%	1
Intra EU exports	43.4	-	1

Source: VCI, 2002.

On a national scale, the industry constitutes a key stone that underpins the German economy. In terms of sales, it is about 10 percent of Germany's manufacturing industry and as of the year 2001, it employed 467,000 people (7.4 percent of total manufacturing industry) and ranked 5th in terms of employment (VCI, 2002). New products and processes are the driving force of the international competitiveness of the chemical industry in Germany. A high trade surplus was realized by this industry in 2001 and together with the automobile industry, the industry contributed decisively to Germany's trade balance surplus in 2001. Despite the aforementioned importance of Germany's chemical industry, it is important to underscore the fact that this industry has inflicted considerable damages to safety, health and the environment. There is therefore a great need to balance the socio-economic value of this industry with the ecological needs. Thus, the next part of this section of the paper is dedicated to product stewardship tools and the fundamental need for product stewardship in Germany's chemical industry.

2.1. Product Stewardship: An Overview of Basic Tools

Germany's chemical industry has a long tradition of being concerned with environmental issues as an umbilical commitment to decrease the negative environmental impacts caused by their organization and activities and to further the process of sustainable development. Product stewardship is one of those tools that has been employed by Germany's chemical industry to sustain this commitment. Within the chemical industrial sector, the first guidance document on product stewardship was issued in 1992 by the US Chemical Manufacturers Association (CMA), soon followed by various other chemical sector organizations (E.g. the European Chemical Industry Council (CEFIC) and the Canadian Chemical Producers Association (CCPA)). Ultimately, product stewardship is all about manufacturers taking on more responsibility for reducing the health and environmental impacts of their products and packaging. It's an incentive to redesign products with fewer hazardous materials, so that they are more durable, reusable and recyclable. Lindhqvist (1992) defines extended product responsibility (EPR) as an environmental protection strategy to reach an environmental objective of a decreased environmental impact of a product, by making the manufacturer of a product responsible for the entire life cycle of a product and especially for the take-back, recycling and final disposal of the product. Lindhqvist argues that while the manufacturer, distributor, user, recycler and final disposal can all influence the environmental impacts of a product, primary responsibility rests with the manufacturer. According to Lindhqvist (1992), this is because the producer is responsible for product development and therefore has the unique opportunity to prevent environmental impacts through the adoption of environmentally conscious product development (or eco-design). Indeed, as Cramer *et al* (1999) further purport, the main purpose of product stewardship is to ensure that an industry systematically structures its safety health and environmental policy at the product level. In order to achieve this goal, a set of tools referred to as Product Stewardship Tools (PST) are necessary for product manufacturers (PPRC, 1999). These tools have been summarized by the United States of America Pacific Northwest Pollution Prevention Resource Centre as follows:

a. Design for the Environment

Design for the Environment (DFE) considers the potential environmental impacts of a product and the process used to make that product, including components and raw materials. It could be considered one facet of life cycle management. DFE principles evaluate facility and local impacts, as well as global impacts and include habitat disturbance, emissions and effluents, chemical releases, inefficient use of water and energy, solid waste, and much more. Application of DFE also considers recovery of the product at the end of its useful life, through Design for Disassembly, Design for Remanufacturing, and Design for Recycling.

b. Eco-Labeling

Eco-label programs are certification programs in which a product, process, or management system is certified to meet specific environmental criteria as established by an assessment body. Manufacturers wishing to achieve certification to an eco-label program may enjoy marketing advantages because the certification positively differentiates their products as environmentally sound. Typically, the assessment organization will certify the product/process and upon approval, allow the applicant to use a logo on the product label, packaging, and/or advertising. Typical examples include Germany's Blue Angel. Independently, a manufacturer can also label their products with non-certified, specific, environmental information, which is another form of eco-labeling. There are many international eco-labeling entities that certify products for their country, and for manufacturers who sell products in their country. Additional programs certify performance-specific features such as the energy efficiency or indoor air quality emissions of various products or processes. Eco-labeling programs have influenced product design in the areas of life-cycle impact, durability, hazards, recyclability, and more. The criteria differ for each program and can cover the product itself, and in addition, the processes used to manufacture the product, or other life-cycle impacts. Examples include efficient use of fossil fuels, reduction of greenhouse gas emission, conservation of resources, concentration restrictions on substances such as heavy metals and toxins.

c. Green Purchasing

These are procurement policies, measures, or preferences that specify environmentally preferable products and services. Businesses and consumers can use their purchasing power to leverage supplier practices, products, and decisions, for better environmental protection. By specifying purchasing preferences for products that have less environmental impact, organizations can spur manufacturers to: use recycled-content, design for recycling/reuse, consume less energy or water, minimize waste, or create less toxic products. Procurement preferences can provide an immediate market for environmentally and socially responsible products.

d. Take Back Programs

These programs are a fundamental tool for product stewardship. They provide manufacturers the physical responsibility for products and/or packaging at the end of their useful lives. Indeed, by accepting used products, manufacturers can acquire low-cost feedstock for new manufacture or remanufacture, and offer a value-added service to the buyer. This play a vital role in limiting the adverse impacts of a product during the cradle to crave process.

2.2. Why the Need for Product Stewardship in Germany's Chemical Industry?

It is my respectful submission that of all the German industries, however, the one in greatest need of sustainable development is the chemical industry. The rationale behind this establishment is that this industry is susceptible to environmental damages on several fronts. Indeed, this industry has a history of ecological damages. More than 50% of all the heavy metals and organic pollutants in the river Rhine are of diffuse origin while the rest emanate from point sources including Germany's chemical Industries located along the Rhine catchment (ICP, 2001).

The river Rhine is an important natural resource in Western Europe and flows through several nations including Germany. Pollution of the Rhine from point sources ends up in the Netherlands at the port of Rotterdam, which is found downstream region of the Rhine. The Concentration of most hazardous substances in the Rhine are slowly decreasing or have stagnated at a low level during the year 1971 to 1998 which illustrate the results of three monitoring stations along the Rhine. Nevertheless, the International Commission for the Protection of the Rhine (ICPR) target values for heavy metals in the suspended matter has not yet been delivered. The last valuations in 2001 shows that the target values for Mercury, Cadmium, Zinc and Copper have not been reached. Elevated heavy metals (Mercury, Cadmium, Zinc and Copper) concentrations are associated with degraded water and sediment quality and can result in potential impacts on human health and the environment. In the eastern harbors of the port of Rotterdam, mainly influenced by inputs from the Rhine, Zinc, Cadmium and Copper frequently exceeds the sea/slufter values in the sedimental dredged material. After the year 2000, it is likely that some 30 million meters of sedimental dredged material will still be contaminated by heavy metals to the extent that it has to be stored in the slufter (Rotterdam Municipal Port Management, 2000). A National Academy of Sciences study in the United States of America suggested that one out of four developmental and behavioral problems in children may be linked to genetic and environmental factors, including exposure to heavy metals like lead and mercury (SFDTRA, 2000). Itai-itai disease is well known as a health hazard induced by Cadmium originating from chemical industries in cadmium-polluted areas. According to Keiko *et al* (1999), the main clinical features of Itai-itai disease are: Multiple proximal renal tubular dysfunctions, hormonal imbalance, and excruciating pain in the bones and all over the body.

At the BASF (One of Germany's largest chemical industry) "Aktiengesellschaft's" site in Ludwigshafen, Germany, approximately 3 Million metric tons of hazardous goods left the site in some 118,000 shipments by road, rail, inland waterways, sea and air in 2001. Four accidents pertaining to the shipment of these hazardous goods occurred (BASF 2001). During the year 1999, many operational disturbances and scenario accidents were recorded in BASF AG. BASF is located in the catchment of the river Rhine and most of the accidents recorded at BASF could pose severe threats to the river Rhine and the environment in general. In January 27, 1999, during the course of cleaning work, a dye was accidentally released into the environment through the wastewater treatment plant. Furthermore, a similar case of industrial contamination was recorded on November 15 of the same year, when a leaky valve caused tertiary butanol to be released accidentally into the environment through an outlet of wastewater treatment plant. These tacit and glaring cases of ecological contamination caused by BASF industrial activities indicate that irrespective of the cumulative efforts to check environmental hazards, plant disturbances and anomalies at the corporate level can never be completely ruled out.

In 1998, a consistent increase in dioxin concentrations in cow milk was observed by the German regional authorities (Malisch, 1998). Contamination was eventually traced back to citrus pulp pellets (CPP) that were contaminated with dioxins. Citrus pulps are used as animal feed. The European Union concluded that lime milk used in the production of CPP was the cause of contamination (EU, 1999). The characteristic of all dioxins is their high melting point, low vapor pressure and good solubility and affinity for non-polar conditions. Moreover, they accumulate and biomagnify in the food chain due to their fat solubility and pronounced resistance to metabolic degradation. As Irwin *et al* (1997) argue, dioxin is lethal and some of its effects to human beings include *inter alia* changes in the human liver, harmful effects on the human kidney, heart, lungs and nervous system, increased rates of human cancer, immunological, reproductive, fetotoxic, genotoxic effects.

3. Research Methodology

In June 2003, a survey was conducted among a series of chemical industries in Germany. A brief questionnaire was prepared and addressed to environmental managers of 122 randomly chosen large, small and medium sized chemical industries from all parts of Germany. This questionnaire based survey was undertaken in order to ascertain the extent to which these industries were implementing Product Stewardship Tools (PST). The raison d′etre for using this method was due to the large sample size and because a questionnaire based survey will undoubtedly enable detailed information about corporate product stewardship to be collected within the shortest possible time frame. All questionnaires were sent by email during the first week of June 2003 and by the end of that month, 63 questionnaires had been received corresponding to 51%. Since June 2003 when the survey was conducted, the sampled industries have undoubtedly accrued experience in the arena of product stewardship and it is therefore possible that if the same questions were posed to them today, their answers will surely be very different. Thus, it is my contention that the results of this study are only valid for the industries sampled during the month of June 2003. The following results were obtained.

4. Survey Results

In this section of the paper, questions raised in the questionnaire are presented and a summary of corresponding results obtained are outlined.

4.1. Have You Adopted Product Stewardship in Your Company?

This first question was formulated to elicit a simple answer (yes or no) to a very crucial question about the adoption of product stewardship as a tool for corporate sustainability. To ensure clarity, the CEFIC definition of product stewardship was given. A total of sixty one (about 97%) industries that responded to the survey had adopted product stewardship. Only two industries (about 3%) answered "no". Industries that answered "no" expressed that they had not fully adopted product stewardship.

4.2. When Was Product Stewardship Initiated in Your Company?

Any assessment of product stewardship needs to begin with the history and evolution of product stewardship and the possible trends for the future. Results obtained indicate that the first chemical industry initiated product stewardship in 1954 (Figure 2). There was a considerable increase in initiation in 1991. A peak was obtained in 1998 (Figure 2).

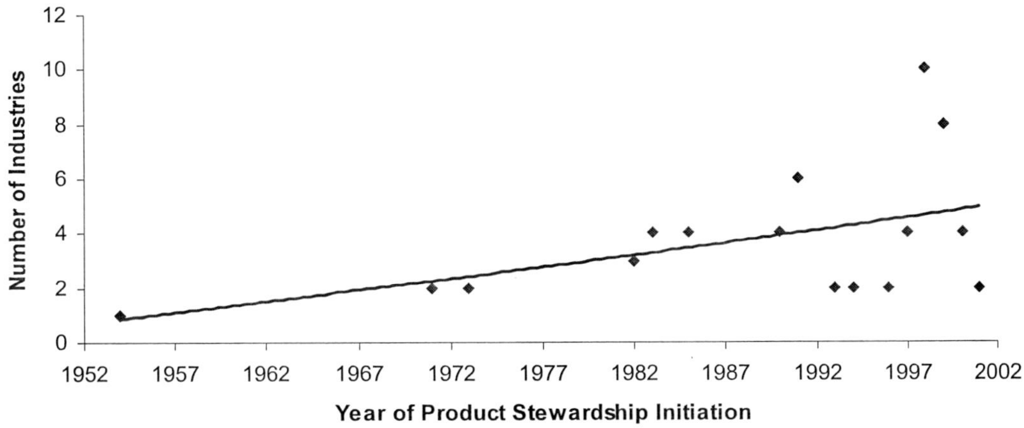

Figure 2. Polynomial trend of product stewardship initiation in some German chemical industry.

4.3. Do You Purchase Environmentally Friendly Raw Material to Be Used in the Design of Your Products? If Yes Please Give Examples

Survey respondents were asked to say whether they were committed to green purchasing. Forty –eight (about 76%) of all the respondents answered affirmatively (Figure 3). Examples of green purchasing were varied but some of the key examples that were mentioned included:

- Soya bean oil (it is used in the manufacture of margarine) in the place of synthetic oil which is less environmentally friendly;
- Formaldehyde free stabilizers (Stabilizers serve as wetting agents , preserve dyes, inhibit bacteria growth and harden films in the film industry);
- Bleaches with biodegradable chelates;
- Non halogenated solvents;
- Bee wax, chalk and dextrin (used in the manufacture of paints and colorants);
- Natural rubber (used in the production of tyres, belts, gaskets etc) instead of synthetic rubber that is synthesized from petroleum;

- A "respimat". This is a novel, reusable, propellant-free (no CFC) multi dose soft mist inhaler that is used in pharmaceutical industries for drug testing. It is an alternative to the conventional metered dose inhaler.

Ten (16%) respondents, however, declared tacitly that they were not committed to green purchasing (Figure 3). The rest did not provide any answers.

Figure 3. Level of implementation of green purchasing, recycling, packaging, labeling and storing of products in some chemical industries in Germany.

4.4. Do You Recycle and Reprocess Used Products?

This question was aimed at documenting recycling and take back programs as one of the fundamental tools of product stewardship. Of the sixty three respondents that responded to the survey, forty six (73%) maintained that they were engaged in take back and recycling programs while seventeen respondents (23%) declared the contrary (Figure 3). In answering affirmatively, major examples of recycled products provided by the respondents included:

- Polyvinyl chloride or PVC (a modern synthetic polymer that is used in the manufacture of plastics , electric cables , pipes etc);
- Sulfur;
- Solvents like Chlorine, butanol and Iso- butanol;
- Polyethylene;
- Styrofoam;
- Glycol;
- Used coolants;

4.5. Are Your Products Packaged, Labeled and Stored According to the Relevant EU Legislation?

Sixty two (about 98%) industries admitted that packaging, storing and labeling of their products was done according to the relevant EU legislation (Figure 3). Only one respondent (about 2%) submitted that packaging, labeling and storing of products as per the requisite EU legislation was not part of their services. This respondent did not state whether this was handled by a third party or not.

4.6. In What Areas Have You Predominantly Reduced Emissions?

In figure five, key arenas where emission reduction was achieved by the sampled industries is presented. Major reduction in emission to water and air were received as some of the answers to this question. Fifty-five (87%) and Fifty-two (82%) of the companies revealed that there had been a considerable reduction in their emission to water (example include reduction in Chemical Oxygen Demand and Biological Oxygen Demand) and air respectively. These respondents sited C0, $C0_2$, and $S0_2$ as key examples of major pollutants whose emission to air had been greatly reduced.

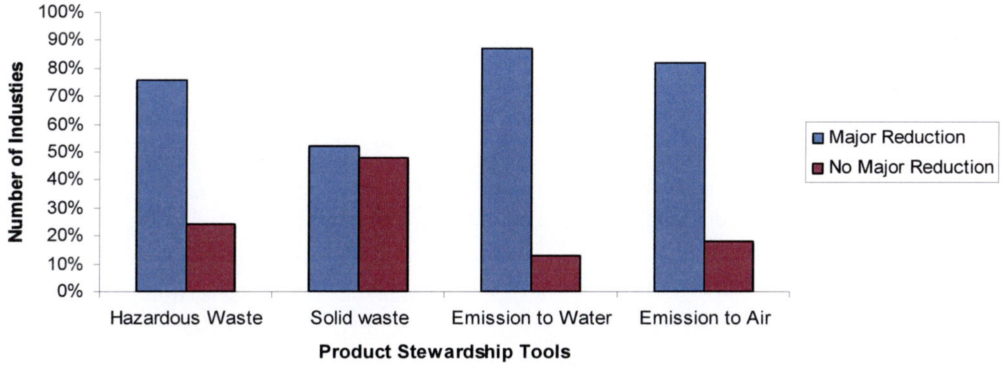

Figure 4. Level of implementation of some product stewardship tools (emission reduction in hazardous and solid waste, emission reduction to air and water) in some chemical industries in Germany.

With regards to hazardous wastes, a total of forty eight (76%) respondents expressed significant reduction in hazardous wastes. The most commonly listed examples of these wastes were: Chlorine, butanol, and polyethylene. Solid waste was sited as an arena where key strides had been made as thirty three (52%) of the responded asserted the fact that there had been a reduction in the emission of solid waste (like glass, and scrap metals) from their operations. Thirty (48%) industries made a contrary declaration.

5. Survey Reflections

The survey demonstrated that initiation of product stewardship in the sampled industries reached its peak in 1998 (Figure 2). A plausible explanation for this trend could be that most industries in Germany might have wanted to boost up their ecological credentials (achieve a "green" image) as a result of political motivation from Germany's coalition government that was formed in 1998 between the Green party and the Social Democratic Party (SPD).

The survey also revealed that almost all the industries that had adopted PST were large and medium sized enterprises. This is consistent with studies obtained by Alemagi *et al* (2006) in a study of environmental management systems in industries along the Atlantic coast of Cameroon. As Alemagi *et al* (2006) argue that larger industries have the requisite knowledge, financial and human resources to design and implement an environmental management system which like product stewardship is a tool for corporate ecological sustainability.

The survey found little differences in the number of industries that declared that they had predominantly reduced emission of solid waste with those that rejected this declaration (figure 4). A possible reason to underpin this observation could be that these industries with little reduction in solid waste may be generating household waste (like used paper) which in most instances could be very difficult to control. Indeed, a closer look at the annual environmental report of some of these industries revealed that solid waste generated during their operation was handled by a licensed third party which in most instances could be the local authority.

It is glaring from the survey responses that most survey respondents were steadfastly committed to the implementation of PST like recycling, green purchasing, packaging, labeling and storing of products (figure 3). This result is consistent with Lewis' (2005) assertion. This author maintains that the most essential element in EPR is the extension of producer responsibility to the post-consumer or end-of-life stage. He argues that this was first mandated in Germany in 1991 for packaging, and is now being applied to packaging and other product sectors in many of the world's industrialized economies. However, comparing our results with the prevailing situation in the United States, it would be fair to make a reasonable inference as to the fact that implementation of product stewardship could be far advanced in Germany's chemical industry as compared to the United States chemical industry. In support of this postulation, a survey of California local government electronic recycling programs revealed that take back programs are still at a premature stage (Boisson and associates 2002). The rationale behind this observation could be that implementation of EPR by Germany's chemical industry is generally regulated by legal obligations which are authorized and enforced by government. In the United States, legal enforcement has generally been avoided in favor of voluntary approaches with shared responsibilities and minimal or no regulation (Ehrenfeld, 2002).

6. The Way Forward

This thesis has provided an insight into the current product stewardship initiatives formulated by some chemical industries in Germany. It submitted that, the efforts made by these industries towards implementing product stewardship and its tools show a right tract to follow but needs to be strengthened. Indeed, further improvements in the arena of product stewardship can be made if these industries become more active in promoting the implementation of all tools for product stewardship particularly in the realm of solid waste reduction. Specifically, the German government needs to investigate various incentives for the implementation of PST. The merits of fundamental and strategic research in the area of solid waste emission could be further investigated and encouraged as this could serve to benefit both the environment and the chemical industry. Furthermore, continued research into PST like recycling and reuse technology with pilot projects in these areas could be given more emphasis. Finally, it is essential that the general public and different stakeholders are enabled and encouraged to become more actively involved in the development of practical implementation of policies and legislation relevant to product stewardship at the corporate level. This means that individuals be given all the relevant information necessary to enable them to exercise informed choices as stakeholders; they must also have access to data related to PST to enable them monitor the performance of industries.

This paper has revealed that, to a large degree, the principle of product stewardship has already been advanced in some Germany's chemical industry through the proactive implementation of PST. The overall conclusion to be drawn from this study is that most of the product stewardship initiatives in place thus far addressed key concept and tools of product stewardship, but in certain cases, particularly in the realm of reduction in emission of solid waste tend to be inadequate. However, important ground work for PST has been laid. Thus it is crucial that both the government and these industries became more proactive in promoting the aforementioned recommendations as this will undoubtedly guarantee that the industries continue along a path of improved corporate social responsibility and environmental management.

References

Alemagi D, Oben PM, Ertel J. (2006). A Survey of Environmental Management System Initiatives in Industries along the Atlantic Coast of Cameroon. *Corporate Environmental Strategy: International Journal for Sustainable Business* (In press).

BASF, (2001). Environment, Safety, Health Report, BASF Group Worldwide.

Boisson and Associates. (2002). Meeting California's E-waste challenge: *A survey of California local government Electronic Recycling Programs*. Prepared for the California Integrated Waste Management board, California.

CEFIC, (2002). The European chemical industry in a world wide perspective. Brussels, Belgium.

Clarkson, R .J. (1998). Industrial uses of chemicals. Available online at: http://www.rjclarkson.demon.co.uk/middle/middle8.htm

Commission of the European Communities. (2001). White Paper on Strategy for a Future Chemical Policy. Brussels, Belgium 2001.

Cramer, J, and Alders G. (1999). The design of a product stewardship management system within Akzo Nobel. *Eco-Management and Auditing* 6: 135 – 139.

Ehrenfeld, J .R. (2002). Sustainability and enterprise: an inside view of the corporation. In: *Extended Producer Responsibility: A Materials Policy for the 21st Century,* Fishbein, B., Ehrenfeld, J., Young, J. (Eds.). Inform, New York.

EU, (1999). Organization of official inspections in the field of animal nutrition: Dioxin contamination of citrus pulp pellets. *EU Report* XXIV/1005/99-MR Final.

ICPR, (2001). Conference of the Rhine Ministers: The state of the Rhine Yesterday – Tomorrow. Strasbourg, France.

Irwin, R.J, Van Mouwerik M, Stevens L, Seese M.D and Basham W. (1997). Environmental contaminants Encyclopaedia. United States of America National Park Service, Water Resources Division, Ford Collins, Colorado, U. S. A.

Lemke, K .W. (2002). Chemistry in North-Rhine Westphalia. Zentrum Für Innovation und Technik in NRW, Germany.

Lewis, H. (2005). Defining product stewardship and sustainability in the Australian packaging industry. *Environmental science and policy* 8: 45 – 55.

Lindhqvist, T. (1992). Extended producer responsibility. Proceedings of the Conference on Extended Producer Responsibility as a Strategy to Promote Cleaner Production, Trolleholm Castle, Sweden.

Malisch, R. (1998). Increase of PCDD/F contamination of milk and butter by use of contaminated citrus pulp as component in feedstuff. *Organohalogen Compounds* 35: 131–134.

PPRC. (1999). Pacific Northwest Pollution Prevention Resource Centre, U.S.A. Available online at: http://www.pprc.org/pprc/howto.html

Rotterdam Municipal Port management, (2000). Dredged Material in the port of Rotterdam: Interface between the Rhine Catchments Area and the North Sea. GKSS Research Centre, Germany.

SFDTRA, (2000). Scientific Frontiers in Developmental Toxicology and Risk Assessment. National Academy of Sciences, National Academy Press, U.S.A.

VCI, (2002). *Branch portrait of the German chemical industry*, Frankfurt, Germany.

In: Environmental and Regional Air Pollution
Editors: D. Gallo, R. Mancini
ISBN: 978-1-60692-893-6

Chapter 10

ANALYTICAL METHOD OPTIMIZATION AND MONITORING OF PRIORITY ORGANIC POLLUTANTS IN URBAN AIR

Senar Ozcan* and Mehmet Emin Aydin
Selcuk University, Department of Environmental Engineering, 42031, Campus, Konya-Turkey

ABSTRACT

In the first phase of the research an analytical method was developed in order to determine three groups of priority organic pollutants simultaneously such as polycyclic aromatic hydrocarbons (PAHs), polychlorinated biphenyls (PCBs) and organochlorine pesticides (OCPs) in atmosphere quantitatively by chromatographic techniques. In the second phase of the research, PAHs, PCBs and OCPs were monitored in gas and particle phases of urban air.

In determining particle phase compounds, the best extraction efficiencies were obtained by ultrasonic extraction method using 25 mL of n-hexane/petroleum ether (1/1, v/v) solvent mixture three times for 15 minutes. For determining compounds in gas phase of air samples soxhlet extraction method was carried out for 16 hours with 150 mL diethyl ether/n-hexane (1/9, v/v) solvent mixture. 10 gram 5% deactivated silica gel column was used for cleaning up the interfering substances while simultaneously fractionation of target compounds prior to GC analyses. Elutions of compounds were carried out as follow: Fraction 1: 70 mL n-hexane, Fraction 2: 3 x 20 mL n-hexane/ethyl acetate (1/1, v/v). PAHs and PCBs compounds were determined with GC/MS systems while OCPs compounds were determined using GC/μ-ECD system. Surrogate and standard spiked filters were analyzed after optimized method. Recovery were between 86%±2 and 104%±3 for PCBs with Relative Standard Deviation (RSD)< ±4, between 66%±6 and 94%±5 for PAHs with RSD<±8, between 67%±5 and 120%±10 for OCPs with RSD<10. Spiked PUF plugs were also analyzed following the optimized method. Recovery ratios obtained from PUF plugs were between 94%±9 and 118%±8 for PCBs with RSD%±9 while the recovery ratios changed from 88%±1 to 120%±2 for PAHs with

* E-mail: sozcan@selcuk.edu.tr

RSD<±8. Recovery ratios of OCP compounds were between 65%±4 and 108%±6 with RSD<±9. In order to determine the matrix effect on optimized method, air sampled filters and PUF plugs were spiked with standards and surrogate standards and analyzed according to the optimized method. Recovery ratios of air sampled and spiked filters and PUF plugs were not statistically significantly different from the recoveries of non air sampled spiked filters and PUF plugs at 95% level.

Gas phase and particle phase PAHs, PCBs and OCPs contents were analyzed in air samples taken from five different locations selected considering residential, industrial and traffic areas between August 2006 and May 2007. Mean PAHs concentrations were between 30 ng m^{-3} and 214 ng m^{-3} while mean PCBs and OCPs were between 0.04 ng m^{-3} and 0.14 ng m^{-3} and 0.81 ng m^{-3} and 6.15 ng m^{-3}, respectively. Total concentration ratios found in gas phase were 61%-63% for PAHs, 63%-89% for PCBs and 45%-70% for OCPs. With increasing in temperature concentration ratios in gas phase were increased while concentration ratios in particle phase decreased for all compounds. The higher concentrations of PAHs were determined in winter seasons while the lower concentrations of PAHs were determined in summer. Results from seasonal monitoring of PCBs in urban air of Konya show that the highest PCBs concentrations were determined in September while the lowest PCBs concentrations in April. The highest total OCPs concentrations were found in autumn seasons that in these months, lands were prepared for planting by farmers. Considering the risk characterization of these compounds determined in Konya urban air it could be concluded that these compounds pose some health risks for human.

Keywords: urban air, polychlorinated biphenyls (PCBs), polycyclic aromatic hydrocarbons (PAHs), organochlorine pesticides (OCPs).

INTRODUCTION

Air pollution could be described as the presence of one or more particulate matters or gases that naturally do not exist in the atmosphere in quantities and/or characteristics for a duration that will have deleterious effect on public health and welfare, natural environmental systems and economy. Main sources of air pollutants are low quality fuel consumption for domestic heating and energy production in power plants. Vehicles exhaust emissions are also important sources of air pollution in cities. A lot of trace level carcinogenic compounds are given to the atmosphere during chemical production, decomposition, vaporization and similar processes. Air pollutants are grouped in two categories as conventional air pollutants such as SO_2, NO_x, CO, CO_2, O_3, particulate matters and lead and carcinogenic toxic air pollutants such as arsenic, benzene, chromium (VI), nickel, PAH, radon, trichloroethylene and toluene (WHO, 2000).

Toxic air pollutants also known as hazardous air pollutants cause cancer, serious acute and chronic health problems on immune, neurological, endocrine and reproduction systems. It is estimated that these pollutants have toxic and harmful environmental effects even in low concentrations. Atmosphere plays an important role on transportation and accumulation of these pollutants. 70%-80% of these compounds brought by winds to atmosphere as attached to the particles and at the end they are deposited on waters or accumulated on other receivers in the environment by dry or wet deposition mechanisms. Then these pollutants enter human body via food chain (US EPA 2005; Karademir 2004).

Some works are carried out by US EPA in order to reduce 188 pollutants that have been determined in environment. Inventor works were carried out about these compounds in atmosphere of a lot of cities of several countries. Works also being carried out for the other mediums such as soil, water, sediments and ashes. Concentrations of these compounds in environmental compartments should be regulated because of their potential toxic effects, interaction between the environmental compartments and they tend to be transported long distances by atmosphere. Therefore, these compounds should be measured and monitored in national and international scales (US EPA, 2005).

Conventional air pollutants could be measured and monitored by automatic instruments located in various places in city centers. However, it is not possible to sample and analyze toxic air pollution compounds called priority organic pollutants by US EPA and WHO such as polychlorinated dibenzo-p-dioxins (PCDDs), polychlorinated dibenzo furans (PCDFs), polycyclic or polynuclear aromatic hydrocarbons (PAHs), polychlorinated biphenyls (PCBs), polybrominated biphenyl (PBBs), organochlorine pesticides (OCPs), polybromodiphenyl ethers (PBDEs), phthalate esters (PEs), polychlorinated naphthalenes (PCNs), volatile organic compounds (VOCs), polychlorinated alkanes (PCAs), chloro benzenes (CBs), phenols, aldehits and ketones by automatic instruments since the complexity of their sampling and analyses as well as their costs. Also, analyses and sample preparation works for organic air pollutants require long time and labour. Because of the need for efficient, repeatable and reliable sample preparation techniques, a lot of attempts were made for rapid and simple method developments without or with minimal organic solvent use.

Sampling method for air analyses usually involves air sampler instruments that vacuum a known amount of air during which organic pollutants are adsorbed on a sorbent medium. During sampling particulate matters are retained on filters while vapor or gas phase compounds are retained on sorbent. Fibre glass and cellulose filters are used for collection of particulate matters while solid adsorbans made of silica gel, carbon and different type of organic polymers are used widely for collecting volatile or semi volatile organic pollutants. Compounds retained on filters and sorbents are then extracted using different extraction techniques. Desorption of organic pollutants from sorbents are carried out either by thermal desorption or using organic solvents. Widely used solvent extraction techniques are soxhlet extraction, ultrasonic extraction, shake flask extraction, supercritical fluid extraction, rapid solvent extraction, microwave assisted extraction, pressured fluid extraction, matrix solid-phase extraction methods. Since the extraction methods are not selective, in order to eliminate interferences the extract is usually cleaned up from interfering compounds by using different clean-up techniques such as gel permeation chromatography, alumina, florosil, silica gel column.

In order to avoid interferences of similar compound or separation of target compound groups fractionation is some times required prior to gas chromatography analyses. Chromatographic techniques such as gas chromatography (GC), liquid chromatography (LC), ion chromatography (IC) etc. are used in qualitative or quantitative analyses of target compounds. The chromatographic techniques will be used for analyses of the target compounds depend on analytes properties such as polarity, volatility and thermo stability (Dean, 2003; Santos and Galceron, 2002).

POLYAROMATIC HYDROCARBONS (PAHS)

Polycyclic aromatic hydrocarbons (PAHs) are by-products of the incomplete combustion of organic matters. Anthropogenic emission sources for PAHs in the atmosphere include emissions from motor vehicles, waste incineration plants, domestic heating, oil refining, and other industrial processes. Besides, an important natural source is the biomass burning that occurs in forest fires (Cecinato et al., 1998; Oros et al., 2001; Junker et al., 2000). PAH molecules are made up of two or eight aromatic rings at least two of which neighboring rings share two adjacent carbon atoms. PAHs form a large group, but the most toxic compounds of this group are PAH molecules that have four to seven rings. Because of their carcinogenic and mutagenic nature, PAHs are investigated more in air pollution works in recent years. Benzo[a]pyrene in particles are accepted as highly carcinogenic. Despite the hundreds of PAH compounds found in environment, 16 PAH compounds are classified as priority pollutants by US EPA. 8 PAH compounds are suggested for investigations in air by WHO. 6 PAH compounds are described as mutagen and carcinogen by IARC (International Agency for Research on Cancer). These PAH compounds are given in Table 1 (Vardar, 1998; Ollivon et al., 2002). The 16 EPA PAH compounds declared as priority organic pollutants are investigated in this work. Chemical structure of the 16 EPA PAHs are given in Figure 1, some physico-chemical properties of PAHs are given in Table 2. Depending on their chemical structures, PAH compounds are easily soluable in organic solvents while their solubility in water decreases with increasing number of aromatic rings. Boiling points of PAH compounds are above 250 ^{o}C and they are in solid form in room temperature.

Table 1. PAH compounds regarded as priority organic pollutants by EPA, WHO and IARC

16 PAH compounds declared as priority pollutants by US EPA	8 PAH compounds that is suggested for investigation by WHO	6 PAH compounds declare as carcinogenic and mutagenic by IARC
Naphthalene	Fluoranthene	Benzo[a]anthracene
Acenaphthalene	Pyrene	Benzo[b]fluoranthene
Acenaphthene	Benzo[a]anthracene	Benzo[k]fluoranthene
Fluorene	Benzo[b]fluoranthene	Benzo[a]pyrene
Phenanthrene	Benzo[k]fluoranthene	Dibenzo[a,h]anthracene
Anthracene	Benzo[a]pyrene	Indeno[1,2,3-cd]pyrene
Fluoranthene	Benzo[g,h,i]perylene	
Pyrene	Indeno[1,2,3,cd]pyrene	
Benzo[a]anthracene		
Chrysene		
Benzo[b]fluoranthene		
Benzo[k]fluoranthene		
Benzo[a]pyrene		
Dibenzo[a,h]anthracene		
Benzo[g,h,i]perylene		
Indeno[1,2,3-cd]pyrene		

Table 2. Some physical and chemical properties of 16-EPA PAH compounds investigated in this study (HSDB, 2006)

Compound names	Molecular formula	Molecular weight	Melting point (^{0}C)	Solubility in water (μg L^{-1}), (at 25 ^{0}C)	Vapor pressure (mm Hg)	Octanol/water partition coefficient (Kow)
Naphthalene	$C_{10}H_8$	128.17	80.2	320	0.085 (25 ^{0}C)	3.30
Acenaphthalene	$C_{12}H_8$	152.20	92-93	390	9.12x10^{-4} (25 ^{0}C)	4.07
Acenaphthene	$C_{12}H_{10}$	154.21	95	190	4.47x10^{-3} (20 ^{0}C)	3.92
Fluorene	$C_{13}H_{10}$	166.21	116-117	190	3.2x10^{-4} (20 ^{0}C)	4.18
Phenanthrene	$C_{14}H_{10}$	178.22	101	120	6.8x10^{-4} (25 ^{0}C)	4.57
Fluoranthene	$C_{16}H_{10}$	202.26	111	100-110	9.22x10^{-6} (25 ^{0}C)	5.16
Anthracene	$C_{14}H_{10}$	178.23	218	76	2.67x10^{-6} (25 ^{0}C)	4.45
Pyrene	$C_{16}H_{10}$	202.26	151.2	77	8.92x10^{-5} (25 ^{0}C)	4.88
Benzo[a]anthracene	$C_{18}H_{12}$	228.29	160	10	2.2x10^{-8} (20 ^{0}C)	5.79
Chrysene	$C_{18}H_{12}$	228.29	258.2	2	6.3x10^{-9} (25^{0}C)	5.73
Benzo[b]flouranthene	$C_{20}H_{12}$	252.32	168	1.5	5x10^{-7} (20 ^{0}C)	6.60
Benzo[k]fluoranthene	$C_{20}H_{12}$	252.32	217	0.8	9.7x10^{-10} (25 ^{0}C)	6.84
Benzo[a]pyrene	$C_{20}H_{12}$	252.32	179	1.6	5.49x10^{-9} (25 ^{0}C)	5.97
Indeno[1,2,3-cd]pyrene	$C_{22}H_{12}$	276.34	163.6	62	10^{-11} (20 ^{0}C)	6.58
Dibenzo[a,h]anthracene	$C_{22}H_{12}$	278.33	266	0.5	1x10^{-10} (20 ^{0}C)	6.50
Benzo[g,h,i]perylene	$C_{22}H_{12}$	276.34	277	0.26	1x10^{-10} (25 ^{0}C)	6.63

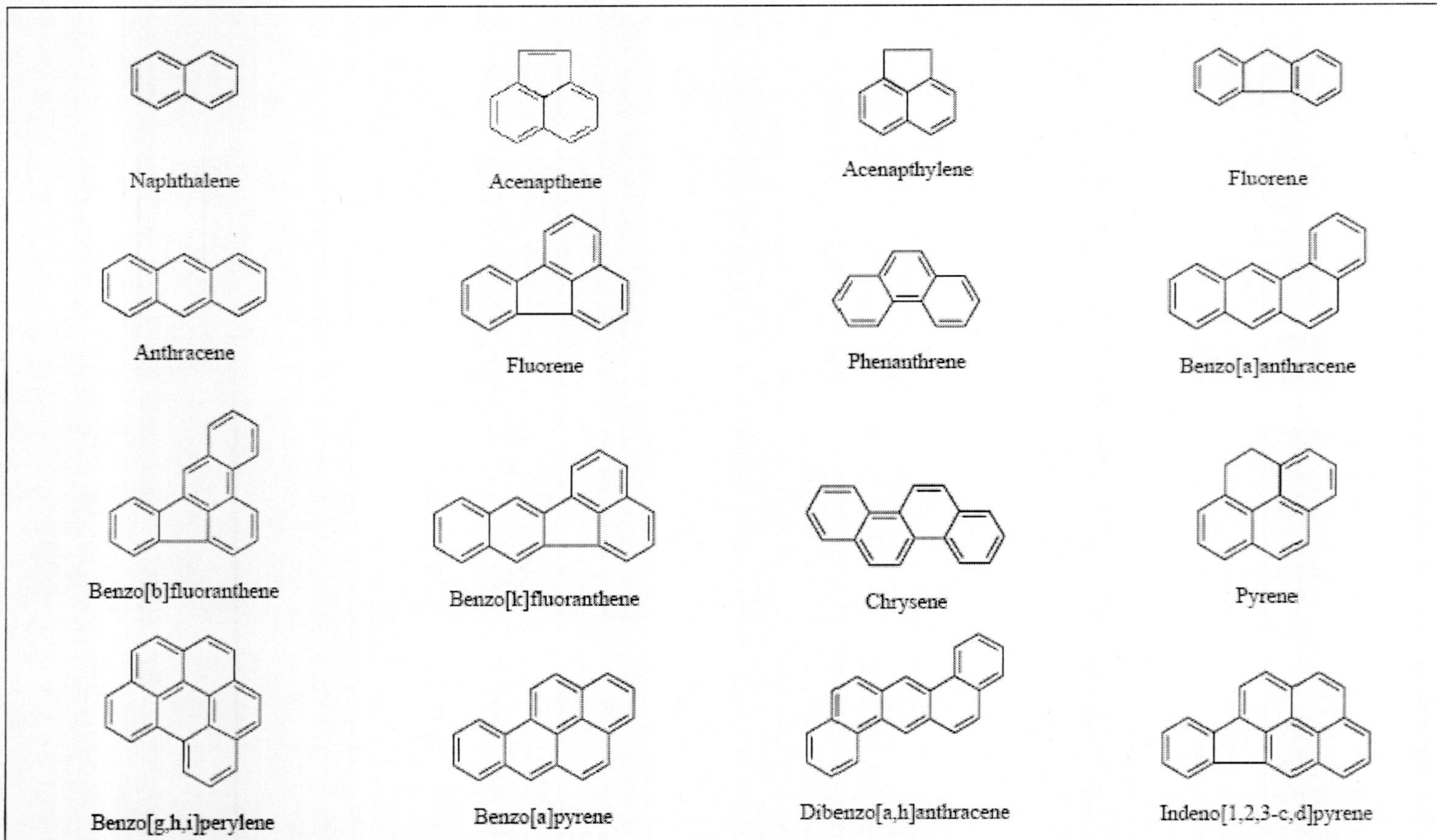

Figure 1. Chemical structures of 16-EPA PAH compounds investigated in this study (HSDB, 2006).

POLYCHLORINATED BIPHENYLS (PCBs)

PCBs consist of chlorine atoms bonding to biphenyls molecules. 209 PCB congeners are possible according to number and location of chlor atom in the molecule. IUPAC (International Union of PURE and Applied Chemistry) named PCB compounds considering the number and location of chlor atom in biphenyl molecules. Of the theoretical 209 congeners about 130 congeners may be found in commercial mixtures. PCB compounds have been commercially produced since early 1930 and they were widely used in a variety of equipment and consumer products such as transformers and capacitors, hydraulic and heat transfer systems, plastics, sealants, adhesives and paints. PCBs spread out in the environment in all over the world since there was no regulation or limitation on their use. However, PCB production and use have been restricted and banned in a lot of countries since mid 1970. Main sources of PCBs in city atmosphere are evaporation of these compounds from were they are used or stored or burning of PCBs containing materials. Because of PCB compounds chemical properties such as stability, low volatility and lipophilicty they tend to be bioaccumulative and have chronic health effects. Also, PCBs may enter the atmosphere from transformers, incinerators, landfills, sludge drying beds (Hermanson and Hites, 1989; Hsu et al., 2003). Therefore, analyses and monitoring of these compounds in ambient air is very important (Barro et al., 2005; Tasdemir, 1997; Yeo et al., 2003). About 213 tons of PCBs are used in Turkey by Electricity Production and Distribution Company. PCB use was banned in 1993 by hazardous chemicals regulation except the use of PCBs in closed systems (Erdogrul et al., 2004).

The most toxic PCBs compounds are non-ortho and mono-ortho congeners. Di-ortho congeners are relatively less toxic comparing the other congeners however, they are more abundant in biological samples. 11 PCB congeners are considered from the toxicological point of view (Alanso et al., 2001). 7 PCB congeners known as ICES (International Council for the Explanation of the Sea) congeners are widely determined congeners in worldwide (Yeo et al., 2004). 12 PCB congeners that have toxicological properties like dioxins and furans called as dioxin like PCB congeners by WHO and they are investigated and monitored systematically in environmental samples and industrial emissions. In addition there are 50 PCB congeners that are found in environmental samples (Aries et al., 2004; Martinez et al., 2000). These PCB congeners are given in Table 3.

Table 3. PCB compounds that are classified as priority pollutants

11 PCB congeners considered from toxicological point of view	7 PCB congeners known as ICES congeners that are determined widely in worldwide	12 dioxin like PCB congeners described by WHO	50 PCB congeners found in environment widely
PCB 8, 18, 28, 52, 44, 118, 128, 180, 195, 206, 209	PCB 28, 52, 101, 118, 138, 153, 180	PCB 77, 81, 105, 114, 118, 123, 126, 156, 157, 167, 169, 189	PCB 8, 18, 15, 16, 31, 28, 33, 22, 52, 49, 47/48, 44, 74, 66/95, 60, 101, 99, 97, 77/110, 149/123, 118, 105, 141, 137, 138/163, 187, 171, 156, 180, 200

Table 4. Some physical and chemical properties of PCB compounds investigated in this study (HSDB, 2006)

Compound names	Compound names (IUPAC)	Molecular formula	Molecular weight	Melting point (^0C)	Solubility in water ($\mu g\ L^{-1}$), (at 25 ^0C)	Vapor pressure (mm Hg)	Octanol/water partition coefficient (Kow)
2,4,4'-Trichlorobiphenyl	PCB 28	$C_{12}H_7Cl_3$	257.54	78	130	43.9	5.9
2,2',5,5'-Tetrachlorobiphenyl	PCB 52	$C_{12}H_6Cl_4$	291.99	87	29	4.92	6.3
2,2',4,5,5'-Pentachlorobiphenyl	PCB 101	$C_{12}H_5Cl_5$	326.43	77	10	1.20	6.8
2,2',3,4,4',5'-Hexachlorobiphenyl	PCB 138	$C_{12}H_4Cl_6$	360.88	79	1.80	0.532	7.4
2,2',4,4',5,5'-Hexachlorobiphenyl	PCB 153	$C_{12}H_4Cl_6$	360.88	103	0.91	0.692	7.8
2,2',3,4,4',5,5'-Heptachlorobiphenyl	PCB 180	$C_{12}H_4Cl_6$	395.32	99	0.63	0.129	-

Table 5. Some physical and chemical properties of OCP compounds investigated in this study (HSDB, 2006)

Compound names	Molecular formula	Molecular weight	Melting point (^0C)	Solubility in water ($\mu g\ L^{-1}$), (at 25 ^0C)	Vapor pressure (mm Hg)	Octanol/water partition coefficient (Kow)
Aldrin	$C_{12}H_8Cl_6$	364.91	104	0.017	$1.2x10^{-4}$	6.50
α-BCH	$C_6H_6Cl_6$	290.85	159-160	2	$4.5x10^{-5}$	3.80
β-HCH	$C_6H_6Cl_6$	290.83	314.5	0.24	$3.6x10^{-7}$ (20 ^0C)	3.78
γ-HCH	$C_6H_6Cl_6$	290.83	112.5	7.3	$4.2x10^{-5}$ (20 ^0C)	3.72
δ-HCH	$C_6H_6Cl_6$	290.83	141.5	10	$3.5x10^{-5}$	4.14
p,p'-DDD	$C_{14}H_{10}Cl_4$	320.04	109-110	0.09	$1.35x10^{-6}$	6.02
p,p'-DDE	$C_{14}H_8Cl_4$	318.03	89	0.04	$6x10^{-6}$	6.51
p,p'-DDT	$C_{14}H_9Cl_5$	354.49	108.5	0.0055	$1.6x10^{-7}$ (20 ^0C)	6.91
Dieldrin	$C_{12}H_8Cl_6O$	380.91	175.5	0.195	$5.89x10^{-6}$	5.40
Endosulfan I	$C_9H_6Cl_6O_3S$	406.93	106	0.51	$6.2x10^{-6}$	3.83
Endosulfan II	$C_9H_6Cl_6O_3S$	406.93	106	0.45	$6.2x10^{-6}$	3.83
Endosulfan sulfate	$C_9H_6Cl_6O_4S$	422.95	181-182	0.48	$1x10^{-11}$	3.66
Endrin	$C_{12}H_8Cl_6O$	380.91	226-230	0.25	$3.0x10^{-6}$ (20 ^0C)	5.20
Endrin aldehyde	$C_{12}H_8Cl_6O$	380.89	235	0.024	$2x10^{-7}$	4.80
Heptachlor	$C_{10}H_5Cl_7$	373.32	95.5	0.18	$4x10^{-4}$	6.10
Heptachlor epoxide	$C_{10}H_5Cl_7O$	389.40	160	0.2	-	4.98
Methoxychlor	$C_{16}H_{15}Cl_3O_2$	345.65	87	0.1	very low	5.08

Chemical structures of PCB compounds investigated in this work are given in Figure 2 while physico-chemical properties of them are given in Table 4. The most important chemical properties of PCB compounds are their inert structures. As the chlorine atom numbers increases in PCB compounds their solubility in water decreases. Solubility of PCB compounds in apolar solvents and accumulation of PCBs in biological tissues are quite high. PCB compounds that low soluble in water and have low vapor pressure are stable compounds with high dielectric conductivity.

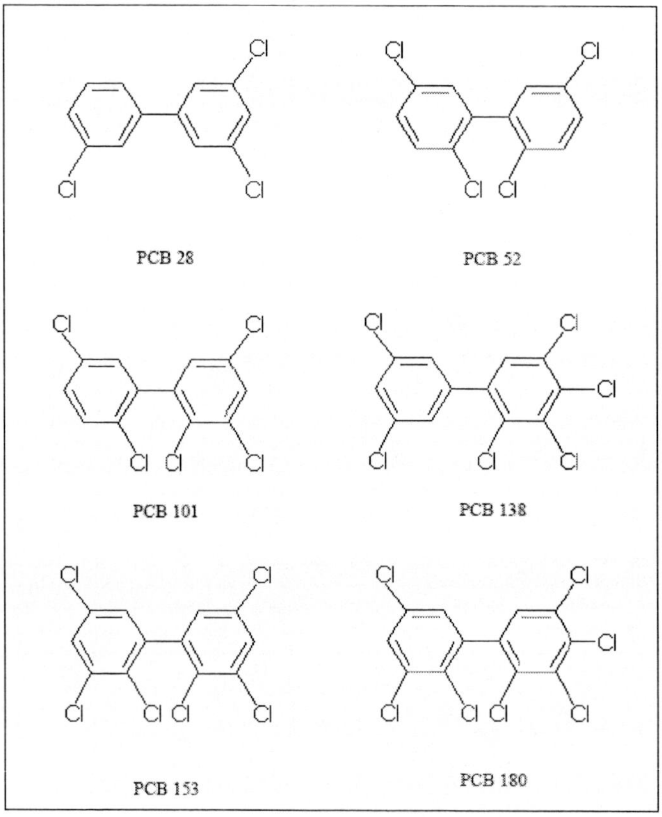

Figure 2. Chemical structures of PCB compounds investigated in this study (HSDB, 2006).

Organochlorine Pesticides (OCPs)

Pesticides are chemicals that are used to protect crops during production and storage against the insects, microorganisms and the other harmful organisms. Pesticides are classified as fungicide, insecticide, herbicide, rodenticide and acaricide according to harmful organisms that are used against. Despite the population increase rapidly in the world, there is no significant increase in food production. This situation forces all countries to increase food production. In order to achieve this goal plant production and protection products especially chemical fertilizers and pesticides are used widely. To protect crops from harmful insects, pathogens and plants thus to prevent danger of hunger, pesticides that are produced during the Second World War have been used widely.

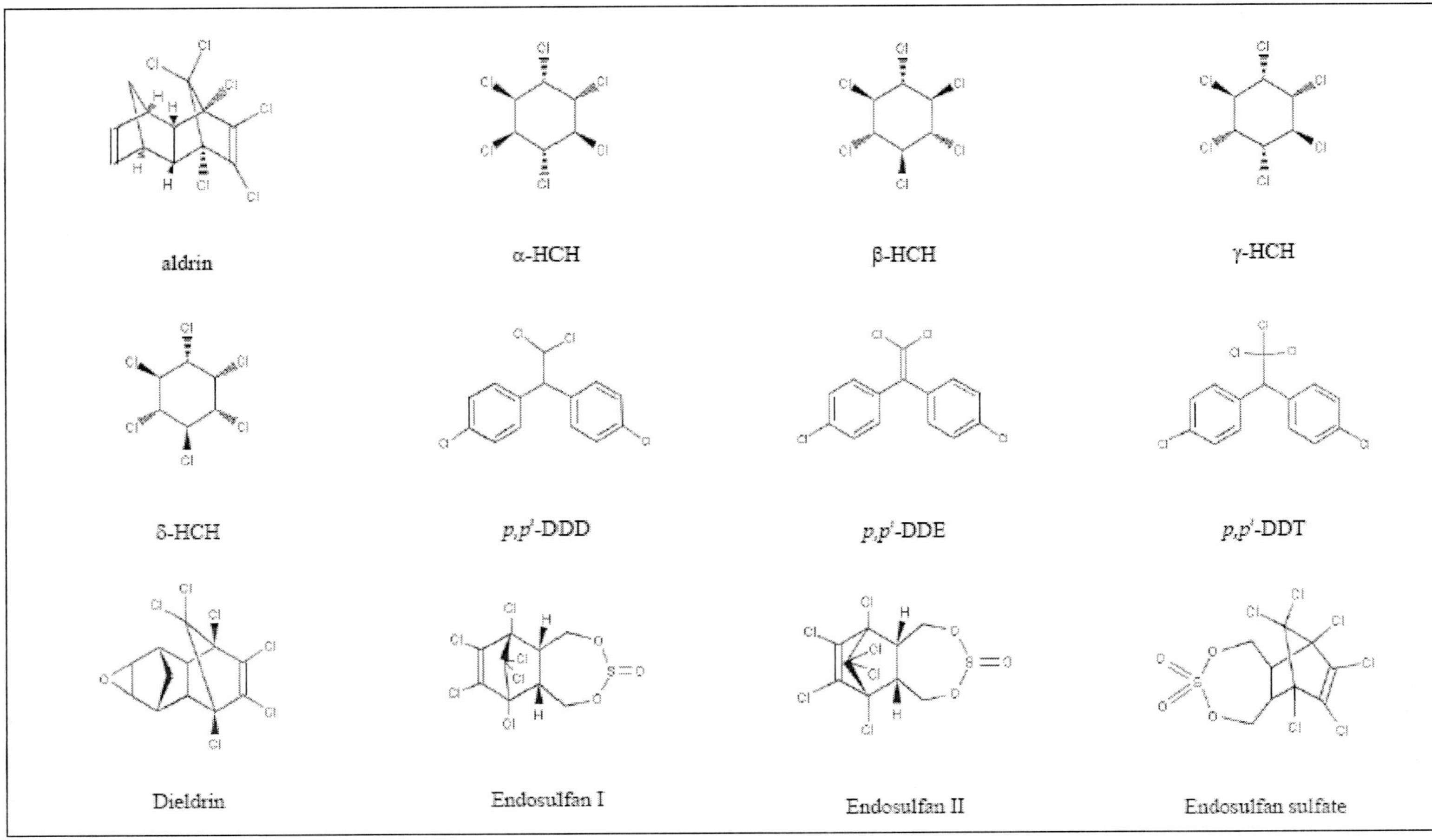

Figure 3. Chemical structures of OCP compounds investigated in this study (HSDB, 2006).

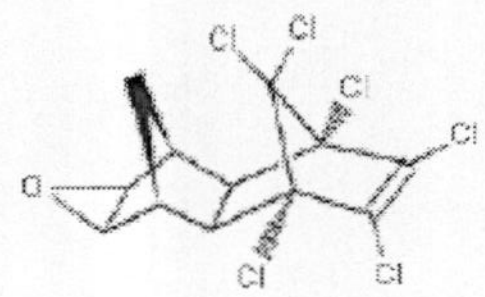

Endrin

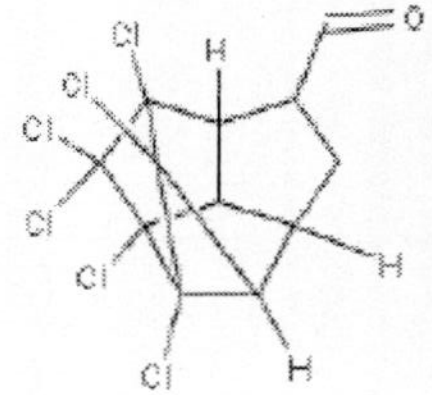

Endrin aldehyde

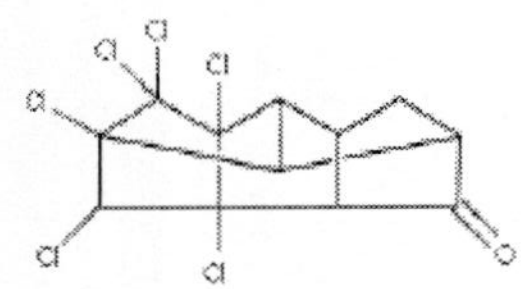

Endrin ketone

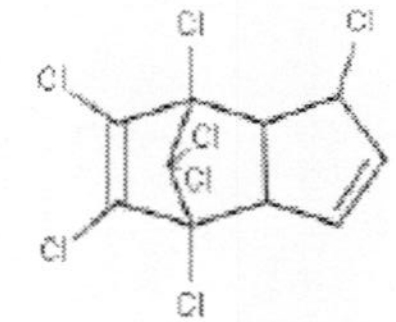

Heptachlor

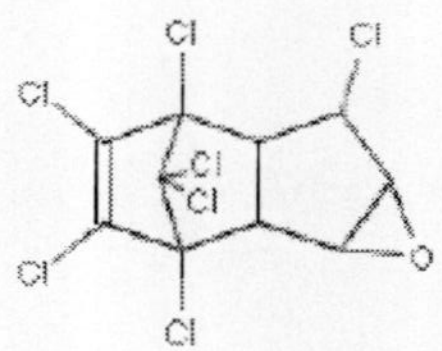

Heptachlor epoxide

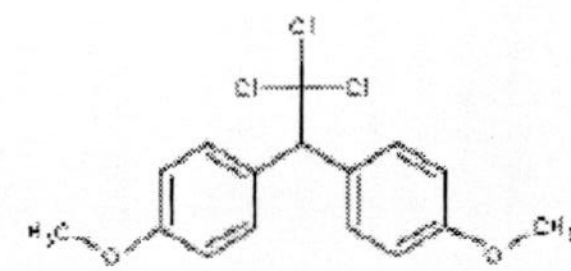

Methoxychlor

Figure 3. Chemical structures of 16-EPA PAH compounds investigated in this study (HSDB, 2006) (continued).

However, pesticides also cause serious problems because of their toxicity, persistency and bioaccumulation in environmental compartments. The pesticides can enter the atmosphere due to the current use, or emission/reemission from the various environmental compartments in which they are already present as residues resulting from historical use (Scholtz and Bidleman, 2006). These compounds are transported in the ecosystem and in atmosphere they could be transported long distances in local and global scales (Breivik et al., 1999). Pesticides were used first time in 1965 in Turkey and achieved considerable increase in crop production. During the pesticides application on soil, plant or seeds, they cause important environmental problems after they are transported with various mechanisms according to their chemical properties. While some part of the pesticides used cause accumulation of persistent toxic compounds in atmosphere by evaporation, some other part of them transformed to toxic or non-toxic compounds by photochemical decomposition. Some part of pesticides polluted soil as they are retained and react in soil in chemical or biochemical mechanisms. Some parts of pesticides reach and pollute surface and ground waters by rain, snow or run off. Some parts of pesticides are taken up by plants and accumulate in and on leaves and fruits. When these plants and fruits consumed by human or animals for long times acute and chronic poising events may occur or they may cause various health problems.

Organochlorine pesticides are important part of semi volatile organic compounds that are widely used in the past as insecticides. Organochlorine pesticides (OCPs) are contaminants of interest because they are relatively ubiquitous and persist in the environment. Most of these compounds have been found in all major environmental sectors and even in pristine environments (Hermanson and Hites, 1989; Hsu et al., 2003). Therefore, they can be transported long distances and can be accumulated terrestrial food chain. Most of the usage of OCPs were banned or restricted in Turkey like in a lot of countries because of their accumulation and persistence in environment. However, some of the OCPs are still being used in some countries. Chemical structures of OCP compounds investigated in this work are given in Figure 5 and some of the physico-chemical properties of them are given in Table 3.

PCBs, PAHs and OCPs partition between the gas and particulate phase based on their concentrations, vapor pressures, the ambient air temperature, and the concentration of particulate matter present in the air. Atmospheric fate, transport, dry and wet deposition of these compounds are influenced considerably by the distribution of substances between gaseous and particles phases (Falconer and Bidleman, 1994; Cotham and Bidleman, 1995; Hoff et al., 1996). PAHs in the atmosphere generally associated to airborne particles. In air, smaller molecular weight (less than 3 ring) PAHs are mostly found in gaseous phase while those of higher molecular weight (more than 5-ring) PAHs are generally associated with particles phase (Venkataraman et al., 1994). Because of the persistence, moderate vapor pressure and lipophilic features of the PCBs and OCPs, it is expected to the same association occurs between these pollutants and airborne particles. PCBs of higher molecular weight (containing 5 or more chlorine atoms) are deposited easily on surfaces of plants, soil and waters, because they have lower vapor pressures than PCBs with lower degree of chlorination (3 or 4 chlorine atoms) (Nielsen et al., 1996; Van der Hoff and Van Zoonen, 1999; ATSDR, 2000).

PAHs, PCBs and OCPs are important three groups of compounds that found simultaneously in urban air. Sampling and analysis procedures are given by EPA method TO-13A for PAHs while method TO-4A is given for the sampling and analysis of both PCBs and

OCPs in air samples. Similar analysis steps for different groups of organic pollutants in air samples are suggested by EPA. However, the procedures in each step are varying according to the physico-chemical properties of target compounds. When it is required to analyze more than one group of pollutants such as PAHs, PCBs and OCPs, it is necessary to use different analysis methods. EPA method TO-13A (1999) described the sampling and analysis procedures for the determination of PAHs. In addition, Method TO-4A (1999) is also used for the sampling and analysis for both PCBs and OCPs in air samples. According to these methods, by using of a high volume air sampler, the air is sucked through a filter, made with fibre-glass or cellulose, and sorbent materials (i.e. activated carbon, organic polymer, silica gel, polyurethane foam etc.). The air borne particles are retained on the filter and pollutants in gaseous phase are sorbed by sorbent materials.

According to EPA methods (TO-4A and TO-13A) PUF plugs and microfibre filters should be extracted by soxhlet extraction using 700 mL solvent mixture of diethylether:*n*-hexane (1/9, v/v) for a minimum of 16 hours. Before clean-up, sample extract is reduced to 1 mL. According to Method TO-4A (1999), if only polar compounds are analyzed, an alumina clean-up procedure should be used. Column is packed with 10 g alumina 10% deactivated and rinsed with approximately 20 mL *n*-hexane. The concentrated sample extract is placed on the column and eluted with 10 mL *n*-hexane. The eluate volume is concentrated to exactly 1 mL and analyzed with GC systems. If both PCBs and common pesticides are analyzed in the same sample, alternate clean-up procedures may be necessary i.e., silicic acid, silica gel. Finally, class separation and improved specificity can be achieved by separation on florosil. According to Method TO-13A (1999), the extract may be cleaned-up by solid phase extraction using activated silica gel. 10 g of silica gel are extracted in a soxhlet extractor with 10% diethyl ether for 6 h and then activated by heating for 16 hours at 150 ^{o}C. Column is packed with 10 g activated silica gel and 1 g anhydrous sodium sulfate, then it is rinsed with diethylether:*n*-hexane (1/9, v/v) at 1 mL min^{-1} for 1 h. It is then pre-eluted with 40 mL of *n*-pentane and the eluate is discarded. The concentrated sample extract is placed on the column and is eluted with 25 mL of *n*-pentane. The eluate volume is reduced to exactly 1 mL and analyzed with GC systems.

Gas chromatographic analysis of the organic pollutants (*i.e.* PCBs, PAHs, OCPs, etc.) follows a tedious sample preparation to extract the pollutants from any matrix. For the isolation of target compounds, different extraction procedures include, i.e. soxhlet (Wobst et al., 1999; Fatoki and Awofolu, 2003; Bakan and Ariman, 2004), shake-flask (Kolb et al., 1995; Pozo et al., 2001; Nawab et al., 2003), sonication (Babic et al., 1998; Castro et al., 2001; Banjoo and Nelson, 2005; Aydin et al., 2006, 2007; Tor et al., 2006), microwave assisted extraction (Ericsson and Colmsjo, 2000; Pino et al., 2000) can be employed. However, these extraction procedures are not specific to each analyte group. Therefore, with these methods, PCBs, PAHs and OCPs are extracted as mixtures from the same filter and sorbent material. However, clean-up procedure is completely different for these groups of pollutants. For instance, activated florosil should be used for PCBs and OCPs while fully activated silica gel is used for analyses of PAHs (US EPA, 1996).

When analyzing more than one group of pollutants in air different analytic methods are necessary. Using different methods means increase in analyses expenses and time. Despite there are various works reported on the analyses of different pollutants in air, either only one or two groups of compounds were analyzed using the same method. In these works, analysis were carried out using different methods (Vasconcellos et al., 2003; Tasdemir et al., 2004a,

2004b; Ozcan et al., 2008). There are some works published for developing common procedures to fractionate samples into chemical groups for Gas Chromatography (GC) analysis. For example, Tor et al. (2004) tested different clean-up procedures with wastewater extracts to design a general method for PAHs and selected PCBs. Their results showed a satisfactory separation of PAHs from PCBs with using a combined column (5 g of aluminum oxide, deactivated with 5% H_2O plus 5 g of fully activated silica gel), but only naphthalene as PAHs family member would be found in the PCBs fraction with recovery of 10%. Aceves et al. (1988) examined various clean-up procedures with sediment extracts to design a general method for PAHs, Aroclor 1254 and some chlorinated pesticides. Their results showed the possibility of separating PAHs from PCBs, but not from chlorinated pesticides. Moreover, their PCB analysis was not specific. Namely, Aroclor 1254 does not contain mono or dichlorinated PCB congeners and only 1% (by weight) of trichlorinated biphenyls. Column chromatography optimization works were carried out simultaneously for clean-up and fractionation of PCBs, PAHs and OCPs in a sample extracted from wastewaters and sediments (Jang and Lee, 2001; Batarseh et al., 2003a, 2003b).

Although there are various studies reporting the determination of the different pollutants in air (Vasconcellos et al., 2003; Tasdemir et al., 2004a, 2004b; Yeo et al., 2003), analytical methods are especially optimized in this research for three groups of PAHs, PCBs and OCPs. Moreover, these analytical procedures used the soxhlet extraction for isolation of target compounds from sample. However, it is intention of this research to get an analytical multimethod for the determination of different groups of the pollutants. Such investigations were performed only occasionally (Mandalakis et al., 2002). Multimethods consisting of a unique sample preparation for different pollutants are meaningful to reduce time-consuming analysis.

The aim of this work is to develop a multi-method for analysis of three groups of pollutants existing together in urban air. These three groups of pollutants were sampled using same filter and sorption materials in order to reduce sampling time, analysis time and cost. In order to be able to make simultaneous analysis of three groups possible a comprehensive clean-up and fractionation procedure is necessary. For the same purposes single extraction procedure will be employed to transfer the pollutants into the same organic phase. Therefore suitable column chromatography for simultaneous clean-up and fractionation of the extracts will be optimized. The extract then will be cleaned-up and fractioned simultaneously employing optimized column chromatography. The optimized column chromatography system will facilitate the use of a one-step clean-up/fractionation to obtain the two fractions, one of which will contain PCBs group, the second one of which will contain PAHs and OCPs groups. Additionally, this fractionation procedure will considerably speed up the analysis procedure. Solvent consumption, sorbent consumption and analysis time also will be reduced. In this way, PAHs, PCBs and OCPs compounds in urban air will be determined using the same optimized method.

In order to assess the potential of priority organic pollutants, gas-particle concentration, temporal changes, and variation relationship with temperature were determined in the urban air of Konya, Turkey. Five sampling locations are determined so as to represent Konya city atmosphere air pollution best considering the residential, industrial areas, traffic and transportation of organic pollutions from outside to the city. Seasonal changes of these compounds in Konya city atmosphere were determined by analyses of samples taken from these locations. Exposure risks for these compounds were determined using the analyses results. As a result of this work problem associated with air pollution and some possible solution will be put forward as well as data will be produced for global emissions data base.

EXPERIMENTAL

Reagents and Standards

All chemicals used were of analytical grade. The standards of PCBs (PCB 28, 52, 101, 138, 153 and 180), 16-EPA PAHs (naphthalene, acenaphthalene, acenaphthene, fluorene, phenanthrene, anthracene, fluoranthene, pyrene, benzo[a]anthracene, chrysene, benzo[b]flouranthene, benzo[k]fluoranthene, benzo[a]pyrene, dibenzo[a,h]anthracene, benzo[g,h,i]perylene, indeno[1,2,3-cd]pyrene), OCPs compounds their metabolites and isomers (α-HCH, β-HCH, γ-HCH, δ-HCH, heptachlor, heptachlor epoxide, dieldrin, aldrin, endrin, endrin aldehyde, endrin ketone, endosulfan I, endosulfan II, endosulfan sulfate, *p,p'*-DDE, *p,p'*-DDD, *p,p'*-DDT, methoxychlor) were from Accustandard. 1,2,3,4-Tetrachloronaphthalene, Chrysene-d_{12}, Phenanthrene-d_{10}, PCB 14, PCB 65, PCB 166 and mixture of tetrachloro-m-xylene and PCB 209 standards were also from Accustandard. Solvents *n*-hexane, acetone, diethyl ether, dichloromethane, petroleum ether (40-60 ^{o}C) used were from Merck. Na_2SO_4, silica gel 60 (0.063-0.2 mm) and florosil (0.15-0.25 mm) were also from Merck.

Chromatographic Analysis

The determinations of PCBs and PAHs were carried out by gas chromatograph (GC, Agilent 6890 N, Agilent Technologies, CA, USA) equipped with mass selective detector (MSD, Agilent 5973, Agilent Technologies, CA, USA). The features and operating conditions of GC-MSD system were as follows: GC, equipped with programmable temperature vaporizer (PTV) inlet, DB-5 MS 5% phenylmethyl siloxane fused silica capillary column and helium was used as carrier gas.

Injection was done into a temperature programmable vaporizer by an Agilent 7683 B Series automatic injector (Agilent Technologies, CA, USA). The temperature of the ion source and mass spectrometer transfer line were maintained at 180 ^{o}C and 280 ^{o}C, respectively. The samples were monitored in the selected ion mode (SIM) the used masses (m/z) of PCBs and PAHs. The monitored m/z ratios were given in Table 6.

The determination of OCP was performed by GC equipped with μ-electron capture detector (μ-ECD, Agilent Technologies, CA, USA). The features and operating conditions of GC/μ-ECD system were as follows: GC Agilent 6890 N installed with HP-5, 5% phenylmethyl siloxane fused silica capillary column.

The split/splitless injector was set at 250 ^{o}C and operated in the splitless mode (purge delay 1 min, purge flow 30.1 mL min^{-1}). The gas chromatographic conditions for the determination and quantification of PCBs, PAHs and OCPs in air samples are given in Table 7.

Table 6. SIM group for analyses of PCB and PAHs by GC/MSD

Compounds	Monitored ions (m/z)	Compounds	Monitored ions (m/z)
TCMX (internal1)	207, 209, 244	Naphthalene	128, 129, 127
PCB 14	222, 152, 224	Acenaphthalene	152, 153, 151
(surrogate1)	256, 258, 186	Acenaphthene	153, 154, 152
PCB 28	292, 220, 290	Fluorene	166, 165, 167
PCB 52	292, 290, 220	Phenanthrene-d_{10} (surrogate1)	188, 187, 189
PCB 65	326, 324, 328	Phenanthrene	178, 176, 179
(surrogate2)	360, 362, 290	Anthracene	178, 179, 176
PCB 101	360, 362, 290	1,2,3,4-tetrachloronaphthalene	266, 264, 268
PCB 153	360, 362, 290	(int)	202, 203, 200
PCB 138	394, 396, 324	Fluoranthene	202, 203, 200
PCB 166	498, 500, 496	Pyrene	228, 226, 229
(surrogate3)		Benzo[a]anthracene	240, 236, 239
PCB 180		Chrysene-d_{12} (surrogate2)	228, 226, 229
PCB 209		Chrysene	252, 253, 250
(internal2)		Bbenzo[b]flouranthene	252, 253, 250
		Benzo[k] fluoranthene	252, 253, 250
		Benzo[a]pyrene	276, 277, 274
		Indeno[1,2,3-cd]pyrene	278, 279, 276
		Dibenzo[a,h]anthracene	276, 277, 138
		Benzo[g,h,i]perylene	

Optimization of Fractionation and Clean-up Column

While extraction and concentration procedures of PCBs, PAHs and OCPs to transfer the pollutants into the same organic phase are similar, the clean-up procedures are very different for these three groups of pollutants. For example, if compounds are PAHs, fully activated silica gel should be used in US EPA Method 3630 C, while if PCBs and OCPs are to be analyzed, 3.3% deactivated silica gel or active florosil should be used in US EPA Method 3620 C. Also, eluation solvent type, volume and elution profile different for every compound group. According to the type and levels of the interfering substances in the sample, additional clean-up procedures may be used as alumina clean-up, gel permeation clean-up, acid-base partition clean-up.

In the first phase of the research a simultaneous clean-up and fractionation method was developed in order to determine three groups of priority organic pollutants such as polycyclic aromatic hydrocarbons (PAHs), polychlorinated biphenyls (PCBs) and organochlorine pesticides (OCPs) simultaneously in urban air. Silica-gel (10 g, fully activated) and florosil (10 g, fully activated) were first tested as column sorbents. Bead size of silica gel used in the study is 0.063-0.2 mm, while bead size of florosil is 0.15-0.25 mm. Silica gel (silicic acid) is a regenerative adsorbent of silica with slightly acidic properties. Silica gel is produced from sodium silicate and sulfuric acid. It can be used in column chromatography for the separation of analytes from interfering compounds of a different chemical polarity.

Table 7. Gas chromatographic conditions for the determination and quantification of PCBs, PAHs and OCPs

	PCBs	PAHs	OCPs
Instrument	HP 6890 Series II	HP 6890 Series II	HP 6890 Series II
Column type	HP-5MS	HP-5MS	HP-5
Length	30 m	30 m	30 m
Internal diameter	250 μm	250 μm	320 μm
Film thickness	0.25 μm	0.25 μm	0.25 μm
Injection	PTV injector, 1 μL auto sampler	PTV injector, 1 μL auto sampler	Split/splitless injector, 1 μL auto sampler
Carrier gas	Helium, 1.9 mL min^{-1}	Helium, 2.0 mL min^{-1}	Helium, 0.8 mL min^{-1}
Make-up gas	-	-	Nitrogen, 47 mL min^{-1}
Detector	Mass spectrometer	Mass spectrometer	μ-Electron capture, ^{63}Ni
Detector temperature	280 ^0C	280 ^0C	300 ^0C
Inlet temperature program	80 ^0C for 0 min, 720 ^0C min^{-1} to 350 ^0C, 350 ^0C for 2 min, 200 ^0C min^{-1} to 80 ^0C	80 ^0C for 0 min, 720 ^0C min^{-1} to 350 ^0C, 350 ^0C for 2 min, 200 ^0C min^{-1} to 80 ^0C	250 ^0C
Oven temperature program	70 ^0C for 2 min, 25 ^0C min^{-1} to 150 ^0C, 3 ^0C min^{-1} to 200 ^0C, 8 ^0C min^{-1} to 280 ^0C, 280 ^0C for 3 min	60 ^0C for 4 min, 15 ^0C min^{-1} to 160 ^0C, 3 ^0C min^{-1} to 300 ^0C, 300 ^0C for 10 min	60 ^0C for 0 min, 15 ^0C min^{-1} to 160 ^0C, 5 ^0C min^{-1} to 300 ^0C, 300 ^0C for 5 min
Total run time	41.87 min	67.33 min	39.67 min

Silica gel clean-up method is carried out for clean-up of sample extracts containing PAHs, derivatized phenolic compounds, OCPs, PCBs as araclors (US EPA, 1996). Florosil is a magnesium silicate with basic properties. It has been used for clean-up of pesticide residues and other chlorinated hydrocarbons. Florosil clean-up procedure is carried out for sample extracts containing phthalate esters, chlorinated hydrocarbons, nitrosamines, nitroaromatics, haloethers, organochlorine and organophosphorus pesticides, PCBs (US EPA, 1996).

Silica gel and florosil were activated at 210 ^{0}C in an oven for 4 hours. After cooling down in a desiccator, they were used. Silica gel was deactivated with 2% H_2O and with 5% H_2O. Before use, they were homogenized and allowed to cool down. Na_2SO_4 was activated at 450 ^{0}C in an oven for 4 hours.

According to US EPA standard methods, these clean-up methods can be carried out either by standard column chromatography techniques or by solid-phase extraction cartridges. Generally, standard column chromatography techniques are used, because they contain larger amounts of adsorbent. So they have a greater clean-up capacity. In standard column chromatography method, the column is packed with the required amount of adsorbent and is conditioned and then loaded with the sample. Elution of the analytes is carried out with suitable solvents that leave the interfering compounds on the column. Then, the eluate can be concentrated and analyzed. Most of the clean-up procedures used for research purposes are modifications of these methods.

The fractionation column, length of 145 mm and 15 mm internal diameter, was prepared with 10 g silica gel according to slurry-packing technique as described in Ref. (Jaouen-Madoulet et al., 2000; Jang and Li, 2001). *n*-hexane was used to make silica gel slurry and florosil slurry. Column was cleaned with acetone and *n*-hexane before filling up with adsorban slurry. 2 g activated Na_2SO_4 was added on the top of the silica gel or florosil according to the same technique. The column was pre-eluted with 40 mL *n*-hexane prior to the addition of the extract. Then, 1 mL of *n*-hexane contained mixed standard including PCBs (each of 100 ng), PAHs (each of 1000 ng) and OCPs (each of 1000 ng) was transferred on the column. The sample vial was rinsed twice 1 or 2 mL *n*-hexane and each rinse was added to the column. *n*-hexane, *n*-pentane and dichloromethane were chosen as mobile phase. The elution was carried out at a rate of 2 mL min^{-1}. At first, elution was carried out with 10 x 10 mL of *n*-hexane and 4 x 10 mL of dichloromethane. After the selection of the most suitable sorbent which separates the PCBs from PAHs and OCPs, optimization was performed by changing the deactivation levels of sorbent and eluent type. The details of optimization are given in Table 8. After determining the fraction cut-off volume, which completely contained the PCBs group, collection of PAHs and OCPs were carried out. Then the eluent was reduced to exactly 1 mL using rotary evaporator at 40 ^{o}C and under a gentle stream of nitrogen and the solvent was exchanged to *n*-hexane. Then internal standards of tetrachloro-m-xylene, PCB209 and 1,2,3,4-Tetrachloronaphthalene (with the concentration of 0.1 ng μL^{-1}) were added in 1 mL of reduced eluents prior to analysis.

Certification of PUF Cartridge

Particles were collected by passing air through the GF/A type glass microfibre filter with a diameter of 90 mm and pore size 1.6 μm. Pollutants in gas phase were collected in PUF (polyurethane foam) plug with length of 5 cm, diameter of 6.5 cm, density of 0.0225 g cm^{-3}.

Table 8. The details of optimization for column chromatography to separate PCBs from PAHs and OCPs

Sorbent type	Elution profiles
10 g, silica gel, fully activated	*n*-hexane (each fraction consisted of 10 mL, total volume: 100 mL), dichloromethane (each fraction consisted of 10 mL, total volume: 40 mL)
10 g, silica gel, fully activated	70 mL of *n*-hexane, (each fraction consisted of 10 mL, total volume: 140 mL), *n*-hexane: ethyl acetate (1/1, v/v) (20 mL of each fraction, total volume: 60 mL)
10 g, florisil, fully activated	70 mL of *n*-hexane, (each fraction consisted of 70 mL, total volume: 70 mL), *n*-hexane: ethyl acetate (1/1, v/v) (20 mL of each fraction, total volume: 60 mL)
10 g, 2% deactivated silica gel	70 mL of *n*-hexane, (each fraction consisted of 70 mL, total volume: 70 mL), *n*-hexane: ethyl acetate (1/1, v/v) (20 mL of each fraction, total volume: 60 mL)
10 g, 5% deactivated silica gel	70 mL of *n*-hexane, (each fraction consisted of 70 mL, total volume: 70 mL), *n*-hexane: ethyl acetate (1/1, v/v) (20 mL of each fraction, total volume: 60 mL)
10 g, silica gel, fully activated	70 mL of *n*-pentane, (each fraction consisted of 70 mL, total volume: 70 mL), *n*-pentane: ethyl acetate (1/1, v/v) (20 mL of each fraction, total volume: 60 mL)

Prior to air sample collection, cleaning procedures were applied to filters and PUF plugs to minimize the interfering substances contamination and to provide acceptable blanks by US EPA (1999). Acceptable levels for multiple components mixtures of PCB and OCP should be less than 100 ng for each pair of filter and adsorbent.

Acceptable levels for Naphthalene is <500 ng/cartridge while other PAH compounds is <200 ng total/cartridge. Each of the PUF plugs were soxhlet extracted with *n*-hexane:diethyl ether (9/1, v/v) for 24 h several times while glass microfibre filters was baked at 450 ^{0}C in muffle furnace to remove the impurities for 5 h. The cleaned PUF plugs and microfibre filters were dried in a desiccator, packed glass cartridge stored in aluminum container until sampling.

The results of the blank tests did not show any interference peaks that would inhibit quantifying the target compounds in PUFs and microfibre filters.

Method and Field Blanks

Blank microfibre filters and PUF plugs were routinely placed in the field to determine possible analytical contaminants during sample handing and preparation for the sampling and analyses. Also, laboratory blank samples were routinely analyzed in same manner as the real samples. Blank levels of individual compounds were determined usually <10% of the amounts in the blank microfibre filters and PUF plugs and in most cases they were not detectable.

Optimization of Extraction

According to US EPA 3500B, the extraction of semivolatile and nonvolatile organic compounds from solid matrix such as soil, waste, filter can be carried out according to soxhlet extraction (US EPA 3540), automatic soxhlet extraction (US EPA 3541), pressurized fluid extraction (ASE) (heat or pressure) (US EPA 3545), ultrasonic extraction (US EPA 3550), supercritical fluid extraction (SFE) (US EPA 3560, 3561). In soxhlet extraction method, solid sample placed into an extraction thimble and extracted using an appropriate solvent in a soxhlet extractor. The extract is concentrated and is exchanged into a solvent and then analyzed. Soxhlet extraction uses relatively inexpensive glassware, provides efficient extraction but it is required rather long time (16 or 24 hours) and uses fairly large volumes of extraction solvents (about 300 mL). According to ultrasonic extraction method, the sample is extracted with solvent using ultrasonic waves. Then, the extract is analyzed. Ultrasonic extraction is fairly rapid method.

The recovery efficiency of the method has been evaluated by the analysis of pre-cleaned glass microfibre filter samples and PUF plug samples spiked with known amounts of PAHs, PCBs and OCPs compounds (McConnell et al., 1998; Alegria et al., 2006; Mandalakis and Stephanou, 2007). There are various studies reporting *n*-hexane:diethyl ether (9/1, v/v) (US EPA, 1999), dichloromethane:petroleum ether (1/4, v/v) (Tasdemir et al., 2004b; Odabası et al., 1999; Sofuoglu et al., 2004), *n*-hexane:dichloromethane (9/1, v/v) (Yeo et al., 2003, 2004), dichloromethane (McConnell et al., 1998; Park et al., 2002; Harrad and Mao, 2004; Gioia et al., 2005) in the extraction of PAHs, PCBs and OCPs from airborne particles and PUF plugs by using soxhlet extraction. At the beginning of this study, the extraction efficiency of *n*-hexane:diethyl ether (9/1, v/v), dichloromethane:petroleum ether (1/4, v/v), and *n*-hexane:petroleum ether (1/1, v/v) were compared. Since dichloromethane was of toxic character, it was not used in the extraction of target compounds investigated in this research. For that, spiked glass microfibre filters and PUF plugs were prepared by adding 10 μL of standard mixture of PCBs (the concentrations are 10 ng μL^{-1} for each compound), 10 μL of standard mixture of PAHs and OCPs (the concentrations are 100 ng μL^{-1} for each compound). While microfibre filters were extracted by ultrasonic method, PUF plugs were extracted by soxhlet extraction.

The spiked glass microfibre filters were sonicated for 45 min with 100 mL of different solvents in an ultrasonic bath. The extracts were reduced to 2 mL with rotary evaporator at 40 ^{0}C and adjusted to exactly 1 mL under a gentle nitrogen stream and the solvent was exchanged to *n*-hexane. Then internal standards of tetrachloro-m-xylene, PCB209 and 1,2,3,4-Tetrachloronaphthalene were added as 0.1 ng μL^{-1} prior to analysis. The amount of extracted target compounds were determined by GC/MS and GC/μ-ECD and the recoveries were calculated. In the second set of experiments, the optimum volume of solvent, optimum sonication time and optimum repetition of extractions were determined. These optimization experiments were carried out by using of *n*-hexane:petroleum ether (1/1, v/v) which gave the highest recovery for the compounds investigated. In order to determine optimum volume of *n*-hexane:petroleum ether (1/1, v/v), spiked glass microfibre filters were sonicated for 45 min with 25, 50, 75, and 100 mL of *n*-hexane:petroleum ether (1/1, v/v). In order to determine the optimum sonication time, spiked glass microfibre filters were sonicated with 75 mL of *n*-hexane:petroleum ether (1/1, v/v) for 15, 30, 45, 60 min. The extraction of spiked microfibre filters with 25 mL of *n*-hexane:petroleum ether (1/1, v/v) for 15 min were repeated three times. The cumulative recoveries were calculated by adding recoveries of compounds determined in each extract. Spiked filters were extracted according to the optimized ultrasonic extraction conditions. Prior to the analysis the extracts were cleaned up as described in section fractionation and clean-up column and recoveries were determined for each group of compounds.

The spiked PUF plugs were extracted for 16 h at 4-6 cycles/hour with 200 mL of different solvents in a soxhlet extractor. The extracts were reduced to 2 mL with rotary evaporator at 40 ^{0}C and adjusted to exactly 1 mL under a gentle nitrogen stream and the solvent was exchanged to *n*-hexane. Then internal standards of tetrachloro-m-xylene, PCB209 and 1,2,3,4-Tetrachloronaphthalene were added as 0.1 ng μL^{-1} prior to analysis. The amount of extracted target compounds were determined by GC/MS and GC/μ-ECD and the recoveries were calculated. These optimization experiments were carried out by using of *n*-hexane:diethyl ether (9/1, v/v) which gave the highest recovery for the compounds investigated. Then, in order to determine optimum volume of *n*-hexane:diethyl ether (9/1, v/v), spiked PUF plugs were extracted for 16 h with 150, 200 and 250 mL of *n*-hexane:diethyl ether (9/1, v/v). In order to determine the optimum extraction time, spiked PUF plugs were sonicated with 150 mL of *n*-hexane:diethyl ether (9/1, v/v) for 8, 16, 24 h. Spiked PUF plugs were extracted according to the optimized soxhlet extraction conditions. Prior to the analysis the extracts were cleaned up as described in section fractionation and clean-up column and recoveries were determined for each group of compounds.

Air Sampling Procedure

Significant quantities of both vapor and aerosol forms of a semi volatile organic compound may be present in air. Also, some gases may adsorb onto other particles in air. Some collected particulate material with relatively high vapor pressure may evaporate during traditional filtration sampling. Polyurethane foam (PUF) samplers are used to overcome these problems in sampling semi volatile organic compounds in air. The PUF sampler is a modification of high volume sampler. The sampling module consists of a quartz fibre filter and a 6 cm plug of foam to concentrate compounds of interest from the air. Polyether type polyurethane foam (density 0.0225 g cm^{-3}) can efficiently collect vapor and particulate forms of pesticides, polychlorinated biphenyls, dioxins and polyaromatic hydrocarbons. A wide verity of sorbents such as Tenax and XAD-2 is added behind the foam plug to enhance the capture of more volatile compounds (Wight, 1994; US EPA, 1999). The modified high volume air sampler of Thermo-Anderson Inc. Model GPS-11 was used for collecting air samples. High volume sampler can pull 325-400 m^3 ambient air through the filter/adsorbent cartridge at a flow rate of approximately 0.225 m^3 min^{-1} for 24 h. The sampler was calibrated using a calibrated orifice flow rate transfer standard and the flow rates were checked before and after sampling by calibrated flow-meters. Particles were collected by passing air through the GF/A type glass microfibre filter with a diameter of 90 mm and pore size 1.6 μm. Pollutants in gas phase were collected in PUF (polyurethane foam) plug with length of 5 cm, diameter of 6.5 cm, density of 0.0225 g cm^{-3}.

Air samples were collected in various volumes in different works carried out previously such as 140 m^3 (Sofuoglu et al., 2001), 150 m^3 (Tasdemir et al., 2004), 223±8 m^3 (Sofuoglu et al., 2004), 250 m^3 (Alonso et al., 2001), 432 m^3 (Vardar et al., 2004), 437 m^3 (McConnell et al., 1998), 600 m^3 (Yeo et al., 2003), 900 m^3 (Mandalakis et al., 2002), 1000 m^3 (Harrad and Mao, 2004), 200-1000 m^3 (Müller et al., 1998), 7200 m^3 (Lodovici et al., 2003), 13000 m^3 (Hung et al., 2005), 47952 m^3 (Mastral et al., 2003). Samplings were carried out generally between 12-24 hours durations. Air samples were collected with an average sampling period of 24 hours in this study. The mean flow rate and sampling volume for each sampling were about 0.25 m^3 min^{-1} and 360 m^3, respectively. Then, samples were kept at 4 ^oC until analyses in the laboratory. Extraction was performed within 7 days of sampling.

Konya is a historical city with about a million population located in central Anatolia region in Turkey. The city is undergoing rapid industrialization, urbanization, and population growth and has high atmospheric emissions due to excessive use of fossil fuels; low quality coal and fuel-oil, and exhaust gases emitted by vehicles. Air pollution is becoming a great environmental concern in the city and it is essential to monitor air pollutants in large areas and take strict measures. Monitoring of air quality has been started in Konya since 1986. Today, air quality measurements are carried on for only PM_{10} and SO_2 parameters, in 5 stations by using full automatic vehicles. CO concentration was measured as exhaust emission while the other pollutants such as NO_x, HC originated from vehicles were not measured in the city. Air samples were taken from August 2006 to May 2007. The sampling program was conducted in urban area. 5 sampling location were determined in Konya city to monitor PCBs, PAHs and OCPs atmosphere considering the population density and spatial distribution. These sampling locations were determined considering the point sources (industrial source), area sources (residential sources), moving sources (vehicles) and outside transportation of target pollutants in the city atmosphere. In addition principles on sampling locations determination for monitoring air quality of city atmosphere given by EU and Turkish Environment and Forestry Ministry were also observed. Dominant wind direction in Konya is north-east direction according to Konya Meteorological District Directorate.

Sampling location 1 (SL1): Selcuk University, Engineering Faculty Garden in Campus, selected to determine outside transportation to the city.

Sampling location 2 (SL2): 1st industrial zone near cement factory, selected to determine point pollution sources from industry.

Sampling location 3 (SL3): Government Waterworks building garden, selected to determine contribution of residential areas to air pollution.

Sampling location 4 (SL4): Garden of provincial Environment and Forestry Directorate building, selected for area sources of pollution from residential areas.

Sampling location 5 (SL5): In front of Kombassan Petrol Station which is on one of main roads in the city, selected to determine vehicular pollution.

Air sampler is placed in locations 1, 2, 3 and 4 in a way that it will not be effected from a direct source, at least 2 m away from any obstacles on air flow and maximum 15 m above ground levels, and contained very little or no traffic load. Sampling locations are selected away from stagnant air places surrounded by high buildings. Parks, gardens, school areas, hospital gardens, sport areas, recreational areas are suggested by Ministry of Environment and Forestry for sampling places. Some other concerns also were considered in selection of sampling location such as easily accessibility, power supply for sampler and security. Vehicular sources pollution sampling location should be away at least 20 m from a function and at least 4 m away from traffic line. Air sampler in sampling location 5 is placed observing these rules. The period of sampling was carried out from 9:00 a.m. to 9:00 p.m and samples were taken in weekdays that pollution will be higher.

Total suspended particles (TSP) were determined weight before and after sampling the microfibre filters. Air temperatures (T), relative humidity (RH), pressure, rainfall (RF), wind speed (WS) and wind direction (WD) were taken from the meteorological station. Conventional pollutants as 10-µm particulate matters (PM_{10}) and sulfur dioxide (SO_2) concentrations were taken from Ministry of Environment and Forestry. Average values of meteorological data and conventional pollutants were calculated for each one of the collected air samples. Results are given in Table 9.

While mean air temperature varied from -6.5 °C (December) to 19.2 °C (May), mean relative humidity varied from 38.4% to 79.3%. It can be seen in Table 9 that rainfalls were determined only three times ranged between 0.1 mm and 8.0 mm. Investigated conventional pollutants, lowest concentrations for PM_{10} and SO_2 were measured during spring while the highest concentrations were observed during winter. Total Suspended Solid concentrations were determined in between 0.05 µg m^{-3} and 0.32 µg m^{-3}.

Table 9. Mean values of air temperature (T), relative humidity (RH), rainfall (RF), pressure, wind speed (WS) and wind direction (WD) and mean concentrations of PM_{10}, SO_2, Total Suspended Solid (TSP)

Date	T (°C)	RH (%)	RF (mm)	Pressure (mbar)	WS (m sn^{-1}) -WD	PM_{10} (µg m^{-3})	SO_2 (µg m^{-3})	TSP (µg m^{-3})
23/08/2006	15.6	73.7	4.7	895.3	4.9 – N	64.1	1.6	0.08
27/09/2006	18.0	59.0	0.0	899.4	1.5 – S	77.1	6.0	0.05
18/10/2006	10.8	79.3	8.0	898.9	1.0 – N	81.0	5.0	0.06
29/11/2006	6.1	69.3	0.0	905.4	1.9 – ESE	148.3	109.3	0.14
27/12/2006	-6.5	67.0	0.1	904.6	7.4 – N	72.0	35.5	0.32
22/01/2007	2.0	73.4	0.0	909.2	1.1 – ESE	251.2	19.2	0.30
20/02/2007	nm	nm	nm	nm	nm	100.0	7.0	nm
17/03/2007	4.8	46.3	0.0	899.9	1.0 – ENE	85.0	8.6	nm
14/04/2007	8.4	38.4	0.0	900.1	2.4 – NNE	41.5	4.8	0.06
09/05/2007	19.2	53.4	0.0	897.0	1.4 – WSW	67.8	8.1	0.08

nm: not measured.

RESULTS

Quality Control and Quality Assurance

The analytical quality of the data was determined using limit of detection (LOD), recovery, reproducibility, linearity and by checking sampling artifacts. The calibration techniques for PAHs, PCBs and OCPs are the internal standard multipoint calibration using at least five concentration levels. Instrument detection limits (IDLs) for all compounds were determined according to published guidelines at a signal-to-noise ratio (S/N) of three (Keith et al., 1983; Sienra et al., 2006). Blank samples (including six pre-cleaned PUF plugs and filters) were prepared, treated and analyzed in the same manner as the real samples. The limit of method detection (MDL) was defined as the mean blank mass + three standard deviations.

No peaks matching the target compounds were found on blank PUFs and filters. Thus, MDLs were normally based on the IDLs.

The detector response was linear in the range of concentration between 0.01-10 ng μL^{-1} for PAHs and OCPs and between 0.001-0.8 ng μL^{-1} for PCBs and the correlation coefficients for the studied compounds ranged from 0.997 to 1.000. MDLs ranged from 0.04 pg m^{-3} (PCB 153) to 0.08 pg m^{-3} (PCB 101) for PCBs. MDLs ranged from 0.05 pg m^{-3} (Naphthalene) to 0.05 ng m^{-3} (Benzo[k]fluoranthene) for PAHs and they ranged from 1.6 pg m^{-3} (α-HCH) to 0.03 ng m^{-3} (Endrin) for OCPs. MDLs correspond for a nominal air volume of 360 m^3 and a final extract volume of 1 mL. The precision was obtained from six injections of a standard with 10 ng mL^{-1} of each congener and relative standard deviation was lower than 2.7% and 5.8% for MSD and μ-ECD, respectively. Collected samples were spiked with surrogate standards before extraction to measure the recovery efficiency of the procedure. PCB 14, PCB 65, PCB 166 congeners were used as surrogate standards for PCBs and OCPs, chrysene-d_{12} and phenanthrene-d_{10} were used as surrogate standards for PAHs. Each congener had a concentration of 10 ng mL^{-1}. The average recovery efficiencies for PCB 14, PCB 65 and PCB 166 were 100%±2, 120%±3 and 117%±1 on filter samples, and 110%±9, 120%±8 and 120%±8 for PUF samples, respectively. The average recoveries of PAH surrogate standards were 89%±1 for chrysene-d_{12}, and 88%±1 for phenanthrene-d_{10} for filter samples and 97%±1 for chrysene-d_{12}, and 84%±1 for phenanthrene-d_{10} for PUF samples.

The materials used in the present study were assessed for possible analytical contaminants. Due to the many experimental materials used in this study, it was necessary to check the background concentration level of target compounds. Glass microfibre filters were wrapped loosely with aluminum foil washed with *n*-hexane and it was baked at 450 ^{0}C in muffle furnace to remove the impurities for 5 h. It was then allowed to cool down to room temperature in a desiccator. PUF plugs were soxhlet extracted with 150 mL *n*-hexane:diethyl ether (9/1) for 16 hours for several times. The results of the blank tests did not show any interference peaks that would inhibit quantifying the target compounds in PUFs and microfibre filters.

Separation of PCBs from PAHs and OCPs

Among the column systems examined in this study, all PCBs were eluted before the PAHs and OCPs. In the adsorption chromatography, the analytes are adsorbed by sorbent material via interactions between the analytes and functional groups of the sorbent. Several different forces, working only at very close distances, contribute to the adsorption process. Their binding energy is low and consequently no strong chemical bond between the analyte and sorbent is formed. The dissociations of these weak binding forces are controlled by the eluent strength (Unger and Weber, 1999). For the presented study, the faster elution of all PCBs might be attributed to the stronger interactions between sorbent and both PAHs and OCPs. These interactions resulted in the longer retention of these higher polarizable pollutants. These findings are in agreement with Castro et al. (2001), which reported the separation of PCBs from PAHs using column chromatography.

As seen in Figure 4, 100 mL of *n*-hexane plus 40 mL of dichloromethane were insufficient for completely eluting of all compounds.

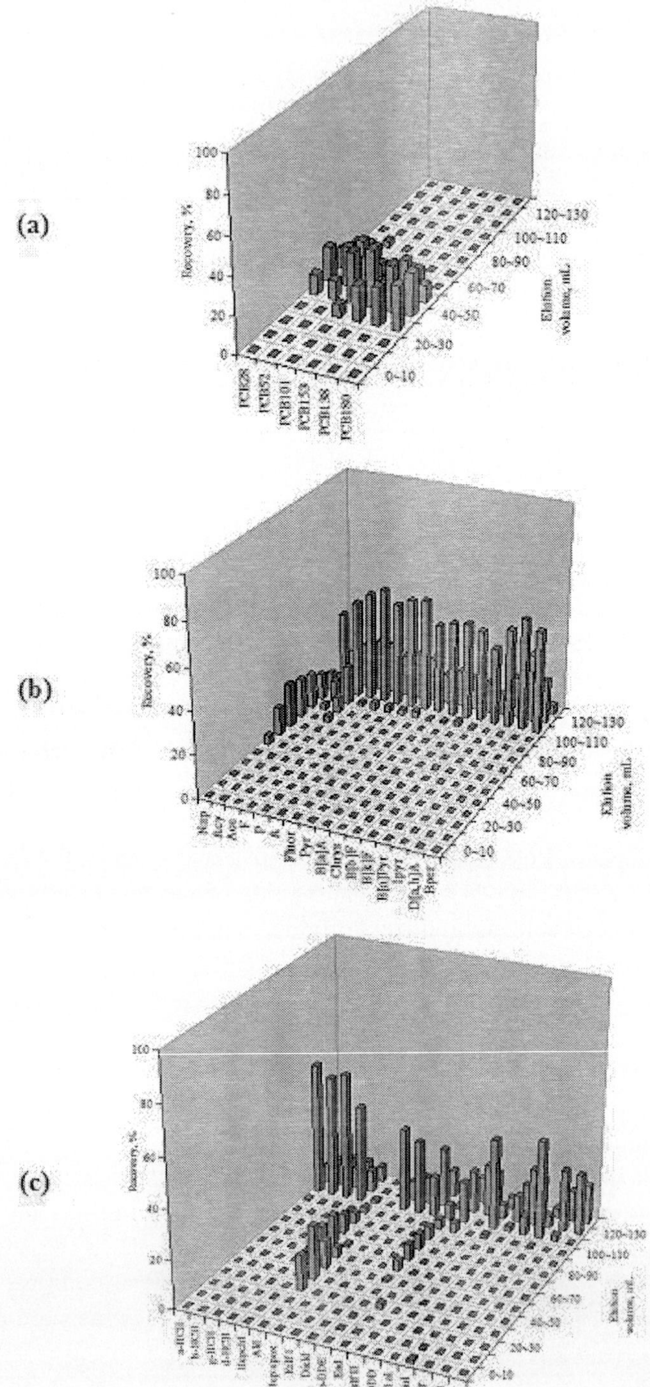

Figure 4. Elution order and recoveries of PCBs (a), PAHs (b) and OCPs (c). (10 g, fully activated silica gel; elution solvent: 10 x 10 mL of n-hexane plus 4 x 10 mL of dichloromethane), [n=4].

Recovery ratios were obtained between 49%-75% for PCBs compounds, 50%-104% for PAHs compounds, and 25%-70% for OCPs compounds. Fraction 1 collected by elution of

100 mL *n*-hexane, Fraction 2 is collected by 40 mL dichloromethane. In fraction 2, there was no PCB congener. Only 67% of naphthalene from PAH compounds eluted in fraction 1 while 33% of naphthalene compounds was eluted in fraction 2. All the rest of the PAH compounds however, were eluted in fraction 2 with 40 mL dichloromethane. Heptachlor, aldrin, *p,p'*-DDE compounds were eluted in fraction 1 in the ratio of 41%, 54% and 25%, respectively. The remaining ratios of these compounds were eluted together with the rest of all OCP compounds in fraction 2. Considering the percentage recovery values it can be suggested that elution of PCBs compounds were completed with 70 mL *n*-hexane.

Moreover, the dichloromethane was of toxic characteristic. For this reason, after 70 mL of *n*-hexane which eluted the PCBs from the column, the elution was carried out by using *n*-hexane:ethyl acetate (1/1, v/v) in order to increase the elution strength of the solvent. After elution with 70 mL of *n*-hexane, it was found that three times elution with 20 mL of *n*-hexane:ethyl acetate (1/1, v/v) was sufficient for the elution of target compounds. Comparing the elution volume of 100 mL of *n*-hexane and 40 ml of dichloromethane with the elution volume of 70 mL *n*-hexane, and 60 mL of *n*-hexane:ethyl acetate (1/1, v/v), it was observed no changes in elution order for PCBs and PAHs, while the recoveries of OCPs increased from 25%-70% to 67%-94%. Recovery ratios were between 84%-95% for PCBs and 50%-88% for PAHs. Considering the elution profile, elution of PCB 166 and PCB 180 congeners completed with fraction 1 (70 mL *n*-hexane) while other PCBs congeners were eluted with fraction 1 (70 mL *n*-hexane) and fraction 2 (60 mL *n*-hexane/ethylacetate (1/1, v/v)). Considering PAHs compounds only 2.5% naphthalene was eluted with fraction 1 while the rest of the all PAHs compounds were eluted with fraction 2. Heptachlor, aldrin, dieldrin and *p,p'*-DDE were eluted 23%, 78%, 3% and 10%, respectively in fraction 1. The rest of these compounds were eluted together with all of the OCPs compounds in fraction 2. Although recovery ratios for PCBs and OCPs were in acceptable range according to the recovery ratio range 60%-120% given by USE EPA (1999) for multiple compounds analyses, recovery ratios obtained for PAHs compound were not in acceptable range.

The volume of *n*-hexane needed for completely eluting the PCBs from fully activated silica gel column was found as 70 mL. In case of fully activated florosil column, the same volume of *n*-hexane (70 mL) caused the overlap between PCBs and PAHs-OCPs. Recovery ratios were changed between 93%-120% for PCBs compounds, 26%-109% for PAHs compounds and 52%-109% for OCPs compounds. Recovery ratios were decreasing as the molecular weight of the compounds increased especially for PAHs compounds. Considering recovery values obtained for PCBs compounds, despite the high recovery ratios they are eluting in both fraction 1 (70 mL *n*-hexane) and fraction 2 (60 mL *n*-hexane/ethylacetate (1/1, v/v). In other words, separation capacity of the silica gel is better than that of florosil column for PCBs and PAHs-OCPs. This situation can be explained by distribution of the sorbents bead sizes. Namely, silica gel had bead size between 0.063-0.200 mm. However, florosil had diameter between 0.150-0.250 mm. As it is well known, smaller bead size means higher surface area and, therefore, the sorption capacity of silica gel is higher than that of florosil.

These results may be explained by considering the "mass transfer" phenomenon between eluent and adsorbent. During the course of a separation, analyte molecules are adsorbed at the surface of the adsorbent and are desorbed again by the fresh eluent. This occurs continuously for all analyte molecules during their passage through the column. These exchange processes between eluent adsorbent do not occur instantaneously but take place with a certain time. During the time needed for analyte transport between phases, the eluent travels further along the column and analyte molecules are retained by new sorbent surface. Because the florosil column has weak sorption capability than that of silica gel, the compounds are easily eluted out from the florosil column. Similar results were reported in Unger and Weber (1999) and Jang and Lee (2001). Therefore, further optimization was focused on silica gel as sorbent, changing its deactivation levels and the elution solvents.

In order to, investigate the effects of eluting solvents on the separation of PCBs from PAHs and OCPs from fully activated (0% water for deactivation) silica gel, *n*-pentane was tested alternatively to *n*-hexane. No considerable change in the elution order and recoveries of compounds were observed. In a run with fully activated silica gel and using *n*-pentane as eluting solvent, recovery ratios were changed from 75% to 93% for PCBs, from 16% to 93% for PAHs while they were changed between 55% and 87% for OCPs. Considering elution profile, all of the PCB 153, PCB 138, PCB 166 and PCB 180 congeners were eluted in fraction 1 (70 mL *n*-pentane) while the PCB 28, PCB 52 and PCB 101 congeners were eluted in the ratio of 67%, 71% and 79%, respectively in fraction 1 (70 mL *n*-pentane). The rest of these PCBs congeners were eluted in fraction 2 (60 mL *n*-pentane/ethylacetate (1/1, v/v)). Only 4% of naphthalene was eluted in fraction 1 while all other PAHs compounds and remaining naphthalene compound were eluted in fraction 2. In addition, heptachlor, dieldrin, *p,p'*-DDE compounds were eluted 12%, 9% and 8%, respectively while aldrin eluted completely in fraction 1. All the rest of the OCPs compounds together with the remaining ratios of the above mentioned compounds were eluted in fraction 2.

Comparing *n*-pentane, with 70 mL of *n*-hexane used as the fraction collecting the PCBs, again only naphthalene was found in the PCBs fraction with 2.5%. Aldrin was found in the PCBs fraction with 78%. Namely, elution strengths of the *n*-pentane and *n*-hexane were nearly the same because the polarity of *n*-pentane is close to that of *n*-hexane.

In addition to fully activated silica gel, 2% and 5% deactivation levels were examined by using the combined standards of PCBs, PAHs and OCPs. For these deactivation levels, 70 mL of *n*-hexane and 60 mL of *n*-hexane:ethyl acetate (1/1, v/v) were consecutively used for the elution of mixed standards from the column. Recovery ratios were changed between 81%-92% for PCBs compounds, 62%-92% for PAHs compounds, 63%-100% for OCPs compounds in 2% deactivated silica gel column. All of the PCBs congeners were eluted in fraction 1 (70 mL *n*-hexane). 64% of naphthalene was eluted in fraction 1 all of the rest of the PAHs compounds were eluted in fraction 2 (60 mL of *n*-hexane:ethyl acetate (1/1, v/v)). Heptachlor, aldrin and *p,p'*-DDE compounds were eluted 66%, 102%, 76%, respectively in fraction 1. The remaining fractions of these compounds were eluted together with the rest of the PAHs compounds in fraction 2.

Increasing moisture contents of the silica gel generally decreased the retention times of all compounds, which mean that higher silica gel activation levels caused longer elution times. These results were in agreement with those of Tor et al. (2004), Aceves et al. (1988), Jang and Lee (2001). They increased the water content of silica gel and found decreased retention times for the separation of PCBs from PAHs. When using fully activated silica gel, the least overlap among PCBs, PAHs and OCPs was found. It could be deduced that lower deactivation levels of silica gel decrease overlapping of PCBs, PAHs and OCPs during elution. The effects of deactivation levels of sorbents on separation efficiency of organic compounds were already extensively investigated by Aceves et al. (1988). They showed that increasing deactivation levels increased the overlap of the compounds, thus complicating the separation of the compounds used. Similar results were observed for silica gel used in this study. Nevertheless, using the mass selective detection and μ-ECD systems eliminated this disadvantage of the 5% deactivated silica gel in this study. Although the deactivation levels of the sorbent are important regarding sorbent-catalyzed solute degradation. The degradation during clean-up is not of primary concern for PCBs, PAHs and OCPs. For these reasons, among the examined column systems, the column filled with 10 g of silica gel deactivated with 5% H_2O was used to separate the PCBs from PAHs and OCPs. The recoveries of the compounds quantified by GC/μ-ECD and GC/MSD analysis for the selected column system are given in Figure 5.

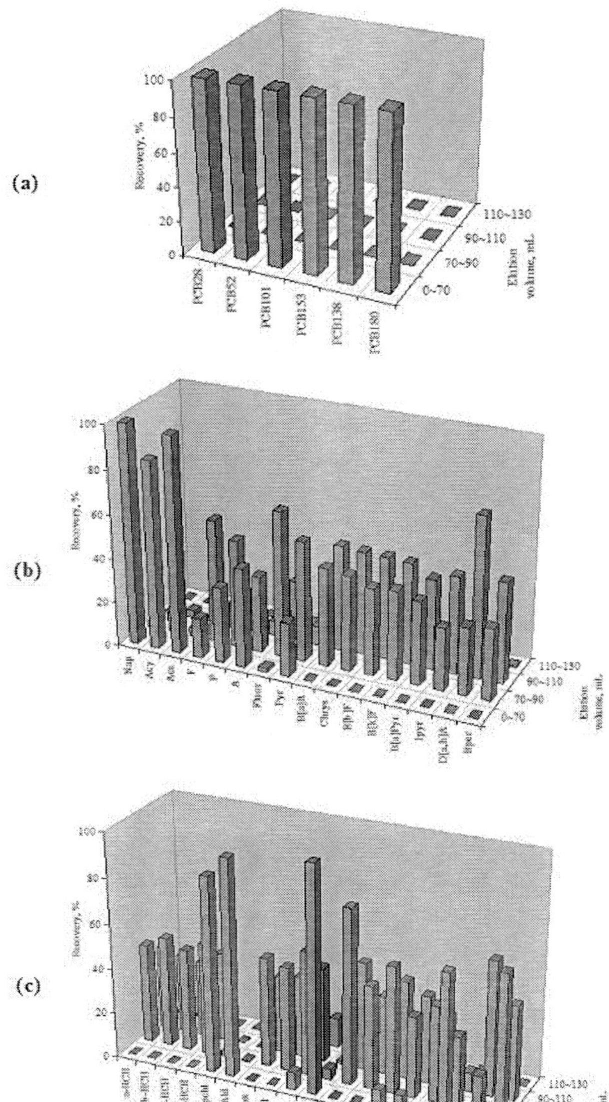

Figure 5. Elution order and recoveries of PCBs (a), PAHs (b) and OCPs (c). (10 g, 5 % deactivated silica gel; elution solvent: 70 mL of n-hexane plus 3 x 20 mL of n-hexane:ethyl acetate (1/1, v/v), [n=4].

Overall, the separation of PCBs from PAHs and OCPs using the column (10 g of silica gel 5% deactivated) was satisfactory. As shown in Figure 5, the recovery values for PCBs, PAHs and OCPs were found as 100%-120%, 73%-108% and 74%-117%, respectively, with RSD values of ≤±5.1%, ≤±6.5% and ≤±6.2%, respectively. All of the PCBs congeners were eluted in fraction 1 (70 mL *n*-hexane). Naphthalene, acenaphthalene, acenaphthene compounds were eluted almost completely with ratios of 104%, 96%, 98% respectively in fraction 1. It was also observed that above 33% of fluorene, phenanthrene, anthracene and pyrene compounds were eluted. The rest of the PAHs compounds were eluted completely in fraction 2 (60 mL of *n*-hexane:ethyl acetate (1/1, v/v)). Considering OCP compounds, heptachlor, aldrin, *p,p'*-DDE, *p,p'*-DDT compounds were eluted 88%, 97%, 100% and 65%, respectively in fraction 1. All the rest of the OCP compounds were eluted from column in fraction 2.

Optimization of Extraction Procedure

There are a lot of works which reported extracting different organic pollutants from airborne particles by using soxhlet extraction (Vasconcellos et al., 2003; Cindoruk and Tasdemir, 2007; Tasdemir and Esen, 2007). However, the main disadvantage of the soxhlet extraction is that it takes long time for extraction (Hartonen et al., 1997). Ultrasonication allows an intensive contact between sample and solvent and it reduces the extraction time. Therefore, ultrasonic extraction can be used to extract target compounds from sample as an alternative to traditional soxhlet extraction (Sporring et al., 2005). Therefore, ultrasonic solvent extraction for the determination of PCBs, PAHs and OCPs from airborne particle holder on glass microfibre filter was aimed in our study.

In the optimization of ultrasonic extraction, different solvent mixtures including *n*-hexane:diethyl ether (9/1, v/v), dichloromethane:petroleum ether (1/4, v/v), and *n*-hexane:petroleum ether (1/1, v/v) were compared for extraction efficiency. Ultrasonic extraction efficiencies of the solvents were checked by calculating of recoveries. The results are given in Figure 6. Statistical significance of the various solvents was determined using the two-tailed unpaired Student's *t*-test (at the 95% confidence interval) on the quantities of the individual PAH, PCB and OCP compounds from four replicate analyses of glass microfibre filters. When statistical evaluation was carried out between extraction efficiencies of *n*-hexane:diethyl ether (9/1, v/v), dichloromethane:petroleum ether (1/4, v/v), and *n*-hexane:petroleum ether (1/1, v/v), no significant differences ($p>0.05$) were observed. The mixture of *n*-hexane:petroleum ether (1/1, v/v) gives the highest recoveries (ranged from 78%±2 to 100%±6) for PAHs among the used solvent mixture. Statistical evaluation of extraction efficiencies of PAHs from filter by *n*-hexane:petroleum ether (1/1, v/v) and *n*-hexane:diethyl ether (9/1, v/v) showed significantly different recoveries for Nap, P, B[a]A, B[a]Pyr.

Moreover, when the same evaluation was carried out between *n*-hexane:petroleum ether (1/1, v/v) and dichloromethane:petroleum ether (1/4, v/v) significant difference was observed for Pyr and B[a]A. For OCPs, *n*-hexane:diethyl ether (9/1, v/v) gave lower extraction efficiency compared to dichloromethane:petroleum ether (1/4, v/v), and *n*-hexane:petroleum ether (1/1, v/v).

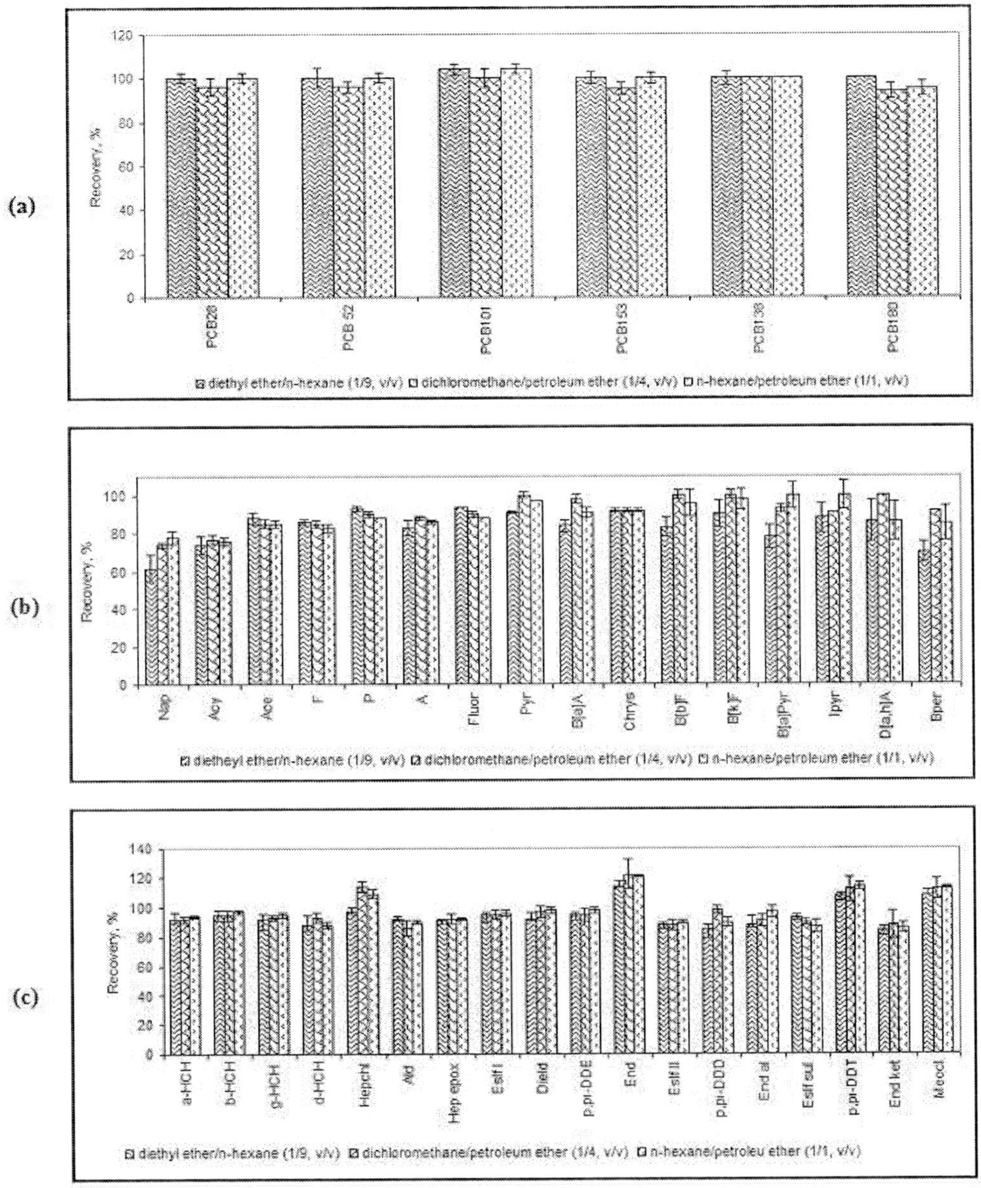

Figure 6. Recoveries of PCBs (a), PAHs (b) and OCPs (c) obtained by ultrasonic extraction with various solvents (extracted once, sonication time: 45 min, solvent volume: 100 mL), [n=4].

Statistical evaluation for the extraction of OCPs from filter by using *n*-hexane:petroleum ether (1/1, v/v) and *n*-hexane:diethyl ether (9/1, v/v) indicated significant difference ($P<0.05$) between the compounds of heptachlor, endrin, *p,p*^{I}-DDT. However, no significant differences were observed between the extraction efficiencies of *n*-hexane:petroleum ether (1/1, v/v) and dichloromethane:petroleum ether (1/4, v/v). However, because dichloromethane is more toxic than *n*-hexane, it was decided that *n*-hexane:petroleum ether (1/1, v/v) should be used for the extraction of OCPs. The mixture of *n*-hexane:petroleum ether (1/1, v/v) was used in further experiments for the extraction of all target compounds.

The aim of the optimization procedure was to improve the extraction efficiency with minimum solvent and time consumption. Therefore, a careful optimization of the other extraction parameters (i.e. solvent volume, extraction time and number of extraction) would be necessary in order to get satisfactory results. Recovery ratios for PCBs by using extraction solvent volume of 25, 50, 75, 100 mL were between 80%±6-100%±4, 90%±0-100%±0, 90%±0-100%±4, 95%±0-108%±2, respectively. By using the same volume of solvents, recovery ratios of PAHs were obtained as between 48%±11-100%±0, 67%±0-100%±0, 70%±7-112%±6, 76%±2-114%±6, respectively. Recovery ratios for OCPs were 56%±6-104%±1, 70%±9-116%±1, 80%±3-118%±0, 86%±3-120%±1, respectively. Generally, increasing the volume of *n*-hexane:petroleum ether (1/1, v/v) from 25 to 75 mL resulted in increasing the extraction efficiencies for all target compounds. However, increasing the volume of solvent mixture from 75 to 100 mL did not significantly improve the extraction efficiencies of PCBs, PAHs and OCPs. Therefore, increasing *n*-hexane:petroleum ether (1/1, v/v) volume was interrupted and further extraction procedures were carried out by using 75 mL of *n*-hexane:petroleum ether (1/1, v/v) (see Figure 7).

The next step was the optimization of the sonication time. Therefore, the sonication time was changed from 15 to 60 min. The effect of sonication time on the extraction of target compounds can be seen in Figure 8. Recovery ratios of PCBs at the end of 15, 30, 45, 60 minutes of sonication were between 88%±7-100%±0, 90%±0-100%±0, 90%±0-100%±0, 90%±0-100%±0, respectively. For the same sonication times, recoveries of PAHs were obtained as between 61%±2-100%±0, 61%±8-100%±0, 70%±7-112%±6, 65%±1-114%±7, respectively. For the compounds of OCPs, recoveries were determined as between 56%±7-107%±4, 56%±17-101%±10, 80%±3-118%±1, 60%±1-120%±15, respectively. Especially for PAHs, it was found that the recoveries increased up to 45 min and there were few fluctuations in the recoveries for increasing time from 45 to 60 min. Additionally, it can be seen in Figure 8 that 45 min was enough for the ultrasonic extraction of PCBs and OCPs. Thus, the optimum sonication time for extraction was selected as 45 min.

If extraction solvent volume divided into two or three and extraction is carried out separately with these solvents twice or three times, the extraction efficiency is increased (Skoog et al., 1996). PCBs, PAHs and OCPs compounds extracted from filters with ultrasonic extraction method using optimum solvent volume of 75 mL, optimum extraction time was determined as 45 minutes. 75 mL extraction solvent was divided into three aliquots and filters were extracted using 25 mL of solvent, 15 minutes sonication time three times sequentially. Extracted solvents were combined and reduced to 1 mL for GC analysis.

When extraction was performed using solvent volume of 75 mL and 45 minutes of sonication, extraction efficiencies of PCB standards were in the range of 90%±0 (for PCB180)-100%±0 (for PCB28). However, when the extraction was carried out three times with solvent volume of 25 mL and 15 minutes of sonication (total volume: 75 mL, total sonication time: 45 minutes), recoveries for all PCBs were obtained closely around 100%. When statistical evaluation was performed between the efficiencies of these two extraction procedures, no significant differences were determined (p>0.05). When PAHs standards were extracted once as mentioned above, recoveries were obtained between 70%±7 (for Acenapthalene) and 112%±6 (for Benzo[k]fluoranthene). While, by application of three times extraction as described above, recoveries were determined in the range of 72%±1 (for Acenapthalene) and 120%±2 (benzo[g,h,i]perylene). Statistical evaluation of both extraction procedures for the recoveries of PAHs showed no significant differences (p>0.05).

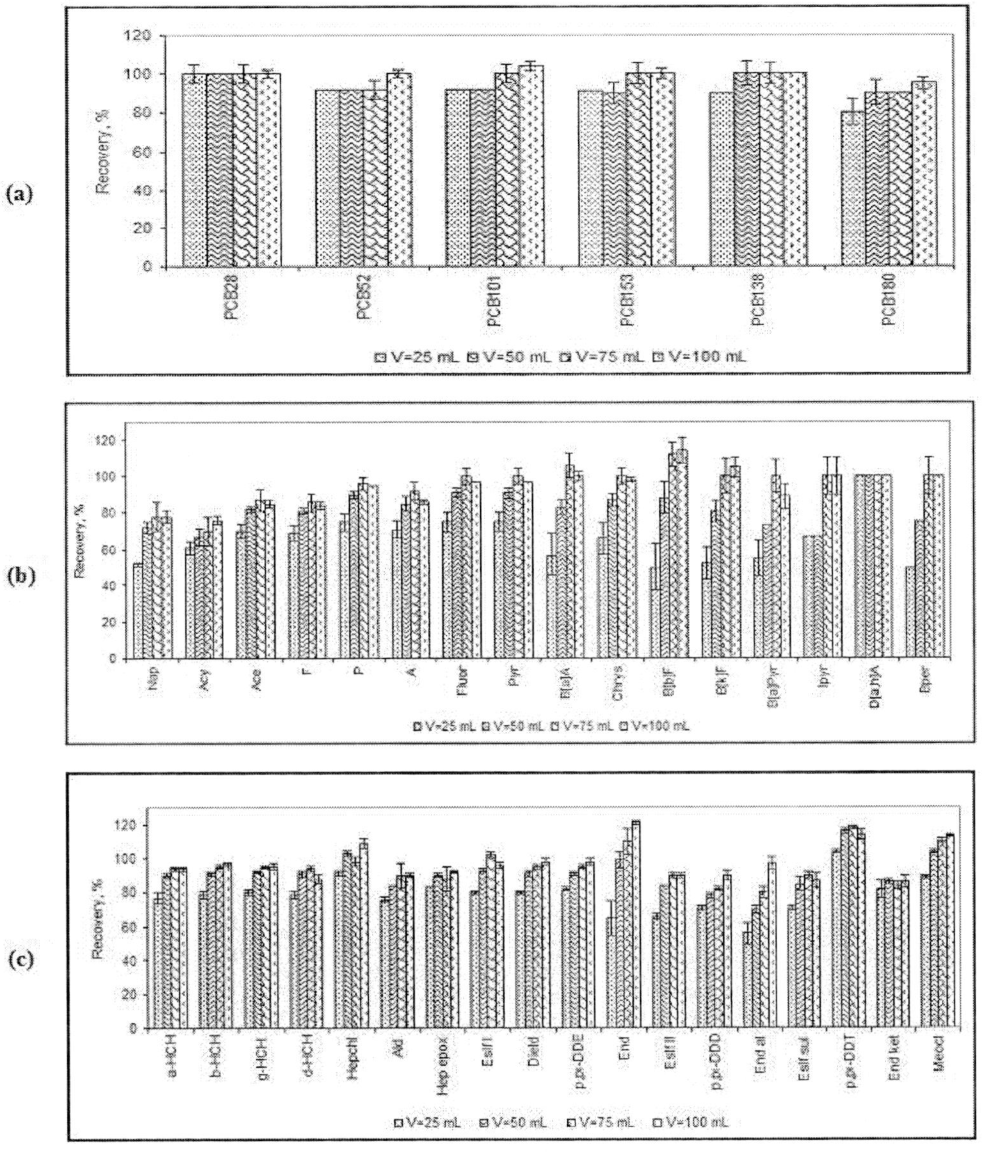

Figure 7. Recoveries of PCBs (a), PAHs (b) and OCPs (c) as a function of the n-hexane/petroleum ether (1/1, v/v) volume (extracted once, sonication time: 45 min), [n=4].

When the same comparison of these two extraction procedures was carried out for the compounds of OCPs, recoveries obtained from once extraction were in the range of 80%±3 (for Endrin aldehyde)-118%±1 (for *p,p'*-DDT), whereas, with application of three times extraction, recoveries ranged from 82%±3 (for Endrin aldehyde) to 120%±6 (Endrin). Statistical evaluation between the efficiencies of these extraction procedures indicated significant difference for the compounds of α-HCH, β-HCH, γ-HCH, heptachlor, dieldrin, *p,p'*-DDE, *p,p'*-DDD, *p,p'*-DDT (p<0.05). Therefore, 25 mL extraction solvent *n*-hexane/petroleum ether (1/1, v/v) was used in the extraction of filters and the extractions were carried out sequentially three times with 25 mL solvent. Extraction time was 15 minutes for each extraction sequence.

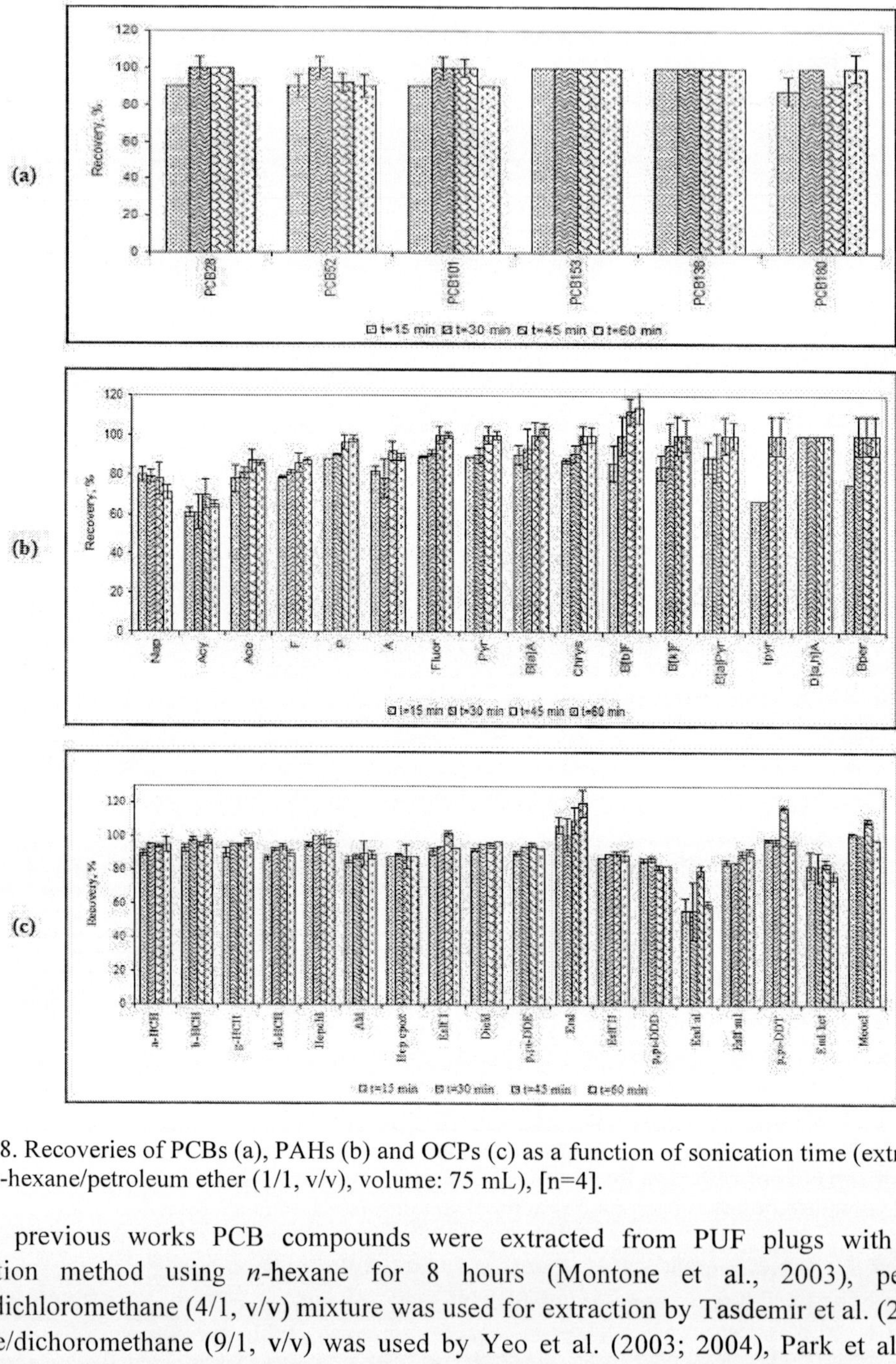

Figure 8. Recoveries of PCBs (a), PAHs (b) and OCPs (c) as a function of sonication time (extracted once, n-hexane/petroleum ether (1/1, v/v), volume: 75 mL), [n=4].

In previous works PCB compounds were extracted from PUF plugs with soxhlet extraction method using *n*-hexane for 8 hours (Montone et al., 2003), petroleum ether/dichloromethane (4/1, v/v) mixture was used for extraction by Tasdemir et al. (2004), *n*-hexane/dichoromethane (9/1, v/v) was used by Yeo et al. (2003; 2004), Park et al. (2002) employed soxhlet extraction for 24 h, Harrod and Mao (2004) used dichloromethane for 16 h extraction, McConnell et al. (1998) carried out soxhlet extraction for 8 h. PAH compounds were extracted also using soxhlet extraction with different solvent mixtures for varying times, Odabası et al. (1999), Sofuoglu et al. (2001) used dichloromethane/petroleum ether (20/80, v/v), Fang et al., (2004) used acetone/*n*-hexane (1/1, v/v), Park et al. (2002) used acetone/*n*-hexane (1/1, v/v) for 24 h extraction, Müller et al. (1998) used dichloromethane for 10 h extraction, Ciganek et al. (2004) used soxhlet extraction method for 8 h. Soxhlet extraction method was also used extracting OCP compounds from PUF plugs with various solvent mixture and for different extraction times by previous researchers. Odabası et al. (1999) used petroleum ether/dichloromethane (4/1, v/v) for 12 h extraction, Sofuoglu et al. (2004) used petroleum ether for 16 h extraction, Algeria et al. (2006) carried out for 24 h, Gioia et al. (2005) used *n*-pentane for 16 h extraction, Yeo et al. (2003) used *n*-hexane/dichloromethane (9/1, v/v) for 24 hours extraction while Louie and Sin (2003) used dichloromethane in soxhlet extraction method.

Spiked standard PCB, PAH, OCP compounds were extracted using 200 mL diethyl ether/*n*-hexane (1/9, v/v), dichloromethane/petroleum ether (1/4, v/v), *n*-hexane/petroleum ether (1/1, v/v) solvent mixtures with soxhlet extraction method for 16 hours from PUF plugs and recovery values obtained are given in Figure 9-a,b,c. Recovery values obtained for PCB compounds were 90%±3-108%±6, 90%±6-100%±6, 77%±5-91%±5 respectively for diethyl ether/*n*-hexane (1/9, v/v), dichloromethane/petroleum ether (1/4, v/v) and *n*-hexane/petroleum ether (1/1, v/v) solvent mixtures. For the same solvent mixtures recovery values for PAH compounds were 74%±3-109%±9, 26%±10-102%±2, 43%±2-1025±2 respectively while the recovery values for OCP compounds also for the same solvent mixtures were 80%-108%±4, 50%±4-120%±5, 61%±2-120%±9 respectively. The highest recovery ratios were obtained with diethyl ether/*n*-hexane (1/9, v/v) solvent mixture for three groups of compounds. Statistical significance of the various solvents in terms of recovery values were determined using two-tailed unpaired students t-test (at 95% confidence interval level) on quantities of individual PAH, PCB and OCP compounds from four replicate analyses of PUF plugs.

When statistical evaluation was carried out between diethyl ether/*n*-hexane (1/9, v/v) with dichloromethane/ petroleum ether (1/4, v/v) solvent mixtures in terms of recovery of PCB compounds only for PCB 101 recovery values were significantly different at 95% level, however when the comparison carried out with *n*-hexane/petroleum ether (1/1, v/v) solvent mixture PCB 14, 28, 52, 65, 101 congeners extraction efficiency values were significantly (at 95% levels) different. When the statistical comparison was carried out between diethyl ether/*n*-hexane (1/9, v/v) and dichloromethane/petroleum ether (1/4, v/v) solvent mixtures in terms of recovery values for PAH compounds, recovery values were significantly different (at 95% level) for Acy, Ace, F, P-d_{10}, A, Fluor, Pyr, B[a]A, Chrys-d_{12}, B[k]F, B[a]Pyr. When the comparison carried out with *n*-hexane/petroleum ether (1/1, v/v), Nap, Acy, P-d_{10}, A, Pyr, B[a]A, Chrys-d_{12}, B[a]Pyr compounds extraction efficiency values were significantly (at 95% levels) different.

Comparing diethyl ether/*n*-hexane (1/9, v/v) with dichloromethane/petroleum ether (1/4, v/v) solvent mixtures in terms of recovery for OCP compounds it could be suggested that recovery values were significantly different at 95% level for Heptachlor epoxide, *p,p*'-DDD, endrin aldehyde however when the comparison carried out with *n*-hexane/petroleum ether (1/1, v/v) solvent mixture recovery values were significantly different (at 95% level) only for endrin aldehyde. The solvent mixtures diethyl ether/*n*-hexane (1/9, v/v), dichloromethane/petroleum ether (1/4, v/v), *n*-hexane/petroleum ether (1/1, v/v) used for extraction of PCB, PAH and OCP compounds from PUF plugs with soxhlet extraction. Recovery for PCB compounds were complying with recovery ratios range of 60%-120% given by EPA for the extraction of multiple compounds. However, only diethyl ether/*n*-hexane (1/9, v/v) solvent mixture was complying with this recovery ratio ranges for PAH and OCP compounds. Therefore, diethyl ether/*n*-hexane (1/9, v/v) solvent mixture was used for the extraction of PCBs, PAHs and OCPs from PUF plugs.

(a)

(b)

(c)

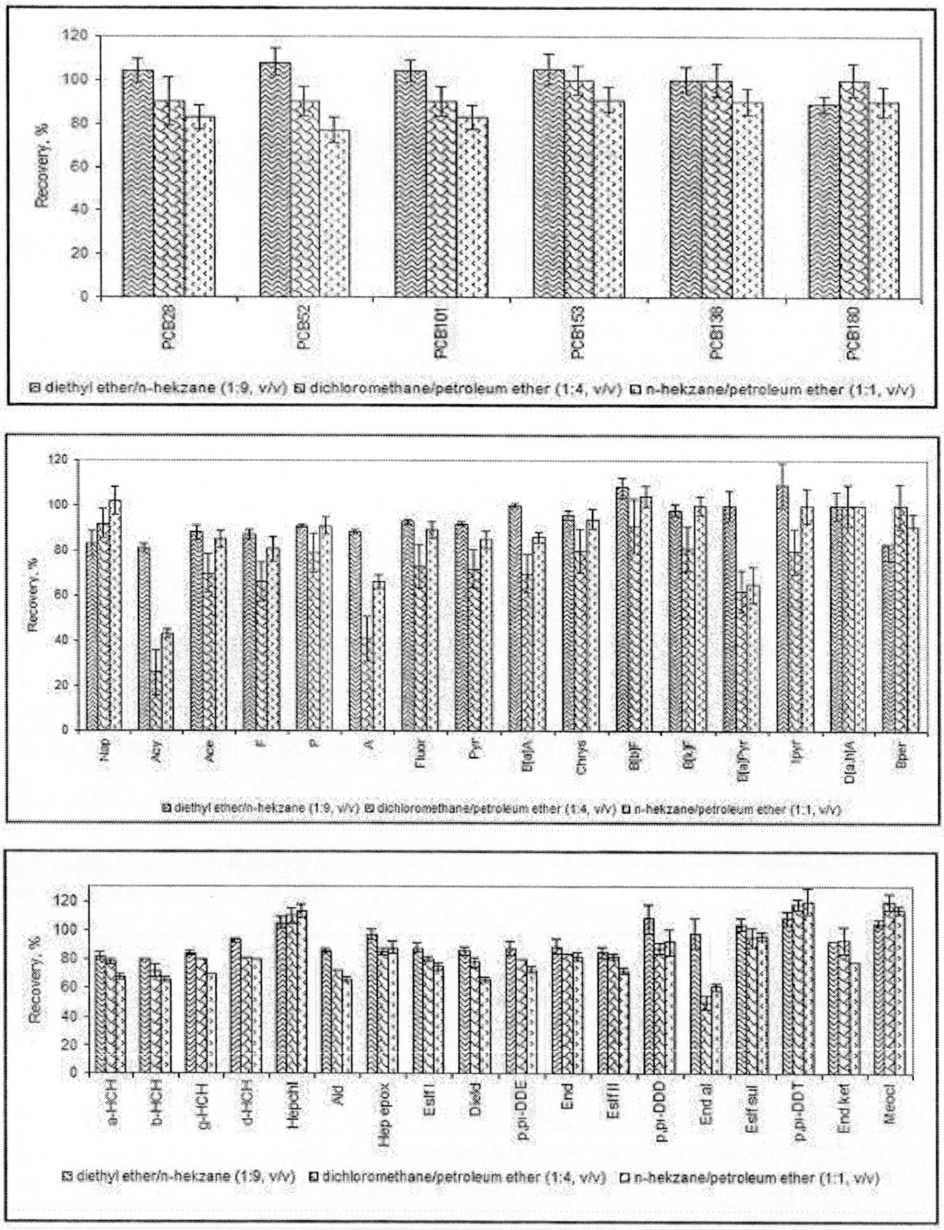

Figure 9. Recoveries of PCBs (a), PAHs (b) and OCPs (c) obtained by soxhlet extraction with various solvents (extraction time: 16 h, solvent volume: 200 mL), [n=4].

The aim of the optimization works carried out was to improve the extraction efficiency with minimum solvent and extraction time. Therefore, optimization of the other extraction parameters such as solvent mixture volume and extraction time would be necessary. After determining the optimum solvent mixture for extraction of PCBs, PAHs and OCPs from PUF plugs, works were carried out to determine optimum solvent volume for soxhlet extraction method using 150 mL, 200 mL and 250 mL of the diethyl ether/*n*-hexane (1/9, v/v). Recovery values obtained from different volumes used in extraction of PCBs, PAHs and OCPs from PUF plugs with soxhlet extraction for 16 hours are given in Figure 10-a,b,c.

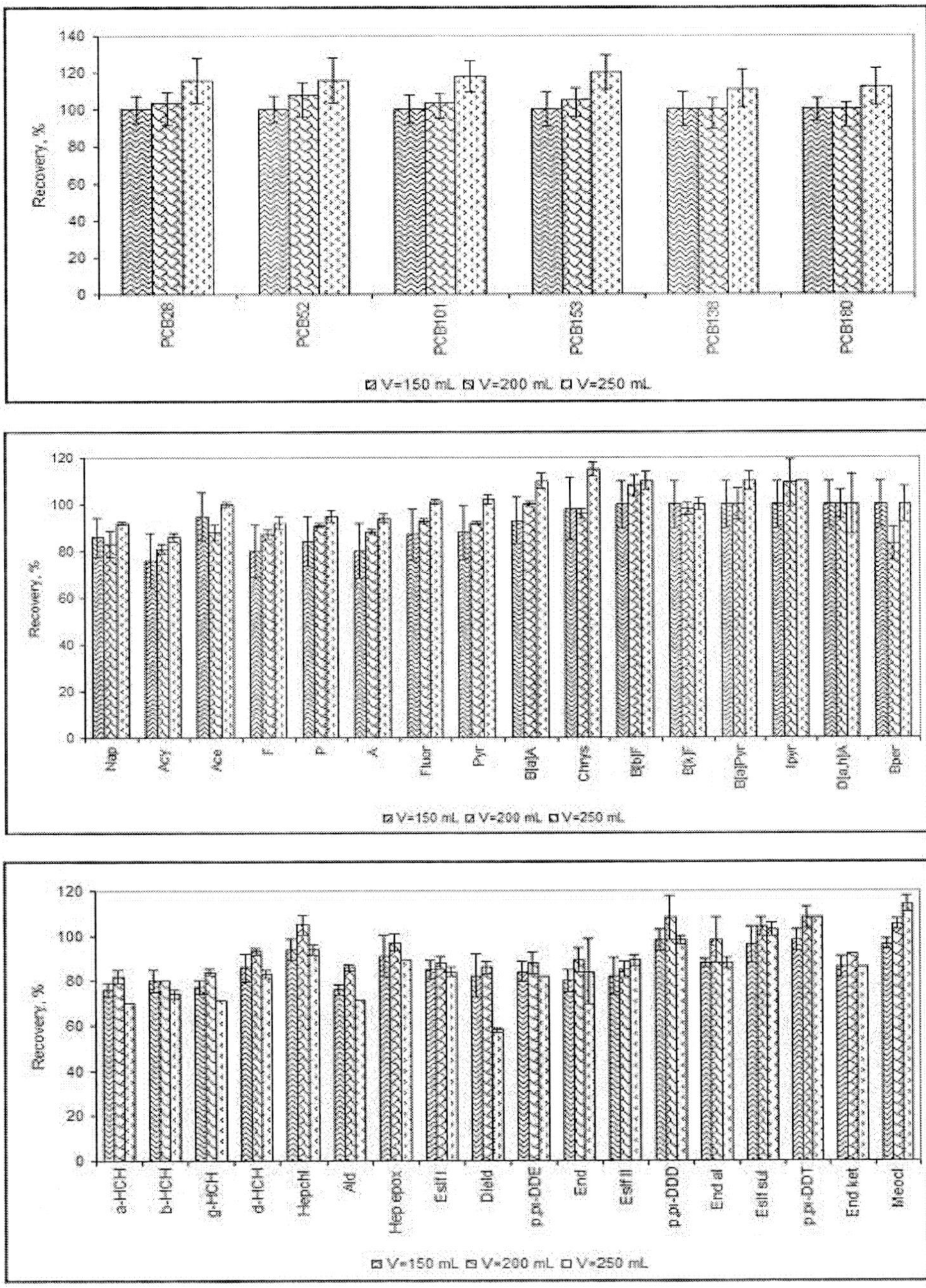

Figure 10. Recoveries of PCBs (a), PAHs (b) and OCPs (c) as a function of the diethyl ether/n-hexane (1/9, v/v) volume (extraction time: 16 h), [n=4].

Recovery values obtained for 150 mL, 200 mL and 250 mL of the extraction solvent volumes are in 100%±7-111%±5, 100%±3-108%±6, 108%±15-120%±9 range respectively for PCB compounds, 76%±11-100%±10, 74%±3-109%±9, 86%±1-115%±3 range respectively for PAH compounds, 76%±2-98%±4, 82%±3-108%±4, 70%-114%±3 range respectively for OCP compounds. Comparing extraction efficiency results of 150 mL solvent mixture volume with the results of 200 mL and 250 mL volumes for PCBs and PAHs statistically it could be suggested that there is no significant difference at 95% level. Although recovery values are in acceptable ranges given by US EPA. However, there was statistically significant difference at 95% level between the recovery values obtained for 150 mL solvent mixture and 200 mL solvent mixture for γ-HCH, heptachlore, aldrin, methoxychlor compounds of OCPs. Extraction efficiencies are also increased with increasing solvent mixture volume for PCB and PAH compounds (see Figure 10). However extraction efficiency was decreased for OCP compounds when solvent mixture volume increased to 250 mL. Considering the higher solvent volume use will increase the analyses cost and extraction efficiencies were in acceptable range 60%-120% for recoveries for all target compounds. Optimum solvent mixture volume was determined as 150 mL for the extraction of PCBs, PAHs and OCPs from PUF plugs with soxhlet extraction method.

After determining the optimum solvent mixture type and volume works were carried out to determine optimum extraction time. 8 h, 16 h and 24 h extraction times were carried out for the extraction of spiked PCB, PAH and OCP compounds to the PUF plugs using 150 mL diethyl ether/*n*-hexane (1/9, v/v) solvent mixture with soxhlet extraction method. Recovery ratios obtained for 8 h, 16 h and 24 h extraction times are given in Figure 11-a,b,c.

Recovery ratios found 100% for PCB compounds for 8 h, 16 h and 24 h extraction times, there was no significant difference observed for different extraction times. Recovery ratios for PAH compounds were in 76%±4-100%±2, 74%±10-100%±10, 78%±1-120%±3 range respectively for 8 h, 16 h and 24 h extraction times. Comparing the results of PAH recovery values statistically it could be suggested that there was no significant difference between the 8 h, 16 h and 24 h results.

Recovery values for OCP compounds were in 64%±4-93%±3, 76%±2-98%±4, 86%±2-120%±1 range respectively for 8 h, 16 h and 24 h extraction times. Comparing 8 hours and 16 hours extraction times results statistically it could be suggested that there was no significant difference between the results of 8 h and 16 h extraction times at 95% significance level. However, comparing 8 h results with 24 h results there was statistically significant difference at 95% confidence range for the extraction efficiency results of α-HCH, β-HCH, δ-HCH, heptachlor, aldrin, heptachlor epoxide, endosulfan I, dieldrin, *p,p'*-DDE, endosulfan II, endrin aldehyde, endosulfan sulfate, *p,p'*-DDT, endrin ketone, methoxychlor compounds comparing 16 h and 24 h extraction times results, it could also be suggested that there was statistically significant (at 95% level) difference between the recovery values for α-HCH, β-HCH, δ-HCH, heptachlor, aldrin, *p,p'*-DDE, endrin, *p,p'*-DDT, methoxychlor compounds.

Despite the improvement of recovery ratios for some of the OCP compounds, the recovery ratios were in the acceptable range of 60%-120% given by US EPA for multiple compounds extractions for all compounds investigated.

There was no difference for the PCBs and PAHs results for different extraction times while there was statistically significant difference between the results of 8 h and 16 h for almost all OCP compounds. Therefore, 16 Hours extraction time was selected as extraction duration with soxhlet extraction method since all of these three groups of compounds will be extracted together from the PUF plugs.

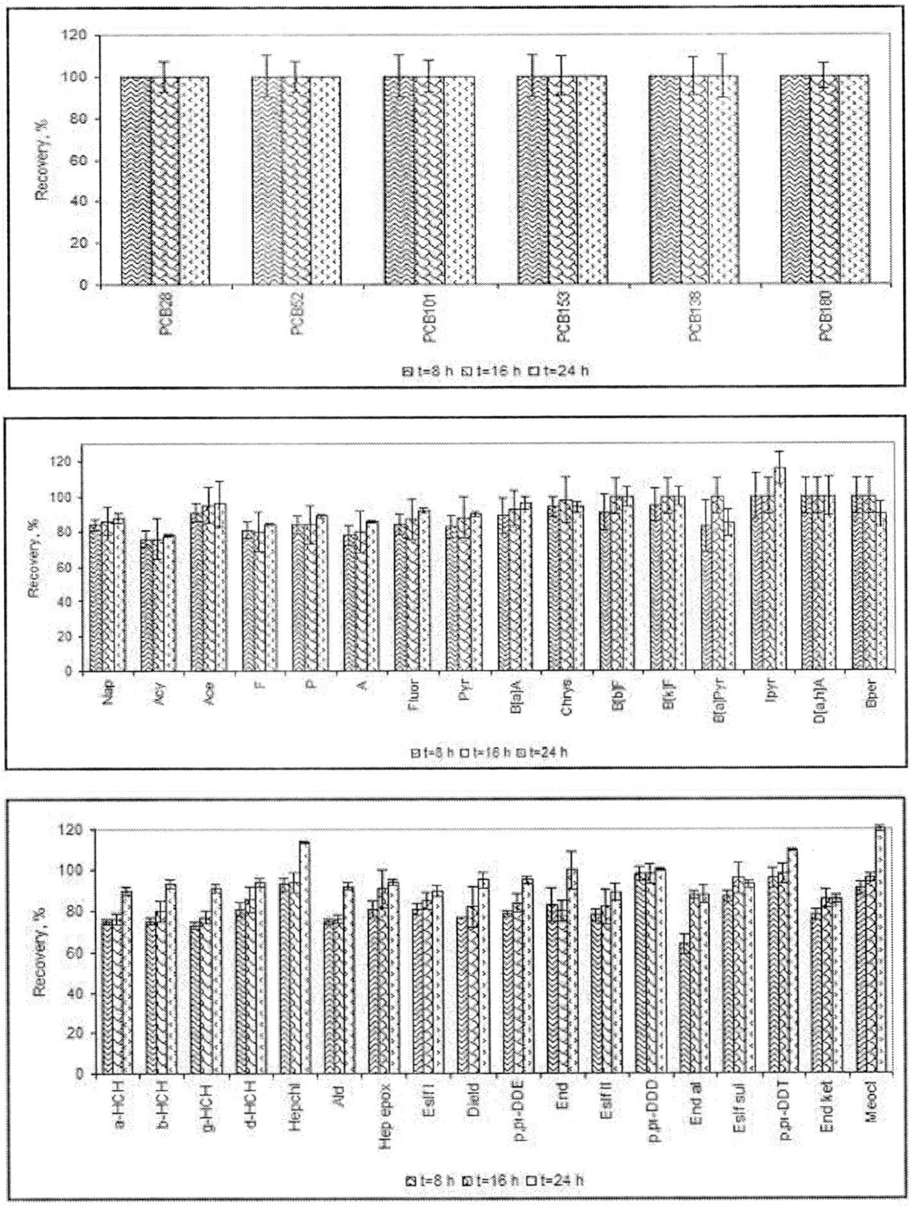

Figure 11. Recoveries of PCBs (a), PAHs (b) and OCPs (c) as a function of extraction time (extracted once, diethyl ether/n-hexane (1/9, v/v), volume: 150 mL), [n=4].

Examination of Matrix Effect

In order to evaluate the matrix effects and to observe the separation performance of the column, sampled microfibre filters and PUF plugs were extracted and analyzed separately. Filters and PUF plugs were extracted as described in section optimization of extraction procedure. Then the extracts volumes were reduced to less than 1 mL and then spiking procedure was carried out by adding 10 μL of standard mixture of PCBs (the concentration is

10 ng μL^{-1} for each compound), 10 μL of standard mixture of PAHs and OCPs (the concentration is 100 ng μL^{-1} for each compound). The volumes were adjusted exactly to 1 mL and the spiked extracts were transferred onto the top of the column and fractionations were carried out consecutively using 70 mL n-hexane and 3x20 mL n-hexane:ethyl acetate (1/1, v/v). Then, the volumes of the elutions were reduced to 1 mL and the internal standards were also added and GC/μ-ECD and GC/MSD analysis were performed. After sampling spiked and non-spiked filters were analyzed simultaneously following the developed method. Total mean (n=5) concentrations were found 2 pg m^{-3} for PCBs, 0.2 ng m^{-3} for PAHs and 0.06 ng m^{-3} for OCPs in filters and 22 pg m^{-3} for PCBs, 1.78 ng m^{-3} for PAHs and 0.44 ng m^{-3} for OCPs in PUF plugs. Recoveries of PCBs, PAHs and OCPs compounds of microfibre filter and PUF plug extracts were calculated after the subtracting corresponding values of PCBs, PAHs and OCPs compounds of microfibre filters and PUF plugs.

The recovery values for the target compounds are given in Table 10 and Table 11 for filters and PUF plugs, respectively. The recovery values (n=5) for the target compounds of the microfibre filters were obtained between 96%±2 (for PCB28)-108%±1 (for PCB138) with RSD≤±9% for PCBs, between 72%±1 (for Acenaphthalene)-101%±1 (for Benzo[k]fluoranthene) with RSD≤±8% for PAHs (except for Naphthalene<31%) while the recoveries were changed from 62%±5 (for Endrin aldehyde) to 110%±1 (for Heptachlor) with RSD≤±6% for OCPs compounds. The recovery values (n=5) for the target compounds of the PUF plugs were obtained between 106%±2 (for PCB180)-120%±5 (for PCB138) with RSD≤±6% for PCBs, between 80%±2 (for Dibenzo[a,h]anthracene)-113%±2 (for Pyrene) with RSD≤±6% for PAHs while the recoveries were changed from 67%±6 (for Endrin aldehyde) to 118%±5 (for p,p^1-DDD) with RSD≤±6% for OCPs compounds.

Table 10. The recoveries of PCBs (a), PAHs (b) and OCPs (c) for the spiked extracts of microfibre filters [*n*=5]

PCBs	Recovery (%)	PAHs	Recovery (%)	OCPs	Recovery (%)
PCB 14	100±2	Naphthalene	31±1	α-HCH	74±1
PCB 28	96±2	Acenaphthalene	72±1	β-HCH	78±1
PCB 52	98±4	Acenaphthene	74±1	γ-HCH	68±3
PCB 65	120±3	Fluorene	80±1	δ-HCH	88±4
PCB 101	102±4	Phenanthrene-d_{10}	89±1	Heptachlor	110±1
PCB 153	104±1	Phenanthrene	84±1	Aldrin	83±1
PCB 138	108±1	Anthracene	87±2	Heptachlor epoxide	69±2
PCB 166	117±1	Fluoranthene	88±2	Endosulfan I	68±1
PCB 180	100±2	Pyrene	94±2	Dieldrin	91±5
		Benzo[a]anthracene	93±3	*p,p'*-DDE	73±4
		Chrysene-d_{12}	88±1	Endrin	109±6
		Chrysene	92±5	Endosulfan II	73±3
		Bbenzo[b]flouranthene	90±2	*p,p'*-DDD	102±4
		Benzo[k] fluoranthene	101±1	Endrin aldehyde	62±5
		Benzo[a]pyrene	98±4	Endosulfan sulfate *p,p'*-	105±6
		Indeno[1,2,3-cd]pyrene	97±3	DDT	82±2
		Dibenzo[a,h]anthracene	87±2	End ketone	75±2
		Benzo[g,h,i]perylene	86±4	Methoxychlor	79±4

Table 11. The recovery values of PCBs (a), PAHs (b) and OCPs (c) for the spiked extracts of PUF plugs [*n*=5]

PCBs	Recovery (%)	PAHs	Recovery (%)	OCPs	Recovery (%)
PCB 14	110±6	Naphthalene	102±1	α-HCH	81±5
PCB 28	110±4	Acenaphthalene	98±1	β-HCH	83±4
PCB 52	111±4	Acenaphthene	99±1	γ-HCH	75±3
PCB 65	120±3	Fluorene	103±1	δ-HCH	105±4
PCB 101	118±4	Phenanthrene-d_{10}	97±1	Heptachlor	112±3
PCB 153	118±4	Phenanthrene	113±1	Aldrin	91±2
PCB 138	120±5	Anthracene	101±1	Heptachlor epoxide	76±5
PCB 166	120±6	Fluoranthene	108±2	Endosulfan I	78±6
PCB 180	106±2	Pyrene	113±2	Dieldrin	103±2
		Benzo[a]anthracene	100±4	*p,p'*-DDE	82±6
		Chrysene-d_{12}	84±1	Endrin	78±4
		Chrysene	100±6	Endosulfan II	77±4
		Bbenzo[b]flouranthene	97±2	*p,p'*-DDD	118±5
		Benzo[k] fluoranthene	96±1	Endrin aldehyde	67±6
		Benzo[a]pyrene	80±4	Endosulfan sulfate	103±4
		Indeno[1,2,3-cd]pyrene	88±3	*p,p'*-DDT	98±6
		Dibenzo[a,h]anthracene	80±2	End ketone	70±5
		Benzo[g,h,i]perylene	90±5	Methoxychlor	86±2

Table 12. Comparison of method recovery ratios of PCBs, PAHs, OCPs standard and surrogate spike (%) with recovery ratios obtained in previous works

Compounds name	Standard/surrogate	Recovery ratios	Reference
PAH PCB OCP PAH PCB OCP	Standard + surrogate	66%±6-94%±5 (filter) (except Naph) 86%±2-104%±3 (filter) 67%±5-120%±10 (filter) 88%±1-120%±2 (PUF plug) 94%±9-118%±8 (PUF plug) 65%±4-108%±6 (PUF plug)	In this study, 2007
PAH	Surrogate	64%±14-79%±12 (filter+PUF plug)	Odabası et al., 1999; Sofuoglu et al., 2001; Vardar et al., 2004
PAH	Standard	81%-96% (filter) 69%-95% (PUF plug plug)	Park et al., 2002
PAH	Standard	50.77%-99.7% (filter)	Omar et al., 2002
PCB	Surrogate	50%-120% (PUF plug)	Montone et al., 2003
PCB	Surrogate	~75% (filter) ~70% (PUF plug)	Tasdemir et al., 2004b
PCB	Standard	~108%±3.2 (filter +PUF plug)	Alonso et al., 2001
PCB	Surrogate	62.9%±7.3-116.6%±10.2	Yeo et al., 2004
PCB	Surrogate	50%-80% (filter +PUF plug)	Harrad and Mao, 2004
PCB	Surrogate	75%-113% (filter +PUF plug)	Park et al., 2002
OCP	Surrogate	51%±7-87%±13 (filter +PUF plug)	Sofuoglu et al., 2001
OCP	Standard	42%±4-128%±22 (filter) 40%±11-128%±22 (PUF plug)	Sofuoglu et al., 2004
OCP	Surrogate	88%-119% (filter +PUF plug)	Yeo et al., 2003
OCP	Surrogate Standard	56%±5-102%±8 (filter+PUF plug) >70% (filter+PUF plug)	Alegria et al., 2006

Recoveries for all three groups of compounds were in acceptable levels comparing the limits given by US EPA (1999) which are between 60%-120% for multiple compounds. The recovery values of PCBs, PAHs and OCPs of the spiked extracts of microfibre filters and PUF plugs were comparable to those of the only standard spiked sample. This means that the deteriorative effects of other co-extracted substances on both fractionation/clean-up process and quantitative analysis of target compounds by GC system were negligible.

The determination of PCBs, PAHs and OCPs in urban air samples was performed according to analytical protocol given in Figure 12. Recovery ratios of investigated compounds for optimized method are compared with results obtained by other researchers and given in Table 12.

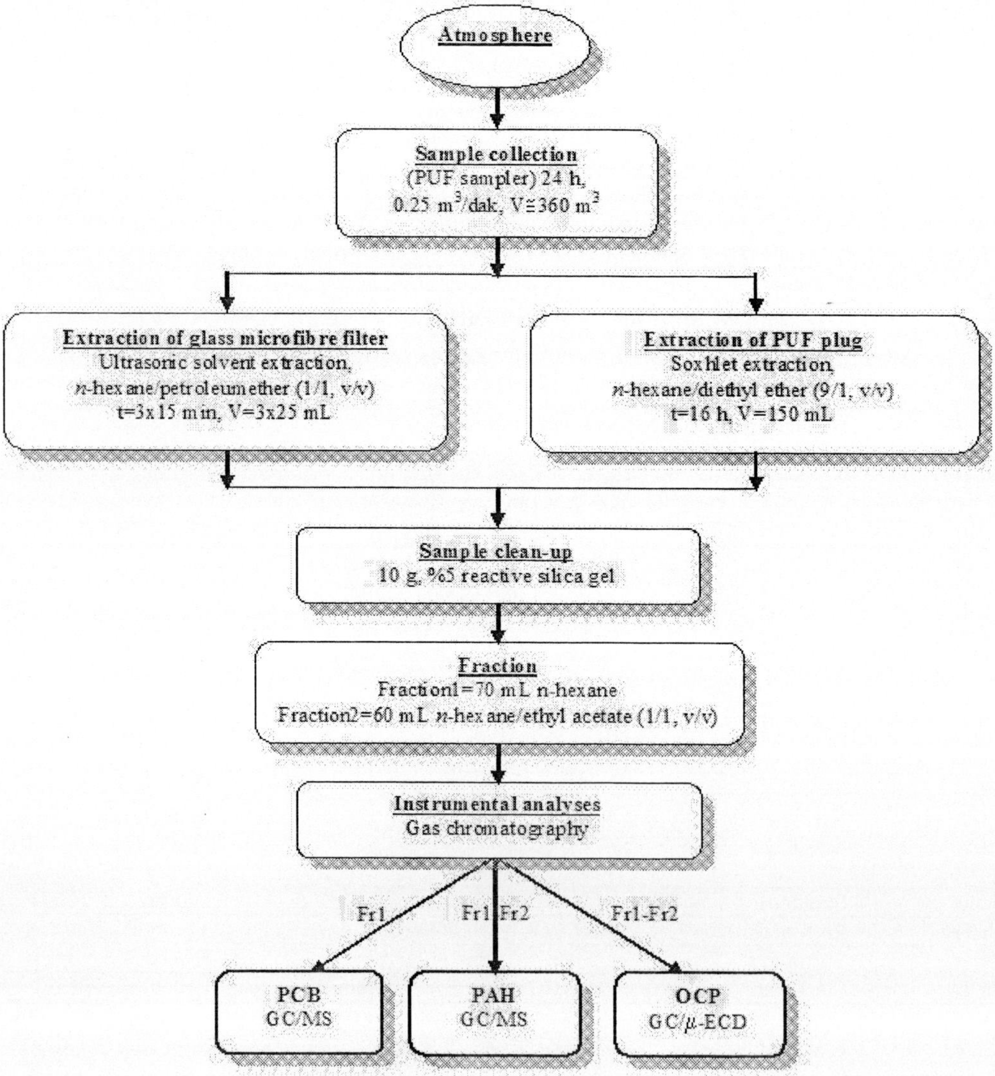

Figure 12. Optimized analytical protocol.

As seen in Table 12, the recoveries obtained in this work were better than the recovery ratios obtained for the same compounds in previous works for the three groups of compounds investigated. The minimum recovery for PCBs after 24 hours soxhlet extraction from filter was reported as 63%, while it was obtained as 86% in this work. In this work the minimum recoveries were obtained as 66% and 67% for PAHs and OCPs, respectively. In the previous works the minimum recoveries were obtained about 50% for PAHs and 40% for OCPs. Minimum method recovery values for PAH, PCB and OCP compounds from PUF plug were obtained 88%, 94%, 65%, respectively. Higher recoveries and lower relative standard deviations were obtained comparing the minimum recoveries in this work with those obtained in the previous works. However, by using the optimized analytical method developed in this work while reducing the hazardous solvent consumption, analysis time and expenses are reduced in the analysis of trace levels organic pollutants.

Air concentrations of PAHs, PCBs and OCPs

Sampling was carried out in five different locations (SL1, SL2, SL3, SL4, SL5) between August 2006 and May 2007 for the analyses of PCBs, PAHs and OCPs in gas and particulate phases. Concentration values of each compound under the pollutants groups were determined in Konya city atmosphere in this work and given in Figure 13. Mean total PCB concentrations were 0.043 ng m^{-3} (0.026 ng m^{-3} in gas phase, 0.016 ng m^{-3} at particulate phase) in SL1, 0.112 ng m^{-3} (0.099 ng m^{-3} in gas phase, 0.013 ng m^{-3} at particulate phase) in SL2, 0.098 ng m^{-3} (0.085 ng m^{-3} in gas phase, 0.013 ng m^{-3} at particulate phase) in SL3, 0.106 ng m^{-3} (0.078 ng m^{-3} in gas phase, 0.029 ng m^{-3} at particulate phase) in SL4, 0.144 ng m^{-3} (0.116 ng m^{-3} in gas phase, 0.028 ng m^{-3} at particulate phase) in SL5. The highest PCB concentration was found in SL5 which was selected to determine the PCB concentrations of traffic while second highest PCB concentration was determined in SL2 which is located in industrial zone. The lowest PCB concentration was observed in SL1 (Selcuk University area) which was selected to determine outside transportation of pollutants to the city considering the dominant wind direction. Higher concentrations of PCBs observed in SL2 and SL5 indicates that the determined PCBs are emitted from traffic and industrial sources in Konya. While investigated all PCBs congeners were detected in the urban air of Konya, PCB 28 and PCB 52 congeners were determined as dominant.

Distributions of PCBs in the atmosphere have been studied by other researchers in various urban, rural and industrialized areas in worldwide. A number of these studies are frequently detected 6 PCB congeners (PCB 28, 52, 101, 153, 138, 180) which were studied in this work. At first time, the atmospheric levels of PCBs were studied Cindoruk and Tasdemir (2007) in Turkey. They measured gas and particulate concentrations of 6-PCBs in an urban area in Bursa (Turkey). They determined characterization and partitioning of 6-PCBs. In the study carried out between August 2004 and May 2005, the mean PCBs concentrations gas and particulate phase were determined as 63 pg m^{-3} and 13 pg m^{-3} in urban location, as 99 pg m^{-3} and 13 pg m^{-3} in industrial location and as 78 pg m^{-3} and 129 pg m^{-3} in traffic location, respectively. The average total PCBs results obtained in this work are lower than average values determined in Bursa. Their sampling point is an urban area with high traffic loading in Bursa which is an industrialized city. Yeo et al. (2004) investigated concentration characteristics of atmospheric PCBs for urban and rural area in Korea. They determined that mean concentrations of total 6-PCBs (gas + particulate phase) in urban and rural air were 39.35 pg m^{-3} and 15.5 pg m^{-3}, respectively.

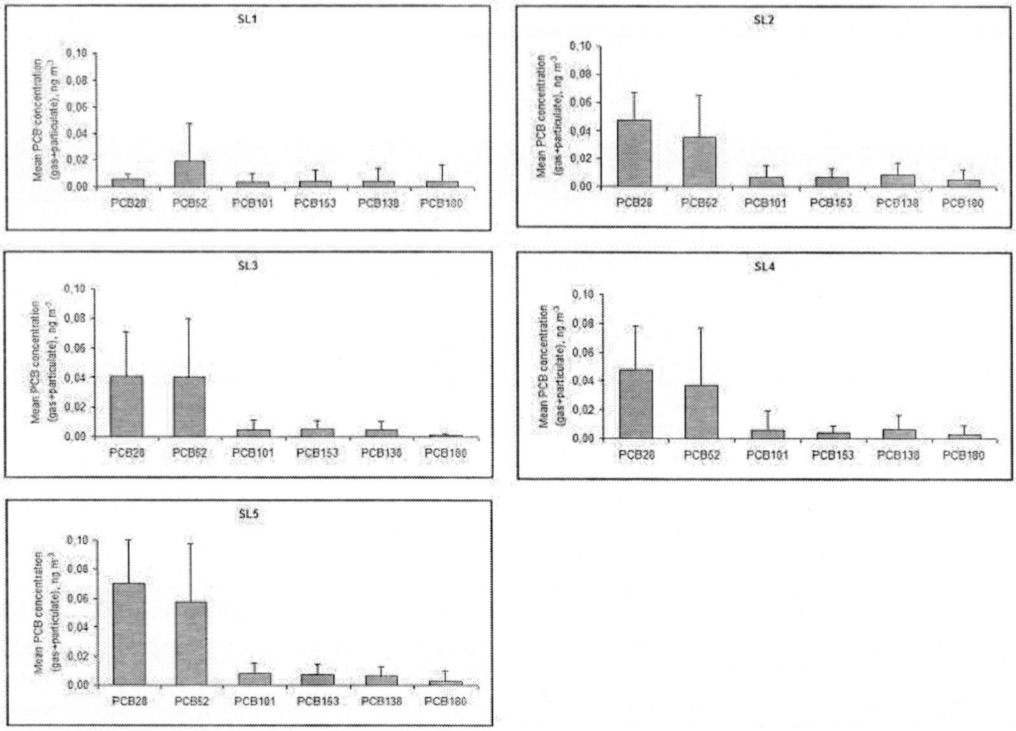

Figure 13. Mean total gas and particulate phases concentrations of each PCB compounds for all sampling locations (ng m-3).

The atmospheric concentrations of the selected PCBs over a summer period at King George Island were investigated by Montone et al. (2003). Mean 6-PCBs congeners in the marine atmosphere were determined 20.4 pg m^{-3}. Mandalakis and Stephanou (2006) measured atmospheric concentrations of PCBs in the rural area in eastern Germany. While the average total concentrations of 6-PCBs congeners were determined as 25 pg m^{-3}, they were predominately observed in the gas phase. Atmospheric total concentrations of 6-PCBs were recorded 64 pg m^{-3} at an urban location in Birmingham, UK between April 1999 and July 2000 by Harrad and Mao (2004). Comparing the PCBs levels measured in our study with other works, measured PCBs concentrations in our study were in the range of their reported results. These differences in between the concentrations can be explained as due to different sources, atmospheric conditions and transportations.

Mean total PAH concentration values are 30.2 ng m^{-3} (18.9 ng m^{-3} in gas phase, 11.3 ng m^{-3} at particulate phase) in SL1, 180.8 ng m^{-3} (110.9 ng m^{-3} in gas phase, 69.9 ng m^{-3} at particulate phase) in SL2, 229.8 ng m^{-3} (147.7 ng m^{-3} in gas phase, 82.1 ng m^{-3} particulate phase) in SL3, 206.2 ng m^{-3} (126.8 ng m^{-3} in gas phase, 79.4 ng m^{-3} at particulate phase) in SL4, 293.7 ng m^{-3} (186.3 ng m^{-3} in gas phase, 107.4 ng m^{-3} at particulate phase) in SL5. The highest PAH compounds were observed in SL5 and the order of PAHs concentration levels from high to low observed in the other sampling locations is SL3 > SL4 > SL > SL1. Sampling locations 3 and 4 are selected for determining PAH compounds given to atmosphere from the burning processes, mostly for domestic heating. From the results obtained in this work it could be suggested that sources of PAHs in Konya city atmosphere are traffic and domestic heating respectively.

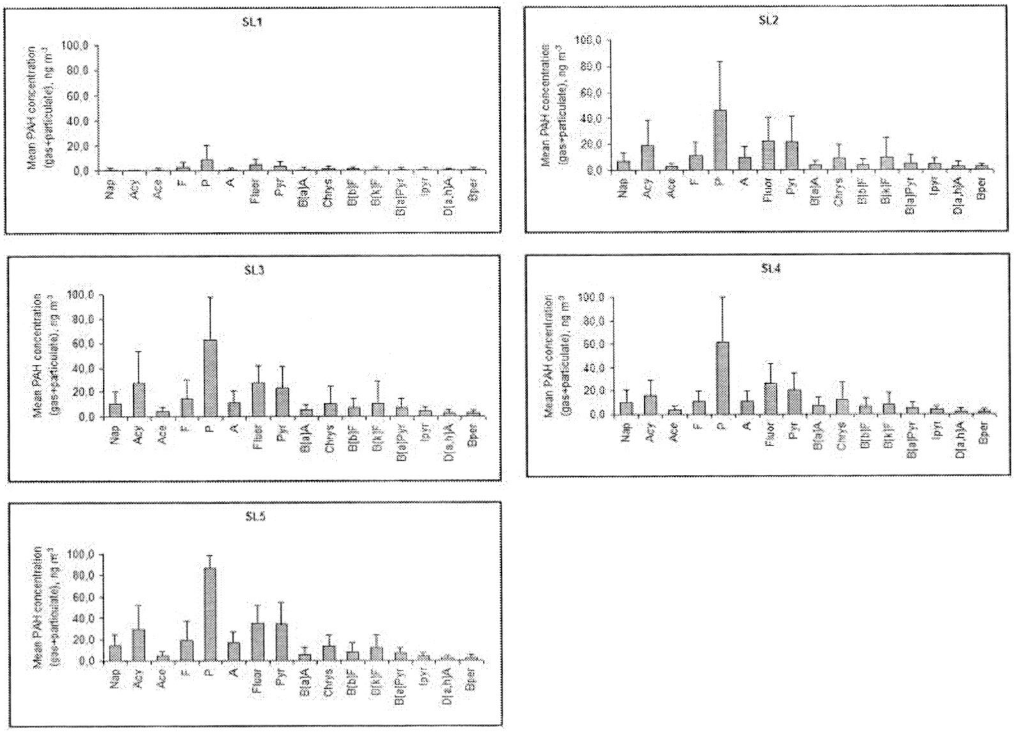

Figure 14. Mean total gas and particulate phases concentrations of each PAH compounds for all sampling locations (ng m-3).

Tasdemir and Esen (2007) reported that the mean total concentration of PAHs in the urban air at a traffic site in Turkey was 456 ng m^{-3}. They found that lighter PAHs compounds were in higher concentrations. In Italy, the mean total PAH concentration in air samples was determined as 59.4 ng m^{-3} and more volatile compounds were dominant (Cincinelli et al., 2007). Tsapakis and Stephanou (2005) determined average total PAH concentration as 79.3 ng m^{-3} in an urban region in Greece. PAH concentrations have been reported for different areas in the world by many researchers (Müller et al., 1998; Odabası et al., 1999; Ollivon et al., 2002; Park et al., 2001; 2002; Poor et al., 2004; Vardar et al., 2004). They found Phenanthrene as the most abundant compounds in all samples while Fluor, Pyr, Anthracene, Chrys were determined high percentage in detected total concentrations. Moreover, results demonstrated that determined PAHs levels display diversity according to sampling procedure, analytical method, the features of sampling area, sampling time and these factors limited comparison of results. Still, our results are in consistent with in the literature. The determined PAH and PCB compounds concentration values were observed in similar pattern in sampling locations. This results show that sources of PCBs and PAHs are similar for industrial or urban areas. Therefore, it could be suggested that their evaporation or transportation paths could be controlled in similar ways.

Mean total OCP concentrations were 0.81 ng m^{-3} (0.62 ng m^{-3} in gas phase, 0.19 ng m^{-3} at particulate phase) in SL1, 3.03 ng m^{-3} (1.36 ng m^{-3} in gas phase, 1.83 ng m^{-3} at particulate phase) in SL2, 3.78 ng m^{-3} (1.53 ng m^{-3} in gas phase, 2.25 ng m^{-3} particulate phase) in SL3, 4.79 ng m^{-3} (1.51 ng m^{-3} in gas phase, 3.28 ng m^{-3} at particulate phase) in SL4, 6.16 ng m^{-3} (1.72 ng m^{-3} in gas phase, 4.44 ng m^{-3} at particulate phase) in SL5. The highest OCP concentration value was determined in SL5 and the following order of OCP concentrations according to sampling locations from high to low was observed as SL4 > SL3 > SL2 > SL1. Mean concentration values of each OCP compounds observed in sampling locations in gas and particulate phases are given in Figure 15. In all sampling points α-HCH, β-HCH, γ-HCH, δ-HCH, Eslf II, *p,p'*-DDT, *p,p'*-DDD, Eslf sul, Eslf II and Meocl were observed as dominant compounds.

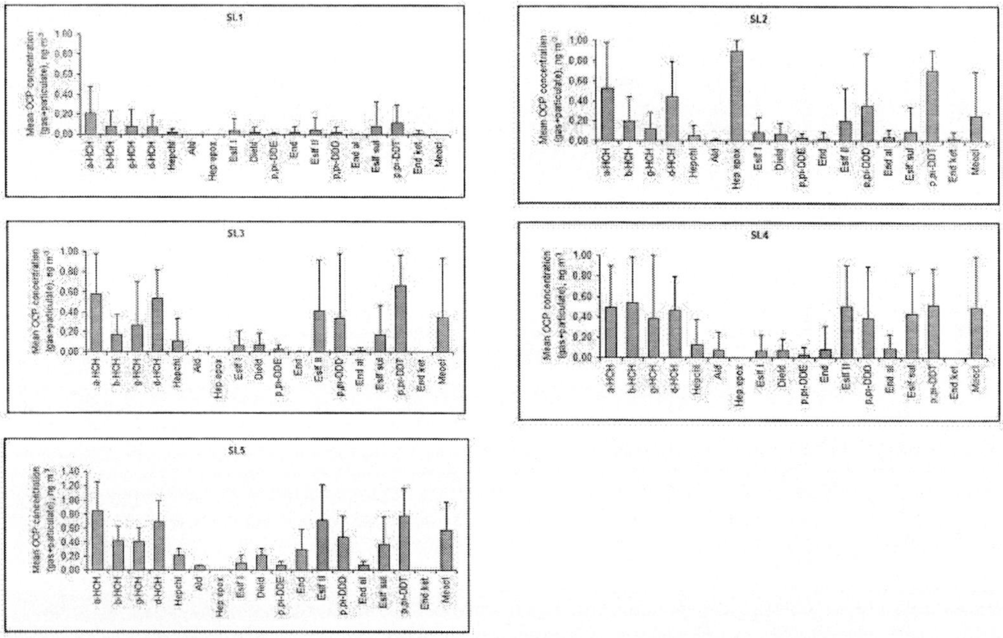

Figure 15. Mean total gas and particulate phases concentrations of each OCP compounds for all sampling locations (ng m-3).

Total mean HCHs (Σ α-HCH+β-HCH+γ-HCH+δ-HCH) were determined 0.44 ng m^{-3}, 1.278 ng m^{-3}, 1.553 ng m^{-3}, 1.89 ng m^{-3}, 2.389 ng m^{-3}, respectively for SL1, SL2, SL3, SL4, SL5. HCH isomers (α+β+γ+δ-HCH) were determined as dominant compounds which accounted for 30% of total OCPs in the investigated organochlorine pesticides. HCHs are one of the most widely used and most readily detected organochlorine pesticides in environmental samples. The use of HCH was restricted in 1978 and was prohibited in 1985 while the use of lindane was banned in 1979 in Turkey. Technical HCH consist of α-HCH (65-70%), β-HCH (7-10%), γ-HCH (14-15%), δ-HCH (~7%), ε-HCH (~1-2%), other components (~1-2%) used widely as a commercial pesticide. Lindane consists of almost 100% γ isomers and is also known as γ-HCH. In the study, other isomers were detected more dominantly than γ-HCH. These results demonstrated that technical HCH was used in Konya area.

Total mean DDTs including *p,p'*-DDT, *p,p'*-DDD, *p,p'*-DDE were determined in urban air of Konya as 0.14 ng m^{-3}, 1.088 ng m^{-3}, 1.02 ng m^{-3}, 0.937 ng m^{-3}, 1.32 ng m^{-3}, respectively for SL1, SL2, SL3, SL4, SL5. Total mean DDTs make up 20% of total OCPs. DDT is the best pesticides and is used for the control of vectors transmitting some diseases as malaria, typhus, plague, trypanosomiasis. Their use were restricted in 1978 and then were banned in

1985 in Turkey. But it is still used in some countries for controling malaria, especially in the tropical region (Hung et al., 2005).

Total mean Endrin including End, End al, End ket were determined as 0.031 ng m^{-3}, 0.084 ng m^{-3}, 0.013 ng m^{-3}, 0.172 ng m^{-3}, 0.366 ng m^{-3} respectively for SL1, SL2, SL3, SL4, SL5. Endrin is used to control mice, moles and other rodents and is spread on the leaves of crops such as cotton and grains. Aldrin is used to kill termites and other soil pests, in grain storage, for vector control. Use of endrin and aldrin were banned in 1979. Dieldrin is used basically to control termites and textile pests. Also it is used to control of insects transmitting some diseases and for insects living in agricultural soils. Aldrin and dieldrin banned in 1971 were determined in very low concentrations in urban air of Konya.

Endosulfan is used to control insects on food and also as a wood preservative. It is considered priority pollutant by the EPA. Total mean (gas + particulate phase) concentrations of Eslf compounds including Eslf I, Eslf II, Eslf sul were determined as 0.158 ng m^{-3}, 0.376 ng m^{-3}, 0.654 ng m^{-3}, 1.004 ng m^{-3}, 1.185 ng m^{-3}, respectively for SL1, SL2, SL3, SL4, SL5. This value accounted for 21% total OCPs. Endosulfan I and Endosulfan II are chlorinated hydrocarbon insecticed that are used against wide variety of insects and acar. They were classified high toxic pesticides by US EPA toxicity classification and their use are limited (Gioia et al. 2005). Mean total gas and particulate phase Endosulfan compound concentration was 722 pg m^{-3} in Konya city atmosphere. This results show that this compound pose a health risk. While Heptachlor are observed in all sampling locations, Hep epox compound was determined only sampling location 2 in 0.9 ng m^{-3} concentration.

Pesticides were used on a large scale after World War II in agriculture and for disease vector control, especially DDT. Their residues have existed in different environment in the world since then. Konya is the biggest Plato and accounted for 17% of all agricultural production in Turkey. Banned pesticides especially DDTs, aldrin, dieldrin were observed at high level in urban area of Konya may be due to wide agricultural areas. Also, they can evaporate from contaminated soils and waters and can be transport long distances from used areas. Some pesticides such as aldrin, toxaphene, DDT, chlordane, dieldrin, endrin, HCB, heptachlor, mirex are listed in the Stockholm Convention on POPs. These pesticides have slow rate degradation, for example half life of DDT is 10-15 years in soil and they have high solubility in fatty substances which causes their bioaccumulation in the food chain. They can be transported over long distances in the atmosphere.

Therefore, pesticides exist in air was investigated by other researches in recent years. Organochlorine pesticides concentrations were investigated between 14 and 23 May 2003 in İzmir, Turkey (Sofuoglu et al., 2004) and mean concentrations of total OCPs were detected 2.2 ng m^{-3}. Comparing their results with total OCPs concentrations determined in May in this work, it can be suggested that values of total OCPs were similar. Sofuoglu et al. (2001) determined organochlorine pesticides in between 0.24 ng m^{-3} and 1.35 ng m^{-3} in the gas phase in Chicago, between June and October 1995. Scheyer et al. (2005) carried out analysis of some orhanochlorine pesticides in an urban air atmosphere in Strasbourg and aldrin and dieldrin were detected in lower concentrations, between 0.01 ng m^{-3} and 0.09 ng m^{-3} while α-HCH and γ-HCH were detected in between 0.01 ng m^{-3} and 4 ng m^{-3}. OCPs concentrations were measured in the ambient air of Chiapas, Mexico during 2001-2001 by Alegria et al. (2006). Aldrin was not detected in air samples while total mean concentrations of other investigated pesticides were detected as 2.26 ng m^{-3}. In the study carried out by Gioia et al.

(2005), average concentrations of gas and particulate phase in the urban/industrial area in New Jersey were determined as 1.5 ng m^{-3} and 17.8 pg m^{-3}, respectively. Levels of OCPs determined in the study were compared with results of different regions by other researcher, it can be said that OCPs concentrations were generally higher than other research results. But total DDT and HCH isomers found dominant like they were determined in this study. Dieldrin is banned in Mexico and all neighboring countries but it is still determined at low levels as about mean 15 pg m^{-3} (Alegria et al., 2006). Yeo et al. (2003) identified in the rural air in between 1999 and 2000 although the use of HCH has been banned since 1979 in South Korea. Even though some of investigated pesticides are legally banned in our country, their residues still present in urban air. Also, Breivik et al. (1999) demonstrated that there was relationship between global use of HCHs isomers and their air concentrations in Europe between 1970 and 1996.

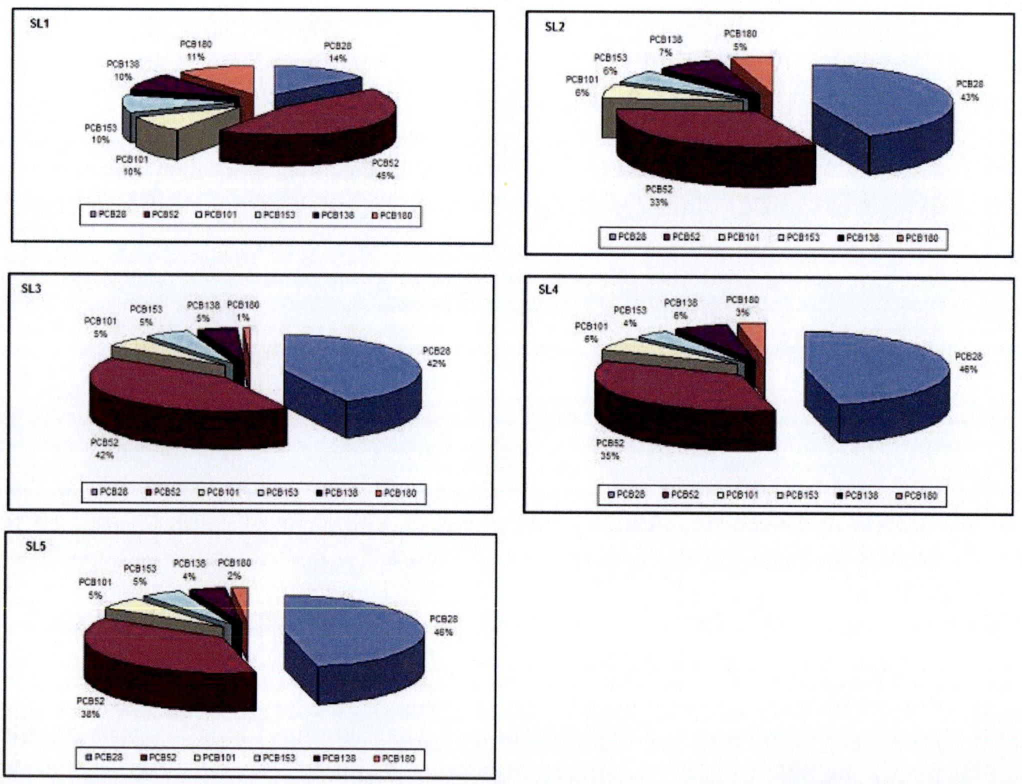

Figure 16. Mean concentration distribution ratios of PCB compounds for all sampling locations.

Percent Contributions of PAHs, PCBs and OCPs in Urban Air

Figure 4, 5, 6 show the contributions of each PCB, PAH, OCP compounds to the total PCB, PAH, OCP concentration. As seen in Figure 4, PCB28 and PCB52 are made up between 14%-46% and 33%-45% of total PCBs, respectively. Other PCB congeners make up between 1% and 10% of total PCBs. As a result, tri-CBs and tetra-CBs homologs of investigated PCB congeners were determined as dominant in all sampling location.

Considering the PCB profiles determined in previous works; Montone et al. (2003) reported low molecular weight PCB congeners such as PCB 101 and smaller molecular congeners were found as dominant PCB compounds in air samples and they were made up about 66.7% of all determined PCB concentration. McConnell et al. (1998) suggested that considering the homolog distribution of PCB congeners tri and tetra chlorinated congeners are made up 70-85% of all determined PCB concentrations. Yeo et al. (2003) observed that tri-CBs and tetra-CBs are the main part of the PCB homologs that they determined in high concentrations in atmosphere.

In the city atmosphere Phenanthrene was found with highest ratio of about 29%, Fluor, Pyr, Acy and F followed Phenanthrene with ratio of 13%, 12%, 10% and 7%, respectively. A, B[a]A, Chrys, B[b]F, B[k]F, B[a]Pyr, Ipyr, D[a,h]A, Bper, Nap, Ace compounds were determined below 5% ratio (see Figure 17). Low molecular weight PAH compounds (with 2 to 4 rings) including P, Fluor, Pyr, Acy, Chrys, F, Nap, B[a]A were dominantly determined than high molecular weight PAH compounds (5 and 6 rings) including B[b]F, B[k]F, B[a]Pyr, Ipyr, D[a,h]A, Bper in the total concentrations. While Phenanthrene, Fluor, Pyr accounted for 29%, 13%, 10% of total PAHs, other compounds formed <5% of total PAHs. 3 and 6 ring PAH compounds profiles were observed as dominant. The PAH compounds observed dominantly show low toxicological properties. Phenanthrene, Fluor, Pyr, Anthracene, Chrys compounds are often found among 16 EPA PAHs in atmosphere (Müller et al., 1998; Halsal et al., 2001; Lin et al., 2002; Ollivon et al., 2002; Polkowska et al., 2000; Park et al., 2001; 2002; Vardar et al., 2004; Hung et al., 2005; Tsapakis and Stephanou, 2005). These reported results are in agreement with the results obtained in this work.

The most abundant compound group found in Konya city atmosphere was total HCH (Σ α-HCH+β-HCH+γ-HCH+δ-HCH) with a ratio of 41% (see Figure 18). The total DDT compound (Σp,p'-DDT+p,p'-DDD+p,p'-DDE) followed the total HCH group with a ratio of 24%. Total Eslf (Σ Eslf-I+ Eslf-II+ Eslf sul) was found 18% followed by End (Σ End+ End al+End ket) with a ratio of 4% while heptachlor and dieldrin compounds were observed 2% and 0.6% ratios, respectively.

22 OCP compounds were determined in gas and particulate phases in six different locations in New York. Mean total chlordane concentration in gas phase were determined as between 83-518 pg m^{-3} while they were determined as between 1-10 pg m^{-3} in particulate phase. Heptachor and heptachlor epoxide were not found in particulate phase while they were found as mean values between 32-304 pg m^{-3}, 1.8-45 pg m^{-3} in gas phase, respectively. Total DDT concentrations were determined between 31-237 pg m^{-3} in gas phase and between 0.59-4.8 pg m^{-3} in particulate phase. In gas phase, aldrin was found between 1.1-77 pg m^{-3} while dieldrin was found between 8.2-75 pg m^{-3}. In particulate phase, aldrin and dieldrin were found only in one location with a concentration of 0.23 pg m^{-3} and 2.8 pg m^{-3}, respectively. Total α-HCH and γ-HCH concentrations in gas phase were found between 37-254 pg m^{-3} however they were not detected in particulate phase. Total Endosulfan concentrations were between 14.9-178 pg m^{-3} in gas phase, while they were changed between 5.9-2.41 pg m^{-3} in particulate phase (Gioia et al., 2005). Concentrations of 11 pesticides in gas and particle phase were investigated in air samples collected from rural and urban atmosphere in France. Pesticides in atmosphere were found in gas phase in atmosphere and these results are important in long distances transportation of pesticides (Sanusi et al., 1999).

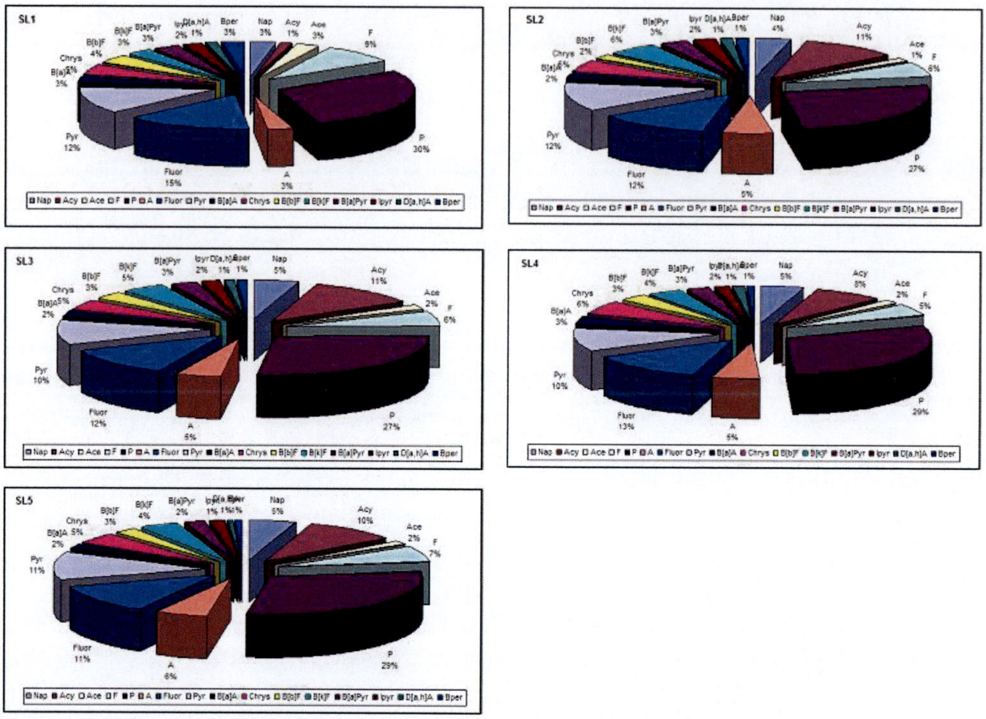

Figure 17. Mean concentration distribution ratios of PAH compounds for all sampling locations.

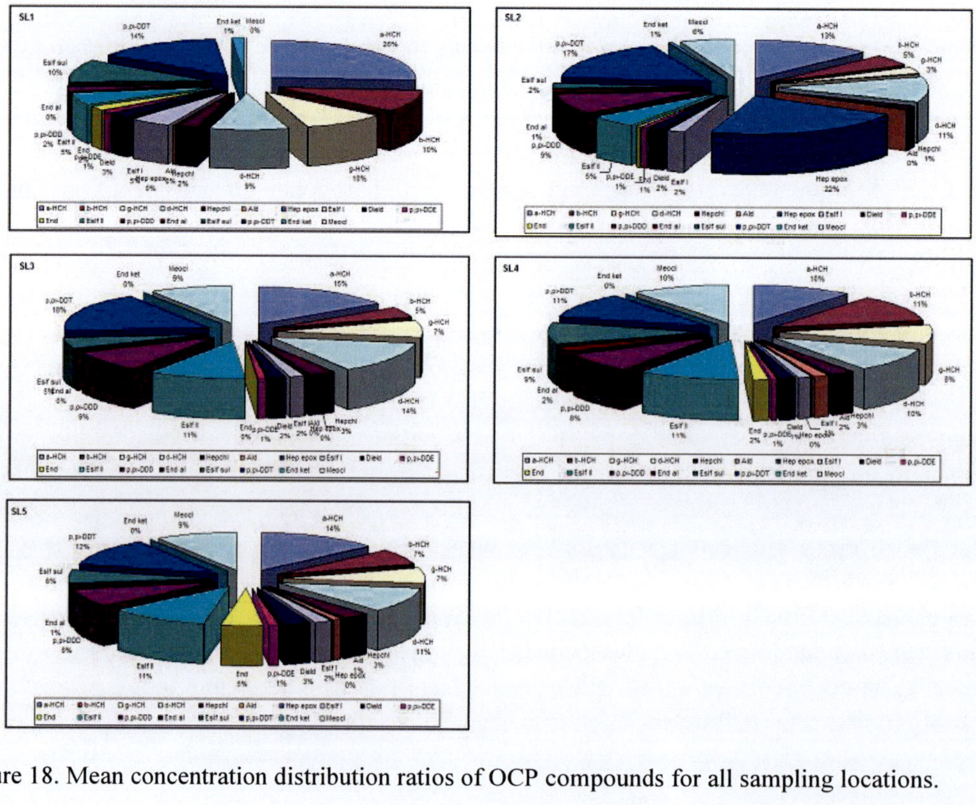

Figure 18. Mean concentration distribution ratios of OCP compounds for all sampling locations.

Seasonal Trends of Total PAHs, PCBs and OCPs in Gas and Particulate Phases

Monthly variations of total PAHs (a), PCBs (b), OCPs (c) in gas and particulate phases for all sampling locations are given in Figure 19, 20, 21. Looking at Figure 19, it can be seen that the highest total PCB concentration was found in September consisting of 0.29 ng m^{-3} (0.26 ng m^{-3} in gas phase, 0.03 ng m^{-3} at particulate phase) in SL2, 0.42 ng m^{-3} (0.37 ng m^{-3} in gas phase, 0.05 ng m^{-3} at particulate phase) in SL3, 0.36 ng m^{-3} (0.22 ng m^{-3} in gas phase, 0.14 ng m^{-3} at particulate phase) in SL4. The highest PCB concentration was observed in March in SL1 (0.14 ng m^{-3} total, 0.03 ng m^{-3} gas phase, 0.11 ng m^{-3} particulate phase) however, in the same sampling location relatively higher value of PCB concentration was determined in September. The highest PCB concentrations in SL5 were determined in January as 0.33 ng m^{-3} (0.25 ng m^{-3} in gas phase, 0.08 ng m^{-3} at particulate phase). In October, February and September lower values were determined.

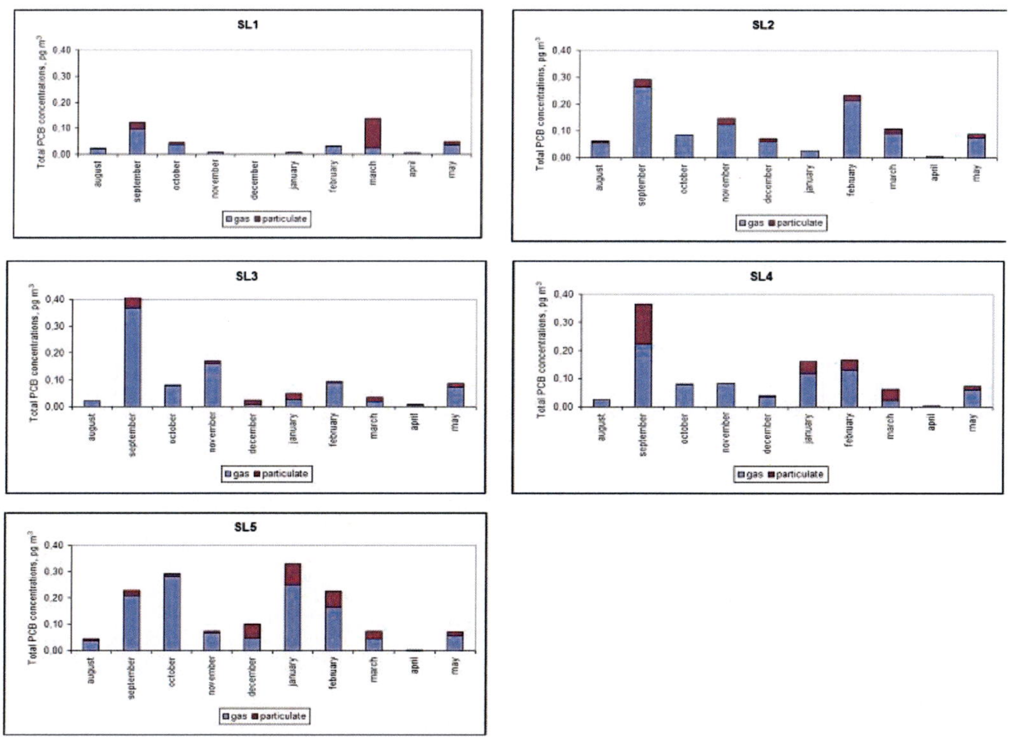

Figure 19. Monthly variations of total PCBs in gas and particulate phases for all sampling locations.

Results from seasonally monitoring of PCBs in urban air of Konya show that the highest PCBs concentrations were determined at September while the lowest PCBs concentrations at April. It can be seen Table 9 that temperature in September is higher than in April. It can be suggested that higher temperature may increase volatilization from contaminated environment in atmosphere. Also, higher wind speed might have caused air transporting of these compounds. Considering Figure 20, the highest PAH concentrations were observed in SL1 and SL2 as 124.9 ng m^{-3} (70.2 ng m^{-3} in gas phase, 74.7 ng m^{-3} at particulate phase) and 442.8

ng m^{-3} (23.1 ng m^{-3} in gas phase, 211.7 ng m^{-3} at particulate phase) respectively in February. In sampling locations SL 3, SL4 and SL5 the highest concentration levels were 586.9 ng m^{-3} (234.6 ng m^{-3} in gas phase, 352.3 ng m^{-3} at particulate phase), 427.5 ng m^{-3} (206.2 ng m^{-3} in gas phase, 221.3 ng m^{-3} at particulate phase), 534.4 ng m^{-3} (335.3 ng m^{-3} in gas phase, 199.1 ng m^{-3} at particulate phase) in January. PAHs concentrations seasonal variations are similar in sampling locations SL1, SL3 and SL4. Therefore, it could be suggested that sources of PAH compounds depend not only variation in temperature and vapor pressure but also emissions from traffic and domestic heating. Reasons of low concentrations observed in summer seasons are the decrease in emissions from domestic heating and traffic load as well as decrease of temperature inversions effects. In addition photochemical decomposition would be increased in summer season sun shine. Observed seasonal variation in concentration of PAHs in the atmosphere show that contribution of domestic heating to the PAH level is significant.

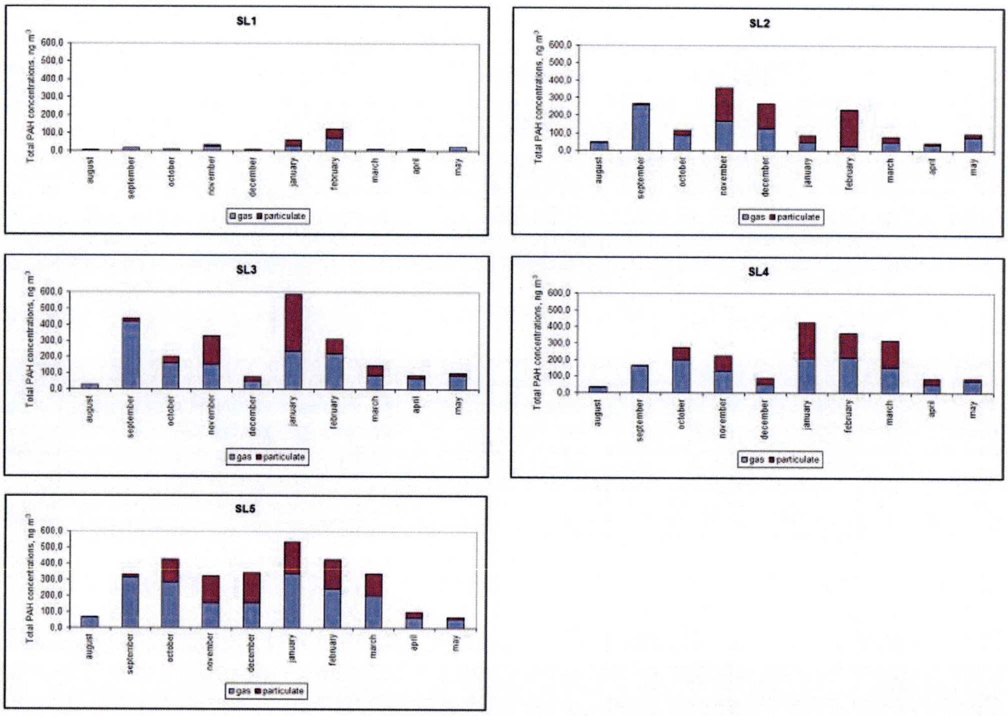

Figure 20. Monthly variations of total PAHs in gas and particulate phases for all sampling locations.

In the sampling location 5, total PAH concentrations were lower in August, April and May comparing the other months values, however, despite the seasonal variation there is no discernible variation in concentration levels. Sampling location 5 was selected for determining PAH compounds in the atmosphere coming from traffic. Sampling location SL1, SL3 and SL4 were selected for determining PAHs in atmosphere originating from domestic heating comparing the results of these sampling locations with SL5 reveals that PAH concentrations decreased in December in SL1, SL3 and SL4 while there is no discernible decrease in SL5. PAH compounds originating traffic add up with the PAH compounds originating from domestic heating and reach the highest levels in this sampling location.

Comparing pollutants concentrations measured in samples taken from different locations in different times is important in determining the seasonal changes. Looking at the seasonal variations, the higher PAH concentrations were observed in cold season sampling times. The highest PAH concentration observed in winter occurred depending on high emissions in winter. The reason why lower concentrations of PAHs observed during summer is coal burning for domestic heating, heavy traffic and the formation of inversion reduced, in addition these compounds were decomposed by means of photochemical reactions. It can be said that domestic heating widely contributed concentrations of PAHs determined in Konya urban air. In a lot of works, PAHs concentrations in the air have been determined higher during the winter season and lower during the summer season.

Looking at Figure 21, it can be seen that the highest OCP concentrations were observed in SL1, SL4, SL5 in October with concentrations of 3.19 ng m^{-3} (2.90 ng m^{-3} in gas phase, 0.29 ng m^{-3} at particulate phase), 9.22 ng m^{-3} (5.05 ng m^{-3} in gas phase, 4.17 ng m^{-3} at particulate phase), 12.29 ng m^{-3} (5.13 ng m^{-3} in gas phase, 7.15 ng m^{-3} at particulate phase), respectively. In sampling location SL2, the OCP concentration was 7.91 ng m^{-3} (1.5 ng m^{-3} in gas phase, 6.41 ng m^{-3} at particulate phase) in November. In sampling location SL3, the OCP concentration was determined as 10.58 ng m^{-3} (1.63 ng m^{-3} in gas phase, 8.95 ng m^{-3} at particulate phase) in January. Higher concentrations observed in October and November could be attributed to pesticides applications during agricultural activities. In these months, lands were prepared for planting by farmers.

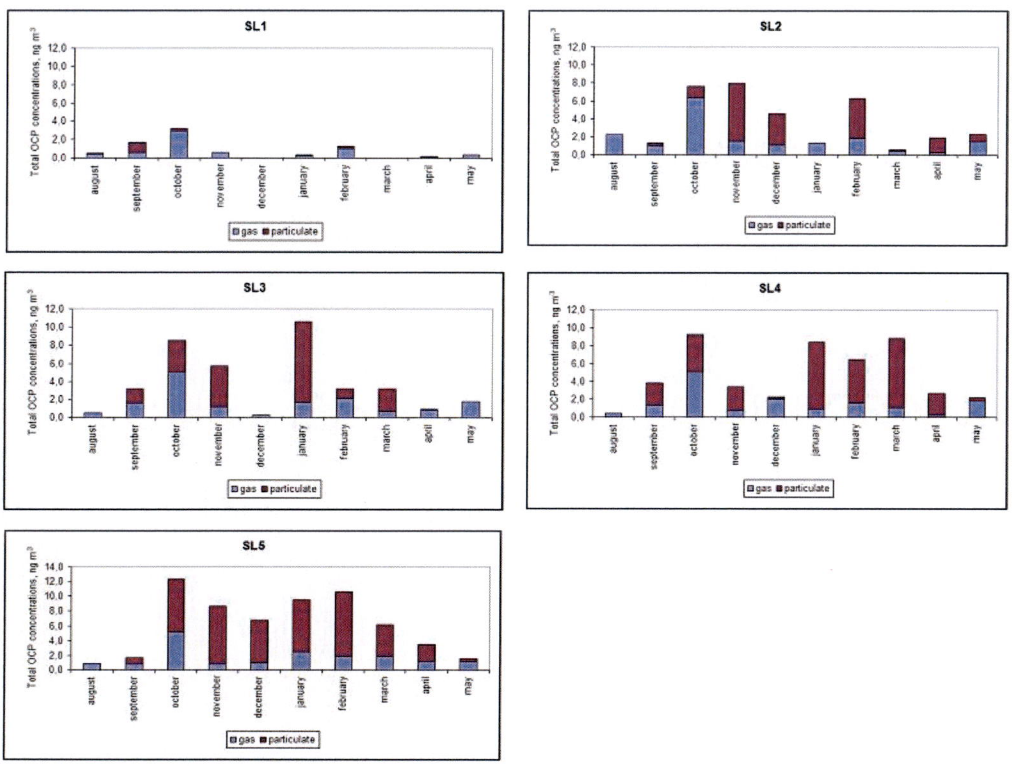

Figure 21. Monthly variations of total OCPs in gas and particulate phases for all sampling locations.

Therefore, as soil is treated the transfer of pesticides increase from soils to the atmosphere. Similar results were recorded by Scheyer et al. (2005). Number of fogy days are quite a lot in winter season according to meteorological data and temperature inversions also occur very often in winter. In stagnant fogy days with no wind, emission from chimneys and motor vehicles exhausts would not disperse in the atmosphere but stay close to the ground. Air pollution in such days rise to quite high levels. Monthly variation of the three groups of pollutants investigated were similar and changing according to atmospheric conditions.

Gas-Particle Partitioning of PAHs, PCBs and OCPs

Distributions of semi volatile organic compounds between gas and particulate phase depend on some properties which are vapor pressure of compounds, characteristics of particulates, ambient temperature. Atmospheric transport and removal of these compounds are strongly influenced by distribution of the substances between these two phases. Also, understanding the gas-particulate partitioning of pollutants may assist estimating health effects related with respiration. (Sanusi et al., 1999; Vardar et al., 1998). In urban air of Konya, as all of the PAHs and PCBs investigated were dominantly detected in gas phase, OCPs were dominantly detected in particulate phase. Total PAHs concentrations in gas phase were 61% of total PAHs concentrations determined in urban air of Konya. While the lower quantity in the particulate phase was recorded in September with 2% value, the highest value was recorded in January with 52%. Total PCBs concentrations in gas phase were 73% of total PCBs concentrations determined in urban air of Konya. Total PCBs and PAHs concentrations in the gas phase were recorded higher than 90% value generally (Mandalakis and Stephanou, 2006; Tasdemir and Esen, 2006; Park et al., 2001). Comparing of the literature values with these results, percentage of existing of PCB and PAH compounds in gas phase in urban air of Konya were lower than that of determined at other cities or countries. This situation can be explained that particulate matter concentrations in the city atmosphere have been raised recently parallel with rapid urbanization and industrialization in Konya region. The dominating wind direction of the city is up to from north-east to south-west and the industrial areas are in the north of the city. Therefore polluted air disperses upon the city. Total OCPs concentrations in the particulate phase were 45% of total OCPs concentrations determined. Similar results were polluted by Lammel et al. (2006) that the particulate bounded fraction of organochlorines compounds were usually higher such as 93% for endrin, 52% for aldrin in air of China.

When molecular weighs of the compounds increases concentrations of the gas phase decreases considering gas and particulate partitioning of the individual compounds. Mean presence of Nap in the gas phase were 97% in the total PAHs concentrations, while compounds such as B[b]F, B[k]F, B[a]Pyr, Ipyr, D[a,h]A, Bper were about 2%. PCB and OCP compounds show similar behavior to PAHs. PCB28 and PCB52 congeners presented in the gas phase were 87% and 68%, respectively while PCB180 was 6% of total PCBs. About 70% of HCHs isomers except for γ-HCH were in gas phase while other pesticides investigated were dominantly in the particulate phase changing between 71% and 99%.

Low chlorinated PCBs were predominately found in the gas phase while high chlorinated PCBs were predominately determined in the particulate phase. The concentrations of PCBs were decreasing with increasing number of Cl in PCBs in the urban air. This type of homolog profiles has been observed by other researchers. Yeo et al. (2004) were observed that the concentration of PCBs homologs in the urban and rural areas decreased with increasing Cl substitutions. Montone et al. (2003) found the lower chlorinated congeners such as below than and PCB 101 predominatly in the air samples. In the study was carried out by Cindoruk and Tasdemir (2006), lower molecular weight PCBs congeners were generally dominant in the urban air samples. These results were also similar to our study results. This situation could be explained by their properties such as $\log K_{oa}$ (octanal-air partition coefficient), $\log K_{ow}$ (octanal-water partition coefficient), vapor pressure, solubility. Bigger congeners than penta-CBs of PCBs homolog deposited easily on surfaces of plants, soil and waters and can evaporate less comparing higher vapor pressure congeners such as tri-CBs and tetra-CBs (Nielsen et al., 1996; Hoff and Zoenen, 1999; ATSDR, 2000). Therefore, examining distribution of PCB homologs for all samples analyzed, the highest levels of PCBs homologs in atmosphere have less chlorine and high vapor pressure such as tri-CBs and tetra-CBs.

In general two or three rings PAHs compounds as Nap, Acy, Ace, F, P are predominately in the gas phase, four rings PAHs compounds as Fluor, Pyr, Chrys exist both in gas and particulate phases, five or more rings PAHs compounds as B[b]F, B[k]F, B[a]Pyr, Ipyr, D[a,h]A, Bper occur predominately in the particulate phase. PCB concentration values determined in gas phase in SL1 were about 63% of total concentration. They were about 89% in SL2, 85% in SL3, 73% in SL4 and 80% in SL5. PCB concentrations determined in Konya city atmosphere were mainly in gas phase. PAH compounds produced during burning processes in high temperatures. Therefore, they are given to atmosphere in gas phase or attached to the micron size particles. The determined PAH concentration values found in this work were in gas phase about 63% in SL1 and SL5, 61% in SL2 and SL4, 64% in SL3 of total concentrations.

In previous works, Mandalakis and Stephanou (2006) found about 95% of PCB concentration they determined in gas phase, Tasdemir and Esen (2006) found about 90% of PAH compounds was in gas phase, Park et al. (2001) more than 95% of PAHs was in gas phase. Lower molecular weight PAH compounds are mainly in gas phase and found in higher ratios in total concentrations. This demonstrates that lower molecular PAH compounds are transported more in the atmosphere. Comparing the ratios of PAH compounds in gas phase results of this work with previous works, the gas phase ratio of PAH compounds was found lower in this work. The possible reason for these results may be that the particle matter (PM) concentration in Konya city atmosphere is much higher comparing with the atmosphere of other cities that previous works carried out.

Investing gas-particle phase changing ratios of total OCP compounds for sampling locations SL1, SL2, SL3, SL4, SL5, about 70% of total OCP compounds determined in sampling locations SL1 was in gas phase, in SL2 and SL3 the ratios were 60% in SL4 and SL5 the ratios were 45%. As a result it could be suggested that the determined pesticides are mainly in gas phase and hence could be transported for long distances. Physico-chemical properties of compounds, temperature, PM content ratios are important in fractionation of OCP compounds in gas and particle phases. In this study, particle fractionation ratios of the three group compounds are changing in similar pattern and gas phase ratios are decreased in winter months.

Variations of PAHs, PCBs and OCPs Concentrations with Atmospheric Temperature

Compounds in gas phases given to atmosphere could cool down and either condensed or attached to the particles. Ambient temperature is an important factor in evaporation or condensation of pollutants in different medias or particles in atmosphere. High temperature in summer increase evaporation of low molecular weight compounds while low temperatures observed in winter decrease or stop this mechanism completely (Vardar, 1998). It was observed in this work that when the temperature increased generally concentration of compounds in gas phase is increased in ambient air depending on the pollutants vapor pressure.

Variation of PCBs, PAHs and OCPs in particle and gas phases with temperatures observed in sampling locations are investigated. There is a weak correlation observed between temperature and PCB concentrations in gas phase (correlation coefficient (r) is between 0.13-0.42) while a weak negative correlation was observed between temperature and PCB concentrations in particle phase (r is between 0.06-0.46). It was also observed that in higher temperatures PAH compounds are also found in higher concentrations in gas phase. The reason for this is that the evaporation is increased from surfaces in high temperatures. While concentrations in gas phases increasing with increasing temperature, particle phase concentration of compounds is decreased. Weak negative correlations were observed between temperature with concentrations in gas and particle phases and also with total concentrations (r is between 0-0.27, 0.49-0.79, 0.32-0.49). These results show that PAH compounds measured are generally emitted from heating sources. In previous researches, weak correlation were found between total PCB concentration and temperature ($r=0.01$) by Montone et al. (2003), between total PAH concentrations and temperatures ($r=0.06$) by Tasdemir and Esen (2006), between PAH concentrations in gas phase and temperature ($r=0.19$) by Odabası et al. (1999), between PAH concentrations in gas phase and temperatures ($r=-055$) by Tsapakis and Stephanou (2005). The similar results obtained in this work indicated that the determined PCB and PAH compounds are coming not from evaporation of sources but from anthropogenic sources such as traffic, industrial activities and domestic heating.

Weak correlations between gas phase concentration of OCP compounds and temperature (r changed between 0.01 and 0.25) were observed while weak negative correlations were observed between particles and total OCP concentrations (r changed between 0.28-0.69 and 0.13-0.52).Observing no strong correlation between gas phase concentration and temperature is an indication that this pollutants are not transported into atmosphere by evaporation from polluting sources. Sofuoglu et al. (2004) investigated the temperature impact on gas phase atmospheric concentrations of OCPs and they determined that it was significant for β-HCH and endosulfan sulfate and was not for other compounds. They also indicated that the wind speed and direction impact on the gas phase were important for some compounds. Correlation between temperature, wind speed and direction, sources with PAHs and PCBs concentrations in Chicago atmosphere were investigated and the variations could not exactly been explained (Sofuoglu et al., 2001).

Sources of PAHs, PCBs and OCPs Determined in Konya Urban Air

In determining PAHs compound emission sources some specific PAH compounds or ratios between certain PAH compounds are used. Especially B[a]Pyr/Bper ratio is used often. B[a]Pyr/Bper ratios between 0.39-0.51 indicates that the source of these compounds are vehicles exhaust gases, if the ratio is above 0.8 than it indicates that sources of PAHs compounds is burning fossil fuels for domestic heating (Lodovoci et al., 2003; Park et al., 2002). B[a]A/(B[a]A+Chrys) ratios are used for differentiating traffic originating compounds formation. If the ratio Fluor/(Fluor+Pyr is between 0.2-0.4 petrol engines, between 0.6-0.9 diesel engines, between 0.4-0.6 fossil fuels burning processes are the main sources in PAH compounds formation (Chen, 2007). Another PAH ratios used in previous works is Ipyr/(Ipyr+Bper). If this ratio is between 0.2-0.3 then the source of PAH compounds formation is petrol engines, between 0.3-0.4 the source is diesel engines, if the ratio however is between 0.4-0.9 than the source is burning processes (Chen, 2007).

PAH ratios are given in Table 13 for sampling locations that PAH concentrations were measured in this work.

The ratios obtained from mean total values determined by the measurements between August 2006 and May 2007 are as follow: B[a]Pyr/Bper ratios are above 0.8 at all sampling locations, Fluor/(Fluor+Pyr) and Ipyr/(Ipyr+Bper) ratios are between 0.4-0.6 for all sampling locations indicating that PAH compounds formation sources are mainly domestic heating processes. B[a]A/(B[a]A+Chrys) ratios are between 0.2-0.4 for all sampling locations indicating that traffic originating PAH compounds coming from mainly petrol engines. Determining PAH compounds formation sources are important in controlling the formation of these compounds. ıt was found that PAH compounds determined in Konya city atmosphere is mainly coming from burning processes, moreover the part of PAH compounds coming from traffic sources mainly from petrol engines.

PCBs may be emitted also from industrial or municipal waste burning processes in addition to the other sources. During gas-particle partition of aliphatic compounds emitted from vehicles exhausts could behave like PCBs (Mandalakis et al., 2002). PCBs are persistent anthropogenic compounds that can stay long time without decomposition since they have resisting physico-chemical properties to photolytic, chemical and biological decomposition. Therefore, PCB compounds determined in Konya city atmosphere assumed to be coming by evaporation from where they are used.

Table 13. PAHs compounds ratios used for identification of source

Ratio	SL1	SL2	SL3	SL4	SL5
B[a]Pyr/Bper	1.08	2.04	2.55	2.71	3.10
B[a]A/(B[a]A+Chrys)	0.40	0.28	0.34	0.36	0.30
Fluor/(Fluor+Pyr)	0.55	0.51	0.54	0.56	0.50
Ipyr/(Ipyr+Bper)	0.47	0.62	0.62	0.64	0.65

In determining the sources of OCP compounds in atmosphere some specific OCP compounds or ratios between some certain OCP compounds are used. *p,p'*-DDT/*p,p'*-DDE ratio could be used as an indicator in determining the time of DDT compounds staying in atmosphere. If the ratio is about 1 indicates the mixture of DDT compounds coming from later use, below 1 ratio indicates only former use of DDT compounds, if this ratio is above 1 than this is an indication of current use of DDT compounds. Alegria et al. (2006) determined the *p,p'*-DDT/*p,p'*-DDE ratio as 1.06 in Chiapas in Mexico, between 2000-2001. Shen et al. (2005) carried out an investigation in USA between 2000-2001 and found *p,p'*-DDT/*p,p'*-DDE ratios as 0.94 in Tapachula city, 1.3 in Cheturnal city, 1.2 in Belmohan and 5.8 in Costa Rica city. Gioia et al. (2005) carried out a research on 22 OCP compounds in 6 different locations of New York from January 2000 to May 2001 and obtained mean *p,p'*-DDT/*p,p'*-DDE ratios between 0.18-0.47. *p,p'*-DDT/*p,p'*-DDE ratios below 1 indicates that recent DDT emissions to the atmosphere is reduced. Technical α-HCH/γ-HCH ratio is between 4-8 range. Lower ratios determined than this ratio indicate continuing lindane (γ-HCH) use in the area. Alegria et al. (2006) found α-HCH/γ-HCH ratio 0.38 in Chiapas in Mexico between 2000-2001. Shen et al. (2005) found α-HCH/γ-HCH ratio as 0.16 in Tapachula, 0.46 in Cheturnal, 0.27 in Belmohan and 0.18 in Costa Rica between 2000-2001. Gioia et al. (2005) found α-HCH/γ-HCH ratios between 1.3 and 2.7 in 6 different locations in New York from January 2000 to May 2001. In this work *p,p'*-DDT/*p,p'*-DDE ratio was found as 20 while α-HCH/γ-HCH ratio was found as 2.55 in Konya city atmosphere. Therefore, by looking at these ratio it could be concluded that use of DDT and lindane (γ-HCH) are continuing during agricultural activities in Konya.

PCBs, PAHs and OCPs Exposure Risk Characterization

Some of the PAH compounds are proved to be as carcinogenic and/or mutagenic compounds and cause for concern with their widely existence in ambient air. IARC (International Agency for Research on Cancer) declared that B[a]Pyr, B[a]A, D[a,h]A compounds are in group 2A class that are possible carcinogens for human, Nap, B[b]F, B[j]F, B[k]F, Ipyr compounds are in group 2B that may be possible carcinogens for human. A, Bper, B[e]Pyr, Chrys, Fluor, F, P, Pyr, Perylene are in group 3 class which are not carcinogen (Iavicoli et al., 2006). It is proved that small particles and PAH compounds (such as B[a]Pyr) attached together and reach to bronchia and alveoli. Gas and particle phase mean total B[a]Pyr concentrations values determined in each sampling location are given in Figure 22. B[a]Pyr concentrations as it can be seen in Figure 22, is found as 0.86 ng m^{-3} in SL1 and in other sampling locations the concentrations were in similar values about 6 ng m^{-3} in this work.

Comparing B[a]Pyr with other PAH compounds in terms of toxicological characteristics, limited information available for other PAH compounds. Therefore, health risks associated with inhalation of PAH compounds evaluated mainly considering B[a]Pyr concentration in air. B[a]Pyr toxic equivalent factor (TEF) is a tool in determining in carcinogenity and potential toxicity of PAH compounds in samples. Potential health risks caused by inhalation of carcinogenic PAH compounds are evaluated considering the B[a]Pyr_{TEF} values given by Castellano et al., (2003) for determined PAH compounds. B[a]Pyr_{TEF} values obtained for each sampling location are given in Figure 23.

Maximum values for B[a]Pyr is given as 1 ng m^{-3} by Italian air quality standards while it was given as 0.25 ng m^{-3} by English air quality standards (Lodovici et al., 2003; Mastrall et al., 2003). Target annual mean B[a]Pyr value is between 0.7-1.3 ng m^{-3} range in some

European countries (Castellano et al., 2003). B[a]Pyr_{TEF} value is determined as 2.01 ng m^{-3} for SL1 while it was about 13 ng m^{-3} for other sampling locations in this study (see Figure 23). The highest B[a]Pyr values were determined in all sampling locations in January and February.

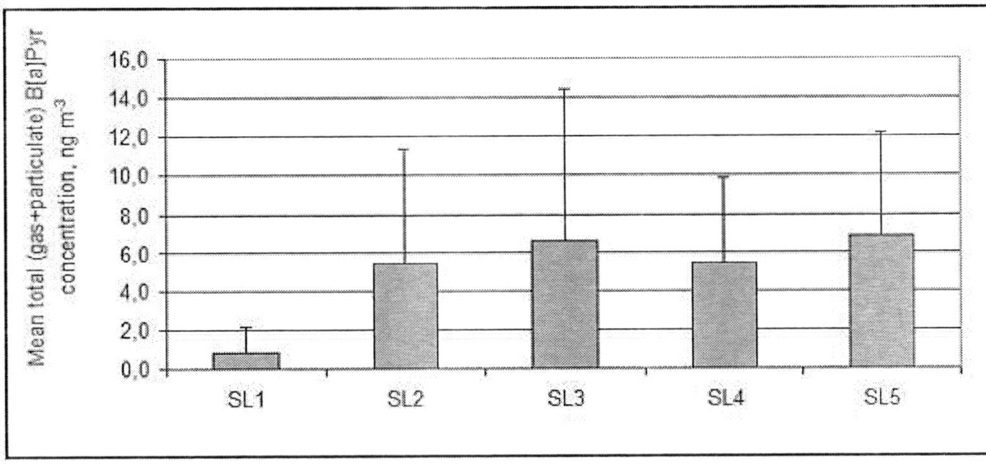

Figure 22. Mean total (gas+particulate) B[a]Pyr concentration (ng m-3).

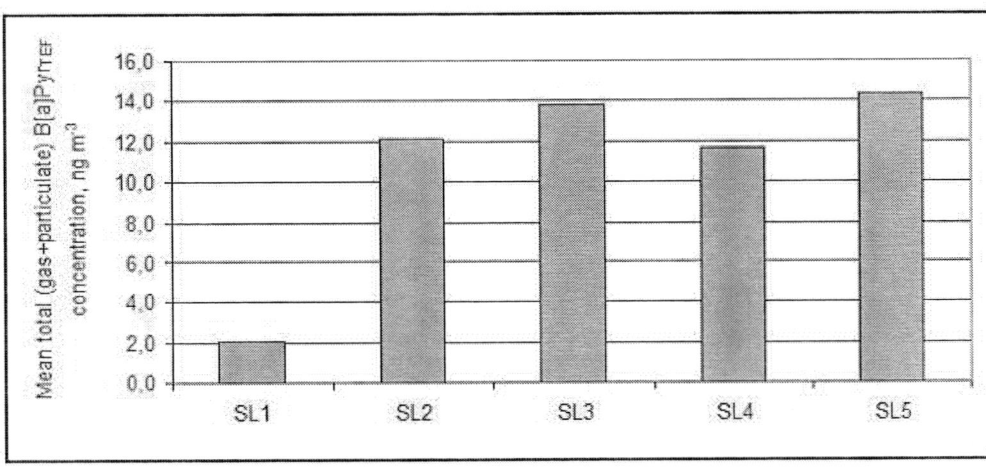

Figure 23. B[a]PyrTEF values determined for each sampling location (ng m-3).

In previous works the highest B[a]Pyr value was 5.8 ng m^{-3} in winter months (Lodovici et al., 2003), 0.605 ng m^{-3} in Spain (Castellano et al., 2003), 0.28 ng m^{-3} in Brazil (Vasconcellos et al., 2003) and 10.13 ng m^{-3} in Brazil (Azevedo et al., 2002), 5 ng m^{-3} in high traffic areas in Australia (Müller et al., 1998). B[a]Pyr_{TEF} values of total determined PAH compounds was between 0.86-2.63 ng m^{-3} (Lodovici et al., 2003) and between 0.32-1.88 ng m^{-3} (Mastral et al., 2003). Comparing the results B[a]Pyr values determined in this work is higher than the values reported in previous works. In addition except sampling location SL1, in other four sampling locations the target value annual mean of 1 ng m^{-3} for B[a]Pyr level were exceeded. This result shows that, there is a health risk especially for people work near these sampling locations. PAH sources in Konya city atmosphere are determined as mainly seasonal burning processes for domestic heating and after that the traffic is important sources for PAHs. In order to reduce B[a]Pyr levels below the annual mean value of 1 ng m^{-3} levels given as target value by some European countries some measures should be taken to reduce PAH emissions from primarily in heating processes and secondarily in traffic and industrial sources.

PCBs, PAHs and OCPs are semi-volatile synthetic organic compounds that evaporate from the areas where they are used and can be transported in atmosphere for long distances. These compounds are directly discharged to aquatic environments or deposited in water mediums after transportation in atmosphere. It is proved that these compounds can be found in very remote places that far away from any residential, industrial or agricultural areas. These compounds are found usually in low concentrations in water and air mediums therefore they do not usually pose threat to human health directly however, they can bioaccumulate in exposed organisms and become dangerous. These compounds have tendency to bioaccumulate in lipid tissues of organism and can be biomagnified through the terrestrial or aquatic food chain. Therefore, they are global problems for human and environmental health. These compounds may cause birth defects and disabilities, various cancers, immune system disruption, reproduction system defects in mammals. In addition long term exposure of humans to these compounds in higher concentrations may cause increase in birth defects, reproduction disabilities, and decrease in IQ levels, endocrinological disruption and some cancers in humans (Ministry of Environment and Forestry, 2006).

Conclusions

A multi-residue analytical method has been developed and validated for the analysis of PCBs, PAHs and OCPs in urban air samples. This method includes the soxhlet extraction of the PUF plugs and ultrasonic extraction of microfibre filters, the clean-up of the extracts and their separations by adsorption chromatography on open columns (10 g of 5% deactivated silica gel collecting PCBs with 70 mL of *n*-hexane and collecting the PAHs and OCPs with 3 x 20 mL of *n*-hexane:ethyl acetate; 1/1, v/v) in two fractions, the first containing all the PCBs and some PAHs and OCPs compounds and the second fraction containing all the rest of PAHs and OCPs. Finally, the target compounds are identified and quantified by GC/μ-ECD and GC/MSD. This method enables the determination of the target compounds at concentrations as low as between 0.04 pg m^{-3} and 0.08 pg m^{-3} for PCBs, 0.05 pg m^{-3} and 0.05 ng m^{-3} for PAHs, 1.6 pg m^{-3} and 0.03 ng m^{-3} for OCPs. The overall recovery values (RV) and their repeatability for PCBs, PAHs and OCPs were found as in acceptable ranges (60% <RV< 120%; RSD < 10%). The matrix effects on separation and recovery of PCBs, PAHs and OCPs were not significant. As a consequence, the advantages of the method developed in this research comparing to EPA Methods TO-13A and TO-4A can be suggested as follows: solvent consumption in the extraction step is reduced about 60% in the analysis of a sample for PCBs, PAHs and OCPs. Sorbent amounts and elution solvents volume consumptions are reduced considerably comparing EPA Methods TO-13A and TO-4A in the clean-up step. In addition, the cost of time and labor will be reduced. The method developed in this work will reduce laboratory expenses without substantial new equipment in addition without compromising from accuracy and precision.

POPs resist photolytic, biological or chemical degradation and can be transported to places where they have never been used. Because of their toxic characteristics, release of persistent organic pollutants must be prevented or reduced. In content of the Stockholm agreement, Turkey carry out national implementation plan to protect human health and the environment by eliminating or restricting the production, use and release of 12 POP chemicals (aldrin, chlordane, DDT, dieldrin, endrin, heptachlor, mirex and toxaphene, hexachlorobenzene, polychlorinated biphenyls, dioxins and furans). Therefore, firstly their concentrations must be identified in environment and then, their possible sources must be determined. Afterwards, in order to reduce or exterminate their possible risk to human health and the environment, precautions must be taken by authorities.

Primary results on 16 EPA PAHs, PCBs which are widely investigated in atmosphere, OCPs which were frequently used during agricultural activity were found between August 2006 and May 2007 in the study. All target compounds were detected in urban air of Konya. According to these results, it can be suggested that Konya is an activly contributing region to persistent organic pollutants in Turkey. Despite banned/restricted from use in Turkey, some organochlorine pesticides were determined in urban air. These results demonstrated that they are illegally being used in the course of agricultural activity and gardens in Konya or they are residues of past use in the environment.

Gas and particulate phase PAHs, PCBs and OCPs were monitored in Konya city atmosphere by sampling and analyses program that has been carried out considering different traffic loads, industrial zones and residential areas for a year. The most abundant compounds in PAHs were Phenanthrene, Fluor, Pyr, in PCBs the most abundant compounds, were PCB28 and PCB52 congeners while total HCH ($\Sigma\alpha$-HCH+β-HCH+γ-HCH+δ-HCH) and total DDT ($\Sigma p,p'$-DDT+p,p'-DDD+p,p'-DDE) were most abundant in OCPs. Considering the relation between these compounds with temperature, there was no strong correlation observed between temperature and PCB, PAH and OCP compounds. Gas phase concentrations of compounds decreased with increasing molecular weights of pollutants. By looking at the results obtained from the sampling and analyses of four different seasons, the highest PCBs, PAHs and OCPs values were found winter and autumn seasons. Compounds concentrations profiles in particle phase were remain about the same in all seasons. The highest concentrations in particle phase were obtained in winter season measurements. Considering the risk characterization of these compounds it could be concluded that these compounds pose health risks for human. So, the long term atmospheric monitoring of persistent organic pollutants must continue in Konya and in other cities in Turkey.

REFERENCES

Aceves, M., Grimalt, J., Broto, F., Comellas, I. (1988). Analysis of hydrocarbons in aquatic sediments. 2. Evaluation of common preparative procedures for petroleum and chlorinated hydrocarbons, *Journal Chromatography,* 436 (3), 503-509.

Agency for toxic substances and disease registry, division of toxicology/toxicology information branch (ATSDR) (2000). *Toxicological profile for polychlorinated biphenyls (PCBs)*, ATSDR, Atlanta, USA.

Alegria, H., Bidleman, T.F., Figueroa, M.S. (2006). Organochlorine pesticides in the ambient air of Chiapas, Mexico, *Environmental Pollution*, 140, 484-491.

Alonso, S.G., Pastor, R.M.P., Cabezas, A.J.Q. (2001). Optimized chromatographic method for the measurement of selected polychlorinated biphenyls in ambient air, *Analytica Chimmica Acta*, 440, 223-230.

Aries, E., Anderson, D.R., Ordsmith, N., Hall, K., Fisher, R. (2004). Development and validation of a method for analysis of "dioxin-like" PCBs in environmental samples from the steel industry, *Chemosphere,* 54, 23-32.

Aydin, M.E., Tor, A., Ozcan, S. (2006). Determination of selected polychlorinated biphenyls in soil by miniaturized ultrasonic solvent extraction and gas chromatography-mass-selective detection, *Analytica Chimica Acta,* 577 (2), 232-237.

Aydin, M.E., Ozcan, S., Tor, A. (2007). Ultrasonic solvent extraction of persistent organic pollutants from airborne particles, *CLEAN-Soil, Air, Water*, Clean 2007 35 (6), 660 – 668.

Azevedo, D.D., Dos Santos, C.Y.M., Neto, F.R.D. (2002). Identification and seasonal variation of atmospheric organic pollutants in Campos dos Goytacazes, Brazil, *Atmospheric Environment*, 36 (14), 2383-2395.

Babic, S., Petrovic, M., Kastelan-Macan, M. (1998). Ultrasonic solvent extraction of pesticides from soil, *Journal Chromatography A,* 823 (1-2), 3-9.

Bakan, G., Ariman, S. (2004). Persistent organochlorine residues in sediments along the coast of mid-Black Sea region of Turkey, *Marine Pollution Bulletin,* 48 (11-12), 1031-1039.

Banjoo, D.R., Nelson, P.K. (2005). Improved ultrasonic extraction procedure for the determination of polycyclic aromatic hydrocarbons in sediments, *Journal Chromatography A,* 1066 (1-2), 9-18.

Barro, R., Ares, S, Garcia-Jares, C., Llompart, M., Cela, R. (2005). Sampling and analysis of polychlorinated biphenyls in indoor air by sorbent enrichment followed by headspace solid-phase microextraction and gas chromatography–tandem mass spectrometry, *Journal of Chromatography A,* 1072, 99-106.

Batarseh, M.I., Kreuzig,R., Al-Nasir, F., Bahadir, M. (2003b). Residue analysis of organic pollutants in sediments from the Amman/Zarqa area in Jordan. Part II: Organochlorine compounds, *Fresenius Environmental Bulletin*, 12 (9), 979-983.

Batarseh, M.I., Kreuzig,R., Bahadir, M. (2003a). Residue analysis of organic pollutants in sediments from the Amman/Zarqa area in Jordan. Part I: Development of analytical methods and distribution patterns of PAHs, *Fresenius Environmental Bulletin*, 12 (9), 972-978.

Breivik, K., Pacyna, J.M., Münch, J. (1999) Use of α-, β- and γ-hexachlorocyclohexane in Europe, 1970–1996, *The Science of The Total Environment*, 239 (1-3), 151-163.

Castro, J., Sanchez-Brunete, C., Tadeo, J.L. (2001). Multiresidue analysis of insecticides in soil by gas chromatography with electron-capture detection and confirmation by gas chromatography-mass spectrometry, *Journal Chromatography A,* 918 (2), 371-380.

Castellano, A.V., Cancio, J.L., Aleman, P.S., Rodriguez, S.R. (2003). Polycyclic aromatic hydrocarbons in ambient air particles in the city of Las Palmas de Gran Canaria, *Environment International,* 29, 475-480.

Cecinato, A., Ciccioli, P., Brancaleoni, E., Brachetti, A., Vasconcellos, P.C., (1997). PAH as candidate markers for biomass burning in the Amazonia forest area, *Annali di Chimica.*, 87, 555-569.

Chen, H.W. (2007). Polycyclic and nitro-polycyclic aromatic hydrocarbons in air particulates from a semiconductor manufacturing are of Taiwan, *Fresenius Environmental Bulletin*, 16 (5), 543-547.

Ciganek, M., Neca, J., Adamec, V., Janosek, J., Machala, M. (2004). A combined chemical and bioassay analysis of traffic-emitted polycyclic aromatic hydrocarbons, *Science of the Total Environment*, 334-335, 141–148.

Cincinelli, A., Bubba, M.D., Martellini,T., Gambaro, A., Lepri, L. (2007). Gas-particle concentration and distribution of *n*-alkanes and polycyclic aromatic hydrocarbons in the atmosphere of Prato (Italy), *Chemosphere*, 68 (3) 472-478.

Cindoruk, S.S., Tasdemir, Y. (2007). Characterization of gas/particle concentrations and partitioning of polychlorinated biphenyls (PCBs) measured in an urban site of Turkey, *Environmental Pollution*, 148, 325-333.

Cotham, W.E., Bidleman, T. (1995). Polycyclic aromatic hydrocarbons and polychlorinated biphenyls in air at an urban and a rural site near Lake Michigan, *Environmental Science Technology*, 29, 2782-2789.

Dean, J. R. (2003). *Methods for environmental trace analysis*, Wiley, North Umbria University, Newcastle, UK.

Erdogrul, O., Covaci, A., Kurtul, N., Schepens, P. (2004). Levels of organohalogenated persistent pollutants in human milk from Kahramanmaraş region, Turkey, *Environmental International*, 30, 659-666.

Ericsson, M., Colmsjo, A. (2000). Dynamic microwave-assisted extraction, *Journal of Chromatography A*, 877 (1-2), 141-151.

Falconer, R.L., Bidleman, T.F. (1994). Vapour pressures and predicted particle/gas distributions of polychlorinated biphenyl congeners as functions of temperature and orthochlorine substitution, *Atmospheric Environment*, 28, 547-554.

Fang, G, C., Wu, Y.S., Chen, M.H., Ho, T.T., Huang, S.H., Rau, J.Y. (2004). Polycyclic aromatic hydrocarbons study in Taichung, Taiwan, during 2002–2003, *Atmospheric Environment*, 38, 3385–3391.

Fatoki, O.S., Awofolu, R.O. (2003). Methods for selective determination of persistent organochlorine pesticide residues in water and sediments by capillary gas chromatography and electron-capture detection, *Journal of Chromatography A*, 983 (1-2), 225236.

Gioia, R., Offenberg, J.H., Gigliotti, C.L., Totten, L.A., Du, S., Eisenreich, S.J. (2005). Atmospheric concentrations and deposition of organochlorine pesticides in the US Mid-Atlantic region, *Atmospheric Environment*, 39, 2309-2322.

Halsall, C.J., Sweetman, A.J., Barrie, L.A., Jones, K.C. (2001). Modeling the behavior of PAHs during atmospheric transport from the UK to the Arctic, *Atmospheric Environment*, 35, 255-267.

Harrad, S., Mao, H. (2004). Atmospheric PCBs and organochlorine pesticides in Birmingham, UK: concentrations, sources, temporal and seasonal trends, *Atmospheric Environmental*, 38, 1437-1445.

Hartonen, K., Bøwadt, S., Hawthorne, S.B., Riekkola, M.L. (1997). Supercritical fluid extraction with solid-phase trapping of chlorinated and brominated pollutants from sediment samples, *Journal of Chromatography A*, 774, 229-242.

Hazardous Substances Data Bank (HSDB), (2006) National Library of Medicine, National Toxicology Information Program, USA.

Hermanson, M.H., Hites, R.A. (1989). Long-term measurements of atmospheric polychlorinated biphenyls in the vicinity of superfund dumps, *Environmental Science Technology*, 23, 1253-1258.

Hoff, R.M., Strachan, W.M.J., Sweet, C.W., Chan, C.H., Shackleton, M., Bidleman, T.F., Brice, K.A., Burniston, D.A., Cussion, S., Gatz, D.F., Harlin, K., Schroeder, W.H. (1996). Atmospheric deposition of toxic chemicals to the Great Lakes: A review of data through 1994, *Atmospheric Environment*, 30, 3505-3527.

Hsu, Y.K., Holsen, T.M., Hopke, P.K. (2003). Locating and quantifying PCB sources in Chicago: Receptor modeling and field sampling, *Environmental Science Technology*, 37 (4), 681-690.

Hung, H., Blanchard, T.P., Halsall, C.J., Bidleman, T.F., Stern, G.A., Fellin, P., Muir, D.C.G., Barrie, L.A., Jantunen, L.M., Helm, P.A., Ma, J., Konoplev, A. (2005) Temporal and spatial variabilities of atmospheric polychlorinated biphenyls (PCBs), organochlorine (OC) pesticides and polycyclic aromatic hydrocarbons (PAHs) in the Canadian Arctic: Results from a decade of monitoring, *Science of the Total Environment,* 342,119-144.

Jang, J.K., Lee, A. (2001). Separation of PCBs and PAHs in sediment samples using silica gel fractionation chromatography, *Chemosphere,* 44, 1439-1445.

Jaouen-Madoulet, A, Abarnou, A., Le Guellec, A.M., Loizeau, V., Leboulenger, F. (2000). Validation of an analytical procedure for polychlorinated biphenyls, coplaner polychlorinated biphenyls and polycyclic aromatic hydrocarbons in environmental samples, *Journal of Chromatography A*, 886, 153-173.

Junker, M., Kasper, M., Roosli, M., Camenzind, M., Kunzli, N., Monn, Ch., Theis, G., Braun-Fahrlander, Ch. (2000). Airborne particle number profile, particle mass distribution and particle-bound PAH concentrations within the city environment of Basel: and assessment as part of the BRISKA Project, *Atmospheric Environment*, 34, 3171-3181.

Karademir, A. (2004). Health risk assessment of PCDD/F emissions from a hazardous and medical waste incinerator in Turkey, *Environmental International,* 30, 1027-1038.

Keith, L.H., Crummett, W., Deegan, J., Libby, R.A., Taylor, J.K., Wentler, G. (1983). Principles of environmental-analysis, *Analytical Chem*istry (Washington, DC, U.S.), *55*, 2210-2218.

Kolb, M., Bohm, H.B., Bahadir, M. (1995). Analytical multimethod for the determination of low volatile organic pollutants in sediments and sewage sludges, *Fresenius Journal Analytical Chemistry,* 351 (2-3), 286-296.

Lammel, G., Ghim, Y.S., Grados, A., Gao, H., Hühnerfuss, H., Lohmann, R. (2007) Levels of persistent organic pollutants in air in China and over the Yellow Sea, *Atmospheric Environment*, 41 (3), 452-464.

Lin, T.C, Chang, F.H., Hsieh, J.H., Chao, H.R., Chao, M.R. (2002). Characteristics of polycyclic aromatic hydrocarbons and total suspended particulate in indoor and outdoor atmosphere of a Taiwanese temple, *Journal of Hazardous Materials*, A95, 1-12.

Lodovici, M., Venturini, M., Marini, E., Grechi, D., Dolara, P. (2003). Polycyclic aromatic hydrocarbons air levels Florence, Italy and their correlation with other air pollutants, *Chemosphere,* 50, 377-382.

Louie, P.K.K., Sin, D.W. (2003). A Preliminary Investigation of persistent organic pollutants in ambient air in Hong Kong, *Chemosphere*, 52, 1397-1403.

Mandalakis, M., Stephanou, E.G. (2007). Atmospheric concentration characteristic and gas-particle partitioning of PCBs in rural area of eastern Germany, *Environmental Pollution*, 147 (1), 211-221.

Mandalakis, M., Tsapakis, M., Tsoga, A., Stephanou, E.G. (2002). Gas-particle concentrations and distribution of aliphatic hydrocarbons, PAHs, PCBs and PCDD/Fs in the atmosphere of Athens (Greece), *Atmospheric Environment,* 36, 4023-4035.

Martinez, M., Ferroro, J.D., Marti, R., Puig, F.B., Comellas, L., Rodriguez-Larena, M.C. (2000). Analysis of dioxin-like compounds in vegetation and soil samples burned in Catalan forest fires. Comparison with the corresponding unburned material, *Chemosphere,* 41, 1927-1935.

Mastral, A.M., Lopez, J.M., Callen, M.S., Garcia, T., Murillo, R., Navarro, M.V. (2003). Spatial and temporal PAH concentrations in Zaragoza, Spain, *The Science of the Total Environment,* 307, 111-124.

McConnell, L.L., Bidleman, T.F., Cotham, W.E., Walla, M.D. (1998). Air concentrations of organochlorine insecticides and polychlorinated biphenyls over Green Bay, WI, and the four lower Great Lakes, *Environmental Pollution*, 101, 391-399.

Ministry of Environment and Forestry (2006). National draft application plan about Stocholm aggrement on POPs, Project UNIDO-POP, Project no: GF/TUR/03/008.

Montone, R.C., Taniguchi, S., Weber, R.R. (2003). PCBs in the atmosphere of King George Island, Antarctica, *The Science of The Total Environment*, 308 (1-3), 167-173.

Müller, J.F., Hawker, D.W., Connell, D.W. (1998). Polycyclic aromatic hydrocarbons in the atmospheric environment of Brisbane, Australia, *Chemosphere,* 37 (7), 1369-1383.

Nawab, A., Aleem, A., Malik, A. (2003). Determination of organochlorine pesticides in agricultural soil with special reference to gamma-HCH degradation by Pseudomonas strains, *Bioresource Technology,* 88 (1), 41-46.

Nielsen, T., Jorgensen, H.E., Larsen, J.C., Poulsen, M. (1996). City air pollution of polycyclic aromatic hydrocarbons and other mutagens: occurrence, sources and health effects, *The Science Total Environment,* 189/190, 41-49.

Odabaşı, M., Vardar, N., Sofuoğlu, A., Taşdemir, Y., Holsen, T.M. (1999). Polycyclic Aromatic Hydrocarbons (PAHs) in Chicago Air, *The Science of the Total Environment*, 227, 57-67.

Ollivon, D., Blanchoud, H., Motelay-Massei, A., Garban, B. (2002). Atmospheric deposition of PAHs to an urban site, Paris, France, *Atmospheric Environment*, 36, 2891-2900.

Omar, N.Y.M.J., Abas, M.R.B., Ketuyl, K.A., Tahir, N.M. (2002). Concentrations of PAHs in atmospheric particles (PM-10) and roadside soil particles collected in Kuala Lumpur, Malaysia, *Atmospheric Environment,* 36, 247-254.

Oros, D.R., Simoneit, B.R.T. (2001). Identification and emission factors of molecular tracers in organic aerosols from biomass burning. Part 2. deciduous trees. *Applied Geochemistry*, 16, 1545-1565.

Ozcan, S., Aydin, M.E., Tor, A. (2008). Chromatographic seperation and analytic procedure for priority organic pollutants in urban air, *CLEAN-Soil, Air, Water*, Clean 2008 36 (8).

Park, J.S., Wade, T.L., Sweet, S.T. (2001). Atmospheric distributions of polycyclic aromatic hydrocarbons (PAHs) and depositions to Galveston Bay, Texas, USA, *Atmospheric Environment,* 35, 3241-3249.

Park, J.S., Wade, T.L., Sweet, S.T. (2002). Atmospheric deposition of PAHs, PCBs and organochlorine pesticides to Corpus Christi Bay, Texas, *Atmospheric Environment,* 36, 1707-1720.

Pino, V., Ayala, J.H., Afonso, A.M., Gonzalez, V. (2000). Determination of polycyclic aromatic hydrocarbons in marine sediments by high-performance liquid chromatography after microwave-assisted extraction with micellar media, *Journal of Chromatography A*, 869 (1-2), 515-522.

Polkowska, Z., Kot, A., Wiergowski, M., Wolska, L., Wolowska, K., Namiesnik, L. (2000). Organic pollutants in precipitation: determination of pesticides and polycyclic aromatic hydrocarbons in GdanH sk, Poland, *Atmospheric Environment*, 34, 1233-1245.

Poor, N., Tremblay, R., Kay, H., Bhethanabotla, V., Swartz, E., Luther, M., Campbell, S. (2004). Atmospheric concentrations and dry depositions rates of polycyclic aromatic hydrocarbons (PAHs) for Tampa Bay, Florida, USA, *Atmospheric Environment,* 38, 6005-6015.

Pozo, O., Pitarch, E., Sancho, J.V., Hernandez, F. (2001). Determination of the herbicide 4-chloro-2-methylphenoxyacetic acid and its main metabolite, 4-chloro-2-methylphenol in water and soil by liquid chromatography-electrospray tandem mass spectrometry, *Journal of Chromatography A,* 923 (1-2), 75-85.

Santos, F.J., Galceran, M.T. (2002). The application of gas chromatography to environmental analysis, *Trends in Analytical Chemistry*, 21, 672-685.

Sanusi, A., Millet, M., Mirabel, P., Wortham, H. (1999). Gas-particle partitioning of pesticides in atmospheric samples, *Atmospheric Environment*, 33, 4941-4951.

Scheyer, A., Graeff, C., Morville, S., Mirabel, P., Millet, M. (2005). Analysis of some organochlorine pesticides in an urban atmosphere (Strasbourg, east of France), *Chemosphere*, 58 (11), 1517-1524.

Scholtz, M.T., Bidleman, T.F. (2006). Modelling of the long term fate of pesticide residues in agricultural soils and their surface exchange with the atmosphere: Part I. Model description and evaluation, *Science of the Total Environment*, 368, 823-838.

Shen, L., Wania, F., Lei, Y.D., Teixeira, C., Muir, D.C.G., Bidleman, T.F. (2005). Atmospheric distribution and long-range transport behavior of organochlorine pesticides in north America, *Environmental Science and Technology*, 39 (2), 409-420.

Sienra, M.D., Rosazza, N.G. (2006). Occurrence of nitro-polycyclic aromatic hydrocarbons in urban particulate matter PM10, *Atmospheric Research*, 81, 265-276.

Skoog, D.A., West, D.M., Holler, F.J. (1996). *Fundamentals of analytical chemistry*, Seventh Edition, Fort Worth [u.a.]: Saunders College Publication.

Sofuoglu, A., Cetin, E., Bozacıoglu, S.S., Sener, G.D., Odabası, M. (2004). Short-term variation in ambient concentrations and gas/particle partitioning of organochlorine pesticides in Izmir, Turkey, *Atmospheric Environment,* 38, 4483–4493.

Sofuoglu, A., Odabası, M., Tasdemir, Y., Nasrin, R.K., Holsen, T.M. (2001). Temperature dependence of gas-phase polycyclic aromatic hydrocarbon and organochlorine pesticide concentrations in Chicago air, *Atmospheric Environment*, 35, 6503-6510.

Sporring, S., Bøwadt, S., Swensmark, B., Bjorklund, E. (2005). Comprehensive comparison of classic soxhlet extraction with soxtec extraction, ultrasonication extraction, supercritical fluid extraction, microwave assisted extraction and accelerated solvent extraction for the determination of polychlorinated biphenyls in soil, *Journal of Chromatography A.,1090*, 1-9.

Tasdemir, Y. (1997). *Modification and evaluation of a water surface sampler to investigate the dry deposition and air water exchange of polychlorinated biphenyls (PCBs)*, Chicago, Illinois, December.

Tasdemir, Y., Esen, F. (2007). Urban air PAHs: Concentration, temporal changes and gaz/particle partitioning at a traffic site in Turkey, *Atmospheric Research,* 84, 1-12.

Tasdemir, Y., Odabasi, M., Vardar, N., Sofuoglu, A., Murphy, T.J., Holsen, T.M. (2004a). Dry deposition fluxes and velocities of polychlorinated biphenyls (PCBs) associated with particles, *Atmospheric Environment*, 38 (16), 2447-2456.

Tasdemir, Y., Vardar, N., Odabası, M., Holsen, T.M. (2004b). Concentrations and gas/particle partitioning of PCBs in Chicago, *Environmental Pollution,* 131, 35-44.

Tor, A., Aydin, M.E., Ozcan, S. (2006). Ultrasonic solvent extraction of organochlorine pesticides from soil, *Analytica Chimica Acta*, 559 (2), 173-180.

Tor, A., Cengeloglu, Y., Aydin, M.E., Wichmann, H., Bahadir, M. (2004). Seperation of some priority organic pollutants in wastewater samples by column chromatography, *Fresenius Environmental Bulletin*, 13 (7), 679-684.

Tsapakis, M., Stephanou, E.G. (2005). Occurrence of gaseous and particulate polycyclic aromatic hydrocarbons in the urban atmosphere: study of sources and ambient temperature effect on the gas/particle concentration and distribution, *Environmental Pollution,* 133, 147-156.

Unger, K.K., Weber, E. (1999). *A Guide to Practical HPLC*, GIT Verl., Germany.

US EPA (1996). Method 3500B, *Organic Extraction and Sample Preparation.*

US EPA (1996). Method 3620B, *Florosil clean-up.*

US EPA (1996). Method 3630C, *Silica gel clean-up.*

US EPA (1999). *Compendium of methods for the determination of toxic organic compounds in ambient air*, 2nd ed., Compendium method TO 4A.

US EPA (1999). *Compendium of methods for the determination of toxic organic compounds in ambient air*, 2nd ed., Compendium method TO 13A.

US EPA, air toxics website, http://www.epa.gov/ttn/atw/allabout.html (2005).

Van der Hoff, G.R., Van Zoonen, P. (1999). Trace analysis of pesticides by gas chromatography, *Journal of Chromatography A,* 843, 301-322.

Vardar, N. (1998). *Characterization of atmospheric concentrations of PAHs and their gas-particle partitioning*, Chicago, Illinois, December.

Vardar, N., Tasdemir, Y., Odabası, M., E.Noll, K. (2004). Characterization of atmospheric concentrations and partitioning of PAHs in the Chicago atmosphere, *Science of the Total Environment*, 327, 163-174.

Vasconcellos, P.C., Zacarias, D., Pires, M.A.F., Pool, C.S., Carvalho, L.R.F. (2003). Measurements of polycyclic aromatic hydrocarbons in airborne particles from the metropolitan area of Sao Paulo City, Brazil, *Atmospheric Environment,* 37, 3009-3018.

Venkataraman, C., Lyons, J.M., Friedlander, S. (1994). Size distributions of aromatic hydrocarbons and elemental carbon. 1. sampling, measurement methods and source characterization, *Environmental Science Technologyi,* 28, 555-562.

Wight, G.D. (1994). *Fundamentals of Air Sampling*, Lewis Publishers, New York.

Wobst, M., Wichmann, H., Bahadir, M. (1999). Surface contamination with PASH, PAH and PCDD/F after fire accidents in private residences, *Chemosphere*, 38 (7), 1685-1691.

World Health Organization (WHO), (2000). *Guidelines for Air Quality*, Geneva.

Yeo. H.G., Choi, M., Chun, M.Y., Kim, T.W., Cho, K.C., Sunwoo, Y. (2004). Concentration characteristics of atmospheric PCBs for urban and rural area, Korea, *Science of the Total Environment*, 324, 261-270.

Yeo. H.G., Choi, M., Chun, M.Y., Sunwoo, Y. (2003). Concentration distribution of polychlorinated biphenyls and organochlorine pesticides and their relationship with temperature in rural air of Korea, *Atmospheric Environment*, 37, 3831-3839.

In: Environmental and Regional Air Pollution
Editors: D. Gallo, R. Mancini

ISBN: 978-1-60692-893-6

Chapter 11

CHARACTERISTICS OF RED RAIN IN THE CILICIAN BASIN, NORTHEASTERN MEDITERRANEAN

***Türkan Özsoy*[a,*] *and Sermin Örnektekin*[b]**
[a]Mersin University, Faculty of Science and Literature,
Chemistry Department, Çiftlikköy Campus, 33343, Mersin
[b]Mustafa Kemal University, Faculty of Science and Literature,
Tayfur Sökmen Campus, 31034, Antakya-Hatay, Turkey

ABSTRACT

The northeastern coast of the Mediterranean Sea is under the influence of sporadic dust transport from North African (The Great Sahara) and Middle Eastern (Arabian Peninsula) deserts, particularly during spring and autumn. Mineral dust from these arid regions typically has a characteristic reddish yellow - reddish brown color owing to its high iron oxide content. Precipitation samples collected at Mersin, on the eastern Mediterranean coast of Turkey, during the period of December 2003 – May 2005 have been analyzed with respect to their red dust contents, and in relation to backward air mass trajectories corresponding to the rainy days to determine potential source regions. Classification of trajectories revealed local sources from Anatolian mainland and in the Cilician Basin, in addition to remote sources in North Africa and the Middle East. Approximately, one fourth of the precipitation samples can be classified as "*red rain*", annually. The mean electrical conductivity and the volume weighted mean (VWM) Aluminium (Al) concentration of red rain were found to be more than four and eight times higher than normal rain, respectively; while the dissolution of calcite in dust was found to result in an order of magnitude decrease in the VWM hydrogen ion concentration of red rain, by neutralizing (buffering) the acidity of rainwater. Due to dissolution of atmospheric CO_2 in rainwater, natural precipitation is slightly acidic (pH ≈ 5.6). The VWM pH values were calculated to be 6.10 for normal rain and 7.21 for red rain. The mean bicarbonate concentration of red rain, estimated from the theoretical relation between pH and HCO_3^-, was sixteen times higher than that of normal rain. Similarly, abundance of the major anions (Cl^-, NO_3^-, SO_4^{2-}) and major cations (Na^+, K^+, Ca^{2+}, Mg^{2+}, NH_4^+) measured in red rain samples were all high, relative to the values

* Turkey, tozsoy@mersin.edu.tr

observed in normal rain samples, including formaldehyde (HCHO), the only organic parameter measured in Mersin precipitation. Similar enhancements were also observed in soluble (Al_s) and particulate (Al_p) aluminium species. Based on these results, red rain can be identified as a rich source of all major ions, formaldehyde and aluminium suggesting mineral dust particles to act as host for all measured species in the northeastern Mediterranean atmosphere.

Keywords: Northeastern Mediterranean, Red rain, Mineral dust, Wet deposition, Aerosol dust flux.

1. INTRODUCTION

1.1. Dust in the Mediterranean Atmosphere

Large quantities of dust, consisting of soil eroded mineral particles originating from arid regions can occasionally be transferred to the atmosphere and transported over thousands of kilometers to other continents. For instance, existence of Saharan dust over the Black Sea (Kubilay et al., 1995), northern Europe; including the Netherlands (Reiff et al., 1986), Scotland (Davies et al., 1992) and Sweden (Franzen et al., 1994), across the Northern Atlantic (Talbot et al., 1986; Kremling and Streu, 1993) and over Florida (Prospero et al., 1987) are all well known events. Desert dust has attracted considerable research interest for more than two decades due to its potential impact on either terrestrial or aquatic biogeochemical cycles when deposited onto the land and sea surfaces at very distant sites. For example, it has been reported that transportation and ensuing deposition of Saharan dust to the Amazon Basin (Swap et al., 1992) and northeastern Spain (Avila et al., 1998; Avila and Penuelas, 1999) has provided ample amount of inorganic nutrients that could contribute to the forest productivity. Quetel et al. (1993) suggested that North African alumino-silicate dominated aerosols are delivered in the form of pulses which may impose non-steady conditions on the biogeochemical processes of iron, which could be limiting nutrient in the marine environment. Donaghay et al. (1991) has proposed different scenarios about marine ecosystem responses to episodic atmospheric inputs of the nutrient type of elements such as iron and nitrogen. Mineral dust contains high amounts of various trace elements (e.g., Mn, Co, Cr, Cu, Ni, Zn) which could also be limiting nutrients for certain species of aquatic organisms; synergistic and antagonistic interactions between multiple trace elements could also be very important on oceanic plankton production and species composition (Bruland et al., 1991). Even though too many studies have been performed on this issue, the impact of aerosol dust on ocean productivity and biogeochemical processes in terrestrial and aquatic ecosystems are still not well understood owing to the limitations in conducting in-situ experiments due to the sporadic nature of transportation events.

North African and Arabian deserts are known to be the most important potential source regions of mineral dust which affect the composition of the western (Bergametti et al., 1989; Martin et al., 1990; Dulac et al., 1992; Chester et al., 1993) and the eastern Mediterranean aerosols (Yaalon and Ganor, 1979; Ganor and Mamane, 1982; Ganor et al., 1991; Ganor 1994; Kubilay and Saydam, 1995). Recently, quite a number of comprehensive studies have been conducted addressing ionic and elemental composition of atmospheric aerosols (Güllü et al., 1998, 2004, 2005; Herut 2001; Herut et al., 2001; Koçak et al., 2004a, 2004b, 2005) and their size distributions in the eastern Mediterranean (Bardouki et al., 2003; Shaka and Saliba, 2004; Koçak et al., 2007; Gerasopoulos et al., 2006; Lazaridis et al., 2008). In a major part of these studies it has been clearly demonstrated that atmospheric aerosol over the eastern Mediterranean is strongly influenced by sporadic dust transport from Saharan Desert, occuring mainly during transitional seasons in March, April, May and October. Geographical dust sources to the Eastern Mediterranean may be variable throughout the year, for example central Sahara is dominating in the spring; eastern Sahara in the summer; the Middle East and Arabian Peninsula in the autumn (Alpert et al., 1990b; Kubilay et al., 2000). In the summer and autumn dust is transported to the region from higher atmospheric altitudes (> 700 hPa) while the lower levels are influenced by anthropogenic aerosols originated either locally or transported from the North (Kubilay et al., 2003). Dust originating from African and Arabian deserts is carried into the Mediterranean region either by the Scirocco and Khamsin local wind systems, or by mid-latitude atmospheric depressions passing along the Mediterranean-European region. A brief information about the climatology and meteorology of the Mediterranean region, related to atmospheric dust transportation is provided in the following section of this chapter.

1.2. Climatic Characteristics of the Mediterranean

The Mediterranean can be regarded as a transitional zone between the continental influences of Europe and Asia, the desert climate of North Africa and the oceanic effects from the Atlantic. The surface pressure field is influenced by the oceanic Azores High, the Siberian winter anticyclone, the Northwestern extension of the South Asian thermal low, as well as more transient anticyclones and travelling depressions (Maheras et al., 1999). Qualitatively, the Mediterranean climate is characterized by generally warm, occasionally cool winter temperatures (November through to February) dominated by rainfall and long warm and dry, often hot and sunny summers (June through to September) (Wallen, 1970). The transitional seasons, spring and autumn are of very variable in length and characteristics. The relatively long spring season (March through May) is noted for periods of unsettled winter-type weather associated with an increased occurence of North African cyclones; the rest of this period is very similar to that in the summer. Autumn usually lasts only one month (October) and is characterized by an abrubt change from the summer to the unsettled weather of winter (Brody and Nestor, 1980; Özsoy, 1981; Milliman et al., 1992). In general, temperature displays negative correlation with latitude while precipitation shows positively high correlation with latitude in the Mediterranean. A positive relationship between precipitation and longitude also exists in all seasons, however only in spring and autumn it is statistically significant (Trewartha, 1961). In terms of annual total rainfall, western Mediterranean receives relatively higher amounts of precipitation and this pattern decreases eastwards, with the exception of the Adriatic Coasts. For example, 757 mm was reported as the long term (42 years, 1958 - 2000) average annual rainfall for Iberia, while the same average was reported as 620 mm for Greece (Goosens, 1985) and 580 mm as 30 years (1963 – 1994) average for the southeastern Mediterranean Coast of Turkey (Özsoy and Saydam, 2001).

The climatic characteristics of the Mediterranean described in the previous paragraph are principally influenced by complex topography and oceanography of the region. Topographic formations interacts with air masses, either intensifying or obstructing their effects, being partly responsible for the Mediterranean type of climate (Barry and Chorley, 1998). Spatial variations in topography may create stormy conditions or intensify pre-existing air systems (Lagouvardos et al., 1996) and the effects of topography are not limited to large-scale climatology, but are also responsible for microclimatic effects that may contribute to the spatially discontinuous trends in both temperature and precipitation (Milliman et al., 1992).

Recently, Harding (2006) has summarized the general characteristics of the Mediterranean climate while investigating the changes in climate extremes. He reported that particularly, in the eastern Mediterranean the dominant pressure system is the Asiatic thermal pressure centre which can be high or low depending on season and this pressure system is characterised by its size and seasonal movement. During winter, the pressure system is centered in cold Siberia and called as Siberian thermal high whereas in summer, it becomes a thermal low and moves south and west to occupy the Northern India, the Middle East and the Eastern Mediterranean, creating landward southerly flow and produces North-easterly winds across Turkey and Greece. The Asiatic summer thermal low is also one of the driving mechanisms of the Indian monsoon (Barry and Chorley, 1998). Cyclogenesis is particularly important for winter rainfall across the basin from west to eastward and the passage of low pressure systems (depressions or cyclones) across the basin, controls intense rain generation and moisture transport. Cyclones are formed in association with the effects of complex topography and the Mediterranean Sea, they are often subsequently steered by the former and reinforced by the latter. Eastern Mediterranean depressions usually form over the Adriatic, the Ionian Sea, Crete and Cyprus Islands during autumn and winter (November through April) whereas during summer cyclogenesis is usually weaker than in winter and many of the tracks particularly in the central and eastern Mediterranean disappear in the warmer months, becoming quasi-stationary and losing their ability to transfer moisture (Alpert et al., 1990a,b). Most likely due to increasing winter temperatures (Serreze et al., 1997) a decline has been reported in the frequency of intense winter cyclones across the Mediterranean (Piervitali et al., 1997). However, a significant increase has also been detected in the intensity of cyclones occuring in both high- and mid-latitudes (McCabe et al., 2001). This type of mid-latitude atmospheric depressions passing along the Mediterranean-European region, in the eastward direction are also responsible for transportation of mineral dust by atmospheric pathways.

Occasionally, when parts of Icelandic low reach as far south as the North coast of Africa, the Mediterranean front, which is created by the boundary between the humid Mediterranean air mass (cyclone) and hotter-drier continental African air mass, is altered and warm African air can be drawn from the Sahara into South-eastern Europe and the counter-clockwise circulation of the low propels the mixed air loaded with red dust across the southern coasts of Europe (Dayan and Miller, 1989). The resulting dry, dusty and hot winds (e.g. the Sirocco, also known as the Ghibli and Khamsin) usually originate from the Arabian or Saharan deserts and catch moisture during their northward excursions when crossing the Mediterranean, causing rainfall in the eastern Mediterranean. The precipitation events coinciding with this type of dust transport, identified by the reddish or yellowish color of membrane filters used in filtering the rainwater are called as "red rain" or "mud rain" (Avila 1996; Avila et al., 1998; Özsoy and Saydam 2000; Özsoy et al., 2008). These events have long been recognized and frequently described in the past 100 years over Europe (Avila and Penuelas 1999). The phenomenon is also identifiable to the naked eye due to the reddish silty marks and stains that raindrops leave on various receiving surfaces (e.g car surfaces, snow covered grounds).

1.3. General Characteristics of the Cilician Basin

Our study region, the Cilician Basin, located in the Northeastern part of the Levantine Basin, Eastern Mediterranean and covers the area between Turkey and the island of Cyprus. Mersin and Iskenderun Bays adjoining wide continental shelves and river deltas of the Göksu, Seyhan, Tarsus and Ceyhan are located along the Turkish coast (Figure 1). Taurus and Amanos mountain ranges bound the Cilician Basin in the north and east, respectively, lined with the narrow coastal plains in those regions excluding the vast delta plains of Seyhan and Ceyhan Rivers.

The climate is typical of the Eastern Mediterranean, with hot, dry (no rain) and humid summers, mild and rainy winters and short transitional seasons. Northerly winds dominate the winter (November to March) while southwesterly winds dominate the summer (April to October). Weather steered by steep mountain ranges but intercepted by valleys along the northern shore, such as the Göksu River Valley and the Gulf of İskenderun. (Reiter, 1979; Özsoy, 1981; Özsoy and Sözer, 2006). Mean annual precipitation and the number of rainy days calculated for a period of 30 years (1963-1994) were found to be 580 mm and 67.2, respectively (Özsoy and Saydam, 2001), which are very close to the annual precipitation average (593.7 mm) and the number of rainy days (62) reported by the Turkish Meteorological Service, Mersin Station. December is the most while August is the least rain receiving months (ÇDR, 2006). According to the last 50 years of records a devastating flood event has occured during December, 2001 following a continuous rainfall of 677.4 mm within only three days. Except this extremity, the highest daily rainfall is 109 mm, recorded on January 24, 1984. No serious droughts have been experienced in the region. The lowest annual rainfall (209.4 mm) was recorded in 1982, which is approximately 65 % lower than the mean annual precipitation of Mersin (ÇDR, 2006).

In general, Cilician Basin consists of mainly terrigenous carbonates, derived from limestone, dolomite and calcretes (Yetiş et al., 1995). Calcareous-karstic Taurus mountains, mostly of Paleozoic and Mesozoic formation, reaching heights of 600-1500 m in the north (Kapur et al., 2000). More than 72 % of the soils in the region is reported to be extremely calcareous, the rest being moderately calcareous (ÇDR, 2006).

During the last few decades, the Mediterranean Coast of Turkey has been subjected to urbanization due to rapid population and industrial growth, agricultural and touristic activities, marine transport linked to the harbours of Mersin, İskenderun and Taşucu, oil storage and pipeline terminals at Yumurtalık, Ceyhan and Dörtyol (including the recently completed Bakü-Tblisi-Ceyhan pipeline transporting oil and gas from the Caspian Sea), leading to significant environmental stresses, with adverse effects on air quality. Rapid urbanization is taking place in the region because of recreational constructions and extensive building of second (vacation) houses on the eastern Mediterranean coastline. This is drastically altering the landscape (NDA, Turkey, 2003). Mersin, where precipitation samples have been collected is the second biggest and industrialized city with 850,000 inhabitants in the studied region and the city has had the highest population increasing rate by internal immigration (% 7) and by birth (‰ 40.6) during the 1980 – 1990 period among the cities in Turkey (ÇDR, 2006).

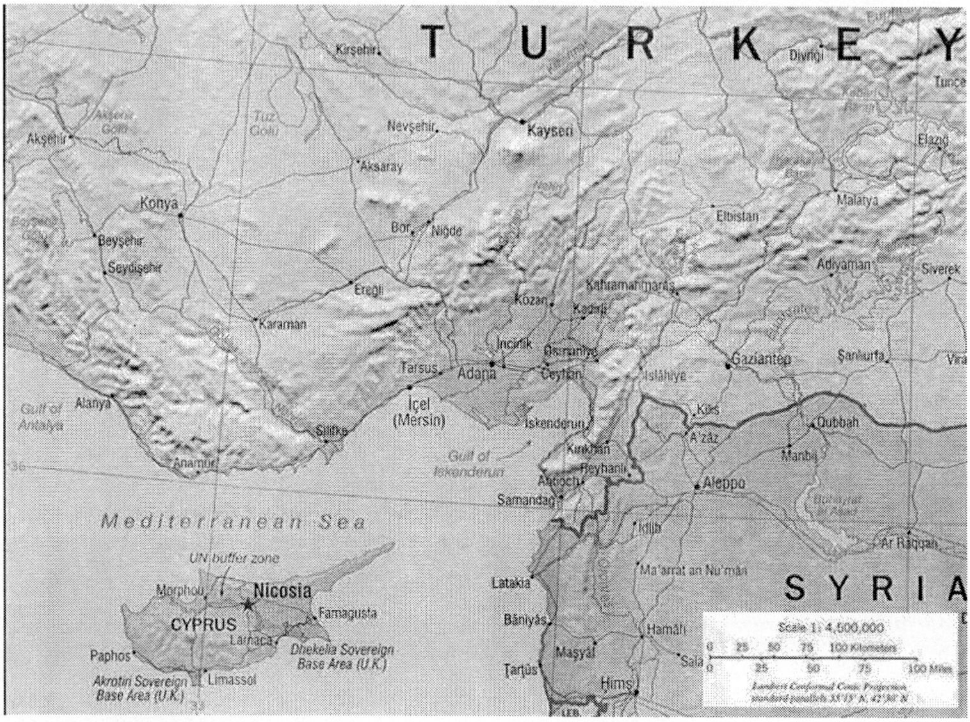

Figure 1. The study area, Cilician Basin.

Internal immigration rate has been mitigated recently and the population increasing rate was reported as ‰ 26.5 during the 1990 – 2000 period. Adana is the most crowded and industrialized city of the region, having 1,130,000 population located inland, 60 km eastward of Mersin. Another area of major environmental concern and pollution hot spots is the Gulf of İskenderun. The population is approximately 2,000,000 around the Gulf and in addition to two international pipe line terminals a huge Iron and Steel Works, a cement factory, two fertilizer plants, a number of smaller steel mills and a welding pipe factory are located along the coastline, contributing air pollution problems to the region. For instance, daily SO_2, NO_x and heavy metal emissions from ISDEMİR, the largest Iron-Steel Work of Turkey, were reported as 270, 12 and 43 ton, respectively (Örnektekin and Çakmaklı, 2003).

1.4. Previous Atmospheric Research Conducted in the Region

One of the most comprehensive study carried out for five years in the northwestern Mediterranean is EROS 2000 (European River Ocean System), has revealed that the material transported to the basin by atmospheric pathways (10 – 27 tons km^{-2} yr^{-1}) is as important as the riverine input and approximately 90 % of this atmospheric input is comprised of Saharan mineral dust (Martin et al., 1989). The results of another study performed within the same region; a systematic atmospheric sample collection and trace metal analyses of 16 month duration, showed that total (dissolved + particulate) atmospheric input of trace metals are comparable to the amount of trace metals carried by Rhone River to the Northwestern Mediterranean Coast (Guieu et al., 1991). Following Western Basin studies which focused on the research from the oceanographic point of view, the first long-term atmospheric research in the Northeastern Mediterranean has been initiated in Erdemli, by the Institute of Marine Sciences of Middle East Technical University (IMS-METU) and the very first results of this land-based atmospheric particulate sampling study during the period of August 1991 – December 1992, have highlighted the major differences in the elemental composition of atmospheric aerosols between the Eastern and the Western Basins (Kubilay and Saydam, 1995):

- The Eastern Basin is much more extently effected by desert particles than the Western Basin. Aerosols collected from the Eastern Basin had higher concentrations of crustal elements (Al, Fe) while mean concentrations of anthropogenic elements (Cd, Pb, Zn) were relatively lower hence, the effect of anthropogenic emissions over the Eastern Basin is much less pronounced.
- Relatively higher Cr and Ni concentrations have been observed over the Eastern Basin and this situation was attributed to the presence of ophiolitic rocks that outcrop in patches on the hinterland of the basin which are relatively enriched with Cr and Ni bearing minerals (Tolun and Pamir, 1975; Çağatay et al., 2002).

Later on, similar type of comparison has also been applied within the Eastern Mediterranean Basin itself by comparing the two contemporaneous data sets, between 1999 and 2001 obtained from Erdemli-Mersin (Turkey) and Tel-Shikmona (Israel) (Koçak et al., 2004). As a result of this study a general increasing trend of the crustal material load from north to south across the Levantine Basin has been determined. Particularly, during Saharan dust outbreaks enhanced concentrations of Al (x1.7), Fe (x1.6) and Mn (x1.9) have been observed at Israeli station compared to Erdemli. Cr was found to be enriched at Erdemli due to the contribution from local ophiolitic rock sources. Even though no significant gradient of Pb concentrations was observed across the Levantine Basin, a decadal decrease of almost 40 % in Pb concentrations of Erdemli aerosols was reported from the comparison of data with the historic Pb values of Erdemli. This was attributed to the diminished emissions from vehicular transport, similar to the observations in Cairo, Egypt (Hassanien et al., 2001), in the Ligurian Sea (Migon and Nicolas, 1998) and in Germany (Lammel et al., 2002).

Recently, it has been suggested that atmospheric inputs of trace (Guerzoni et al., 1999) and nutrient type of elements (particularly, nitrogen) (Krom et al., 2004) are quantitatively exceeding the riverine inputs to the Mediteranean. Apparently, much more information is avilable on atmospheric fluxes of trace elements to the Western Basin compare to that of the Eastern Basin. Therefore, in order to obtain reliable data for dry deposition flux calculations for the Eastern Basin, aerosol sampling period at Erdemli has been prolonged to four years (January 1999 – December 2002) and aerosol trace metal concentrations have been used for flux calculations of elements (Al, Fe, Mn, Cu, Pb, Zn, Cd) by taking into account both seasonal and inter-annual variations over the northeastern Mediterranean (Koçak et al., 2005). Previously, the importance of atmospheric dry deposition of mineral dust on an annual basis at Erdemli has been emphasized by Kubilay et al. (2000). Estimations were based on aerosol Al measurements, assuming crustal material has an Al abundance of 8.2 %. Total deposition flux of mineral dust in 1992 at Erdemli was estimated at 13 tons km^{-2} and approximately 30 % of the annual deposition of dust supplied by only two Saharan episodes. Dry deposition accounted for 56 % of total deposition and this contribution increased to 93 % during summer (Kubilay et al., 2000).

Major ionic composition of Erdemli aerosols have been investigated for the samples collected in two different sampling periods: a) October 1991 – December 1992, totally 303 samples (Özsoy et al., 2000); b) A four year period between 1996 and 1999, totally 610 samples (Koçak et al., 2004). The results of the first study revealed that the mean non sea salt sulfate (nss-SO_4^{2-}) concentration was one of the highest among the reported values for rural areas worldwide. The same results have also primarily highlighted the biogenic source contribution to the very high atmospheric sulfate levels measured in Erdemli aerosols, particularly from enhanced coccolithophorid bloom in the Black Sea. Özsoy et al. (2000) have utilized the nitrate concentration as an indicator of anthropogenic contribution and the crustal Enrichment factors (Ef_{crust}) of mainly anthropogenic elements for tracing changes in chemical aerosol composition instead of lacking methanesulfonate (MSA) concentrations, which may have served as a tracer of the biogenic dimethylsulfide (DMS) contribution to aerosol nss-sulfate. Following this study, MSA and nss-sulfate concentrations in bulk aerosol samples collected during January 1996- December 1999 period at two stations along the Turkish and Cretan coastline of the eastern Mediterranean have been measured by Kubilay et al. (2002) and it has been verified that the majority of the biogenic contribution at Erdemli is originated from summer coccolithophorid *Emiliania huxleyi* blooms developed in the Black Sea.

Since the eastern Mediterranean marine environment is known as one of the least productive area in the world (Krom et al., 1992; Bethoux et al., 1998; Guerzoni et al., 1999; Herut et al., 1999; Psarra et al., 2000) and having a high N:P molar ratio ranging from 24 to 28, compared to the standard N:P Redfield ratio of 16, (Krom et al., 1991; Bethoux et al., 1998; Yılmaz and Tuğrul, 1998) soluble macro-nutrient contents of wet deposition at Erdemli have been quantified with respect to their contribution to biological productivity (Özsoy, 2003). Most likely, due to the low solubility of crustal phosphorus, dust episodes were not found to be important sources of reactive P, in terms of wet deposition. Additionally, the N:P molar ratio of the wet deposition was found to be quite higher (52.5) than the mean ratio reported for surface seawater. Similar study has also been conducted at Erdemli and on the Crete Island in both rainwater and aerosol samples collected almost contemporaneously (Markaki et al., 2003) and a quite higher N:P molar ratio was reported for the dry deposition samples of Erdemli (261) compared to the ratio observed for the wet deposition samples (63). These high N:P ratios reported for both wet and dry atmospheric samples are most probably due to the higher influence of nitrogen sources over the region and suggest a stronger reinforcement for arguments concerning the P-limited nature of the northeastern Basin.

In the present study, precipitation samples have been collected on event basis at four different stations in Mersin during the December 2003 - May 2005 period. In order to identify potential source regions of red dust and the general characteristics of eventual red rain, pH, electrical conductivity and total Al concentrations of precipitation samples have been evaluated in conjunction with air mass back trajectories calculated for the corresponding rainy days.

2. EXPERIMENTAL

2.1. Precipitation Sampling

Precipitation samples have been collected from Mersin, which is an industrialized city with almost one million inhabitants and a site for various industries (e.g. petroleum refinery which was active until mid 2004, thermal power plant, soda, chromium, fertilizer, cement and glass production), almost all situated in the industrial district, NE of the city. Four sampling stations (two urban, two suburban) were selected within the residential area of the city, located at roofs of private buildings based on practical and safety considerations. St.1 is located in Çiftlikköy Campus of Mersin University, at an appoximate height of 60 m from sea-level, on the slopes of Taurus Mountains. The campus is surrounded by agricultural land, mainly lemon groves. Sampling buckets of 30 cm diameter made out of HDPE (high density polyethylene) were deployed at the roof of Mersin Vocational School, a building with central heating and therefore no active chimney. St.2 is located on the roof of a private building in the newly developing residential area of Yenişehir, surrounded by a low density of small private houses and gardens, far from active chimneys. A rain-gauge (Akım Elektronik, Pluviograph, Model PHD5-02) with a special facility for rainwater sampling were deployed at St.2. The urban stations St.3 and St.4 were located at roofs of a private house and a business center, respectively, both heated by electricity, near the highly populated center of the city under the direct influence of other residential heating and dense traffic activity in the area (Özsoy et al., 2008).

Rainwater samples were transferred to the laboratory located in Çiftlikköy Campus of Mersin University, following cessation of the particular rain event. In order to avoid any probable contamination during transportation, the buckets were tightly sealed with a clean plastic cover. Samples were filtered through a 0.45 μm pore size membrane filter (MFS, cellulose acetate, 47 mm diameter) after measuring their volume and then they were divided into aliquotes for different analyses. Subsamples of precipitation reserved for trace element analysis have been preserved in rigorously cleansed PE (polyethylene) bottles at +4°C temperature after acidification with 0.1 M HNO_3 before storage. Sampling buckets, bottles and all glassware used in the laboratory have been washed first with detergent and rinsed with double distilled deionized water. Laboratory equipment have been additionally soaked in 1 M HNO_3 washing solution for 48 h and rinsed with distilled, deionized water several times prior to usage. Utmost care have shown to avoid contamination of samples due to difficulty in the analysis of trace elements in rainwater samples.

2.2. Electrical Conductivity (EC) and pH Measurements

The pH in rainwater was measured by a Mettler Toledo MP120 pH-meter which has an accuracy of 0.01 pH unit. The instrument was calibrated against the buffer solutions of pH 4.01, 7.01 and 10.01 prior to every pH measurements. EC measurements were performed with DIST H198300/3 (mS cm^{-1}) and DIST H198300/4 (μS cm^{-1}) Conductivity/TDS Meters with automatic temperature compensation. Prior to every EC measurements the instruments were also calibrated against a Standard solution which has a constant EC of 1413 μS cm^{-1}. Quality assurance of the data was provided by participating in semi-annual WMO/GAW (World Meteorological Organisation/Global Atmosphere Watch) Precipitation Chemistry Laboratory Intercomparison. For this purpose simulated rainwater samples of the 30th intercalibration programme, distributed by QA/SAC (Quality Assurance/Science Activity Center) of the same institution have been analysed for pH and EC and analytically the accuracy was found to be better than 3.7 % for both of the parameters (Özsoy et al., 2008).

2.3. Al Analyses

A Varian Liberty II Model Inductively coupled plasma atomic emission spectroscopy (ICP-AES) at the Central Research Laboratory of Mustafa Kemal University was used for the analysis of Al in both soluble and insoluble phases of precipitation. The ICP-AES with axial torch was controlled by PC Pentium III. Plasma power was 1.0 kW and integration time was 1.5 sec. with three replicates. The analytical detection limit, defined as three times the standard deviation of five replicate blank measurements and it was 2.0 μg L^{-1} for Al analysis in soluble fraction (Al_s).

Membrane filter samples which contain insoluble fraction of precipitation have been preserved at -18°C in a deep-freeze until analysis time. These samples were digested according to the procedure given in UNEP (1995), using a microwave digestion oven (MARS 5, Microwave Accelerated Reaction System, CEM Microwave Technology Ltd.) at the Central Research Laboratory of Mustafa Kemal University. Total (HF) decomposition method has been applied with little modification in the adjustment of the microwave oven's power and timing. In digestion of samples HF and Aqua Regia (1:3 v/v HNO_3 and HCl) mixture have been used. Digested samples were diluted to 25 mL volume by double distilled de-ionized water and then transferred into PE bottles in order to preserve them at +4°C until analysis time. Al mesurements in insoluble phase of precipitation were performed by a teflon torch of ICP-AES instrument. The detection limit was found to be 0.7 μg L^{-1} for Al in insoluble fraction (Al_p).

Standard solutions for calibration have been prepared daily from high purity MERCK CertiPUR ICP 1000 mg L^{-1} multi-element Stok Standard Solution IV and they were run before every analysis. Standard addition method has been applied in order to check for matrix interferences. Reproducibility (variation of coefficient %) of results was better than 4 % for both Al_s and Al_p analysis. Five field blanks have been collected during the sampling period and they have been analysed by the same procedures applied to the samples. The concentration of field and laboratory blanks measured throughout the study were all below the detection limits of Al_s and Al_p.

2.4. Air Mass Back Trajectory Calculations

In order to interpret atmospheric measurements such as to characterize the mean air flow direction for a specific geographic location and to understand long range transportation of chemical constituents and/or dust from remote sources, one of the most widely utilized technique is to calculate air mass trajectories through running an atmospheric trajectory model (Dulac et al., 1987; Martin et al., 1987; 1990). These models are usually based on an assumption that the tracked air particle follows the mean air flow and hence this allows the calculation of transport pathways forward or backward in time from a source location or from a receptor, respectively. In any trajectory model, selection of the spatial and temporal resolution of the meteorological data, which is going to be interpolated both in space and time is important. For example, coarse resolution of input data reduces the accuracy of the model and the magnitude of the interpolation errors on the location of trajectories increases. Since the estimates of errors range from 350 to 495 km after 72 hrs, trajectory calculations are usually performed for maximum three days (Whelpdale and Moody 1990). On the other hand, selection of a fine resolution means an expense of much longer computer time.

In the present study the three-dimensional trajectory model developed at the European Center for Medium Range Weather Forecasts (ECMWF) in Reading, England by Ray McGrath of the Irish Meteorological Service has been used with a spectral truncation of t-123, a time step of 1 h going backward in time, for each trajectory. The model uses direct calls to the ECMWF atmospheric data archive system MARS to retrieve global model derived wind fields and generated its own internal index to mimic the original wind database. Three-dimensional, 3-day backward trajectories of air masses subjected to scavenging by precipitation and arriving at Mersin Station #2 (36° 47′ 25″ N – 34° 37′ 36″ E) at the beginning of every particular rain event, at 1000, 900 and 850 hPa pressure levels (supposed to be within the planetary boundary level; PBL) and 700 and 500 hPa (supposed to be within the free troposphere) have been calculated (Özsoy et al., 2008). Back trajectories of different pressure levels display both the geographical origin of precipitation carrying air masses and their vertical motion within the atmospheric column during their excursions until they arrive at the sampling point. However, they don't provide information about meso-scale meteorological events e.g. precipitation, that the air mass might have subjected to during its travel.

3. Results and Discussion

3.1. Precipitation Data

Precipitation has been recorded at 0.1 mm sampling increments by a rain-gauge (Akım Elektronik, Pluviograph, Model PHD5-02) installed at St.2. The data belonging to the sampling period of approximately 18 months (from January 2004 to June 2005) were compared with the monthly mean precipitation of 14 years (1990 – 2003) obtained from the Turkish Meteorological Service, Mersin Station, located 4 km south of the sampling St.2 in Figure 2. There is not much coherence between precipitation recorded at St.2 and the mean climatological rainfall in Mersin. Cumulative precipitation for a total of 74 rainy days during December 2003 – June 2005 period was 735.4 mm. Minimum and maximum event-based rainfall values were 0.2 and 61.8 mm, respectively. According to the 14 years of climatological data from Mersin, mean annual precipitation and the number of rainy days were 587.5 mm and 67, respectively. The annual cumulative precipitation corresponding to the period of January 2004 – December 2004 was 530.6 mm and the number of rainy days was 52 during the same period, which both are quite lower than the mean climatological values.

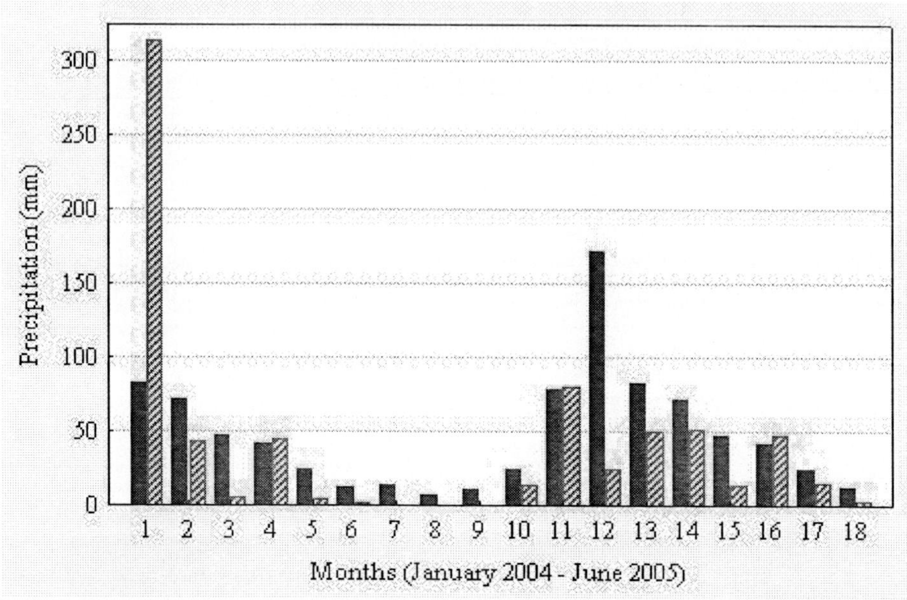

Figure 2. Precipitation data for the sampling period of January 2004 – June 2005 (striped columns) at Mersin (St.2) and the climatological monthly mean precipitation of 14 years (1990 – 2003) for Mersin, obtained from the Turkish Meteorological Service (dark columns).

Most likely, due to heat islands over the city center annual rainfall in the central stations (St.4: 468 mm; St.3: 469 mm) was less than the rainfall measured at suburban stations (St.2: 530.6 mm; St.1: 591 mm). These results might be attributed to the high temporal and spatial variability of precipitation events throughout the year except in summer over most of Turkey (Kutiel et al., 2001; Türkeş and Erlat, 2003; 2005). However, lower amount of precipitation observed in Mersin during 2004, might as well be part of the decrasing trend of rainfall in the Mediterranean over the last half century (Tsonis, 1996; Zhang et al., 1997; Buffoni et al., 1999; Xoplaki et al., 2004), or else it might partly come from the temporal variability itself. It has been reported that in Spain rainfall has reduced by 10 – 20 % over a 30 year period (1964 – 1993) (Romero et al., 1998). A decrease of 26 % has also been reported in some parts of the South Mediterranean, such as Tunisia, southern Spain and southern Italy over the 1951 – 1995 period (Alpert et al., 2002). Long term temporal variabilities of precipitation have been investigated for the 1930 – 2002 period in Turkey and an apparent decreasing trend in the winter precipitation totals was detected which is particularly strongest over the Mediterranean and the inland of the western Anatolia, whereas a general increasing trend was reported in the precipitation totals of spring, summer and autumn seasons (Türkeş et al., 2008). In another study, Partal and Kahya (2006) have utilized non-parametric methods in order to determine trends in long-term annual mean and monthly total precipitation series of Turkey for the 1929 – 1993 period and they observed a noticeable decrease in the annual mean precipitation mostly in western and southern Turkey. From the monthly precipitation variables January, February and especially September were determined to have strong decreasing trends, similar to the findings presented in Türkeş et al. (2008). Downward trend in annual precipitation has also been attributed to the strong decreasing trends of precipitation during those winter months (Partal and Kahya, 2006).

3.2. Mean Directions of the Precipitation Carrying Air Masses

In order to identify the mean airflow directions of the air masses carrying precipitation to the sampling site, 3-days back trajectories of the air masses have been classified according to 90° sectors of arrival. For this purpose, the wind rose was divided into four geographical sectors of 90° with respect to St.2 at the center (36° 47′ 25″ N-34° 37′ 36″ E) and each trajectory was assigned to the sector in which it spent most of its three days of travel. According to this partitioning the geographical regions of which each sector encompasses the potential emmission sources are as follows:

Sector 1: Eastern Anatolia and Common Wealth of Independent States (NE).
Sector 2: Western Anatolia and Europe, Atlantic Ocean (NW)
Sector 3: The Mediterranean Sea and North Africa (SW)
Sector 4: The Middle East and Arabian Peninsula (SE)

The trajectory classification for rainy days was performed at 1000, 900, 850 hPa levels within the Planetary Boundary Layer (PBL) and at 700, 500 hPa barometric levels within the free troposphere, respectively and the results are presented in Table 1, where the percentage in each direction are given for a total of 78 trajectories during rainy days, of the period December 2003 – May 2005 (Özsoy et al., 2008).

Table 1. Percentage frequencies of mean airflow directions of precipitation carrying air masses at five different (1000, 900, 850, 700, 500 hPa) pressure levels for the period of December 2003 – May 2005 in Mersin (Özsoy et al., 2008)

hPa	Sector 1-NE	Sector 2-NW	Sector 3-SW	Sector 4-SE
1000	% 10	% 18	% 44	% 28
900	% 12	% 23	% 28	% 37
850	% 9	% 23	% 45	% 23
700	% 3	% 27	% 69	% 1
500	% 2	% 35	% 63	% 0

As can be seen from Table 1, the highest percentage frequencies of the air masses belong to the SW, Sector 3. The secondary distinctive trajectory direction is SE, Sector 4, particularly for the air masses move within the PBL. These results are consistent with the long term observation of Turkish Meteorological Service, Mersin Station, indicating prominent wind directions of S-SW and SE during rainy days. The industrial district of Mersin is located on the NE of the city (see Section 2.1) and this direction has the lowest percentage frequencies of the air masses on almost all barometric pressure levels, which means that any possible pollutant transport towards the city from the industrial district might take place at its minimum level (Özsoy et al., 2008).

3.3. Red Rain in Mersin

In order to determine red rain events, filter samples have been classified according to their colors and Al_t contents. Air mass back trajectories of the corresponding samples having reddish or dark yellowish color of membrane filters and very high Al_t concentrations revealed 28 dust transport episodes associated with 31 red rain events, during the entire sampling period. Among these episodes the most intense one has occured on 1April 2004 and following this transport, red rain samples have been collected from three different stations, at around 8:00 am on the next day, 2 April 2004. Satellite images recorded on 1 April 2004 are presented in Figure 3a,b while back trajectories of air masses arrived to the St.2 at 8:00 am on 2 April 2004 are given in Figure 4.

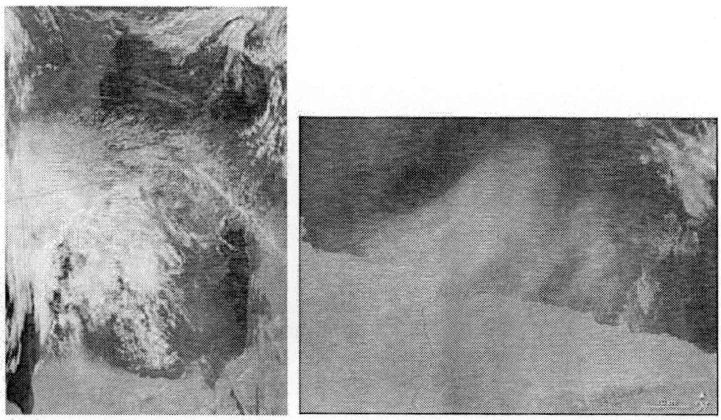

Source: http://earthobservatory.nasa.gov/Newsroom/NewImages/images_index.php3).

Figure 3a,b. The satellite images of dust transport episode occured on 01/04/2004.

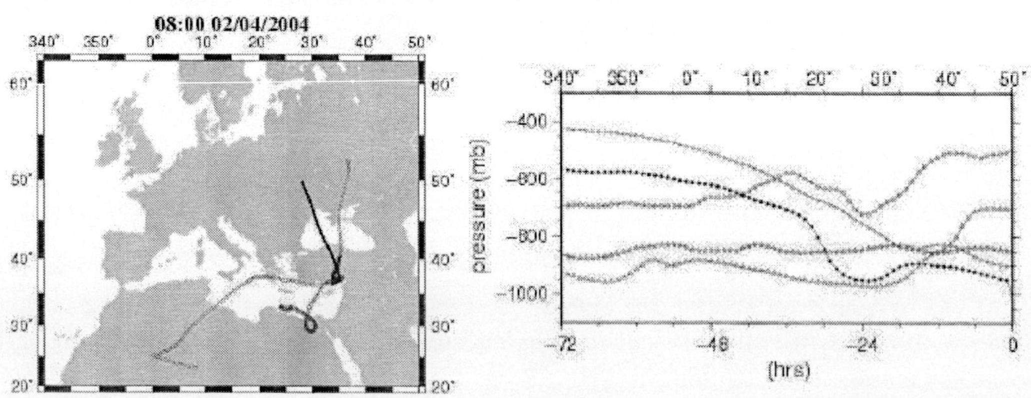

Figure 4. Horizontal and vertical projections of air mass back trajectories arriving at 1000, 900, 850, 700 and 500 hPa levels on 2 April 2004.

Dust plume over the eastern Mediterranean can be seen clearly in Figure 3a and b. On 2 April, air masses ending at 500 and 700 hPa levels have originated from north Africa in accordance with the satellite images while air masses at 1000, 900 and 850 hPa levels have swept over Anatolian mainland thus having dust carrying capacity from local sources as well.

Precipitation samples collected from three different stations on the same day have all displayed the characteristic color of red rain. Electrical conductivity of the St.1, St.2 and St.3 samples (1150, 812, 1456 $\mu S\ cm^{-1}$) were the highest values among the data set and the pH values were measured as 7.95, 8.03, 8.05, respectively.

Trajectory classification of 28 dust episodes showed that 17 of these events were quite outstanding and they have transported material from distant sources, mainly from the North Africa while the remaining 11 cases originated from the local sources, from the Anatolia and the Cilician Basin (Özsoy et al., 2008). The fraction of the episodes originated from Arabian Peninsula and the Middle east was relatively small. Trajectories of 17 events, plotted for 500 hPa pressure level and the trajectories of 11 events plotted for 900 hPa level are presented in Figure 5 and Figure 6, respectively.

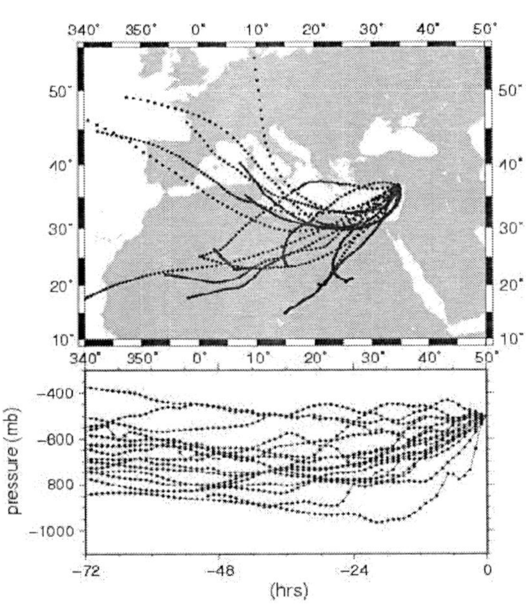

Figure 5. Trajectories of air masses for 17 episodes, arriving 500 hPa pressure level at Mersin, having red dust carrying capacity from remote sources, North Africa.

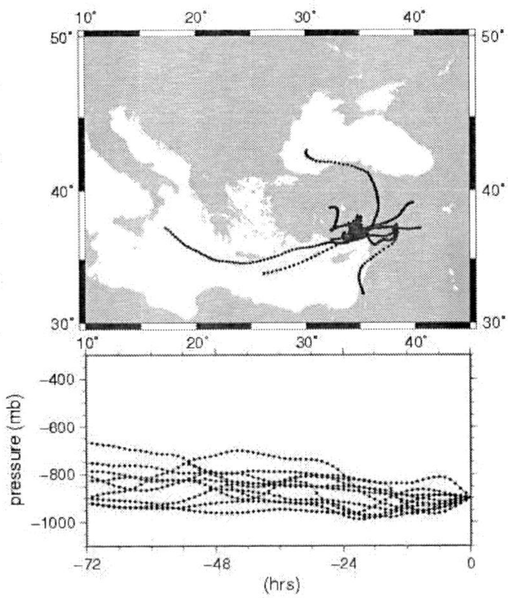

Figure 6. Trajectories of air masses for 11 episodes, arriving 900 hPa pressure level at Mersin, having dust carrying capacity from local sources, Anatolian mainland and the Cilician Basin.

3.4. pH, EC and Al Concentrations of Mersin Precipitation

Calcite ($CaCO_3$) content of mineral dust usually buffer the acidity of rain water and generally very high pH values are measured in red rain samples. Mineralogical analysis of aerosol samples collected from Erdemli had revealed that the calcite content varies from 8 to 31 % (Kubilay et al., 1997) and this range had previously been reported as 5 - 30 % with some gypsum in aeolian Saharan dust collected from Corsica, western Mediterranean (Loye-Pilot et al., 1986). In the present study, the highest pH and EC values have been measured in red rain samples. In order to understand the differences in between red and normal rain samples the VWM H^+ and Al concentrations and the geometric mean of EC values associated with normal versus red rain events are presented in Table 2.

Table 2. Comparison of VWM and Standard deviation of H^+, Al_s, Al_p and Al_t ($\mu g\ L^{-1}$) concentrations in normal and red rain events. Standard deviations of the VWM values were calculated using the formula given by Galloway et al. (1984). Rainfall is in mm unit, EC is in $\mu S\ cm^{-1}$ unit and geometric mean of the data is presented in the table. Factor refers to Red rain/Normal rain ratio and n refers to number of the samples

Parameter	Red rain (n:96)	Normal rain (n: 150)	Factor
Rainfall	635	2034	
pH	7.21	6.10	
EC	121.4 (2.6)	26.2 (2.8)	4.6
H^+	0.0615 ± 0.0171	0.792 ± 0.222	0.078
Al_s	12.2 ± 2.4	4.9 ± 1.2	2.5
Al_p	1546 ± 198	185.0 ± 20.3	8.4
Al_t	1555 ± 199	189.9 ± 20.6	8.2
Al solubility	% 0.785	% 2.58	0.3

The geometric mean EC and the VWM Al_t concentration of red rain were found to be more than four and eight times higher than normal rain, respectively; while the dissolution of calcite in dust was found to result in more than an order of magnitude decrease in the VWM H^+ concentration of red rain, by neutralizing (buffering) the acidity of rainwater. Since Al is mainly a crustal element its solubility in aerosol samples is quite low. However, a large variation of Al solubility appears in the literature and this variability is usually attributed to the pH of the final rainwater solution as well as ligands, ionic strength, amount of precipitation and the type of aerosols scavenged from the atmosphere during wash-out processes (Losno et al., 1993; Spokes et al., 1994). For the whole data set (n: 246) the mean solubility of Al in Mersin precipitation is calculated as 1.34 ± 0.26 %. This value is significantly lower than the values reported for Ankara (14.0 %, Kaya and Tuncel, 1997); Cap Ferrat (17.0 %, Chester et al., 1997) and Mexico City (42.56 %, Baez et al., 2007), while it is higher than the value reported for a suburban site of Istanbul (0.24 %, Başak and Alagha, 2004). Additionally, a quite big difference is observed between the mean solubilities of Al calculated for normal and red rain data sets, most likely due to pH effect. Spokes and Jickells (1996) have investigated the solubility of aerosol Al in different pH conditions and they observed high solubilities in low pH values: 0.9 ± 0.4 % at pH: 5.5 and 7.1 ± 0.6 % at pH: 2.0. The mean solubility of Al in normal rain samples is found to be more than three times higher than the value calculated for red rain. Similar differences are also valid for the mean major ion concentrations which have been previously measured in the same samples (Özsoy et al., 2008). Approximately 5 times of enhancement in Mg^{2+}; 4 times of enhancements in Cl^-, Na^+, K^+ and Ca^{2+} and 2 times of enhancements in NO_3^-, SO_4^{2-}, NH_4^+ and HCHO concentrations have been observed in red rain compared to normal rain samples.

3.5. Wet Deposition Flux of Aluminium and Mineral Dust

Wet deposition flux of an element can be calculated by using the following equation:

$$F_w = C_w \, . \, P$$

where F_w is the wet deposition flux (mg m^{-2}), C_w is the element concentration in precipitation and P is the amount of precipitation. In the literature usually, the volume weighted mean element concentrations are multiplied by the annual mean precipitation. However, this approach would cause greater uncertainties in wet deposition flux estimates. Because every measured concentration deviates from the mean concentration either in negative or positive direction and the amount of precipitation is not the same for every events. Due to non-linear variability of these two parameters, up to 10^2 order of magnitude difference has been observed between the fluxes calculated by two different methods in a study performed in the North Sea (Baeyens et al., 1990). Therefore, in order to obtain a more realistic elemental flux, the amounts of precipitation during each event were converted into daily quantities and the annual fluxes were calculated as the sum of the daily fluxes measured for the year 2004 as given by the equation

$$F_w = \sum_{i=1}^{N} C_i \, . \, P_i \; (i = 1 \ldots N)$$

where, C_i is the element concentration in rain event (i), P_i is the amount of precipitation of the corresponding event and N is the number of samples. The total annual wet deposition flux of Al were 153.8, 131.2, 224.3 and 201.0 mg m^{-2} respectively for stations 1, 2, 3 and 4. The mean annual flux for Mersin was calculated as 174.5 mg m^{-2} (4.0 mg m^{-2} was in Al_s and the rest, 170.5 mg m^{-2} was in Al_p form) and this value is compared to the other values given in the literature in Table 3.

Table 3. Comparison of annual wet deposition flux of Al_t (mg m^{-2} yr^{-1}) with the data reported from various locations around the world

Location	Al_t	Reference
Mersin	174.5	This study
Erdemli	870	Özsoy and Saydam (2000)
İskenderun	8.8	Örnektekin and Çakmaklı (2003)
Antalya	92.0	Al-Momani et al. (1998)
Ankara	3.4	Kaya and Tuncel (1997)
Singapur	47.8	Hu et al. (2003)
Massachusetts	29.4	Golomb et al. (1996)
Hiroshima	0.09	Takeda et al. (2000)

Wet deposition flux of Al calculated in this study is almost five times lower than the previous Erdemli value, but is still considerably higher than the rest of the values presented in

Table 3. The big difference between Al fluxes of Mersin and Erdemli stations which are very close to each other can be attributed to high interannual variation of dust load in the eastern Mediterranean atmosphere. Aluminium is the most abundant metallic element in the Earth's crust comprising about 8.2 % of its mass at the surface (Rudnick and Gao, 2003) and it is often used as an indicator of mineral dust load either in aerosols or the particulate fraction of precipitation. In order to estimate total dust flux through wet deposition to the region, Al_t content of rainwater samples were used and annual wet depositional dust flux is calculated as 2.13 tons km^{-2} in Mersin for the year 2004 which is five times lower than the estimation (10.6 tons km^{-2}) based on previously reported Al flux at Erdemli (Özsoy and Saydam, 2000). This result shows that interannual variation in wet depositional dust flux over the eastern Mediterranean is quite high as it has already been observed in dry deposition. A recent study conducted on dry samples by Koçak et al. (2005) represents a good example to this situation. Average dry deposition fluxes of Al for the Northern (Erdemli) and the Southern (Tel Shikmona) Levantine Basin, during a four year sampling period were reported as 320 and 520 mg m^{-2} yr^{-1}, respectively (Koçak et al., 2005). These values correspond to the annual dry deposition dust flux of 3.9 tons km^{-2} for the Northern Levantine and 6.3 tons km^{-2} for the Southern Levantine Basin which both of them are lower than the previous estimation (7.3 tons km^{-2}) made by Kubilay et al. (2000). Similarly, a very high variability in atmospheric dust deposition has been reported for the western Mediterranean Basin, e.g., NE Spain, the average annual dust deposition for the 11 year period was given as 5.3 tons km^{-2}, with a very high variability (S.D.= 8.7 tons km^{-2}) (Avila et al., 1996).

CONCLUSION

The current study has presented general properties of red rain through the analysis of precipitation samples collected at Mersin, on the eastern Mediterranean coast of Turkey, during the period of December 2003 – May 2005, with respect to their red dust contents, and in relation to backward air mass trajectories corresponding to the rainy days to determine potential source regions. Approximately, one fourth of the precipitation samples was classified as red rain, annually and in addition to remote sources in North Africa and the Middle East, local sources such as Anatolian mainland and the Cilician Basin have been determined as potential source regions of red dust. Red rain samples have been found to be enriched with respect to almost all analyzed parameters except H^+, e.g., the geometric mean EC and the VWM Al_t concentration of red rain were found to be more than four and eight times higher than normal rain, respectively; while the dissolution of calcite in dust was found to result in more than an order of magnitude decrease in the VWM H^+ concentration of red rain, by neutralizing the acidity of rainwater. The mean bicarbonate concentration of red rain, estimated from the theoretical relation between pH and HCO_3^-, was 16 times higher than that of normal rain. Similarly, up to 4 or 5 times of enhancements in Cl^-, Na^+, K^+, Ca^{2+} and Mg^{2+} and 2 times of enhancements in NO_3^-, SO_4^{2-}, NH_4^+ and HCHO concentrations have been observed in red rain compared to normal rain samples. Comparison the two wet deposition dust fluxes; (*i*) 2.13 tons km^{-2} at Mersin obtained from this study and (*ii*) 10.6 tons km^{-2}, estimated from previosly reported Al flux for Erdemli suggests that interannual variation in wet deposition dust flux over the eastern Mediterranean is quite high as it has already been observed in dry deposition. In order to have an idea about the total dust flux to the region, if it is assumed that the wet deposition percentage relative to the total (wet+dry) deposition is 44 % (Kubilay et al., 2000) then 2.13 tons km^{-2} of wet deposition would supply a total deposition dust flux of 4.84 tons km^{-2} which is far from the previous estimation of 13 tons km^{-2} made by Kubilay et al. (2000) for the year 1992. In fact, the most reliable estimation of annual dust flux can only be made by simultaneous measurements of Al concentrations in both precipitation and aerosol samples collected from the same location. However, no such study has ever been conducted so far. Due to dynamic meteorological components, seasonal and interannual variability in atmospheric transportation events is too high and dust loading in the Eastern Mediterranean atmosphere is highly variable both in space and time. Therefore, any intermittent monitoring programme not based on daily (or event) sampling protocol would lead underestimations in dust flux calculations as well as any results obtained from continuous and simultaneous measurements of Al in wet and dry atmospheric samples would be specific to that of sampling period.

ACKNOWLEDGEMENT

This study was financially supported by the Turkish Scientific and Technical Research Council, TÜBİTAK, through the project 102Y144 and by the Scientific Research Projects Fund of Mersin University, through the Project BAP-FEF-KB 2004-3. We thank Dr. Abdullah Öksüz, Dr. Kemal Sangün and the technical personnel of the Central Research Laboratory of Mustafa Kemal University; Mehmet Bayrakçıoğlu, Hidayet Duman, Atilla Çekiç and Seher Mısırlıoğlu, for their help and valuable support concerning the ICP-AES analyses. Prof. Emin Özsoy (IMS-METU) is acknowledged for his valuable contribution for providing access to the ECMWF trajectory data.

REFERENCES

Al-Momani, İ. F., Aygün, S. and Tuncel, G. (1998). Wet deposition of major ions and trace elements in the Eastern Mediterranean Basin. *Journal of Geophysical Research*, 103, 8287-8299.

Alpert, P., Neeman, B. U. and Shay-El, Y. (1990a). Climatological analysis of Mediterranean cyclones using ECMWF data. *Tellus*, 42A, 65-77.

Alpert, P., Neeman, B. U. and Shay-El, Y. (1990b). Intermonthly variability of cyclone tracks in the Mediterranean. *Journal of Climate*, 3 (12), 1474 - 1478.

Alpert, P., Ben-Gai, T., Baharad, A., Benjamini, Y., Yekutieli, D., Colacino, M., diyotato, L., Ramis, C., Homar, V., Romero, R., Michaelides, S. and Manes, A. (2002). The paradoxical increase of Mediterranean extreme daily rainfall in spite of a decrease in total values. *Geophysical Research Letters*, 29(11), art. # 1536, doi:10.1029/2001GL013554.

Avila, A. (1996). Time trends in the precipitation chemistry at a mountain site in Northeastern Spain for the period 1983-1994. *Atmospheric Environment*, 30 (9), 1363-73.

Avila, A., Queralt, I., Gallart, F. and Martin-Vide, J. (1996). African dust over Northeastern Spain: mineralogy and source regions. In S. Guerzoni and R. Chester (Eds.), *The Impact of Desert Dust Across the Mediterranean*: (pp. 201-205). Dordrecht, Kluwer Academic Publishers.

Avila, A., Alarcon, M. and Queralt, M. I. (1998). The chemical composition of dust transported in red rains-Its contribution to the biogeochemical cycle of a Holm Oak Forest in Catalonia (Spain). *Atmospheric Environment*, 32 (2), 179-191.

Avila, A. and Penuelas, J. (1999). Increasing frequency of Saharan rains over northeastern Spain and its ecological consequences. *Science of the Total Environment*, 228, 153-156.

Baeyens, W., Dehairs, F. and Dedeurwaerder, H. (1990). Wet and dry deposition fluxes above the North Sea. *Atmospheric Environment*, 24A (7), 1693-1703.

Baez, A., Belmont, R., Garcia, R., Padilla, H. and Torres, M. C. (2007). Chemical composition of rainwater collected at a southwest site of Mexico City, Mexico. *Atmospheric Research*, 86 (1), 61 -75, doi:10.1016/j.atmosres.2007.03.005.

Bardouki, H., Liakakou, H., Economou, C., Sciare, J., Smolik, J., Zdimal, V., Eleftheriadis, K., Lazaridis, M., Dye, C. and Mihalopoulos, N. (2003). Chemical composition of size resolved atmospheric aerosols in the eastern Mediterranean during summer and winter. *Atmospheric Environment*, 37 (2), 195-208.

Barry, R. G. and Chorley, R. J. (1998). *Atmosphere, weather and climate*. London/New York, Routeledge Pres.

Başak, B. and Alagha, O. (2004). The chemical composition of rainwater over Büyükçekmece Lake, Istanbul. *Atmospheric Research*, 71, 275-288.

Bergametti, G., Dutot, A. L., Buat-Menard, P., Losno, R. and Remoudaki, E. (1989). Seasonal variability of the elemental composition of atmospheric aerosol particles over the North-western Mediterranean. *Tellus,* 41B, 353-361.

Bethoux, J. B., Morin, P., Chaumery, C., Connan, O., Gentili, B. and Ruiz-Pino, D. (1998). Nutrients in the Mediterranean Sea, mass balance and ststistical analysis of concentrations with respect to environmental change. *Marine Chemistry*, 63, 155-169.

Brody, L. R. and Nestor, M. J. R. (1980). Handbook for Forecasters in the Mediterranean, Part 2. Regional Forecasting Aids for the Mediterranean Basin. *Naval Environmental Prediction Research Facility,* Tech. Rep. TR 80-10, Monterey, California, VII-1; VII-13.

Bruland, K. W., Donat, J. R. and Hutchins, D. A. (1991). Interactive influences of bioactive trace metals on biological production in oceanic waters. *Limnology and Oceanography*, 36 (8), 1555-1577.

Bucher, A. (1989). Fallout of Saharan Dust in the NW Mediterranean Region. In M. Liene and M. Sarnthein (Eds.), *Paleoclimatology and Paleometeorology* (pp. 565-584). Dordrecht, Kluwer.

Buffoni, L., Maugeri, M., Nanni, T., 1999. Precipitation in Italy from 1833 to 1996. *Theoretical and Applied Climatology* 63 (1-2), 33-40.

Chester, R., Nimmo, M., Alarcon, M., Saydam, C., Murphy, K. J. T., Sanders, G. S. and Corcoran, P. (1993). Defining the chemical character of aerosols from the atmosphere of the Mediterranean Sea and surrounding regions. *Oceanologica Acta*, 16, 231-246.

Chester, R., Nimmo, M. and Corcoran, P. A. (1997). Rain water-aerosol trace metal relationships at Cap-Ferrat: A coastal site in the Western Mediterranean. *Marine Chemistry*, 58, 293-312.

Çağatay, N., Yiğitbaş, E. and Güney, A. (2002). Sources and distribution of chromate mineral concentrations in alluvial and coastal sands in the region of Mersin and Tarsus. *Report of the Faculty of Mine,* Technical University of Istanbul, pp. 33 (in Turkish).

ÇDR (Çevre Durum Raporu), 2006. 2005 Yılı Mersin İl Çevre Durum Raporu. T.C. Çevre ve Orman Bakanlığı, Mersin Valiliği İl Çevre ve Orman Müdürlüğü, pp 263.

Davies, T. D., Tranter, M., Jickells, T. D., Abrahams, P. W., Landsberger, S., Jarvis, K. and Pierce, C. E. (1992). Heavily-contaminated snowfalls in the remote Scottish highlands: A consequence of regional scale mixing and transport. *Atmospheric Environment*, 26, 95-112.

Dayan, U. and Miller, J. (1989). Meteorological and climatological data from surface and upper air measurements for the assessment of atmospheric transport and deposition of pollutants in the Mediterranean Basin. *MAP Tech. Rap*. Ser. 30, 130.

Donaghay, P. L., Liss, P. S., Duce, R. A., Kester, D. R., Hanson, A. K., Villareal, T., Tindale, N. W. and Gifford, D. J. (1991). The role of episodic atmospheric nutrient inputs in the chemical and biological dynamics of oceanic ecosystems. *Oceanography*, 4 (2), 62-70.

Dulac, F., Buat-Menard, P., Arnold, M. and Ezat, U. (1987). Atmospheric input of trace metals to the western Mediterranean Sea: 1. Factors controlling the variability of atmospheric concentrations. *Journal of Geophysical Research*, 92 (D7), 8437-8453.

Dulac, F., Tanre, D., Bergametti, G., Buat-Menard, P., Desbois, M. and Sutton, D. (1992). Assessment of the African airborne dust mass over the Western Mediterranean Sea using meteosat data. *Journal of Geophysical Research*, 97 (D2), 2489-2506.

Franzen, L. G., Hjelmroos, M., Kallberg, P., Brorstrom-Lunden, E., Juntto, S. and Savolainen, A. L. (1994). The "Yellow-Snow" episode of Northern Fennoscandia, March 1991-A case study of long-distance transport of soil, polen and stable organic compounds. *Atmospheric Environment*, 28, 3587-3604.

Galloway, J. N., Likens, G. E. and Hawley, M. E. (1984). Acid precipitation: Natural versus anthropogenic components. *Science*, 226, 829-831.

Ganor, E. and Mamane, Y. (1982). Transport of Saharan dust across the Eastern Mediterranean. *Atmospheric Environment*, 16, 3, 581-587.

Ganor, E., Foner, H. A., Brenner, S., Neman, E. and Lavi, N. (1991). The chemical composition of aerosols settling in Israel following dust storms. *Atmospheric Environment,* 25A, 2665-2670.

Ganor, E. (1994). The frequency of Saharan dust episodes over Tel Aviv, Israel. *Atmospheric Environment*, 28, 2867-2871.

Gerasopoulos, E., Kouvarakis, G., Babasakalis, P., Vrekoussis, M., Putaud, J. P. and Mihalopoulos, N. (2006). Origin and variability of particulate matter (PM_{10}) mass concentrations over the Eastern Mediterranean. *Atmospheric Environment*, 40, 4679-4690.

Golomb, D., Ryan, D., Eby, N., Underhill, J. and Zemba, S. (1996). Atmospheric deposition of toxics onto Massachusetts Bay-I. Metals. *Atmospheric Environment*, 31, 1349-1359.

Goosens, C. (1985). Principal component analysis of Mediterranean rainfall. *Journal of Climatology,* 5, 379-388.

Guerzoni, S., Chester, R., Dulac, F., Loye-Pilot, M. D., Migon, C., Molinaroli, E., Moulin, C., Rossini, P., Saydam, C., Soudine, A. and Ziveri, P. (1999). The role of atmospheric deposition in the biogeochemistry of the Mediterranean Sea. *Progress in Oceanography*, 44, 147-190.

Guieu, C., Martin, J. M., Thomas, A. J. and Elbaz-Poulichet, F. (1991). Atmospheric versus river inputs of metals to the Gulf of Lions. Total concentrations, partitioning and fluxes. *Marine Pollution Bulletin*, 22, 176-183.

Güllü, G. H., Ölmez, İ., Aygün, S. and Tuncel, G. (1998). Atmospheric trace element concentrations over the eastern Mediterranean sea: Factors affecting temporal variability. *Journal of Geophysical Research*, 103, 21943-21954.

Güllü, G. H., Ölmez, İ. and Tuncel, G. (2004). Source apportionment of trace elements in the eastern Mediterranean atmosphere. *Journal of Radioanalytical and Nuclear Chemistry*, 259 (1), 163-171.

Güllü, G., Doğan, G. and Tuncel, G. (2005). Atmospheric trace element and major ion concentrations over the eastern Mediterranean Sea: Identifications of anthropogenic source regions. *Atmospheric Environment*, 39, 6376-6387.

Harding, A. E. (2006). *Changes in Mediterranean Climate Extremes: Patterns, Causes and Impacts of Change*. Ph D Thesis, University of East Anglia, Norwich, pp. 549.

Hassanien, M. A., Riewerts, J., Shakour, A. A. and Bitto, A. (2001). Seasonal and annual variations in air concentrations of Pb, Cd and PAHs in Cairo, Egypt. *International Journal of Environmental Health Research*, 11, 13-27.

Herut, B., Krom, M. D., Pan, G. and Mortimer, R. (1999). Atmospheric input of nitrogen and phosphorus to the southeast Mediterranean: Sources, fluxes and possible impact. *Limnology and Oceanography*, 44, 1683-1692.

Herut, B. (2001). Levels of aerosol trace metals along the Mediterranean coast of Israel (1996-97). *Fresenius Environmental Bulletin*, 10, 89-93.

Herut, B., Nimmo, M., Medway, A., Chester, R. and Krom, M. (2001). Dry deposition of trace metals at the Mediterranean coast of Israel (SE Mediterranean): *Sources and fluxes. Atmospheric Environment*, 35, 803-813.

Hu, G. P., Balasubramanian, R. and Wu, C. D. (2003). Chemical characterization of rainwater at Singapore. *Chemosphere*, 51, 747-55.

Kapur, S., Saydam, C., Akça, E., Çavuşgil, V.S., Karaman, C., Atalay, I. and Özsoy, T. (2000). Carbonate Pools in Soils of the Mediterranean: A Case Study from Anatolia. In R. Lal, J. M. Kimble, H. Eswaran, and B. A. Stewart (Eds.), *Global Climate Change and Pedogenic Carbonates* (pp. 187-212). Florida: Lewis Publishers.

Kaya, G. and Tuncel, G. (1997). Trace element and major ion composition of wet and dry deposition in Ankara, Turkey. *Atmospheric Environment*, 31, 3985-98.

Koçak, M., Kubilay, N. and Mihalopoulos, N. (2004a). Ionic composition of lower tropospheric aerosols at a northeastern Mediterranean site: Implications regarding sources and long-range transport. *Atmospheric Environment*, 38, 2067-2077.

Koçak, M., Nimmo, M., Kubilay, N. and Herut, B. (2004b). Spatio-temporal aerosol trace metal concentrations and sources in the Levantine Basin of the Eastern Mediterranean. *Atmospheric Environment*, 38, 2133-2144.

Koçak, M., Kubilay, N., Herut, B. and Nimmo, M. (2005). Dry atmospheric fluxes of trace metals (Al, Fe, Mn, Pb, Cd, Zn, Cu) over the Levantine Basin: A refined assessment. *Atmospheric Environment*, 39, 7330-7341.

Koçak, M., Mihalopoulos, N. and Kubilay, N. (2007). Contributions of natural sources to high PM10 and PM2.5 events in the Eastern Mediterranean. *Atmospheric Environment*, 41, 3806-3818.

Kremling, K. and Streu, P. (1993). Saharan dust influenced trace element fluxes in deep North Atlantic subtropical waters. *Deep Sea Research*, 40, 1155-1168.

Krom, M. D., Kress, N., Brenner, S. and Gordon, L. I. (1991). Phosphorus limitation of primary productivity in the eastern Mediterranean Sea. *Limnology and Oceanography*, 36, 424-432.

Krom, M. D., Brenner, S., Kress, N., Neori, A. and Gordon, L. I. (1992). Nutrient dynamics and new production in a warm-core eddy from the eastern Mediterranean Sea. D*eep Sea Research*, 39, 467-480.

Krom, M. D., Herut, B. and Mantoura, R. F. C. (2004). Nutrient budget for the Eastern Mediterranean: Implications for P limitation. *Limnology and Oceanography*, 49, 1582-1592.

Kubilay, N. and Saydam, A. C. (1995). Trace elements in atmospheric particulates over the Eastern Mediterranean: Concentrations, sources and temporal variability. *Atmospheric Environment,* 29, 2289-2300.

Kubilay, N., Yemenicioğlu, S. and Saydam, A. C. (1995). Airborne material collections and their chemical composition over the Black Sea. *Marine Pollution Bulletin*, 30, 475-483.

Kubilay, N., Saydam, A. C., Yemenicioğlu, S., Kelling, G., Kapur, S., Karaman, C. and Akça, E. (1997). Seasonal, chemical and mineralogical variability of atmospheric particles in the coastal region of the Northeast Mediterranean. *Catena,* 28, 313-328.

Kubilay, N., Nickovic, S., Moulin, C. and Dulac, F. (2000). An illustration of the transport and deposition of mineral dust onto the Eastern Mediterranean. *Atmospheric Environment* 34 (8), 1293-1303.

Kubilay, N., Koçak, M., Çokaçar, T., Oğuz, T., Kouvarakis, G. and Mihalopoulos, N. (2002). Influence of Black Sea and local biogenic activity on the seasonal variation of aerosol sulfur species in the eastern Mediterranean atmosphere. *Global Biogeochemical Cycles*, 16 (4), 1079, doi:10.1029/2002GB001880.

Kubilay, N., Çokaçar, T. and Oğuz, T. (2003). Optical properties of mineral dust outbreaks over the northeastern Mediterranean. *Journal of Geophysical Research*, 108 (D21), 4666, doi:10.1029/2003JD003798.

Kutiel, H., Hırsch-Eshkol, T. R. and Türkeş, M. (2001). Sea level pressure patterns associated with dry or wet monthly rainfall conditions in Turkey. *Theoretical and Applied Climatology,* 69, 39-67.

Lagouvardos, K., Katroni, V., Dobricic, S., Nickovic, S. and Kallos, G. (1996). The storm of October 21-22, 1994, over Greece: Observations and model results. *Journal of Geophysical Research – Atmospheres* 101 (D21), 26217-26226.

Lammel, G., Rohrl, A. and Schreiber, H. (2002). Atmospheric lead and Bromine in Germany-post abatement levels, variabilities and trends. *Environmental Science and Pollution Research,* 9, 397-404.

Lazaridis, M., Dzumbova, L., Kopanakis, I., Ondracek, J., Glytsos, T., Aleksandropoulou, V., Voulgarakis, A., Katsivela, E., Mihalopoulos, N. and Eleftheriadis, K. (2008). PM_{10} and $PM_{2.5}$ levels in the Eastern Mediterranean (Akrotiri Research Station, Crete, Greece). *Water, Air and Soil Pollution*, 189, 85-101.

Losno, R., Colin, J. L., Le Bris, N., Bergametti, G., Jickells, T. and Lim, B. (1993). Aluminium solubility in rainwater and molten snow. *Journal of Atmospheric Chemistry*, 17, 29-43.

Loye-Pilot M. D., Martin, J. M. and Morelli, K. (1986). Saharan dust: Influence on the rain acidity and significance for atmospheric Input to the Mediterranean. *Nature*, 321, 427-428.

Maheras, P., Xoplaki, E. and Kutiel, H. (1999). Wet and dry monthly anomalies across the Mediterranean basin and their relationship with circulation, 1860 – 1990. *Theoretical and Applied Climatology,* 64 (3-4), 189-199.

Markaki, Z., Oikonomou, K., Koçak, M., Kouvarakis, G., Chaniotaki, A., Kubilay, N. and Mihalopoulos, N. (2003). Atmospheric deposition of inorganic phosphorus in the Levantine Basin, eastern Mediterranean: Spatial and temporal variability and its role in seawater productivity. *Limnology and Oceanography*, 48 (4), 1557-1568.

Martin, D., Mithieux, C. and Strauss, B. (1987). On the use of the synoptic vertical wind component in a transport trajectory model. *Atmospheric Environment*, 21, 45-52.

Martin, J. M., Elbaz-Poulichet, F., Guieu, C, Loye-Pilot, M. D. and Han, G. (1989). River versus atmospheric input of material to the Mediterranean Sea: An overview. *Marine Chemistry,* 28, 159-182.

Martin, D., Bergametti, G. and Strauss, B. (1990). On the use of the synoptic vertical velocity in trajectory model: Validation by geochemical tracers. *Atmospheric Environment*, 24A, 8, 2059-2069.

McCabe, G. J., Clark, M. P. and Serreze, M. C. (2001). Trends in Northern Hemisphere surface cyclone frequency and intensity. *Journal of Climate*, 14 (12), 2763-2768.

Migon, C. and Nicolas, E. (1998). Effects of antipollution policy on anthropogenic lead transfers in the Ligurian Sea. *Marine Pollution Bulletin*, 36, 775-779.

Milliman, J. D., Jeftic, L. and Sestini, G. (1992). The Mediterranean Sea and Climate Change-An Overview. In L. Jeftic, J. D. Milliman and G. Sestini (Eds.), *Climatic Change and the Mediterranean*: (pp. 1-14), UNEP.

National Diagnostic Analysis (NDA) Turkey, 2003. UNEP/MAP, pp. 67.

Örnektekin, S. and Çakmaklı, S. (2003). Chemical composition and acidity o rain at the Gulf of Iskenderun, North-East Mediterranean. *Water, Air and Soil Pollution: Focus*, 3, 151-166.

Özsoy, E. (1981). On the Atmospheric Factors Affecting the Levantine Sea. ECMWF, European Centre for Medium Range Weather Forecasts, *Technical Report* No. 25, 29.

Özsoy, E. and Sözer, A. (2006). Forecasting circulation in the Cilician Basin of the Levantine Sea. *Ocean Science Discussions*, 3, 1481-1514.

Özsoy, T. and Saydam, A. C. (2000). Acidic and alkaline precipitation in the Cilician Basin, North-Eastern Mediterranean Sea. *Science of the Total Environment*, 253, 93-109.

Özsoy, T. and Saydam, A. C. (2001). Iron speciation in precipitation in the North-Eastern Mediterranean and its relationship with Sahara dust. *Journal of Atmospheric Chemistry*, 40, 41-76.

Özsoy, T. (2003). Atmospheric wet deposition of soluble macro-nutrients in the Cilician Basin, north-eastern Mediterranean Sea. *Journal of Environmental Monitoring*, 5, 971-976.

Özsoy, T., Türker, P. and Örnektekin, S. (2008). Precipitation chemistry as an indicator of urban air quality in Mersin, North-Eastern Mediterranean Region. *Water, Air and Soil Pollution,* 189, 69-83.

Partal, T. and Kahya, E. (2006). Trend analysis in Turkish precipitation data. *Hydrological Processes,* 20, 2011-2026.

Piervitali, E., Colacino, M. and Conte, M. (1997). Signals of climatic change in the central-western Mediterranean basin. *Theoretical and Applied Climatology*, 58 (3-4), 211-219.

Prospero, J. M., Nees, R. T. and Uematsu, M. (1987). Deposition rate of particulate and dissolved aluminum derived from Saharan dust in precipitation at Miami, Florida. *Journal of Geophysical Research*, 92, 14723-14731.

Psarra, S., Tselepides, A. and Ignatiades, L. (2000). Primary productivity in the oligotrophic Cretan Sea (NE Mediterranean): Seasonal and interannual variability. *Progress in Oceanography,* 46, 187-204.

Quetel, C. R., Remoudaki, E., Davies, J. E., Miquel, J. C., Fowler, S. W., Lambert, C. E., Bergametti, G. and Menard, P. B. (1993). Impact of atmospheric deposition on particulate iron flux and distribution in Northwestern Mediterranean waters. *Deep Sea Research*, 40, 898-1002.

Reiter, E. R. (1979). Handbook for Forecasters in the Mediterranean, Weather Phenomena of the Mediterranean Basin, *Part 1: General Description of the Meteorological Processes*, Tech. Pap. 5-75, pp. 344 pp, Environmental Prediction Research Facility, Naval Postgraduate School, Monterey, California.

Reiff, J., Forbes, G. S., Spieksma, F. TH. M. and Reynders, J. J. (1986). African dust reaching Northwestern Europe: A case study to verify trajectory calculations. *Journal of Climatology and Applied Meteorology*, 25, 1543-1567.

Romero, R., Guijarro, J. A., Ramis, C. and Alonso, S. (1998). A 30-year (1964 – 1993) daily rainfall data bas efor the Spanish Mediterranean regions: First exploratory study. *International Journal of Climatology*, 18 (5), 541-560.

Rudnick, R. L. and Gao, S. (2003). The Composition of the Continental Crust. In H. D. Holland and K. K. Turekian (Eds.), *Treatise on Geochemistry*, (Vol 3, pp. 1-64).

Serreze, M. C., Carse, F., Barry, R. G. and Rogers, J. C. (1997). Icelandic low cyclone activity: Climatological features, linkages with the NAG and relationships with recent changes in the Northern Hemisphere circulation. *Journal of Climate*, 10 (3), 453-464.

Shaka, H. and Saliba, N. A. (2004). Concentration measurementsand chemical composition of $PM_{10-2.5}$ and $PM_{2.5}$ at a coastal site in Beirut, Lebanon. *Atmospheric Environment*, 38, 523-531.

Spokes, L. J., Jickells, T. D. and Lim, B. (1994). Solubilisation of aerosol trace metals by cloud processing: A laboratory study. Geochim. *Cosmochim. Acta,* 58 (15), 3281-3287.

Spokes, L. J. and Jickells, T. D. (1996). Factors controlling the solubility of aerosol trace metals in the atmosphere and on mixing into seawater. *Aquatic Geochemistry*, 1, 355-374.

Swap, R., Garstang, M., Greco, S., Talbot, R. and Kallberg, P. (1992). Saharan dust in the Amazon Basin. *Tellus,* 44B, 133-149.

Takeda, K., Marumoto, K., Minamikawa, T., Sakugawa, H. and Fujiwara, K. (2000). Three-year determination of trace metals and the lead isotope ratio in rain and snow depositions collected in Higashi-Hiroshima, Japan. *Atmospheric Environment*, 34, 4525-4535.

Talbot, R. W., Haris, R. C., Browell, E. V., Gregory, G. L., Sebacher, D. I. and Beck, S. M. (1986). Distribution and geochemistry of aerosols in the tropical North Atlantic troposphere: Relationship to Saharan dust. *Journal of Geophysical Research*, 91, 5173-5182.

Tolun, N. and Pamir, A. N. (1975). Explanatory text of the geological map of Turkey, Hatay sheet. *Publ. Min. Res. And Explor. Inst*. Publ., Ankara, pp. 99 and appendices.

Trewartha, G. T. (1961). *The Earth's problem climates*. University of Wisconsin Press, Wisconsin.

Tsonis, A. A. (1996). Widespread increases in low-frequency variability of precipitation over the past century. *Nature,* 382 (6593), 700-702.

Türkeş, M. and Erlat, E. (2003). Precipitation changes and variability in Turkey linked to the North Atlantic Oscillation during the period 1930 – 2000. *International Journal of Climatology*, 23, 1771-1796.

Türkeş, M. and Erlat, E. (2005). Climatological responses of winter precipitation in Turkey to variability of the North Atlantic Oscillation during the period 1930 – 2001. *Theoretical and Applied Climatology*, 81, 45-69.

Türkeş, M., Koç, T. and Sariş, F. (2008). Spatiotemporal variability of precipitation total series over Turkey. *International Journal of Climatology*, 28, published online, doi: 10.1002/joc.1768.

UNEP (United Nations Environment Programme) (1995). Manual for the Geochemical Analyses of Marine Sediments and Suspended Particulate Matter, *Reference Methods for Marine Pollution Studies* No. 63.

Wallen, C. C. (1970). Introduction In C. C. Wallen (Ed.), Climates of Northern and Western Europe. *World Survey of Climatology*. Amsterdam/Oxford/New York, Elsewier Publishing,.

Whelpdale, D. M. and Moody, J. L. (1990). Large Scale Meteorological Regimes and Transport Processes. In A. H. Knap (Ed.), *The Long-Range Atmospheric Transport of Natural and Contaminant Substances,* (pp. 3-39), Netherlands, Kluwer Academic Publishers.

Xoplaki, E., Gonzales-Rouco, J. F., Luterbacher, J. and Wanner, H. (2004). Wet season Mediterranean precipitation variability: influence of large-scale dynamics and trends. *Climate Dynamics*, 23 (1), 63-78.

Yaalon, D. H. and Ganor, E. (1979). East Mediterranean trajectories of dust-carrying storms from the Sahara and Sinai. In C. Morales (Ed.), *Saharan Dust*, (Chapter 9, pp. 183-187), New York, John Wiley and Sons.

Yetiş, C., Kelling, G., Gökçen, S. L. and Baroz, F. (1995). A revised stratigraphic framework for later Cenozoic sequences in the Northeastern Mediterranean Region. *Geol. Rund*., 84, 794-812.

Yılmaz, A. and Tuğrul, S. (1998). The effect of cold- and warm-core eddies on the distribution and stoichiometry of dissolved nutrients in the NE Mediterranean. *Journal of Marine Systems*, 16 (3-4), 253-268 .

Zhang, X. B., Wang, X. L. L. and CorteReal, J. (1997). On the relationships between daily circulation patterns and precipitation in Portugal. *Journal of Geophysical Research-Atmospheres*, 102 (D12), 13495-13507.

In: Environmental and Regional Air Pollution
Editors: D. Gallo, R. Mancini
ISBN: 978-1-60692-893-6

Chapter 12

DEVELOPMENT OF FORECASTING MODELS FOR PREDICTING HOURLY OZONE CONCENTRATIONS AND EXCEEDANCES FOR DAYTON, OH, USING MM5 REAL-TIME FORECASTS

Ashok Kumar, Raga Kalapati and Akhil Kadiyala
Department of Civil Engineering, University of Toledo,
Toledo, OH 43606

ABSTRACT

The objective of this research was to develop and evaluate models for predicting hourly ozone concentrations, ozone exceedances and hourly air quality index (AQI) in Dayton, OH. As the hourly ozone concentrations are closely related to the meteorological conditions, three variables - temperature, wind speed, and dew point temperature were chosen for this study. The ozone data were extracted from the EPA' s Aerometric Information Retrieval System (AIRS) database for the period between 1996 and 2003. The meteorological data were taken from the National Climatic Data Center (NCDC) for the same period.

An analysis of variations in hourly ozone concentrations and ozone episode occurrences was carried out for the period Apr.-Oct. for the years 1996-1999. Also, analysis of the long-term trends in annual means of ozone concentrations, temperature, wind speed, and dew point temperature was performed using the same data set. Based on this analysis, the ozone data was divided into pre-summer (Apr.-Jul.) and post-summer (Aug.-Oct.) seasons, to account for seasonal variations, and each season was further divided into three regimes, namely, stable period (hours: 1-8), ascent period (hours: 9-16), and descent period (hours: 17-24). The Kolmogorov – Zurbenko (KZ) filter technique was used to reduce the scatter in the time series, and models were developed for the three regimes for each season by regression, using the corresponding independent parameter values.

A total of twelve models were developed to predict ozone concentrations for pre-summer and post-summer periods. Six models considered temperature, wind speed, and dew point temperature as the independent variables (three-parameter models), and the other six considered temperature, and wind speed as variables (two-parameter models).

Also, three models each for pre-summer and post-summer season were developed for predicting the ozone exceedances. The performance of the models was evaluated using statistical indicators in three ways:

a) Initial evaluation (or validation) of the models was conducted using 2002 data.
b) The effectiveness of these models was further evaluated using available MM5 (a mesoscale meteorological forecasting model) real-time forecasts from the Ohio State University for the months of Aug.-Oct., 2003.
c) Finally, the performance of the three-parameter models was compared with that of the two-parameter models.

The study shows that the forecasts of hourly ozone concentrations made by the models based on KZ filters are reliable only to a limited extent. However, the models performed well in predicting AQI values reported by the EPA. Also, the three-parameter models performed better in predicting the peak concentrations when compared to the two-parameter models.

1. Introduction

Ground-level ozone is considered to be one of the most important noxious air pollutants in view of its concentrations and its effects on human health. The U.S Environmental Protection Agency (EPA) has identified ozone as one of the six common air pollutants that cause adverse effects to public health and environment. Ozone is linked to a wide range of health impacts, including respiratory distress, resulting in increased emergency hospital visits, hospital admissions, and even premature deaths. Elevated levels above the national standard may cause lung and respiratory disorders. Recent scientific evidence indicates that there is no lower threshold for the health effects of ground-level ozone. Due to its dependence on weather conditions, ozone is typically a summer-time pollutant and a chief component of summer-time smog.

High concentrations of ozone occur during the summer months i.e., from May through September (often termed as "ozone season") in the northern hemisphere. These high concentration episodes are now proclaimed as "Ozone Action Days" (OADs) by many urban areas. The photochemical processes leading to ozone formation in the atmosphere are complex and nonlinear. Ozone is a secondary component, and is created by chemical reactions between its precursors which include nitrogen oxides (NO_x) and volatile organic compounds (VOCs). Emissions from industrial facilities and electric utilities, motor vehicle exhaust, gasoline vapors, and chemical solvents are some of the major sources of NO_x and VOCs. In the presence of sunlight (ultra-violet radiation) and suitable meteorological conditions, the precursors react photo-chemically to produce ozone.

The current National Ambient Air Quality Standards (NAAQS) for ozone promulgated by the EPA are given in Table 1. Based on the analysis of monitored ozone data, all the states have developed a list of counties to be designated as non-attainment under the current standards. Many states will have to implement some control strategies to bring these areas into compliance with the standards. Accurate O_3 forecasts can be useful tools for air pollution control officials and public health officials in preventing the detrimental health effects of this pollutant through emergency air pollution control actions and public health warnings (Ojha et al., 2002).

Table 1. Current National Ambient Air Quality Standards for Ozone (U.S EPA)

Standard Type	Standard
1-hour Average, Primary and Secondary	0.12 ppm (235 $\mu g/m^3$)
8-hour Average, Primary and Secondary	0.08 ppm (157 $\mu g/m^3$)

Research in the development of ozone forecasting models for urban areas around the world has been continuous over the last two decades. Efforts have been made to forecast ground-level ozone based on various meteorological parameters such as temperature, wind speed, relative humidity, background and previous ozone concentrations, and ozone precursor (volatile organic compounds and oxides of nitrogen) levels. Use of so many predictors for forecasting ozone necessitates the development of parallel forecasting models for each of them, ultimately making the process too complicated and unrealistic.

A literature review of ozone forecasting models indicated that most of the recent research studies used various methods or their combinations and different meteorological parameters for predicting ozone concentrations. Different methods of forecasting ozone concentrations include artificial neural networks (Salazar-Ruiz et al. (2008), Wang et al. (2003), Pastor-Barcenas et al. (2005), Ballester et al. (2002), Dutot et al. (2007), Gomez-Sanchis et al. (2006), Ruiz-Suarez (1995), Prybutok et al. (2000)), non-linear multiple regression (Aneiros-Perez et al. (2004), Bordignon et al. (2002)), time-series forecasting (Bauer (2001)), and fuzzy logic (Mintz et al. (2005)). Some other studies that have used a combination of these methods to predict ozone concentrations are Ryan (1995), Lin and Cobourn (2007), Heo and Kim (2004) while a comparison of forecasting results was provided by Schlink and Volta (2000), Jorquera et al. (1998), Chaloulakou et al. (2003), and Sharma et al. (2005). The primary objective of the researchers has been to develop a forecasting model to predict the maximum ozone concentration. There has been limited work in the area of developing models to predict the ozone concentration for the next hour.

The objective of this study was to develop simple ozone forecasting models to predict hourly maximum ozone concentration and air quality index for the next hour for Dayton, Ohio, using the Kolmogorov – Zurbenko (KZ) filtering technique and regression analysis. The KZ filter technique was used to reduce the scatter in ozone data, and regression analysis was used to develop the models following the method approached by Kumar et al. (2000). Some recent studies that have used KZ technique to adjust ozone trends are Wise and Comrie (2005), Lu and Chang (2005), Ibarra-Berastegi et al. (2001). The dataset for the years 1996-1999 was used for the development of the models. Temperature, wind speed, and dew point temperature were chosen as independent variables. The variations in ozone concentrations, ozone episode occurrences, and long-term trends in annual means of ozone concentrations, temperature, wind speed, and dew point temperature were analyzed for the period of 1996-1999. The ozone concentrations were correlated to hourly temperature, wind speed and dew point temperature. The models obtained were three-parameter models. The efficiency of these models was compared to the two-parameter models that were obtained using temperature and wind speed as independent variables. Separate models for accounting for peak ozone concentrations were developed.

The data for the year 2002 was used for validation of the models. The forecasting ability of the models was further analyzed using MM5 (a mesoscale meteorological forecasting model) real-time forecasts for temperature, wind speed, and dew point temperature for the year 2003. These evaluations using real-time forecasts are generally not reported by researchers for air quality predictions. All the evaluations and comparisons are based on statistical parameters, such as fractional bias (FB), normalized mean square error (NMSE), factor of two (Fa2), and forecasting skill (FC) discussed later.

2. Database Development

The ozone concentrations in the state of Ohio are monitored between the months of April through October every year. The hourly ozone concentration measurements were obtained from the EPA's AIRS database (AIRS/AFS). A discussion on air monitoring networks is given by Kumar et al. (2005). Detailed information of the monitoring site is summarized in Table 2.1. The data were obtained from this station for the time period 1995-2003.

Table 2.1. Details of the Ozone Monitoring Station in Dayton, OH

Station ID	39-113-0019
City	Dayton
County	Montgomery
Monitor type	SLAMS
Site address	2100 Timber Lane
Land use	Suburban

In this research, temperature, wind speed, and dew point temperature were considered as variables for development of the model. This data for a selected station (refer Table 2.2 for station details) were obtained from the National Climatic Data Center (NCDC) on a CD-ROM for the time period 1995-2002. The obtained ozone and meteorological data were disordered formatted files and required a lot of processing to get it into the desired format. The files were exported to an MS Access database, where the data was sorted in separate tables by year in the chronological order with the help of built in queries.

Table 2.2. Details of the Station Monitoring Meteorological Data in Dayton, OH

Station ID	724290
Station Name	Dayton International Airport
City	Dayton
County	Montgomery
Cooperative station number	338598
WBAN number	93815

Each table had the details of monitoring site, year being monitored, month, day and the hour of the day (at which the concentration was monitored).

Equipment defects or failure can cause problems resulting in missing and erroneous data. There can be periods of data missing either completely at random or missing at random. These periods of missing data invalidate the assumption of equally spaced observations. A method to overcome this limitation is to develop schemes for interpolating the missing values.

Linear interpolation through averaging of two adjacent bracketing values has been frequently employed to replace missing values in air pollution data (Aramburu, 1983). This technique can be used with a measure of confidence by professionals who are confronted with the task of filling missing data for air pollution data (Terry et al., 1986). In this study, the method of linear interpolation was adopted for filling the missing data.

Table 2.3 below shows the number and percentage of missing values in ozone concentration, temperature, wind speed, and dew point temperature in the model development period, Apr.1996–Oct.1999 and the evaluation period Apr.2002–Oct.2002.

Finally, once the missing values are filled, the ozone data and the meteorological data are stored together in a single table according to the year.

The MM5 real time forecasted data for temperature, wind speed, and dew point temperature were required for the model analysis. The data were extracted from the Polarmet website maintained by the Polar Research Group at Ohio State University (Bromwich, 2003). The real-time numerical weather prediction model of the polar research group uses the PSU/NCAR Mesoscale Model Version 3 (MM5).

Table 2.3. Number and Percentage of Missing Values in the Period Apr. 1996-Oct.1999 and Apr. 2002-Oct.2002

Year Apr. -Oct.	Total Data Points	Ozone		Temperature		Wind Speed		Dew Point Temperature	
		Missing Data	% Missing	Missing Data	% Missing	Missing Data	% Missing	Missing Data	% Missing
1996	5136	106	2.06	226	4.40	226	4.40	226	4.40
1997	5136	32	0.62	47	0.91	47	0.91	47	0.91
1998	5136	26	0.50	63	1.22	63	1.22	63	1.22
1999	5136	18	0.35	46	0.89	46	0.89	46	0.89
2002	5136	26	0.50	8	0.15	8	0.15	8	0.15

This model runs at 0000UTC/1200UTC as its initial time twice a day and produces a 72-hour prediction for the State of Ohio and surrounding areas. The initial fields and boundary tendencies are interpolated from the NWS Early Eta model output with a bi-linear interpolation scheme. The water surface temperatures used in the model were obtained from a climatological data set from the National Center for Atmospheric Research.

The forecasted data for the year 2003 were available only for the months of August to October for Dayton area. This data were modified to a desired format and used for model evaluation.

3. Data Analysis

Knowledge of spatial and temporal variations of ozone is crucial for developing improved models of predicting ozone patterns and establishing effective strategies to reduce the amount of ambient ozone. A comprehensive analysis of the variations in hourly ozone concentrations, ozone episode occurrences, and long-term trends in annual means of ozone concentrations, temperature, wind speed, and dew point temperature was performed for the period Apr.-Oct. for the years 1996-1999. The results are discussed below:

3.1. Annual Trends in Hourly Ozone Concentrations

Figure 3.1 statistically summarizes the annual averages of hourly ozone concentrations in the study region from 1996-1999 using box plots, in which the percentiles (5th, 10th, 25th, 50th, 75th, 90th, and 95th), means, and maxima are indicated. The figure shows that all the maxima values exceeded the standard of 85 ppb and the 95th percentiles were found below 85 ppb for all the years. Table 3.1 shows that the annual mean and 50th-95th percentiles of the hourly ozone concentrations over the period of interest exhibit decreasing trends (-0.2%/yr) with the average values in the range of 30-35 ppb.

Table 3.1. Change in Annual Means and Percentiles of Hourly Ozone Concentrations in the City of Dayton, 1996-1999

Mean	Percentile			
	50th	75th	90th	95th
-0.20	-1.63	0.81	0.81	-2.45

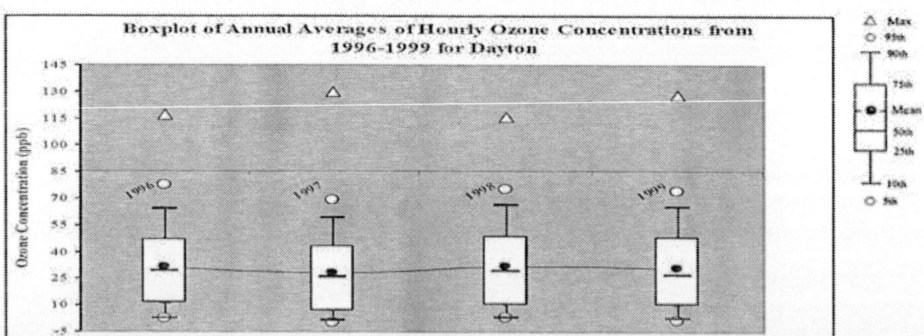

Figure 3.1. Box plot of annual averages of hourly ozone concentrations from 1996-1999 for Dayton.

3.2. Long-term Trends in the Annual Means of Hourly Ozone Concentrations, Temperature, Wind Speed, and Dew Point Temperature

The long-term trends in the annual means of hourly ozone concentrations, temperature, wind speed, and dew point temperature from 1996-1999 are presented in the Figure 3.2. The trend lines, obtained by the least square regression method, are indicated by dotted lines for all the variables. The corresponding slopes are given in Table 3.2. The negative and positive values indicate decreasing trends of ozone, wind speed, and dew point temperature and increasing trend of temperature respectively. This can also be observed in Figure 3.2.

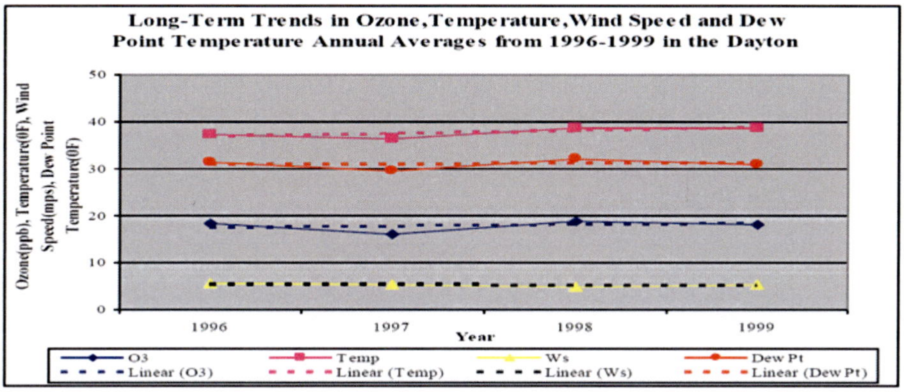

Figure 3.2. Long-term trends in ozone, temperature, wind speed and dew point temperature annual averages from 1996-1999 in Dayton.

Table 3.2. Slopes of the Trend Lines of Ozone, Temperature, Wind Speed, and Dew Point Temperature from 1996-1999

Parameter	Slope
Ozone	-0.023
Temperature	0.358
Wind Speed	-0.032
Dew point Temperature	-0.122

3.3. Seasonal Trends in Hourly Ozone Concentrations

The monthly averages of hourly ozone concentrations are summarized in Figure 3.3. The figure shows that all of the maxima exceed the standard value of 85 ppb, except for the month of April. A study in eastern Canada by Fuentes and Dann (1994), which considered seasonal variations in monthly means of the ozone concentrations of the urban and rural sites, concluded that the site under considerable anthropogenic influence recorded the highest concentrations from Jul. to Aug. (summer months). Similar results were obtained in this study. The monthly means of hourly ozone concentrations exhibit maximum variation in the summer months (May.–Aug.).

3.4 Ozone Episodes

The number of monthly ozone episodic days shown in Figure 3.4 indicates that the maximum exceedance occur during the months of June to August for all the years. Late May and early September were also found to be potential in the years 1998 and 1999. 91 episodic days were found over a 4-year period of around 736 days (Apr. to Oct.). That is, on an average, one day in a week was episodic in the years 1996-1999. Conversely, in a study carried out in Taiwan by Chen et al. (2004), which analyzed the meteorological effects on high ozone episodes, it was seen that the ozone episodes occurred most often in autumn and least frequently in summer.

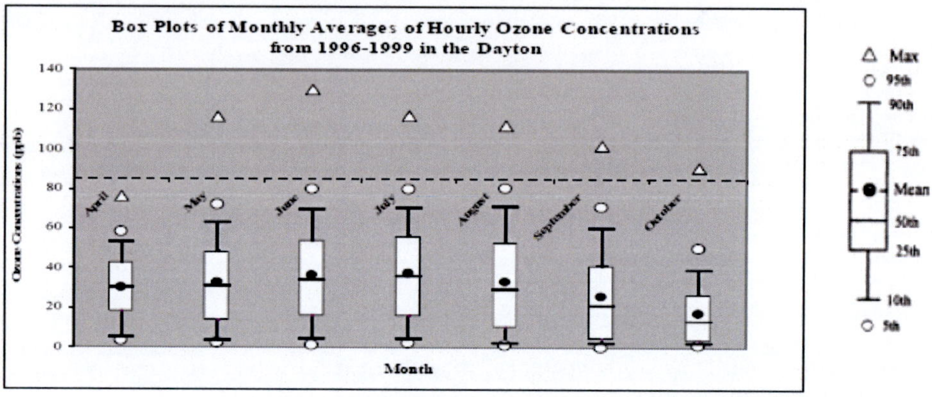

Figure 3.3. Box plot of monthly averages of hourly ozone concentrations from 1996-1999 in the Dayton.

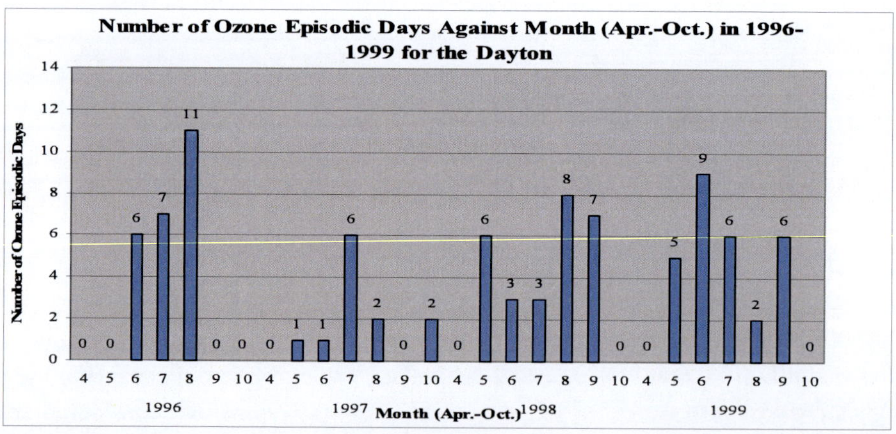

Figure 3.4. Number of ozone episodic days against month (Apr.-Oct.) in 1996-1999 for the Dayton.

3.5. Meteorological Effects on Ozone Episodes

Figures 3.5, 3.6, and 3.7 plot the hourly ozone concentrations that exceeded 85 ppb with the corresponding meteorological parameter values. High values of temperature (70-90 °F), dew point temperature (50-70 °F), and low wind speeds (<8 mph) were found during most of these days in all the years. Hence, it is clear from the figures that meteorological conditions play an important role in the occurrence of ozone exceedances.

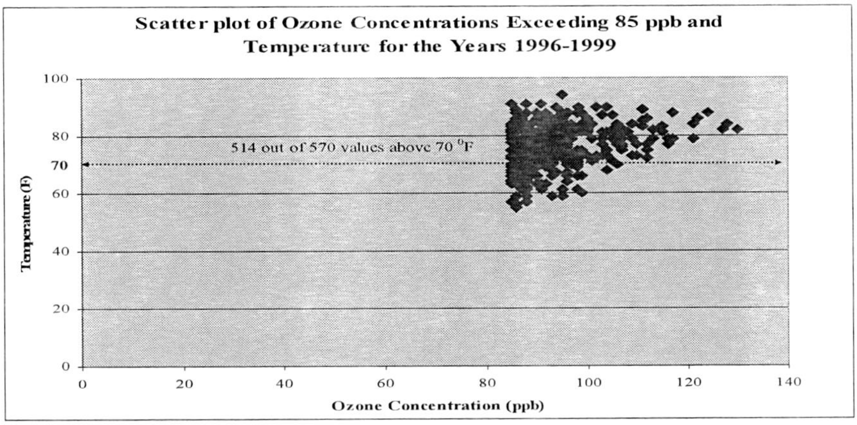

Figure 3.5. Scatter plot of ozone concentrations exceeding 85 ppb and temperature for the years 1996-1999.

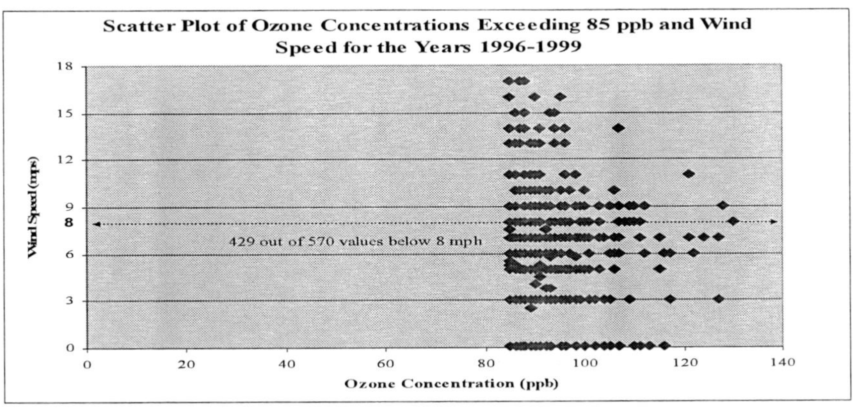

Figure 3.6. Scatter plot of ozone concentrations exceeding 85 ppb and wind speed for the years 1996-1999.

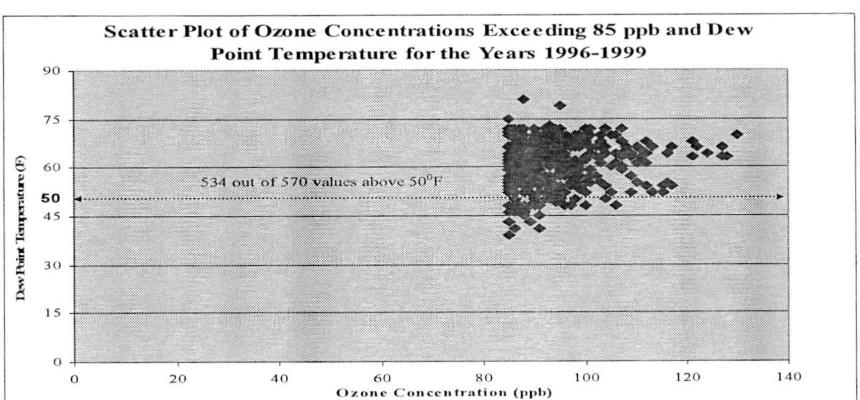

Figure 3.7. Scatter plot of ozone concentrations exceeding 85 ppb and dew point temperature for the years 1996-1999.

3.6. Conclusions from Data Analysis

The variations in ambient hourly ozone concentrations in the city of Dayton were analyzed for the period 1996-1999. The following conclusions were drawn from the analysis:

- Most of the annual and monthly means of the maxima exceeded 85 ppb.
- All of the 95th percentiles of annual and monthly means were below 85 ppb.
- The long-term trends in the annual means depict decreasing trends of ozone, wind speed, and dew point temperature and an increasing trend of temperature.
- The seasonal pattern of surface hourly ozone concentrations showed maximum variation during the summer months (May.-Aug.).
- Ozone episodes occurred on days with high temperatures and dew point temperatures and low wind speeds, which are typical in the months of Jun.-Aug.

From the analysis, it can be concluded that one out of seven days exceeded the ozone standard. The air quality in Dayton is a serious issue and, hence, accurate predictions are needed in order to take precautionary measures to avoid ozone episodic days. The following sections of the paper discuss the development of a model to forecast the hourly ozone concentrations for the city of Dayton.

4. METHODOLOGY

This section deals with the development of the model and statistical evaluation. The regression method was chosen for this study, as it is easy to develop and operate. This method has been well documented and widely used in various disciplines. It has been successfully used for ozone forecasting in many parts of the country.

4.1. KZ Filtering Technique

The models were developed using the KZ filtering technique coupled with regression analysis. This technique, developed by Rao et al. (1995), is a statistical method of moderating the influence of meteorology on surface ozone concentrations. In this study, the temperature, wind speed, and dew point temperature were used as independent variables and KZ filter was used to reduce the fluctuations. According to this method, time series of the log of the ozone concentration (Oz) can be represented by the following equation:

$$X(t) = e(t) + S(t) + W(t) = \text{Baseline}(t) + W(t) \qquad (4.1)$$

where:

$X(t)$ = Original ozone concentration time series
$e(t)$ = Long-term trend component
$S(t)$ = Seasonal variation
$W(t)$ = Short-term variation, and

t is the time.

The baseline is separated from the short-term variation using the Kolmogorov-Zurbenko ($KZ_{m,p}$) filter. The $KZ_{m,p}$ is a low-pass filter produced by repeated iterations of a simple moving average (Zurbenko, 1991). Each iteration of the moving average, Y_i is defined as follows:

$$Y_i = (1/m)\sum_{j=-k}^{k} X_{i+j} \tag{4.2}$$

where,

m = 2k +1.

Y_i becomes the input for the second pass, and the process continues accordingly.

Determination of the final low-pass filter (specifying "m" and the number of passes "p") is an iterative process in which the user determines if the white noise (short-term variation) has been removed. The output time series, Y_t, is the low frequency part of X_t denoted as:

$$Y_t = KZ_{m,p}(X_t) \tag{4.3}$$

Y_t contains both the long-term trend and seasonal effects.

Based on the above theory, simple forecasting models to predict the hourly ozone concentrations for different seasons in the city of Dayton, Ohio have been developed in this study.

4.2. Model Development

A total of twelve models were developed in this study. Six models used temperature, wind speed, and dew point temperature as independent parameters (here after referred to as three-parameter models). The other six considered temperature, and wind speed as independent variables (here after referred to as two-parameter models). Initially the entire dataset (Apr.-Oct.) was considered for the model development and it was seen that the seasonal variations were not accounted for.

Based on the analysis performed on the raw ozone data, it was seen that the seasonal patterns of the hourly ozone concentrations show maximum variation during the summer months. Hence, the dataset was divided into two groups namely pre-summer (Apr.-Jul.) and post-summer (Aug.-Oct.). It was noticed that this classification did not improve the model performance, as the hourly concentrations varied over a wide range. Hence, following the method adopted by Kim and Kumar (2005), it was determined to split each day into three regimes. According to this study, the observed concentrations against hour in a day, shows that the first 8-hours of the day is the stable period of the ozone levels, the second 8-hours is the ascent period, and the remaining 8-hours is the descent period.

A similar trend was observed with the data considered in this study (An example is shown in Figure 4.1.). The ozone data for pre-summer and post-summer was divided into three regimes, namely, hours: 1-8, hours: 9-16 and hours: 17-24. A more or less uniform dataset was obtained from this classification. An overview of the developed models is given in Table 4.1.

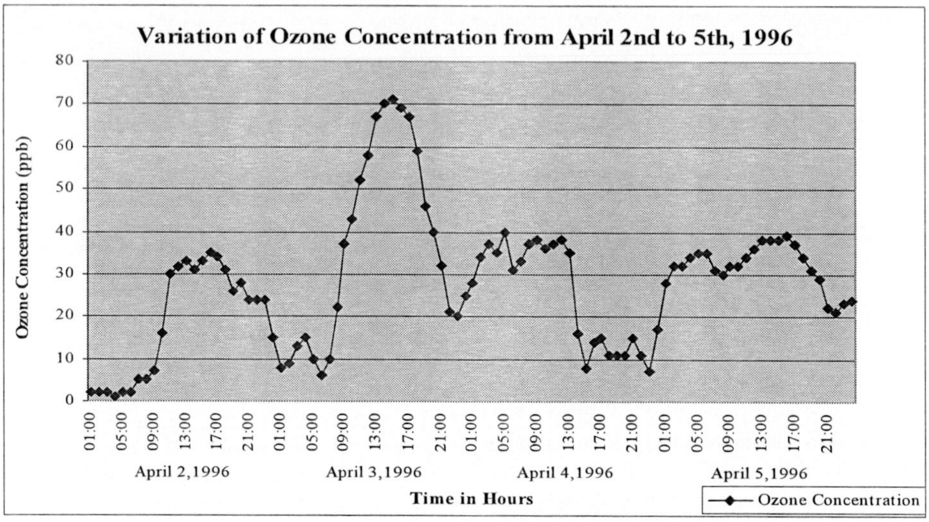

Figure 4.1. Variation of ozone concentration from April 2nd to 5th, 1996.

Table 4.1. Summary of the Models Developed

Season	Hour of the day	Years of Data Considered	Variables Involved	Number of Models
Pre-Summer	1-8, 9-16, 17-24	1996-1999	Temperature, Wind Speed, Dew point Temperature	3
			Temperature, Wind Speed	3
Post- Summer	1-8, 9-16, 17-24	1996-1999	Temperature, Wind Speed, Dew point Temperature	3
			Temperature, Wind Speed	3

Overview of the KZ Model

The development of the model involved two steps:

1) Application of the KZ Filter

The data were separated year-wise and then into two periods namely April to July and August to October considering the seasonal variations of ozone during these periods and each season was further divided into three regimes (hrs:1-8, 9-16, and 17-24). The theory of moving averages was applied to both the ozone and meteorological data. The concept of moving averages is to moderate the fluctuations in the parameter values.

Different values of 'm' (sample size) and 'p' (number of iterations) were investigated. The values of 'm' studied were 50, 100, 200, 275 and 300 depending on the number of data points. The values of 'p' investigated were 1, 2, and 3.The optimal $KZ_{m,p}$ filter was chosen by examining coefficients of determination.

Tables 4.2 and 4.3 list the values of R^2 obtained for pre-summer and post-summer for both three-parameter and two-parameter models. The filters that provided the best R^2 values were chosen for the model development (also highlighted in these tables).

Table 4.2. Coefficient of Determination for Different Combinations of KZ Filters for Pre-summer (Apr.-Jul.) and Post-summer (Aug.-Oct.), Three-parameter Model

Pre-summer					
Hour: 1-8		*Hour: 9-16*		Hour: 17-24	
Type of Filter	R^2	Type of Filter	R^2	Type of Filter	R^2
100,1	0.42	100,1	0.69	100,1	0.41
100,2	0.38	100,2	0.73	100,2	0.45
100,3	0.36	100,3	0.76	100,3	0.48
200,1	0.31	200,1	0.76	200,1	0.44
200,2	0.36	200,2	0.83	200,2	0.46
200,3	0.32	200,3	0.88	200,3	0.47
275,1	0.31	275,1	0.82	275,1	0.39
275,2	0.32	275,2	0.89	275,2	0.31
Post-Summer					
Hour: 1-8		Hour: 9-16		Hour: 17-24	
Type of Filter	R^2	Type of Filter	R^2	Type of Filter	R^2
50,1	0.54	50,1	0.79	50,1	0.80
50,2	0.55	50,2	0.77	50,2	0.84
50,3	0.56	50,3	0.76	50,3	0.85
100,1	0.61	100,1	0.84	100,1	0.87
100,2	0.66	100,2	0.83	100,2	0.91
100,3	0.64	100,3	0.84	100,3	0.92
200,1	0.71	200,1	0.86	200,1	0.93
200,2	0.75	200,2	0.90	200,2	0.95
275,1	0.76	275,1	0.89	275,1	0.95
300,1	0.76	300,1	0.91	300,1	0.96

2) Development of the Model

With the KZ filter chosen, a linear regression of the filtered natural log of ozone concentration [$O_{KZ}(t)$] on filtered temperature [$T_{KZ}(t)$], filtered wind speed [$W_{KZ}(t)$], and filtered dew point temperature [$D_{KZ}(t)$] was performed with the dataset. Three-parameter and two-parameter models were developed for the periods Apr.-Jul. and Aug.-Oct. Each model was evaluated by comparing the model with an independent dataset collected from that site. The relation between the independent variables and the dependent parameter can be expressed in the following linear equation:

$$O_{KZ}(t) = aT_{KZ}(t) + bW_{KZ}(t) + cD_{KZ}(t) + d + \varepsilon(t) \tag{4.4}$$

where, $O_{kz}(t)$ is the filtered natural logarithm of the hourly ozone concentration for the chosen KZ filter,

$T_{kz}(t)$ is the filtered hourly temperature for the chosen KZ filter,
$W_{kz}(t)$ is the filtered hourly wind speed for the chosen KZ filter,
$D_{kz}(t)$ is the filtered hourly dew point temperature for the chosen KZ filter,
a is the coefficient of $T_{kz}(t)$,
b is the coefficient of $W_{kz}(t)$,
c is the coefficient of $D_{kz}(t)$,
d is the regression constant, and
$\varepsilon(t)$ is the residual and represents the variation in ozone concentrations attributable to factors other than temperature, wind speed, and dew point temperature.

Excluding the residual, the relationship between the variables in the above equation may be expressed as in Equation 4.5.

$$O_{KZ}(t) = aT_{KZ}(t) + bW_{KZ}(t) + cD_{KZ}(t) + d \tag{4.5}$$

Three other equations were developed using the dataset. Equation 4.6 is a relationship between the filtered temperature and the raw temperature. This relationship was obtained by linear regression using $T_{kz(m,p)}$ as the dependent variable.

$$T_{KZ(m,p)}(t) = a_1T(t)+c_1 \tag{4.6}$$

where,
$T_{KZ(m,p)}(t)$ is the filtered temperature for the chosen KZ filter,
$T(t)$ is the hourly temperature,
$a_1(t)$ is the coefficient of the hourly temperature, and
c_1 is the constant.

Equation 4.7 is a relationship between the filtered wind speed and the raw wind speed. This relationship was obtained by linear regression using $W_{kz(m,\ p)}$ as the dependent variable.

$$W_{kz(m,p)}(t)=a_2W(t)+c_2 \tag{4.7}$$

where,
$W_{kz(m,p)}(t)$ is the filtered hourly wind speed for the chosen KZ filter,
$W(t)$ is the hourly wind speed,
$a_2(t)$ is the coefficient of the hourly wind speed, and
c_2 is the regression constant.

Equation 4.8 is the relationship between filtered dew point temperature and raw dew point temperature. This relationship was obtained by linear regression using $D_{kz(m,p)}$ as the dependent variable.

$$D_{kz(m,p)}(t)=a_3D(t)+c_3 \tag{4.8}$$

where, $D_{kz(m,p)}(t)$ is the filtered hourly dew point temperature for the chosen KZ filter,

D(t) is the raw hourly dew point temperature,
$a_3(t)$ is the coefficient of raw hourly dew point temperature, and
$c_3(t)$ is the regression constant.

Finally, Equation 4.9 is the relationship between the natural log of the ozone concentration and the filtered natural log of the ozone concentration. This relationship was obtained by linear regression using $O_{kz(m,p)}$ as the predictor variable.

$$\ln O(t) = a_4 O_{kz(m,p)}(t) + c_4 \quad (4.9)$$

where, $O_{kz(m,p)}$ is the filtered natural logarithm of the hourly ozone concentration for the chosen KZ filter,

O(t) is the hourly ozone concentration,

a_4 is the coefficient of the filtered natural logarithm of the hourly ozone concentration for the chosen KZ filter, and c_4 is the regression constant.

Table 4.3. Coefficient of Determination for Different Combinations of KZ Filters for Pre-summer (Apr.-Jul.) and Post-summer (Aug.-Oct.), Two-parameter Model

Pre-summer					
Hour: 1-8		Hour: 9-16		Hour: 17-24	
Type of Filter	R^2	Type of Filter	R^2	Type of Filter	R^2
100,1	0.38	100,1	0.52	100,1	0.33
100,2	0.33	100,2	0.57	100,2	0.37
100,3	0.29	100,3	0.59	100,3	0.41
200,1	0.23	200,1	0.62	200,1	0.37
200,2	0.22	200,2	0.68	200,2	0.44
200,3	0.18	200,3	0.72	200,3	0.45
275,1	0.15	275,1	0.69	275,1	0.35
275,2	0.11	275,2	0.75	275,2	0.37
275,3	0.09	275,3	0.86	275,3	0.34
Post-Summer					
Hour: 1-8		Hour: 9-16		Hour: 17-24	
Type of Filter	R^2	Type of Filter	R^2	Type of Filter	R^2
50,1	0.52	50,1	0.69	50,1	0.74
50,2	0.53	50,2	0.67	50,2	0.77
50,3	0.53	50,3	0.65	50,3	0.79
100,1	0.57	100,1	0.77	100,1	0.80
100,2	0.58	100,2	0.76	100,2	0.82
100,3	0.55	100,3	0.77	100,3	0.83
200,1	0.60	200,1	0.82	200,1	0.85
200,2	0.56	200,2	0.86	200,2	0.86
275,1	0.62	275,1	0.87	275,1	0.87
300,1	0.61	300,1	0.88	300,1	0.88

By substituting Equations 4.6, 4.7, and 4.8 in Equation 4.4, and the resultant expression in Equation 4.9, the natural logarithm of hourly ozone concentration may be expressed as a function of hourly temperature, wind speed, and dew point temperature. This relationship, expressed in Equation 4.10, is essentially the final forecasting model.

$$\ln O(t) = A * T(t) + B * W(t) + C * D(t) + E \quad (4.10)$$

where ,

O(t) is the hourly ozone concentration,
T(t) is the hourly temperature and *A* is coefficient of *T(t),*
W(t) is the hourly wind speed and *B* is coefficient of *W(t),*
D(t) is the hourly dew point temperature and C is the coefficient of D(t), and
E is the regression constant.

These models were developed using the time-series data between 1996 and 1999 (both inclusive) for Dayton, Ohio.

4.3. Model Evaluation

Evaluation of the models developed in this study was done in three different ways:

1) The models were validated using the data for the year 2002. Observed temperature, wind speed, dew point temperature, and ozone concentrations were used. The dataset for the year 2002 was separated into pre-summer and post-summer, and each period was again divided into three sets based on the hour of the day. The predicted ozone values for each hour of each set were obtained using observed meteorological parameter values. These predicted ozone concentration values, obtained by using the corresponding model, were compared against observed values, using the statistical parameters. As this model was checked using the observed ozone data for 2003, this does not lead to a perfect evaluation. However, the evaluation gives us a good idea of the performance of the model, if perfect forecasted meteorological variable values were available.
2) This evaluation also reflects the meteorological forecasting ability of the MM5 model.
3) The models developed using temperature, and wind speed as variables (two-parameter models) were evaluated for 2002 using observed data and for 2003 using MM5 data in the similar manner as discussed in step 1 and 2. These two- parameter models were then compared with the three-parameter models using evaluation parameters.

The models were also checked for their performance in predicting hourly AQI values, and the results were plotted for the observed and predicted values for the models.

The statistical evaluation parameters such as coefficient of determination (r^2), fractional bias (FB), normalized mean square error (NMSE), and fraction of 2 (Fa2) are used to determine the model performance of both the models. In addition forecasting skills of the model was also computed. Each parameter is discussed below:

1) Coefficient of Determination (R^2)

R^2 is a measure of the correlation between the dependent and independent variables in a regression analysis. It is calculated as the square of the correlation coefficient, r, and varies between 0 and 1. An ideal model would have an R^2 of 1.

2) Fractional Bias (FB)

The bias is normalized to make it non-dimensionless. This fractional bias (FB) varies between +2 and -2 and has an ideal value of zero for an ideal model. It is written in symbolic form as:

$$FB = 2*(\overline{Co} - \overline{Cp}) / (\overline{Co} + \overline{Cp}) \qquad (4.11)$$

3) Normalized Mean Square Error (NMSE)

This statistic emphasizes the scatter in the entire data set and is known as Normalized Mean Square Error (NMSE). The normalization by the product Cp*Co assures that the NMSE will not be biased towards models that over predict or under predict. Smaller values of NMSE denote better model performance. The expression for the NMSE is given by:

$$NMSE = \overline{(Cp - Co)^2} / \overline{(CoCp)} \qquad (4.12)$$

4) Factor of x (Fa_X)

The factor of x (Fa_X) is defined as the percentage of the predictions within a factor of 'x' of the observed values. The ideal value for the factor of 'x' should be 1 (100%).

$$Fa_X = \text{Fraction of data which } (1/x) <= C_p/C_o <= x \qquad (4.13)$$

In the present study the value of 'x' was considered '2'.

5) Forecasting Skill

Another indicator of performance of a forecasting model is forecasting skill (FC). Chenevez and Jensen (2001) proposed an indicator for calculating the performance of a model while making peak forecasts. This indicator is being used for the evaluation of performance of the forecasting systems currently operating in Denmark.

They have defined forecasting skill (FC) of a model as:

$$FC = 100*N_r/N \qquad (4.14)$$

where,

N_r is the number of forecast hits inside an uncertainty interval r, and

N is the total number of forecasts.

Table 4.6. Evaluation Parameters for a Perfect Model

R^2	FB	NMSE	Fa_2	FC
1.0	0	0	1.0	100%

R^2 values of the models were not calculated as it is considered unreliable in judging the performance of a forecasting model. It was used only to fit a model and not to check for the model performance.

5. Results and Discussion

The final forecasting model equations and their evaluations are provided in this section. The application of the model evaluation techniques and the calculation of specific performance parameters based on air quality evaluation studies are also presented.

5.1. Models Developed

The final model equation obtained for predicting ozone concentration, is in the form

$$\ln (Oz) = (A*T) + (B* W) + (C*D) + E \qquad (5.1)$$

where,

T = hourly temperature in °F,
W = hourly wind speed in mph,
D = hourly dew point temperature in °F,
Oz = hourly ozone concentration in ppb,
A = coefficient of temperature T,
B = coefficient of wind speed W,
C = coefficient of dew point temperature D, and
E = regression constant.

Tables 5.1 and 5.2 summarize the coefficients of the three-parameter and two-parameter models developed for pre-summer and post-summer seasons.

The values of coefficients for temperature from the tables indicate that the KZ model has shown a positive relationship between the hourly ozone concentrations and hourly temperature for pre-summer and post-summer seasons.

The co-efficient of wind speed was negative for hr: 9-16 models for pre-summer, indicating a negligible effect of wind speed during this period in which high concentrations of ozone occur. During other periods (hr: 1-8, 17-24), wind speed has shown more significance on the ozone concentration than temperature.

Table 5.1. Summary of Coefficients of the KZ Regression Models Developed for Pre-summer: Apr.-Jul

Type of Model	Hour of the Day	Coefficient of Temperature (A)	Coefficient of Wind Speed (B)	Coefficient of Dew point Temperature (C)	Regression Constant (E)
Three-parameter Model	1-8	-0.008	0.132	0.037	0.82
	9-16	0.049	-0.071	-0.038	3.535
	17-24	0.040	0.042	-0.015	1.132
Two-parameter Model	1-8	0.014	0.128	-	0.609
	9-16	0.010	-0.117	-	4.231
	17-24	0.025	0.043	-	0.931

Table 5.2. Summary of Coefficients of the KZ Regression Models Developed for Post-summer: Aug.-Oct

Type of Model	Hour of the Day	Coefficient of Temperature (A)	Coefficient of Wind Speed (B)	Coefficient of Dew point Temperature (C)	Regression Constant (E)
Three-parameter Model	1-8	0.015	0.095	0.026	-0.829
	9-16	0.069	0.102	-0.018	-0.496
	17-24	0.037	0.139	0.027	-2.869
Two-parameter Model	1-8	0.031	0.054	-	-0.349
	9-16	0.056	0.117	-	-0.640
	17-24	0.052	0.079	-	-1.940

The wind speed was seen to have more impact on ozone concentrations than temperature during all periods of post-summer. This may be due to the presence of low temperatures in the post-summer season. It has been reported in the literature that wind plays a very complex role in the transport of ozone and ozone pre-cursors in and out of an area, and the ozone-wind relationship depends on the historical data at a specific site (Hubbard and Cobourn, 1998).

A clear idea of how the dew point temperature affects the ozone concentrations could not be found from the values obtained. However, in most cases it was seen to have less impact on ozone concentrations than temperature and wind speed. Moreover, during morning times (hr: 1-8), dew point temperature was noticed to have more impact than temperature for both pre-summer and post-summer. A detailed examination of the predicted results indicates that the peak values were predicted well qualitatively by including dew point temperature as a variable.

5.2. Models Developed for Pre-summer (Apr.-Jul.)

The hourly concentrations for Apr.-Jul., 2002 were predicted. The model evaluation parameters discussed in Section 4.3 were calculated and the results are summarized in the Table 5.3 for pre-summer. The results of these statistical analyses meet the criteria specified by Kumar et al. (1993) for an acceptable air quality model. NMSE and Fa_2 are close to the ideal values (zero). Forecasting skill, FC, was less than 50 percent for all the models and needs further improvement. It can be seen that all the FB values were negative, indicating the tendency of the model to over-predict. As a further attempt to examine the usefulness of the models, AQI generated from the observed concentrations and predicted concentrations were compared. It can be seen from Table 5.4 for pre-summer that the models are acceptable in predicting ranges rather than discrete values, and extreme values in particular. It was noticed that the models are predicting most of the values in the 41-80 ppb range.

The results of statistical analysis and AQI forecasting of the two-parameter model for the year 2002 are also given in Tables 5.3 and 5.4 respectively. NMSE and Fa_2 were close to ideal values. The 9-16 hour model seemed to over-predict, but the other models were under-predicting. This is evident from the values of FB in Table 5.3. The models meet the requirements of an acceptable air quality model, but, unlike the three-parameter models, these models were unable to predict most of the peak concentrations. It is noticed that considering dew point temperature, as an additional variable, accounted for the majority of the peak values, although the models tend to over-predict most of the time.

Table 5.3. Summary of Results for Pre-summer and Post-summer

Pre-summer (Apr.-Jul.)					
Type of Model	Hour of the Day	Evaluation Parameters			
		FB	NMSE	Fa_2	FC
Three-parameter Model	1-8	-0.46	0.194	0.61	44.32
	9-16	-0.17	0.028	0.89	47.69
	17-24	-0.17	0.029	0.75	42.58
Two-parameter Model	1-8	0.40	0.149	0.85	61.9
	9-16	-0.06	0.003	0.88	35.2
	17-24	0.34	0.111	0.86	35.9
Post-summer (Aug.-Oct)					
Type of Model	Hour of the Day	Evaluation Parameters			
		FB	NMSE	Fa_2	FC
Three-parameter Model	1-8	-0.04	0.001	0.57	81.42
	9-16	-0.01	0.0001	0.88	37.50
	17-24	-0.06	0.002	0.68	58.96
Two-parameter Model	1-8	0.31	0.081	0.64	80.60
	9-16	-0.17	0.023	0.82	37.22
	17-24	0.22	0.032	0.71	59.09
Post-summer (Aug.-Oct.) Using MM5 Data					
Type of Model	Hour of the Day	Evaluation Parameters			
		FB	NMSE	Fa_2	FC
Three-parameter Model	1-8	-0.25	0.045	0.44	74.45
	9-16	-0.23	0.051	0.83	64.13
	17-24	-0.23	0.041	0.58	45.45
Two-parameter Model	1-8	-0.09	0.007	0.43	79.34
	9-16	-0.84	0.749	0.49	17.02
	17-24	0.69	0.439	0.75	48.72

Table 5.4. AQI Forecasting for Pre-summer and Post-summer

Type of Model	AQI	Corresponding Ozone (ppb)	Hr: 1-8		Hr: 9-16		Hr: 17-24	
			Obs.	Pre.	Obs.	Pre.	Obs.	Pre.
Pre-summer (Apr.-Jul)								
Three-parameter Model	0-50	0-40	93%	74%	31%	24%	59%	42%
	51-100	41-80	7%	24%	62%	58%	38%	57%
	101-150	81-120	0%	2%	7%	14%	3%	1%
	151-200	121-160	0 %	0 %	0 %	4 %	0 %	0 %
Two-parameter Model	0-50	0-40	93%	99%	31%	39%	59%	93%
	51-100	41-80	7%	1%	62%	47%	38%	7%
	101-150	81-120	0%	0%	7%	10%	3%	0%
	151-200	121-160	0 %	0 %	0 %	4 %	0 %	0 %
Post-summer (Aug.-Oct.)								
Type of Model	AQI	Corresponding Ozone (ppb)	Hr: 1-8		Hr: 9-16		Hr: 17-24	
			Obs.	Pre.	Obs.	Pre.	Obs.	Pre.
Three-parameter Model	0-50	0-40	99%	100%	48%	49%	79%	79%
	51-100	41-80	1%	0 %	42%	37%	19%	18%
	101-150	81-120	0 %	0 %	10%	13%	2%	3%
	151-200	121-160	0 %	0 %	0 %	1 %	0 %	0 %
Two-parameter Model	0-50	0-40	99%	100%	48%	41%	79%	92%
	51-100	41-80	1%	0 %	42%	36%	19%	8%
	101-150	81-120	0 %	0 %	10%	21%	2%	0%
	151-200	121-160	0 %	0 %	0 %	2 %	0 %	0 %
Post-summer (Aug.-Oct.) Using MM5 Data								
Type of Model	AQI	Corresponding Ozone (ppb)	Hr: 1-8		Hr: 9-16		Hr: 17-24	
			Obs.	Pre.	Obs.	Pre.	Obs.	Pre.
Three-parameter Model (Using MM5 Data)	0-50	0-40	99%	99%	52%	36%	87%	84%
	51-100	41-80	1%	1 %	46%	48%	12%	14%
	101-150	81-120	0 %	0 %	2%	16%	1%	2%
	151-200	121-160	0 %	0 %	0 %	0%	0 %	0 %
	201-300	161-240	0 %	0 %	0 %	0%	0 %	0 %
	301-500	241-400	0 %	0 %	0 %	0%	0 %	0 %
Two-parameter Model (Using MM5 Data)	0-50	0-40	99%	100%	52%	15%	87%	100%
	51-100	41-80	1%	0%	46%	26%	12%	0%
	101-150	81-120	0 %	0 %	2%	28%	1%	0%
	151-200	121-160	0 %	0 %	0 %	16%	0 %	0 %
	201-300	161-240	0 %	0 %	0 %	10%	0 %	0 %
	301-500	241-400	0 %	0 %	0 %	5%	0 %	0 %

5.3. Models Developed for Post-Summer (Aug.-Oct.)

A similar evaluation was performed for post-summer. The performance characteristics for the year 2002 are provided in Table 5.3 for post-summer. All the values obtained were close to the ideal values. Again, negative FB values were obtained, indicating the over-predicting tendency of the models. A better forecasting ability of the models was seen except for the 9-16 hour model, which needs improvement. The AQI forecasting for post-summer, as seen in Table 5.4, shows a tendency of the models to over-predict AQI most of the time. This is generally acceptable in declaring an ozone action day for avoiding non-attainment status of an area.

The result of the evaluation of the two-parameter model (refer to summary Tables 5.3 and 5.4) shows that 1-8 hour and 17-24 hour models were under-predicting. Although the statistical parameter values were close for three-parameter and two-parameter models, the three-parameter models performed better in predicting the peak concentrations.

The results of the evaluation of the models for Aug.-Oct., 2003 using MM5 forecasted data for temperature, wind speed, and dew point temperature are also given in the summary tables. The forecasting skill of the models improved when forecasted data was used. Theoretically, forecasted data may or may not improve the forecasting ability of a model. All the models over-predicted most of the values, which can be concluded from the negative FB values. AQI forecasting, as seen in Table 5.4, shows that the results are in the acceptable range. The observed and the predicted values were fairly close.

The results of the two-parameter models as seen in the summary tables show that the 9-16 hour model did not predict well when the forecasted data was used. The rest of the models were under-predicting.

The three-parameter models performed well in predicting most of the peak concentrations. Previous studies on the forecasting (Ghose and Kumar (2003), Kumar et al., 2004), which considered temperature and wind speed or temperature alone, failed to simulate most of the peak concentrations. Considering dew-point temperature as an additional parameter improved the performance of the model in predicting these values.

6. Exceedance Model

Accurate prediction of the ozone episodic days is required for implementing precautionary measures so as to avoid ozone threats. From the previous discussion, it was seen that the three-parameter model performed well in predicting the trend of most of the peak concentrations, but the forecasting ability to predict peaks within ± 10 ppb of the observed values was found to be low in the range 40-50% (See Table 6.3). Therefore, models to accurately predict peak concentrations were developed for pre-summer and post-summer seasons by examining all the observed values of ozone concentrations above 85 ppb.

6.1 Development of Exceedance Model

The ozone exceedances (> 85ppb) during the years 1996-1999 and the corresponding values of temperature, wind speed, and dew point temperature were extracted and stored in the separate table. The data was divided into pre-summer and post-summer seasons, and each season was further separated based on hour of the day (Hr: 1-8, 9-16, and 17-24).

The number of ozone exceedances during the first eight hours of the day (Hr: 1-8) is usually low because of low pre-cursor emissions and low temperatures. Only 0.02% of the total values were found in this period for pre-summer and post-summer seasons in the years 1996-1999. Therefore, models were developed only for Hr: 9-16 and Hr: 17-24 periods.

Linear regression of the natural logarithm of the ozone concentrations and three variables was performed, and four models - two for both pre-summer and post-summer were developed. An overview of the developed models is shown in the Table 6.1.

Table 6.1. Summary of the Models Developed

Season	Hour of the day	Years of Data Considered	Variables Involved	Number of Models
Pre-Summer	9-16, 17-24	1996-1999	Temperature, Wind Speed, Dew point Temperature	2
Post-Summer	9-16, 17-24	1996-1999	Temperature, Wind Speed, Dew point Temperature	2

Table 6.2. Summary of Coefficients of the Exceedance Models for Pre-summer and Post-summer

Pre-summer (Apr.-Jul.)				
Hour of the Day	Coefficient of Temperature (A)	Coefficient of Wind Speed (B)	Coefficient of Dew point Temperature (C)	Regression Constant (E)
1-8	-	-	-	-
9-16	4.079	0.006	-0.002	-0.002
17-24	4.244	0.004	-0.011	-0.001
Post-summer (Aug.-Oct)				
Hour of the Day	Coefficient of Temperature (A)	Coefficient of Wind Speed (B)	Coefficient of Dew point Temperature (C)	Regression Constant (E)
1-8	-	-	-	-
9-16	4.369	0.002	-0.0006	0.0002
17-24	4.116	0.004	-0.008	0.0017

The models were evaluated for the year 2002. The ozone exceedances in the year 2002 were extracted, and the corresponding meteorological variables were obtained. This data was divided into pre-summer and post-summer seasons. The observed meteorological data and the appropriate models were used to predict the ozone concentrations. These predictions were compared with the observed ozone data for the year 2002.

The performance of the model was studied using the evaluation parameters NMSE, FB, and Fa_2. The forecasting ability of the models to predict the peaks within $\pm$ 10 ppb of observed values was also found.

6.2. Results and Discussion

The final model is of the form discussed in Equation 5.1 in the section 5.1:

$$\ln(Oz) = (A*T) + (B*W) + (C*D) + E \quad (5.1)$$

where,

T = hourly temperature in °F,
W = hourly wind speed in mph,
D = hourly dew point temperature in °F,
Oz = hourly ozone concentration in ppb,
A = coefficient of temperature T,
B = coefficient of wind speed W,
C = coefficient of dew point temperature D, and
E = regression constant.

The coefficients of the models are given in Table 6.2 for pre-summer and post-summer seasons. The evaluation parameters of the model predictions for the year 2002 are summarized in the Table 6.3 for both pre-summer and post-summer seasons.

The temperature was seen to have a strong positive relation with the peak ozone concentration, which is evident from the values of coefficients of temperature in Table 6.2. The wind speed and dew point temperatures have shown a negligible effect on ozone concentrations. The values of NMSE, FB and Fa_2, as seen in the Table 6.3, are close to ideal values and hence meet the requirements of an acceptable air quality model. The forecasting ability of the models for pre-summer was found to be more than 85% and for post-summer it was more than 80%. An increase of approximately 40-60% of the forecasting ability was observed when compared to that of the general three-parameter model described in the previous section.

Table 6.3. Evaluation Parameters for the Exceedance Models for Pre-summer and Post-summer

Pre-summer (Apr.-Jul.)					
Hour of the Day	Evaluation Parameters				
	FB	NMSE	Fa_2	FC	FC (Three-parameter Model)
1-8	-	-	-	-	-
9-16	-0.48	0.002	1.0	89	40
17-24	0.25	0.001	1.0	100	40
Post-summer (Aug.-Oct)					
Hour of the Day	Evaluation Parameters				
	FB	NMSE	Fa_2	FC	FC (Three-parameter Model)
1-8	-	-	-	-	-
9-16	0.39	0.001	1.0	82	40
17-24	0.38	0.001	1.0	82	50

The model can be used to predict ozone episodes when a possibility of occurrence of high pre-cursor emissions coupled with high temperatures (> 70^0F) is forecasted.

Conclusions

The first part of this study presents the development and evaluation of simple forecasting models using the KZ regression approach for predicting hourly ozone concentrations for

Dayton, Ohio, considering temperature, wind speed, and dew point temperature as independent variables. The performances of the developed models meet the general criteria specified by Kumar et al., (1993) for air quality models. The forecasting skills (FC= ± 10 ppb of observed concentration) of the models were found to be low. The models accounted for most of the peak concentrations. Further analysis of forecasted data indicates that predicted AQI indices are either close to observed values or higher than the observed values. This prediction serves well for public use.

The following conclusions can be drawn from this part of the study:

1) The predictions made by the models are reliable only to a limited extent.
2) The analysis of the raw ozone and meteorological data was helpful in assessing the variations and relations between the parameters.
3) High temperature (70-90 °F) and dew point temperatures (50-70 °F), and low wind speeds (<8 mph) were found to be responsible for an ozone episode most of the time. However, temperature was a critical parameter influencing its occurrence.
4) Developing different models for the three regimes (Hr: 1-8, 9-16, 17-24) provided more accurate predictions and also helped in accounting for most of the peak concentrations.
5) The three-parameter models showed an over-predicting tendency, and thus performed well in predicting peak concentrations. The two-parameter models were under-predicting in most cases.
6) The KZ model has shown a positive relationship between the ozone concentration and the temperature for all the models. This is in accordance with most previous studies on forecasting of ozone with temperature.
7) The wind speed was seen to have a negligible effect during 9-16th hour during pre-summer in which high concentrations of ozone occur. However, wind speed had a more significant impact than temperature for all other periods of pre-summer (Hr: 1-8, 17-24) and during post-summer.
8) The dew point temperature was noticed to have more impact than temperature during 1-8th hour for both pre-summer and post-summer. Inclusion of dew point temperature improved the ability to predict high ozone concentrations.

The second part of the study focused on developing exceedance models using linear regression analysis. The exceedance models developed for predicting peak concentrations performed well with a forecasting ability (within ±10ppb) in the range of 80-100%, and can be used to forecast an ozone episode when the predicted temperature is more than 70°F.

In summary, it is suggested to use three parameter models along with the ozone exceedance models to predict hourly ozone concentrations for Dayton area.

In this research, the performance of the model considering dew point temperature as an independent variable along with temperature and wind speed was studied. The models showed a tendency to over-predict the observed ozone concentrations most of the time. There are many other meteorological variables like wind direction, relative humidity, cloud cover that can be explored for improving forecasting models for ozone. Similarly, other methods of forecasting -such as artificial neural networks, non-linear multiple regression, time-series forecasting, and fuzzy-logic based- can also be employed to develop models which can effectively forecast the ozone concentrations.

REFERENCES

Aneiros-Perez, G.; Cardot, H.; Estevez-Perez, G.; Vieu, P. Maximum ozone concentration forecasting by functional non-parametric approaches. *J.Environmetrics*. 2004, 15, 675-685.

Aramburu, J. C.; Parkhurst, W. J. Clarifying a Cloudy Issue: Time Series Analysis of Atmospheric Deposition. *Presented at the 76th Annual Meeting of the APCA*, 1983.

Ballester, E.B.; Valls, G.C.; Carrasco-Rodriguez, J.L.; Olivas, E.S.; Valle-Tascon, S. Effective 1-day ahead prediction of hourly surface ozone concentrations in eastern Spain using linear models and neural networks. *J.Ecol. Model*. 2002, 156, 27-41.

Bauer, G.; Deistler, M.; Scherrer, W. Time series models for short term forecasting of ozone in the eastern part of Austria. *J.Environmetrics*. 2001, 12, 117-130.

Bromwich, D.; Hadad, C.; Ma, Y.; Berkman, P.; Long, D. Central Ohio air quality modeling: achievements and future plans. Paper No. OS-02-02, *Proceedings of Ohio Air Pollution Research Symposium*, 2002.

Bordignon, S.; Gaetan, C.; Lisi, F. Nonlinear models for ground-level ozone forecasting. *J.Stat Meth Appl*. 2002, 11, 227-245.

Chaloulakou, A.; Saisana, M.; Spyrellis, N. Comparative assessment of neural networks and regression models for forecasting summertime ozone in Athens. *J.Sci. Total Environ*. 2003, 313, 1-13.

Chen, K.; Ho, Y.T.; Lai, C.H.; Tsai, Y.A. Trends in Concentration of ground-level ozone and meteorological conditions during high ozone episodes in the Kao-Ping airshed, Taiwan. *J. Air Waste Manage. Assoc*. 2004, 54, 36-48.

Chenevez, J.; Jensen, C. Operational ozone forecasts for the region of Copenhagen by the Danish Meteorological Institute. *J.Atmos. Environ*. 2001, 35, 4567-4580.

Dutot, A.L.; Rynkiewicz, J.; Steiner, F.E.; Rude, J. A 24-h forecast of ozone peaks and exceedance levels using neural classifiers and weather predictions. *J.Environ Modell Softw*. 2007, 22, 1261-1269.

Fuentes, J.D.; Dann, T.F. Ground level ozone in eastern Canada: seasonal variations, trends, and occurrences of high concentrations. *J. Air Waste Manage. Assoc*. 1994, 44, 1019-1026.

Ghose, S.; Kumar, A. Forecasting daily maximum ozone concentrations at Columbus, Ohio. Paper No.OS-03-09, *Proceedings of Second Annual Ohio Air Quality Research Symposium*. 2003.

Gomez-Sanchis, J.; Martın-Guerrero, J.D.; Soria-Olivas, E.; Vila-France, J.; Carrasco, J.L; Valle-Tasco, S. Neural networks for analysing the relevance of input variables in the prediction of tropospheric ozone concentration. *J.Atmos. Environ*. 2006, 40, 6173-6180.

Heo, J.S.; Kim, D. A new method of ozone forecasting using fuzzy expert and neural network systems. J.Sci. Total Environ. 2004, 325, 221-237.

Hubbard, M. C.; Cobourn, W. G. Development of a regression model to forecast ground-level ozone concentrations in Louisville, KY. *J.Atmos. Environ*. 1998, 32, 2637-2647.

Ibarra-Berastegi, G.; Madariaga, I.; Elıas, A.; Agirre, E.; Uria, J. Long-term changes of ozone and traffic in Bilbao. *J.Atmos. Environ*. 2001, 35, 5581-5592.

Jorquera, H.; Perez, R.; Cipriano, A.; Espejo, A.; Letelier, M.V.; Acuna, G. Forecasting ozone daily maximum levels at Santiago, Chile. *J.Atmos. Environ*. 1998, 32, 3415-3424.

Kim, S. E.; Kumar, A. Accounting seasonal nonstationarity in time series models for short-term ozone level forecasts. *J.Stoch. Environ. Res. Risk Assess*. 2005, 19, 241-248.

Kumar, A.; Luo, J; Bennett, G. Statistical evaluation of lower flammability distance (LFD) using four hazardous release models. *J.Process Saf. Prog*. 1993, 12, 1-11.

Kumar, A., Rao, H. G., Jampana, S., and Varadarajan, C. (2005). Air quality monitoring and associated instrumentation, Chapter 2, *Environmental Technology Handbook* (edited by N.P. Cheremisinoff), Government Institutes, Rockville, MD, 99-136, 2005.

Kumar, A.; Tandale, A.; Kalapati, R. Forecasting models for daily maximum ozone concentrations for Cleveland and Akron Areas, Paper No 162, *Proceedings of AWMA Conference*, 2004.

Kumar, A.; Vedula, S; Sud, A. Development of an ozone forecasting model for non-attainment areas in the State of Ohio. *J.Environ. Monit. Assess*. 2000, 62, 91-111.

Lin, Y.; Cobourn, W.G. Fuzzy system models combined with nonlinear regression for daily ground-level ozone predictions. *J.Atmos. Environ*. 2007, 41, 3502-3513.

Lu, H.C.; Chang, T.S. Meteorologically adjusted trends of daily maximum ozone concentrations in Taipei, Taiwan. *J.Atmos. Environ*. 2005, 39, 6491-6501.

Mintz, R.; Young, B.R.; Svrcek, W.Y. Fuzzy logic modeling of surface ozone concentrations. *J.Comput. Chem. Eng*. 2005, 29, 2049-2059.

Ojha, S.; Coutinho, J.; Kumar, A. Developing Systems to Forecast Ozone and Particulate Matter Levels. *Environ. Prog*. 2002, 21, J5-J10.

Pastor-Barcenas, O.; Soria-Olivas, E.; Martın-Guerrero, J.D.; Camps-Valls, G.; Carrasco-Rodrıguez, J.L.; Valle-Tascon, S. Unbiased sensitivity analysis and pruning techniques in neural networks for surface ozone modeling. *J.Ecol. Model*. 2005, 182, 149-158.

Prybutok, V.R.; Yi, J.; Mitchell, D. Comparison of neural network models with ARIMA and regression models for prediction of Houston's daily maximum ozone concentrations. *J.Eur J Oper Res*. 2000, 122, 31-40.

Rao, S.T.; Zalewsky, E.; Zurbenko, I.G. Determining temporal and spatial variations in ozone air quality. *J. Air Waste Manage. Assoc*. 1995, 45, 57-61.

Ruiz-Suarez, J.C. Short-term ozone forecasting by artificial neural networks. *J.Adv Eng Softw*. 1995, 23, 143-149.

Ryan, W.F. Forecasting severe ozone episodes in the Baltimore metropolitan area. *J.Atmos. Environ*. 1995, 29, 2387-2398.

Salazar-Ruiz, E.; Ordieres, J.B.; Vergara, E.P.; Capuz-Rizo, S.F. Development and comparative analysis of tropospheric ozone prediction models using linear and artificial intelligence-based models in Mexicali, Baja California (Mexico) and Calexico, California (US). *J.Environ Modell Softw*. 2008, 23, 1056-1069.

Schlink, U.; Volta, M. Grey Box and component models to forecast ozone episodes: A comparison study. *J.Environ. Monit. Assess*. 2000, 65, 313-321.

Sharma, P.; Masuraha, A.; Kumar, A. Development and evaluation of forecasting models for daily maximum ozone concentrations for Detroit, Michigan. *Proceedings of Environmental Data Analysis - Assessing Health and Environmental Impacts*, Developing Policy, and Achieving Regulatory Compliance Conference, Air and Waste Management Association, 2005, Oak Brook, IL.

Terry, W. R.; Lee, J. B.; Kumar, A. Time series analysis in acid rain modeling: evaluation of filling missing values by linear interpolation. *J.Atmos. Environ*. 1986, 20, 1941-1945.

U.S EPA Air Releases (AIRS/AFS), http://www.epa.gov/enviro/html/airs/airs_query.html, Site Visited on December 19, 2003.

U.S. EPA National Ambient Air Quality Standards (NAAQS), http://www.epa.gov/air /criteria.html , Site Visited on December 17, 2003.

Wang, W.; Lu, W.; Wang, X.; Leung, A.Y.T. Prediction of maximum daily ozone level using combined neural network and statistical characteristics. *J.Environ. Int.* 2003, 29, 555-562.

Wise, E.K.; Comrie, A.C.Meteorologically adjusted urban air quality trends in the Southwestern United States. *J.Atmos. Environ*. 2005, 39, 2969-2980.

Zurbenko, I. G. Spectral analysis of non stationary time series. *J.Int. Stat. Rev*. 1991, 59, 163-168.

In: Environmental and Regional Air Pollution
Editors: D. Gallo, R. Mancini ISBN: 978-1-60692-893-6

Chapter 13

ON THE POTENTIAL HIGH SURFACE OZONE CONCENTRATIONS IN SHANGHAI, CHINA

Xuexi Tie[1] and Fuhai Geng[2]
[1]National Center for Atmospheric Research, Boulder, CO, USA
[2]Shanghai Meteorological Bureau, Shanghai, China

High surface ozone concentrations have harmful effects on human's health and crop production. To prevent high ozone concentration event become an important environmental issue. Shanghai Meteorological Bureau (SMB) has established an observational network for VOC, NOx and O_3 measurements in the Shanghai region. The results show that O_3 concentrations are higher in rural area than in center of the city, suggesting that at the present, there are O_3 depression processes in center of city and the air pollutant emissions are not favorable for O_3 chemical formation. However, this situation may quickly change in the future. Chemical O_3 formation is mainly controlled by two factors which are related to human activities, such as (1) photochemical activities and (2) the ozone precursors emitted from the surface. In eastern China, the economical development is very rapid.

For example, the number of automobiles increases from 0.47 million (1996) to 2.54 million (2007), accounting for 540% increase during the last 10 years in Shanghai. At the present, air pollution, especially high aerosol concentrations, become one of the major environmental problems in this region. The high aerosol concentrations play important roles in reduction of sunlight, resulting in decrease in photochemical activities and ozone chemical productions. However, efforts of clean air actions in China could lead to reduction in aerosol concentrations, and enhancements in photochemical activities and ozone production.

The fast economical development also intends to change the emission ratio of NOx/VOCs. Because the emissions of VOCs are strongly related to the usage of oil fuels, the rapid increase in automobiles will lead to decrease in the emission ratio of NOx/VOCs, and the reduction of the ratio is favorable for the enhancement of ozone chemical production nearby cities. Overall, the future conditions are more favorable for ozone chemical production as the reduction of aerosols and emission ratio of NOx/VOCs, and could lead to potential high ozone concentrations in eastern China, including the city of Shanghai.

Rapidly increasing urbanization will be a major environmental driving force in the 21st century. For example, in Shanghai of China, the areas of building have expanded from 172.56 to 702.82 million m^2 between 1990 and 2006. The numbers of automobiles have increased by 540% between 1996 and 2007 (see Figure 1). The rapid growth of urbanization cause wide-ranging potential consequences related to urban environmental problems (such as air pollutions), human health, and for cultivated and natural ecosystems. One of the important effects is the increases in ozone concentrations in the city of Shanghai. The surface ozone is not directly emitted, and it is mainly chemically formed in the lower troposphere when nitrogen oxides (NOx), carbon monoxide (CO) and volatile organic compounds (VOCs) react in the atmosphere with the presence of sunlight. Thus, CO, NOx and VOCs are called ozone precursors, which are mainly released from motor vehicle exhaust, industrial emissions, and chemical solvents. As a result, the rapid increase in urbanizations and industrial activities could lead to changes in ozone concentrations in the city of Shanghai.

Long-term exposure in high ozone concentrations affects human's health. Scientists have found that ozone primarily affects the respiratory system. According to the study by Santa Barbara County Air Pollution Control District, California, roughly one out of three people in the U.S. is at risk of experiencing ozone-related health effects.

Expose in high ozone concentrations normally causes the following effects: (1) Ozone can irritate human's respiratory system. (2) Ozone can reduce lung function. And (3) Ozone can make asthma symptoms worse. Another important environmental effect of ozone is that high concentrations of ozone damage vegetation and crops. The study by the Fifth Framework Programme of the European Communities within the Programme Energy, Environment and Sustainable Development, (the EU FP5 BIOSTRESS Programme) shows that high ozone concentrations affect the surface of plant leaves. Corrosive effects of ozone on the cuticles (a waxy layer that covers the surfaces of plant leaves) have been observed in trees which were artificially fumigated with ozone.

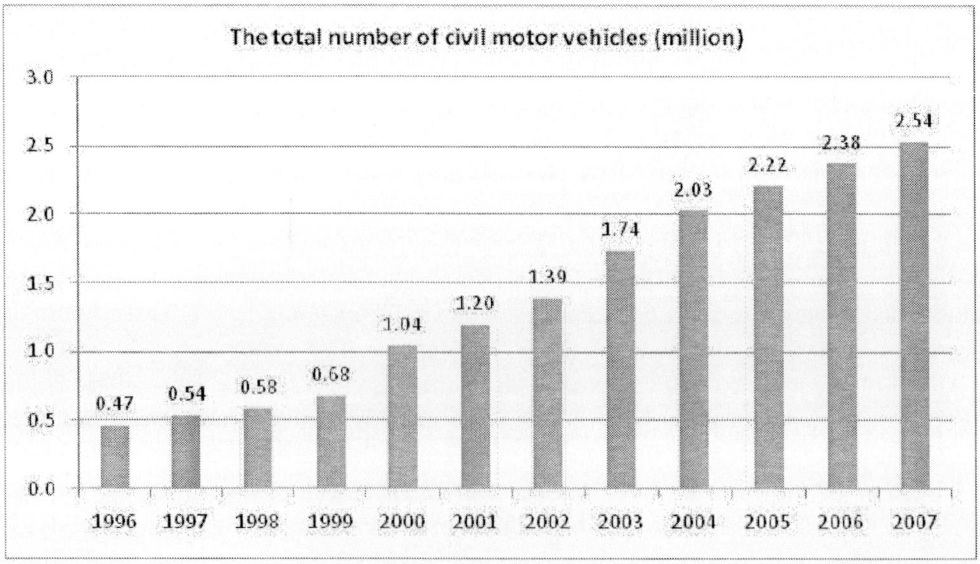

Figure 1. The trends of total number of civil automobiles (million) from 1996 to 2007 in Shanghai.

Study by the CHINA-MAP project [Zhou et.al, 2004] suggests that high ozone concentrations affect rice production in the Yangtz River Delta (YRD) region.

In order to prevent the harmful effects of high ozone concentrations, Shanghai Meteorological Bureau (SMB) has established a VOC laboratory and a monitoring network for VOC, NOx and O_3 measurements in the Shanghai region. The monitoring net work includes 9 measurement locations.

The locations of measurement sites are carefully selected to represent different categories of the land surfaces, including (1) an urban-center site which is under strong influence by urban air pollutant emissions; (2) an urban site which is located in urban area, but is also close to a large city park; (3) a sub-urban site which is located in south of the city with a big complex of chemical industrial zone; (4) a sub-urban site which is located in north of the city with a big complex of steel industrial zone; (5) a rural site which is located in an island surrounding by farm lands; (6) a sub-urban site which is located in west of the city; (7) a sub-urban site which is located in east of the city; (8) a remote site which is located in the edge of an island, and surrounded by a large wet land area; (9) a background site which is located in an small island. The detailed geographical distribution of the 9 sites is indicated in Figure 2.

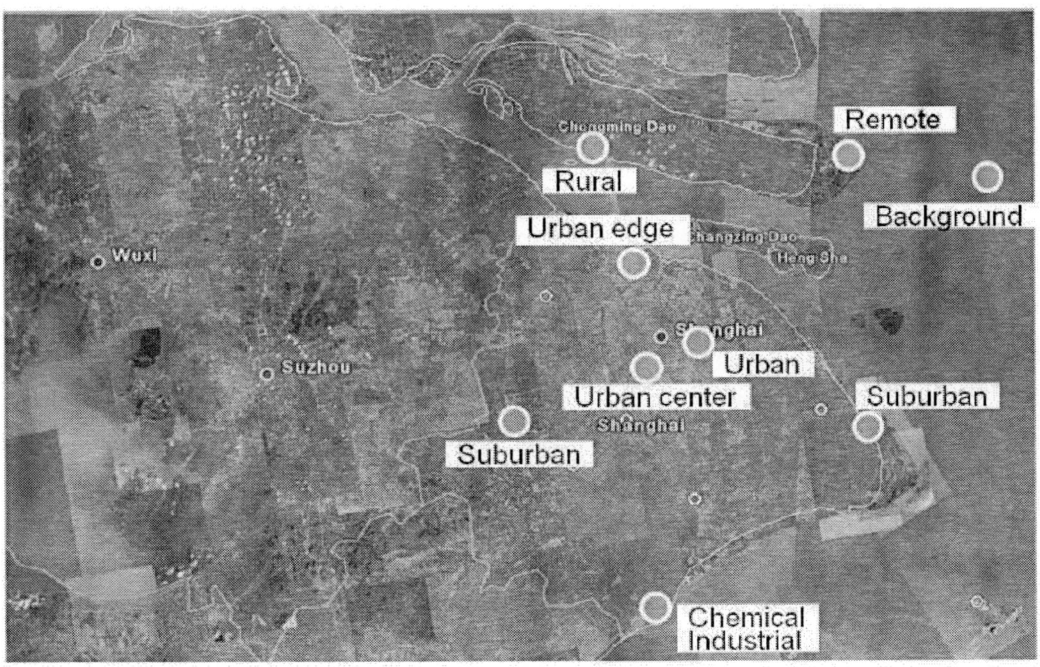

Figure 2. The monitoring net work for ozone and its precursors in the Shanghai region.

The different measurement locations provide insights of O_3 concentrations in various geographical locations with different surface emission characterizations.

As shown in Figure 3, the averaged O_3 concentrations range from 12 to 35 ppbv in 5 locations with relative long-term measurements. The maximum O_3 concentration (35 ppbv) occurs at the rural location, and the minimum O_3 concentration (12 ppbv) takes places at the urban center location.

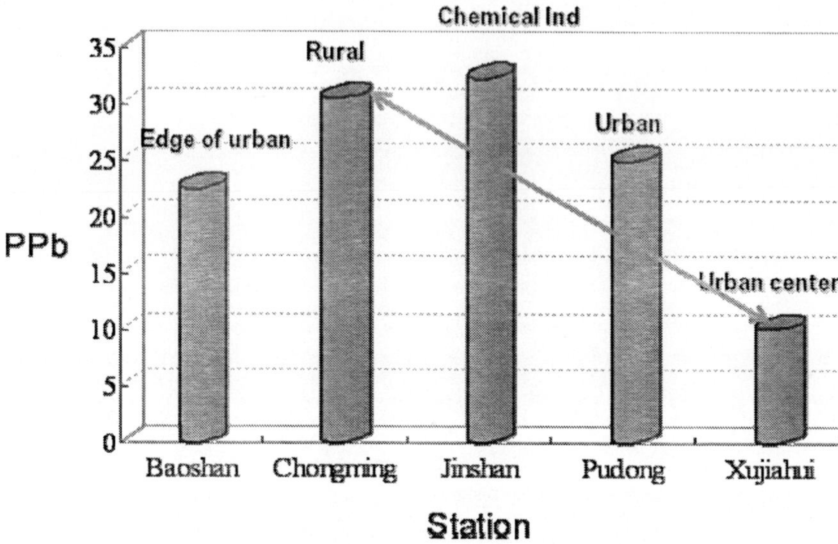

Figure 3. Measured ozone concentrations (ppbv) in the Shanghai region, averaged between January/2006 and September/2007.

The result indicates that in the urban center, which is strongly influenced by urban air pollutant emissions, the O_3 concentrations have lowest values, suggesting that there are O_3 depression processes at this urban site, and the air pollutant emissions are not favorable for O_3 production in the center of the city. As we mentioned before, O_3 is mainly photochemical produced and initiated by the reaction between hydroxyl radical (OH) and VOCs. This O_3 production process continues with the photolysis of NO_2 to finally form O_3 molecular. As a result, the present of both NOx and VOCs is required for the O_3 chemical formation. However, NOx molecular also acts to depress the O_3 chemical formation. For example, under much high NOx conditions, the chemical loss of OH concentrations is mainly due to the reaction of $NO_2 + OH \rightarrow HNO_3$ [Sillman, 1995]. Because OH is a key chemical oxidant to initiate the O_3 chemical formation, the loss of OH due to high NOx concentrations reduces the O_3 chemical formation. At the present, the higher NOx concentrations in the urban area lead to lower OH concentrations than in the rural area, resulting in smaller O_3 production at the urban area in the Shanghai region. Thus, it is important to understand the sensitivity of O_3 concentrations to the changes in NOx and VOC concentrations. This O_3 sensitivity is often defined as either NO_x-sensitive or VOC-sensitive regimes. If O_3 formation is under NO_x-sensitive regime, the increase of NOx emissions will lead to increase in O_3 concentrations. However, if O_3 formation is under VOC-sensitive regime, the increase of NOx emissions will lead to decrease in O_3 concentrations. The O_3 sensitivity in Shanghai can be determined by NOx and VOC measurements with an O_3-isopleth diagram as shown in Figure 4.

The measured results suggest that at the present, the city of Shanghai is in the VOC-limited regime for O_3 formation (at position 1 of Figure 4). Under this condition, increase in NOx concentrations will lead to decrease O_3 concentrations (as pointed by position 2). By contrast, increase in VOC concentrations will enhance O_3 concentrations (as pointed by position 3). From this study, we learn the following scientific aspects; (1) The high NOx concentrations in the urban area of Shanghai depress O_3 concentrations, leading to lower O_3 concentrations in the urban area than in the rural area (as shown in Figure 3). (2) The chemical O_3 formation is strongly determined by the ratio of NOx/VOCs. The chemical O_3 formation is more favorable when the ratio of NOx/VOCs is low. For example, O_3 concentrations are lower with the ratio of about 0.3 (at the position 2 of Figure 4), while are higher with the ratio of 0.2 (at the position 3 of Figure 4).

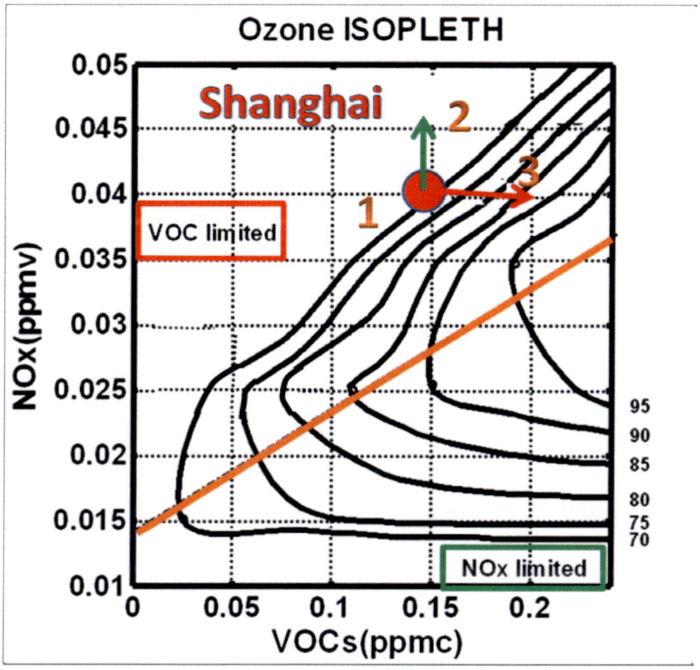

Figure 4. The O3-isopleth diagram. The red dot indicates the measured values of O3 (ppbv), NOx (ppbv), and total VOCs (ppbC) in Shanghai at the present. The brown line separates the O3 sensitivity to VOC-limited regime (above the brown line) and NOx-limited regime (below the brown line).

A recent study suggests that automobile emission is a major VOC emission source in Shanghai [Geng et al., 2007]. As a result, increase in automobiles in Shanghai could lead to lower NOx/VOCs ratio, and higher O_3 concentration in the future.

As we mentioned before, the chemical O_3 formation is initiated by the reaction of OH with VOCs/CO. Thus, the level of OH is a key factor in controlling O_3 formation. The formation of OH is mainly through the following reactions:

$$O_3 + hv \rightarrow O(^1D) + O_2$$

$$O(^1D) + H_2O \rightarrow 2\ OH$$

where hv represent photo-association of O_3 at UV wave length, which is a controlling factor for the formation of OH. Thus, the chemical O_3 formation is strongly determined by the strength of UV radiation, and the level of UV radiation can be considered as "photochemical-limitation" factor for the chemical O_3 formation. The level of UV radiation at the surface of the earth is affected by several important factors, including zenith angle from sun, absorption and scattering of sunlight due to gases, clouds, and aerosol particles. Among these factors, aerosol particles are mainly resulted from anthropogenic emissions. As a result, in high aerosol pollution areas, the photochemical activity is reduced, and the chemical O_3 formation is decreased. Figure 5 illustrates (by schematic pictures) the reduction of O_3 formation due to aerosol particles released from human activities.

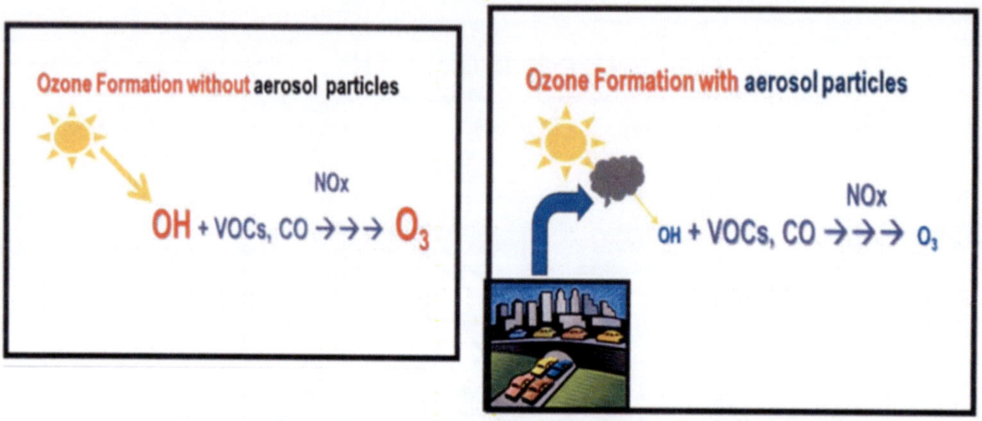

Figure 5. Effects of aerosol particles on O3 formation shown by schematic pictures. The left panel shows that without aerosol particles, the radiation of UV is higher, resulting in higher OH and O3 concentrations. The right panel shows that with aerosol particles released by human activities, the radiation of UV is reduced, and the concentrations of OH and O3 are decreased.

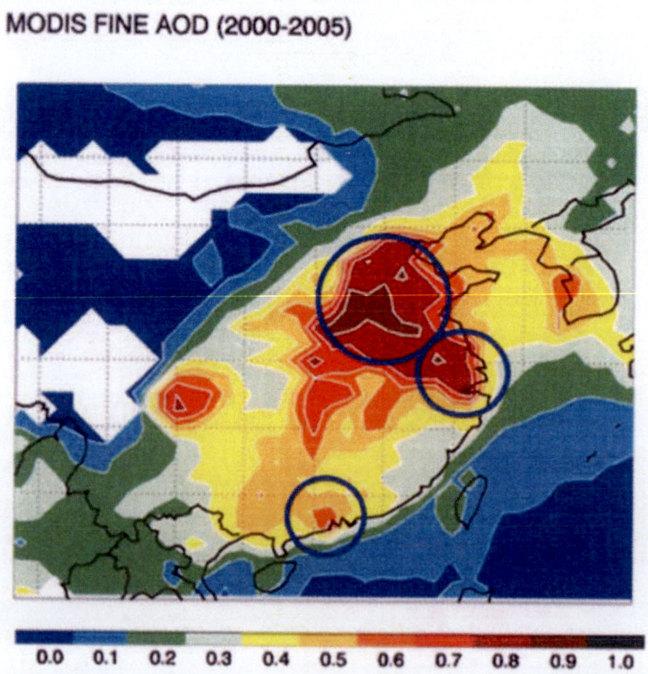

Figure 6. Satellite measurements (MODIS) of aerosol optical depth (AOD) in eastern China for small particles (particle radium less than 0.5 um). The values are averaged between 2000 and 2005. The blue circles show the three important regions (CEC-top circle, YRD-middle circle, and PRD-lower circle).

At the present, the aerosol pollutions due to human activities in eastern China are very heavy. Satellite measurement (MODIS - MODerate resolution Imaging Spectroradiometer) shows that aerosol optical depth (AOD) in the eastern China is about 2-10 times higher than in the US [Tie et al., 2006]. In eastern coast of China, there are three most industrialized regions, including (1) the Central-Eastern China (CEC), (2) the Yangtz-River Delta (YRD) region, and (3) the Pearl-River Delta (PRD) region. In these three regions (see the circled areas in Figure 6), the aerosol loadings are generally high. For example, in the YRD region, the AOD values are often higher than 0.8-0.9, which is about 4-5 times higher than the AOD values in the US. The high aerosol loadings in the YRD region lead to reduction of UV radiation and photochemical activities. As a result, the small photochemical activity limits the chemical formation of O_3.

In summary, the chemical O_3 production is depressed by the high NOx/VOCs ratio and heavy aerosol loadings in the YRD region (or the Shanghai region). However, this region is under rapidly economical developments. The automobiles and oil usages increase dramatically (see Figure 1). Meanwhile, the local government recognizes the environmental problems, such as poor visibility due to aerosol pollutions, and makes efforts to reduce aerosol particles. These changes could have important impacts on the O_3 concentrations in this region. For example, if VOC emissions (mainly due to automobiles and oil usages in Shanghai) increase, it will decrease the NOx/VOCs ratio. The reduction of the NOx/VOCs ratio could lead to higher O_3 production. In addition, the efforts of clean air actions could lead to reductions in aerosol concentrations as well as the enhancements in photochemical activities and chemical O_3 production. Overall, the future conditions are more favorable for ozone chemical productions as the reductions of aerosols and the emission ratio of NOx/VOCs, and could lead to potential high ozone concentrations in eastern China, including the city of Shanghai.

References

Geng, F.H., C.S. Zhao, X. Tang, G.L. Lu, and X. Tie (2007), Analysis of ozone and VOCs measured in Shanghai; *A Case study*, *Atmos. Environ.*, 41, 989-1001.

Sillman, S. (1995), The use of NOy, H_2O_2, and HNO_3 as indicators for ozone-NOXhydrocarbon sensitivity in urban locations, *J. Geophys. Res.*, *100*, 14175-14188.

Tie, X., G. Brasseur, C. Zhao, C. Granier, S. Massie, Y. Qin, P.C. Wang, GL Wang, PC, Yang (2006), Chemical Characterization of Air Pollution in Eastern China and the Eastern United States, *Atmos. Environ., 40.*, 2607-2625.

Zhou et al. (2004), Interaction Between Physical-Chemical Processes in the Lower-level Atmosphere at YRD, *China Metrological Press*.

In: Environmental and Regional Air Pollution
Editors: D. Gallo, R. Mancini
ISBN: 978-1-60692-893-6

Chapter 14

TOXICITY OF SMOKE IN AN AMPHIBIAN MOLLUSC FROM INDIA

Sajal Ray*, Mitali Ray and Niladri Sekhar Bhunia
Department of Zoology, University of Calcutta,
35 Ballygunge Circular Road, Kolkata-700019,
West Bengal, India

ABSTRACT

Agricultural field of India supports a wide variety of biodiversity including specific members of invertebrate phylum Mollusca. Few of the molluscan species inhabit in the regional environment of the subsoil ecosystem of the agricultural field. Microenvironment of the subsoil compartment shelters characteristic infaunas including *Pila globosa* (Mollusca: Gastropoda: Prosobranchia), an edible amphibian snail. *P.globosa* is abundantly distributed both in the terrestrial and freshwater ecosystem of Indian subcontinent. This species is suitably adapted for respiration both in land and water by pulmonary sac and ctenidium respectively. In the post-harvestaion period of winter, *P. globosa* remain in a state of physiological inactivity in its subsoil habitat for months. During post-harvestation period, incineration of dried stems of paddy and dry hay stock is a traditional procedure of manuring in India. During this activity, *P.globosa*, an important member of terrestrial ecosystem is exposed to field smoke for a considerable period of time. In this chapter, toxicity of smoke was examined in depth in relation to hemocyte function of *P. globosa*. Hemocytes are the circulating blood cells of molluscs and are capable of eliciting immunological response under the challenge of xenobiotics and pathogenic infection. They exhibit diverse physiological functions including phagocytosis of nonself particulates, cell aggregation and generation of intracellular cytotoxic molecules. Total hemocyte count is a marker of pollution which was determined both in the control and smoke exposed *P. globosa*. Exposure initiated an alteration in total hemocyte density. In *P. globosa*, discrete subpopulations of hemocyte were identified as lymphoblast, granulocyte, agranulocyte, asterocyte and hyalinocyte. Exposure of *P. globosa* to smoke for 24, 72 and 96 hours (h) in controlled laboratory condition resulted a disruption in subpopulation ratio which never recovered in the

---

* E.mail: raysnailmail@ rediffmail.com

pre-exposed animals maintained in smoke free environment for 7 days. Cell-cell aggregation is a metabolic behavior which was examined in hemocyte population under the exposure of smoke for different spans of exposure. Smoke resulted a disruption in aggregation pattern indicating a state of physiological stress. Phagocytosis of nonself particulates is an important immunological response which was studied in hemocyte under the challenge of yeast in short term culture. Exposure to smoke for 96 hours resulted an increase in phagocytic index (PI) upto a value of 463.33 against a control value 214 which was never restored in smoke free environment. Increment in PI under smoke exposure is considered as perturbation in immunological status of *P.globosa*. Capability of generation of intrahemocyte cytotoxic molecules like nitric oxide (NO), phenoloxidase (PO) and superoxide anion is reported as potential strategy of immunological defence in mollusc. Exposure to smoke yielded a marked inhibition of NO, PO, and superoxide anion generation. Highest inhibition of NO was determined as 0.0386 μM nitrite/ min/ 10^6 cells under the exposure of smoke for 24 hours against a control value of 0.62 μM nitrite/ min/ 10^6 cells. A maximum inhibition of PO was recorded as 1.316 unit/min/ mg of protein under 96 hours of smoke exposure against a control value of 14.11 unit/min/mg of protein. Similar inhibition of generation of superoxide anion was recorded in specimens exposed to smoke in comparison to control. None of these three cytotoxic molecules were restored in 100% when the smoke exposed specimens were maintained in smoke free environment for 7 days. Data indicated a trend of irreversible damage of immunity caused by smoke. Depression in intrahemocyte cytotoxicity is a strong indication of immuno-suppression resulted by smoke in *P.globosa*. Disruption of hemocyte density and inhibition of cytotoxic response may lead to a state of alteration of immunity reported for the first time in *P.globosa*. Such a situation in natural environment may lead to a possible decline in population density of the species in its habitat.

Keywords: *Pila,* Hemocyte, Smoke, Cytotoxicity, Immunity.

INTRODUCTION

Mollusca constitutes a large and wide ranging invertebrate phylum of nearly 100,000 species. It includes varied forms of shelled and shelless animals including snail, clam, scallop, oyster, squid, octopus etc. Many of the species of Mollusca are aquatic and around 10,000 species are areally respiring forms adapted to terrestrial environment. Characteristic diversity of habitat of this group lies parallel to multifarious physiological adaptation which ensures successful reproduction and perpetuation of this group in land and water. Basic body plan of molluscs is unique and their size ranges from few millimeters to 15 meters [1]. In most of the forms, internal visceral organs are situated between ventral muscular foot and dorsally positioned shell. Hard shell is calcareous in nature and a product of secretion of epidermal sheath called mantle. Space between mantle and visceral mass constitutes mantle cavity which houses the gill or ctenidium. Body of the mollusc is unsegmented and the main body cavity is a hoemocoel filled with hemolymph – the blood fluid of both aerial and aquatic respiring forms [2]. *Pila globosa* (Mollusca : Gastropoda : Prosobranchia) is an important amphibian snail of Indian subcontinent capable of respiring both in land and water. This species is popularly known as 'apple snail' and constitutes an important dietary item of human and poultry. *P. globosa* is a prolific breeder and capable of surviving in air and freshwater environment for a long period of time. This unique species bears both ctenidium or gill and lung as respiratory organs in water and air respectively [3]. India with nearly 43.41% of agricultural land supports a wide range of biodiversity in soil and subsoil ecosystem. Subsoil ecosystem is a preferred habitat of diverse fauna including *P. globosa*. This species is abundantly distributed in agricultural field both in rainy and dry seasons. In the dry season of winter, *P. globosa* hibernates deep into the soil around 1- 2 feet beneath the surface. In this characteristic microenvironment, pulmonary sac or lung acts as major respiratory surface of the animal for a period of nearly 4 months. Pulmonary sac of mollusc arises as a fold of body wall in which pulmonary vessels appear during development. Membranous expansion of pulmonary sac overlies the antero-dorsal region of the visceral hump. Large blood vessels develops in the sac which participates in the process of gaseous exchange. Floor of pulmonary sac is thin and moist [3-5]. The respiratory surface of pulmonary sac is suitably adapted for gaseous exchange even in oxygen depleted regional environment. Inspite of its dietary importance, *P.globosa* is yet not cultured scientifically in India. Species is indiscriminately collected from its natural habitat manually for human consumption. India being an important agricultural country of the world practices several traditional methods of manuring for higher yield of crop. In the post-harvestation time period of September and October, the rudiments of dried stems of paddy plants attached to soil are routinely incinerated for increment of soil fertility [6]. This age old process of manuring is practiced in various regions of India as an easy and inexpensive procedure. Incineration of hay in bulk in open agricultural field often cause a severe regional air pollution affecting many aerially respiring soil dwelling species including *P. globosa*. During incineration of hay, *P. globosa* are subjected to intermittent or continuous field smoke pollution resulting a varied degree of physiological stress. In the current chapter, the smoke induced physiological stress was examined in *P. globosa* for the first time. Smoke is a seasonal air pollutant of open agricultural fields of India. During burning of dried plant parts including hay, soil inhabiting invertebrates are exposed to smoke. During tilting of soil in post-hervastation period, *P.globosa* are surfaced and exposed to field smoke pollution. Field burning is a common agricultural process which emits substantial amount of smoke and soot in the regional environment. In India, combustion of fossil fuel and their products are responsible for substantial amount of air pollution in urban area. Considerable degree of air pollution is resulted from burning of noncommercial fuels. Burning of noncommercial fuel emits particulate matter, sulphur dioxide, carbon monoxide and organic substances including hydrocarbon at amount of 7.54, 4.67, 1.97, 2.05 and 6.19 millions of tonnes annually in India respectively [6]. Toxic components of natural smoke initiates adverse health effects in human and higher mammals. No report is available regarding the effect of smoke pollution in amphibian molluscs of India.

Hemolymph, the blood of *P.globosa* is composed of cellular and fluid components. Hemocytes are considered as blood cells of molluscs which circulate in the hemolymph following a definite route. Hemocytes are essential cellular components of molluscan blood and capable of performing diverse physiological functions. These circulatory cells are actively involved in carriage of nutritive agents, gases and various bioactive substances. Moreover, hemocyte is a fundamental component of cell mediated immunity in molluscs and performs diverse immunological functions i.e. phagocytosis, encapsulation, aggregation, generation of cytotoxic molecules and recognition of nonself particulates. Hemocytes are generated from hematopoetic tissues distributed in different locations of molluscan body.

However, in *P.globosa,* the specific location of hematopoietic tissue is yet to be reported. Hemocytes are diverse in relation to their morphology and function. Morphological and functional diversity of molluscan hemocytes are in report. Nomenclature and classification of molluscan hemocytes reported by various workers are not uniform and often raised debates. George and Farguson (1950) classified the hemocytes of gastropods as lymphoid cell, macrophage, and granular amoebocytes [7]. Chakrabarty *et al* (2008) reported morphologically distinct hemocyte subpopulation in an Indian freshwater bivalve [8]. Hemolymph of bivalve contains diverse subpopulations of hemocytes namely agranulocytes, granulocytes, hyalinocytes, with varying size and shape. Total and differential hemocyte counts are considered as potential biomarker of physiological stress of mollusc exposed to xenobiotics and parasites.

Functional status of hemocyte, the chief immunoeffector cells of molluscs was examined in depth under the experimental exposure of natural smoke. Cell density, cell aggregation and phagocytosis of nonself particulates were studied under the exposure of smoke in controlled laboratory conditions. Generation of cytotoxic agents in hemocytes is considered as an immunological response in invertebrates. Nitric oxide, superoxide anion and phenoloxidase are fundamental cytotoxic elements which were estimated in hemocyte of *P. globosa* exposed to smoke. Subsequent analyses of physiological recovery of stress parameters were estimated in post-treated animals maintained in smoke free environment. Data are indicative of substantial dysfunction of hemocytes of the species inhabiting in toxic environment.

MATERIAL AND METHOD

Adult *P. globosa* of uniform size group (5-7 cm) were collected manually from selected agricultural fields of the state of West Bengal of India. These fields did not have the previous history of smoke pollution. After collection, specimens were transported to laboratory in closed containers to avoid possible exposure of ambient air pollution. After proper acclimatization in the laboratory, specimens were exposed to a continuous flow of smoke in a specially designed smoke chamber built for this purpose. The smoke chamber was made of seasoned wood and of rectangular in shape with an internal dimension of 12 X 10 X 10 inches. Chamber had two perforated racks placed horizontally. In the lowest rack of the chamber finely chopped dehydrated hay of paddy of 50 gram weight was placed for incineration. On the top rack, the experimental animals were placed and the hay bed is ignited as a semi closed source of "field smoke" [6]. In every two hour of smoke generation, burnt and semiburnt paddy hay had been replaced routinely by fresh supply of hay of same weight. For control, identical specimens were kept in a dry space devoid of smoke pollution effect. Span of exposure to smoke continued for a period 24, 72 and 96 hours. For recovery experiment, smoke exposed specimens were placed in identical chamber without smoke for a period 7 days before examination.

For cellular analyses of hemocytes, hemolymph of treated and control *P. globosa* were collected aseptically from the heart [9] using sterile needle and syringe and stored in prechilled glass vials. The collected hemocytes were resuspended in sterile snail saline (5mM HEPES, 3.7mM NaOH, 36mM NaCl, 2mM KCl, 2mM $MgCl_2.2H_2O$ and 4 mM $CaCl_2.H_2O$ at pH 7.8) and stored at 4^0C [8]. A part of the freshly collected hemolymph was placed on glass slide in moist chamber and the hemocytes were allowed to settle. Supernatant of the hemolymph was carefully removed by micropipet and hemocytes were transferred to microfuge vial by gentle jetting of sterile snail saline. For all the experiments, the unit density of hemocyte was maintained by adjusting the blood volume with sterile snail saline. Hemocytes are adherent in nature and studied microscopically as glass adhered cells. Viability of hemocytes were routinely screened by staining the cells with 2% trypan blue for 2 minutes following the principle of dye exclusion [10]. Cell suspension with more than 95% hemocyte viability were considered for immunological and biochemical experiments. Glass adhered viable hemocytes were stained with Giemsa's stain [8] for 10 minutes. Enumeration of hemocytes was carried out microscopically with modified Neubaur hemocytometer. Morphological analyses of control and treated hemocytes were performed microscopically (Olympus BX51, Germany) in depth. Phase contrast image analyses of live cells were carried microscopically (Olympus BX51, Germany). Digital documentation of cellular image was carried out using the camera (ProgRes C5, Jenoptik, Germany) attached with the microscopes.

Cell cell aggregation is considered as important metabolic behaviour of hemocytes. Magnitude of hemocyte aggregation was estimated under the controlled exposure of smoke. For aggregation assay, at least two aliquots with same number of viable hemocytes were used. To one aliquot, mixed equal volume of 2.5% glutaraldehyde (Sigma,USA) to obtain 100% free cell suspension. The resulted fixed hemocytes generated the value of "no aggregation" and considered as control. The other aliquot was not treated with glutaraldehyde and contained normal aggregated cells. Total number of free cells were enumerated in both the aliquots and degree of aggregation was determined using a specific formula [11]. Phagocytosis is an important biomarker of environmental pollution and was estimated *in vivo* and *in vitro*. A fixed number of cultured yeast in a concentration of 52.26 X 10^6 cells/0.5 ml PBS was aseptically injected into the foot muscle of *P.globosa* prior to smoke exposure along with control. The entry of yeast resulted an encounter with circulating hemocyte. Hemocytes are reported as potential phagocytes and are capable of performing phagocytosis of nonself particulates under toxic exposure. Phagocytosis of particulates involved several steps i.e. chemotaxis, attachment, ingestion etc. For assay of phagocytosis of yeast *in vivo*, both the treated and control specimens were bled aseptically and hemocytes were isolated. Different steps of phagocytic response including ingestion of yeast was enumerated microscopically and recorded digitally. For assay of phagocytosis *in vitro*, both the control and smoke treated specimens were bled for isolation of hemocytes. Isolated hemocytes were maintained in a short term culture and were challenged with yeast particles at a ratio of 1:10 for engulfment. The cell suspension was incubated in moist chamber for 4 hours for completion of phagocytosis. Post-incubated hemocytes of control and treated specimens were fixed and stained with Giemsa's stain. Engulfment of yeast was estimated microscopically and phagocytic index (PI) was determined [12-14].

Freshly prepared live hemocytes were collected in prechilled glass vials and treated with Triton X- 100 (0.1%) for 30 minutes on ice. Cells were routinely checked for completion of lysis. Lysate supernatant of hemocytes were stored at 4°C and used for biochemical analyses. Estimation of activity of phenoloxidase (EC 1.14.18.1) enzyme was carried out spectrophotometrically (Shimadzu UV–1700, Japan) by monitoring the rate of formation of dopachrome from L-3, 4 DOPA [15]. Estimation involved incubation of cell lysate with 10% SDS (Merck, India) for 15 minutes at 25°C. Post-incubated sample was reacted with L–

DOPA for monitoring the rate of formation of dopachrome whose absorbance was measured at 475 nm. Activity of phenoloxidase was expressed as unit / mg protein/min. Generation of intrahemocyte superoxide anion was estimated spectrophotometrically by nitro blue tetrazolium (NBT) reduction reaction. Estimation involved incubation of hemocytes with 0.3% NBT for 30 minutes at 37°C. Post-incubated hemocytes were washed with 70% methanol and coated with KOH (2M) and DMSO for dissolution of resulting formazan. Blue coloured formazan was measured spectrophotometrically (Shimadzu UV–1700, Japan) at 630 nm and activity of superoxide anion was expressed as OD/ min/10^6 cells [16]. Generation of intrahemocyte nitric oxide was assayed spectrophotometrically (Shimadzu UV – 1700, Japan) at 550 nm following the principle of Griess reaction [17]. Reaction process involved optical measurement of pink coloured nitrite, stable and non-volatile breakdown product of nitric oxide. This reaction system is capable of generating nitrite in biological materials including cell suspension. The molar concentration of nitrite in the sample was determined by using sodium nitrite as standard. Activity of nitric oxide generation was expressed as μM nitrite / min / 10^6 cells. Protein was estimated from hemocyte lysate supernatant using bovine serum albumin as standard [18].

Result and Conclusion

Hemocytes are the chief immunoeffector cells of gastropod molluscs which constitutes the cellular components of circulating blood fluid – the hemolymph [19]. Gastropods are distributed in varied ecosystems which often bear the risk of contamination by xenobiotics of diverse chemistry and toxicity. In India, toxicological screening of pollutants is not being carried out thoroughly in respect to biology of amphibian molluscs. Amphibian molluscs like *P. globosa* are often distributed in the land – water interface. This unique habitat of the species is a less studied area which demands a special attention to toxicologists and ecologists. *P. globosa* bears a wide range of humidity tolerance as evident from its adaptabilities in both dry and wet environment. They are capable of performing gaseous exchange both by pulmonary sac and ctenidium in terrestrial and aquatic habitat respectively. Field smoke pollution is an important anthropogenic activity which adversely affects the terrestrial habitat of various invertebrates including *P. globosa*. In this chapter, the immunobiological response of hemocytes was examined in *P. globosa* exposed to smoke generated from common biofuel in controlled laboratory condition. Molluscs under the exposure of xenobiotics and parasites are able to elicit immunological response through hemocytes distributed in blood. Functional attributes of hemocytes includes modulation of density, cell cell aggregation, phagocytosis, surface adhesion and generation of diverse intracellular molecules of cytotoxicity. A detailed functional analyses of hemocytes were done in depth under the exposure of smoke. Sources of air pollutants are diverse and numerous. Air pollution may be of natural or anthropogenic in origin. In rural India, one of the important anthropogenic sources of air pollution is 'refuse burning' or 'field burning'. The typical pollutants originate from refuse burning include smoke, flyash and soot. Rao (1991) reported [6] a spectrum of commercial and noncommercial sources of air pollution in India which included pollution sources due to burning of firewood, dry cattle dung, vegetable waste and refuse burning.

Morphological description and nomenclature of molluscan hemocyte had been raised a debate among scientists. Malacologists, immunologists and cell biologists proposed several schemes of nomenclature of hemocytes which initiated contradiction [20, 21]. In this chapter, total and differential hemocyte count and hemocyte aggregation were determined in *P. globosa* following its exposure to smoke. Possible restoration of normal cell density, aggregation were examined by maintaining the treated snails in smoke free environment. This experiment was aimed to screen the pattern of recovery of these cell biological parameters [8]. In the hemolymph, specific cells such as lymphoblast cells (Figure IA), agranulocytes (Figure IB), granulocytes (Figures IC,ID, IE and IM), hyalinocytes (Figures IF,IG and IH) and asterocytes (Figures II, IJ, IK and IL) were reported using bright field and phase contrast microscopes.

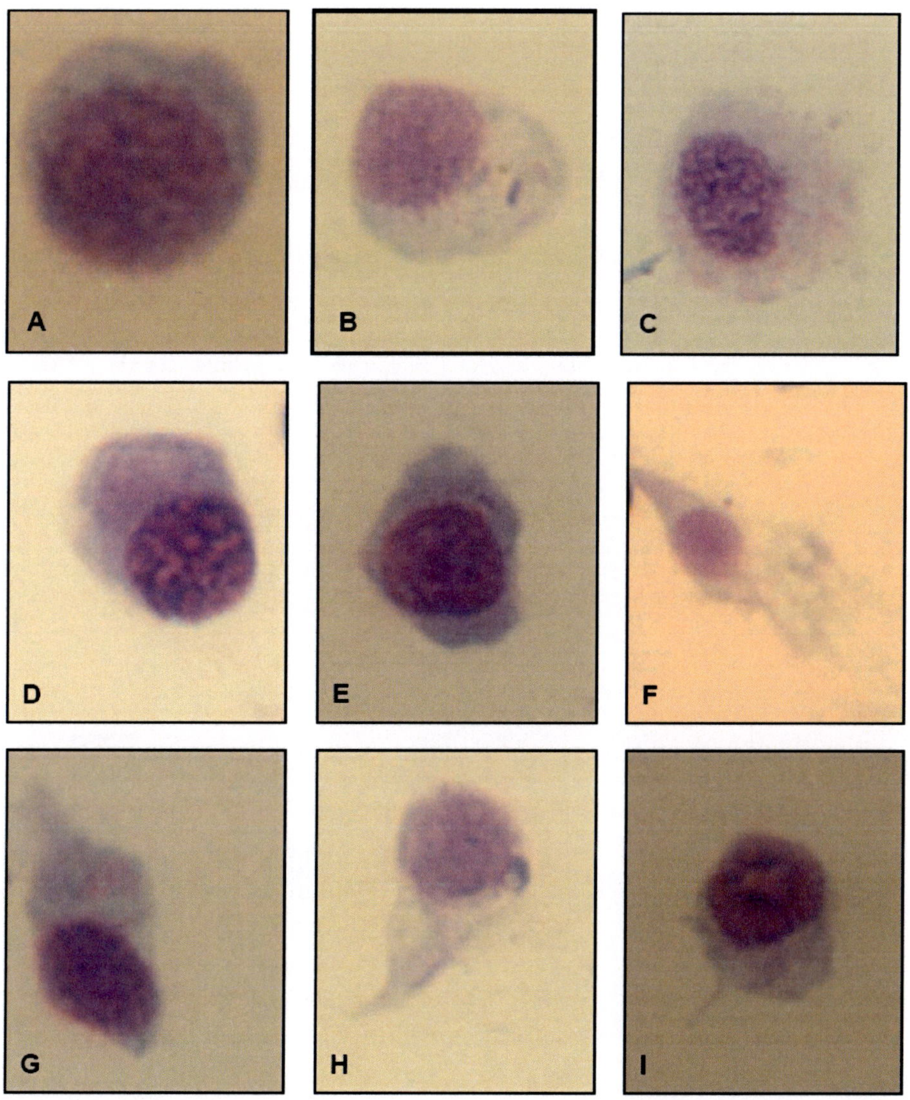

Figure I. (Continued).

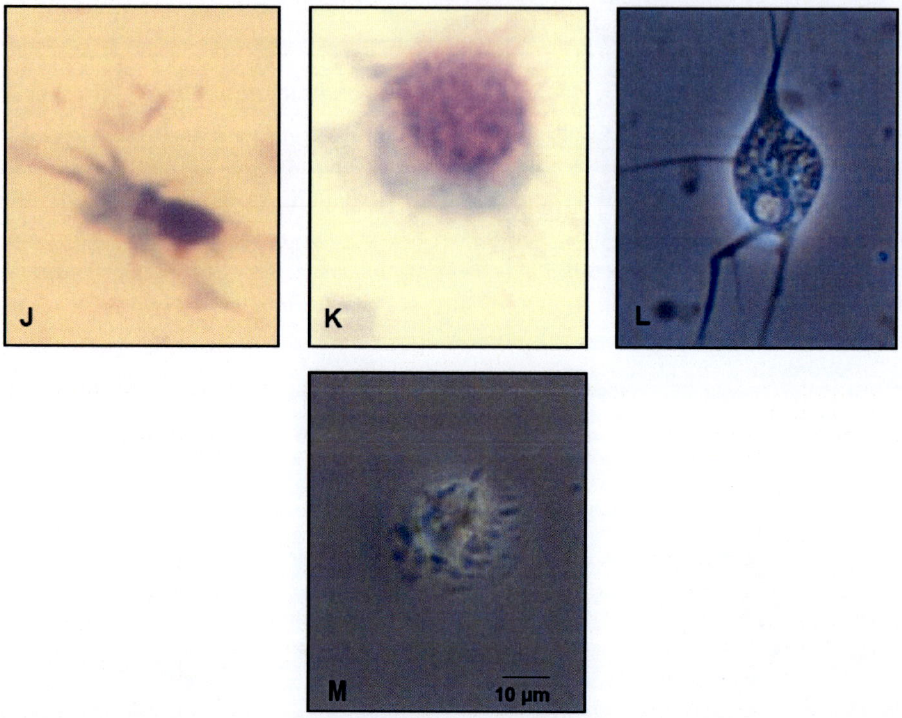

Figure I. Bright field and phase contrast micrograph of the hemocyte of *P. globosa*. (A)Lymphoblast. (B) Agranulocyte. (C), (D), (E) and (M) Granulocytes. (F), (G) and (H) Hylinocytes. (I), (J), (K) and (L) Asterocytes. Scale =10µm.

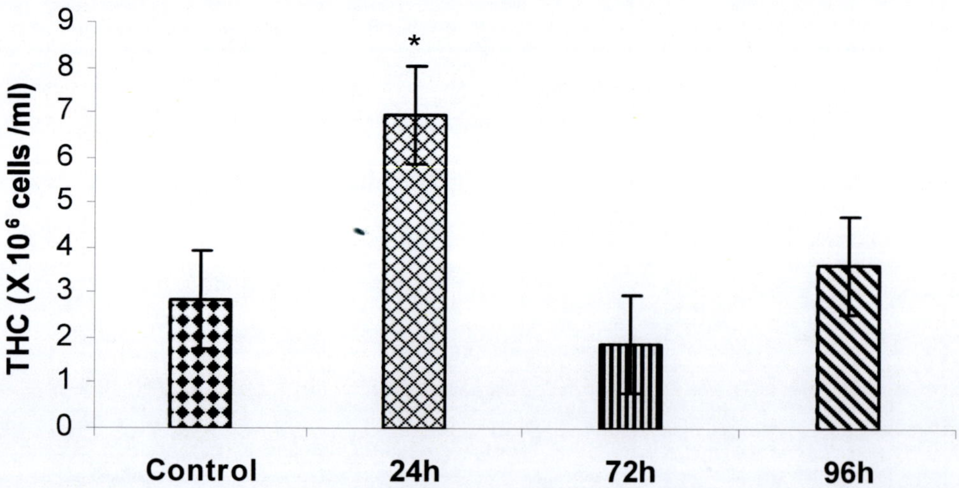

Figure II. Alteration of total hemocyte count (THC) of *P.globosa* under the exposure of smoke for 24h,72h and 96 h .Data expressed as mean ± standard error. *$p < 0.05$. ($n = 3$).

Morphology of these hemocyte variants were recorded as oval, round, star shaped with or without pseudopodial extensions with differential tinctoriality. Morphology of nucleus and cytoplasm were not uniform due to difference in functional status and maturation stage. Total hemocyte count (THC) was determined in *P.globosa* exposed to smoke for 24, 72 and 96 hours of duration along with control. Exposure to smoke for 3 days resulted an increase in THC as 6.93 X 10^6 cells/ ml against a control value of 2.82 X 10^6 cells/ml (Figure II). Exposure to smoke for 72 and 96 hours yielded a value of 1.88 X 10^6 cells/ml and 3.62 X 10^6 cells/ml respectively. Exposure to smoke initiated a fluctuation of hemocyte density which is indicative of destabilization of hematological homeostasis. For analyses of possible recovery of normal density of hemocytes, the pre-exposed animals were kept in smoke free environment for 7 days and THC were determined. A 74.71% of restoration of hemocyte density was achieved by the animals which were exposed to smoke for 96 hours. The groups of animals exposed to smoke for 72 hours of duration did not recover the normal hemocyte density. A partial recovery of 39.53% of THC were recorded in the groups exposed to smoke for 72 hours (Table 1).

Data are suggestive of smoke induced alteration of THC which was irreversible and detrimental for *P.globosa*. Analyses of subpopulation densities of circulatory hemocytes were carried out extensively. In the control specimen, subpopulation maintained a definitive cell ratio (Figure III). Exposure of *P.globosa* to 24, 72 and 96 hours of smoke resulted a marked alteration in control ratio. Result is suggestive of smoke induced disruption of cellar dynamics in the blood of experimental specimen. Partial restoration of control ratio was observed in the specimens maintained in the smoke free environment for 7 days of span.

Cell - cell aggregation is considered as an important metabolic behaviour [11] which is often interfered with alteration of immunological status caused by xenobiotics and parasites. Hemocyte aggregation (Figures IVA and IVB) was studied in depth under the exposure of smoke for 24 and 96 hours along with control. Exposure to 24 and 96 hrs yielded no difference in aggregation response as compared to control (Figure V). *P.globosa* exposed to smoke for 24 and 96 hours exhibited 31.59% and 32.65% of recovery of hemocyte aggregation when the specimens were maintained in smoke free environment for 7 days respectively. Recovery appears to be incomplete for the studied period of time (Table 2). Phagocytosis of nonself particulates is a classical immunological response expressed by discrete population of hemocytes [22, 19].

Table 1. Percent restoration of total hemocyte count of smoke exposed *P. globosa* maintained in smoke free environment for 7 days

Duration of smoke treatment (hour)	% restoration of THC in smoke free environment
24	71.38
72	39.53
96	74.71

Table 2. Recovery of hemocyte aggregation of smoke exposed *P. globosa* maintained in smoke free environment for 7 days

Duration of smoke treatment (hour)	% restoration of aggregating cells in smoke free environment
24	31.59
96	32.65

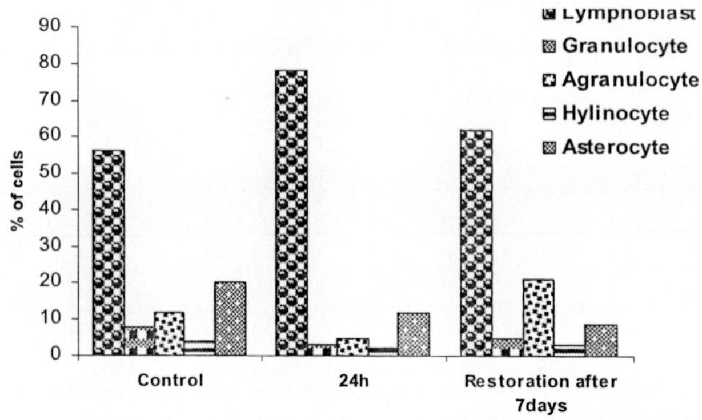

A

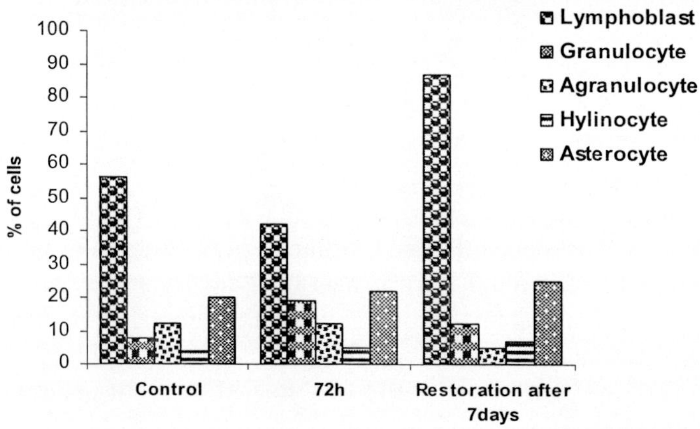

B

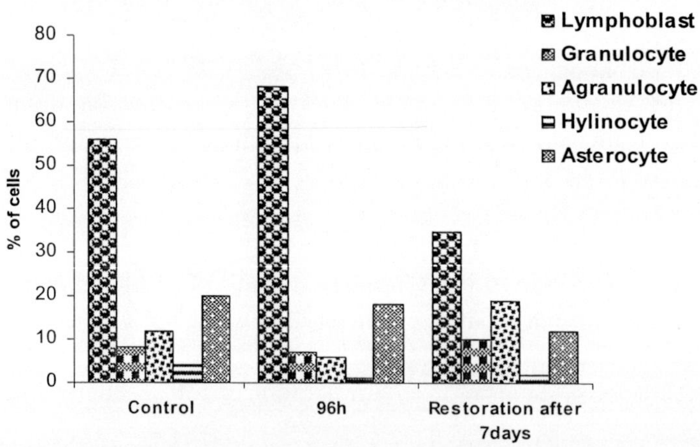

C

Figure III. Hemocyte spectrum of *P.globosa* exposed to smoke for (A) 24h, (B)72h and (C) 96 h and subsequent recovery after 7 days of maintenanance in smoke free environment.

Asterocytes (Figures VIA and VIB) and agranulocytes (Figure VIC) expressed higher degree of phagocytic response in comparison to other cell types. Under the challenge of yeast, both the cells extended pseudopodial extensions for particle entrapment. Within specified period of incubation, yeast particles were engulfed by both types of hemocytes. Asterocytes were characterized by their extended star shaped pseudopodial projections, whereas the agranulocyte exhibited single sided extensions of pseudopodium capable of engulfment of yeast.

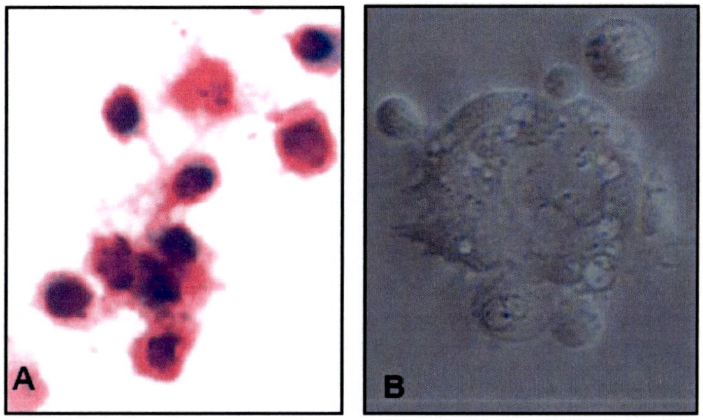

Figure IV. Bright field (A) and phase contrast (B) micrograph of the aggregating hemocytes of *P. globosa*.

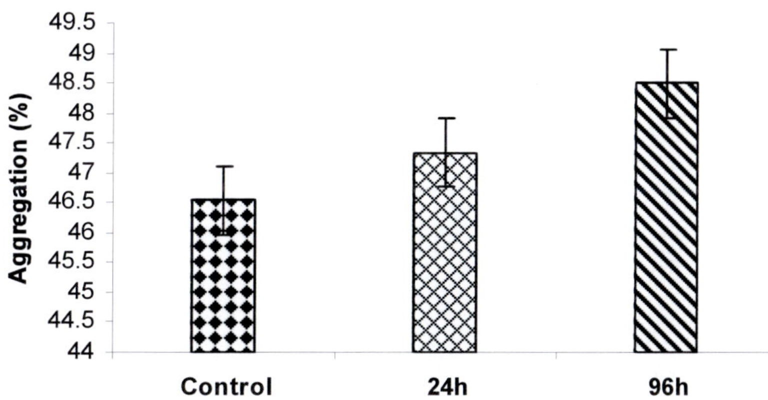

Figure V. Modulation of aggregation of hemocyte of *P. globosa* under the exposure of smoke for 24h, 72h and 96 h. Data expressed as percentage. * $p<0.05$ ($n = 3$).

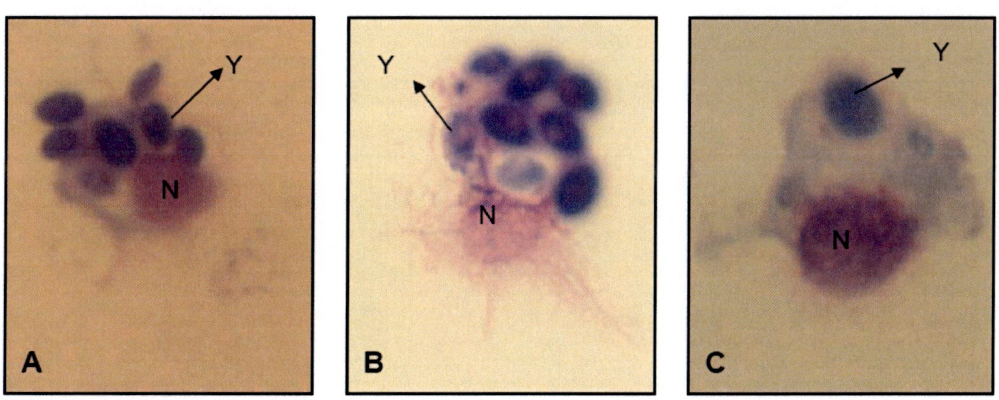

N : Nucleus. Y : Yeast.

Figure VI. Phagocytosis of yeast by asterocyte (A, B) and agranulocyte (C) of *P.globosa*.

Phagocytic efficiency of hemocytes was expressed as phagocytic index [12-14] which was determined in *P.globosa* exposed to smoke for 24, 72 and 96 hours along with control. A progressive increase (Figure VII) of PI was recorded with a maximum PI of 463.33 in hemocyte exposed for 96 hours against a control value of 214. Data exhibited an exposure dependent response. For recovery analyses, pre-exposed snails were kept in smoke free chamber upto a maximum period of 96 hours and PI was determined at different intervals of 24, 72 and 96 hours. None of the specimen exhibited full recovery of phagocytic efficiency (Table 3). A maximum recovery of 64.24% was recorded in the group maintained in smoke free environment for 7 days. Smoke induced depression of PI of hemocyte indicated an irreversible impairment within the studied span of exposure.

Table 3. Percent restoration of phagocytic index (PI) of smoke exposed *P.globosa* maintained in smoke free environment for 7 days

Duration of smoke treatment (hour)	% restoration of PI in smoke free environment
24	11.27
72	50.64
96	64.24

Table 4. Percent recovery of generation of intrahemocyte cytotoxic agents of *P. globosa* maintained in smoke free environment for 7 days

Duration of smoke treatment (hour)	Generation of intrahemocyte cytotoxic agents in smoke free environment		
	Nitric Oxide (μM/min/10^6cells)	Phenoloxidase (Unit/ min/ mg protein)	Superoxide anion (OD/ min)
24	8.26	47.45	49.54
72	42.67	47.54	33.47
96	64.68	56.24	34.62

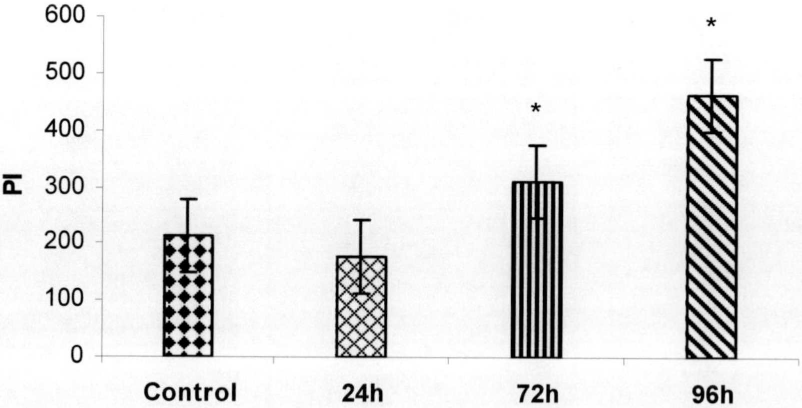

Figure VII. Dynamics of phagocytic response of hemocyte of *P. globosa* under the challenge of yeast exposed to smoke for 24h, 72h and 96 h. Data expressed as mean ± standard error. *p< 0.05.(n = 3).

Generation of intrahemocyte nitric oxide was decreased significantly in exposed group. Highest depression of nitric oxide was recorded against 24 hours of exposure as 0.0386μM nitrite/ min/10^6 cells against the control value of 0.62μM nitrite/ min/ 10^6 cells (Figure VIII).

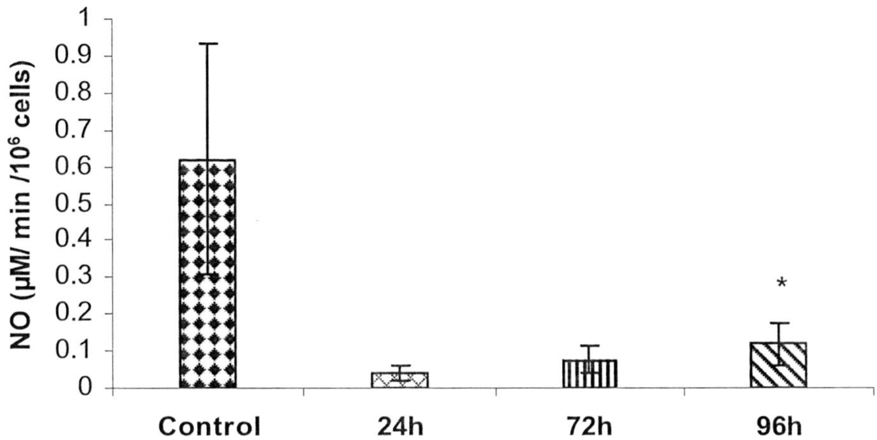

Figure VIII. Alteration of intrahemocyte nitric oxide (NO) activity of *P. globosa* under the exposure of smoke for 24h, 72h, 96 h. Data expressed as mean ± standard error. *$p < 0.05$. (n =3).

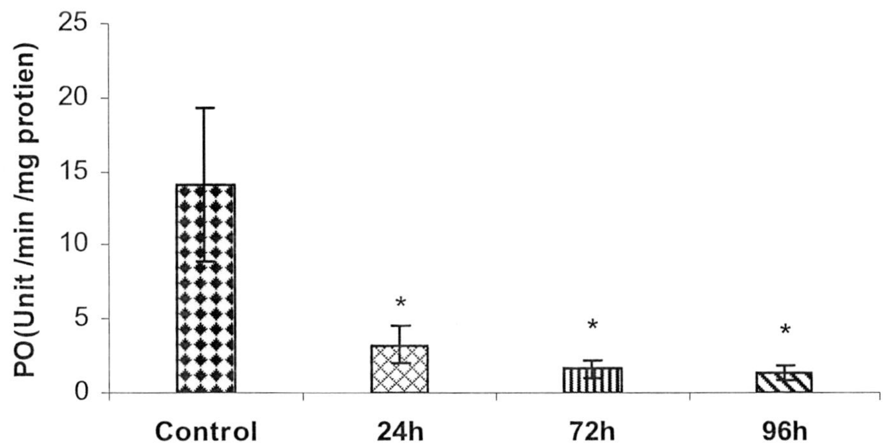

Figure IX. Alteration of intrahemocyte phenoloxidase (PO) activity of *P. globosa* under the exposure of smoke for 24h, 72h and 96 h. Data expressed as mean ± standard error. * $p < 0.05$. (n =3).

Percent of recovery of nitric oxide did not exceed 64.68 % for all the groups maintained in smoke free chamber upto 7 days (Table 4). Like NO, the intrahemocyte phenoloxidase activity was inhibited significantly (Figure IX) under the exposure of smoke. The maximum level of inhibition was recorded as 1.316 unit/min/mg protein against a control value of 14.11 unit/min/mg protein. Activity of PO did not recover more than 56.24% when the pre-exposed snails were maintained in smoke free environment (Table 4). Both NO and PO are principal cytotoxic agents engaged in destruction of intracellular parasites or microbes. These cytotoxic agents are generated as innate immune response of molluscs and considered as important defence molecules [23-25]. Exposure to smoke resulted a decrease in the activity of both these agents whose activity did not recover fully in smoke free environment where the treated specimens were maintained for 7days of span. Generation of superoxide anion in the hemocytes is aimed at intracellular destruction of invading pathogens and is a well studied parameter of invertebrate immunity [26]. Exposure to smoke resulted severe depression of superoxide anion generation under all three spans of exposure (Figure X). Activity of superoxide anion did not recover beyond 49.54% after their maintenance in smoke free environment (Table 4).

Primary defence reaction in invertebrates is identified as "cellular response" [21]. Cellular response plays a pivotal role in molluscan immunity under the challenge of parasites and xenobiotics. Molluscs lack immunoglobulins and mostly depend on circulatory hemocytes to elicit immunological reactions towards xenobiotics and invading pathogens. Phagocytosis, aggregation, generation of cytotoxic agents are discrete cell biological reactions offered by gastropod molluscs under adverse physiological state. Report on toxicity of smoke in *P.globosa* in relation to hemocyte function is absent in current literature. Disruption of hemocyte density and subpopulation ratio is indicative of irreversible breakdown of blood cell homeostasis. Acute depression in cytotoxic response was recorded under smoke exposure which did not restore after maintaining the smoke exposed specimens in smoke free environment for 7 days. Invertebrates including molluscs are dependent on both oxygen dependent and oxygen independent modes of killing of intracellular pathogens [25, 26]. In the biologically unsafe environment, the molluscs are capable of eliciting phagocytic response against invading pathogens [27] and can generate nitric oxide, phenoloxidase and superoxide anion as principal defence molecules. This innate immune response of molluscs help the animals to combat with toxic pathogens abundantly distributed in their natural habitat. Smoke induced decline in activity of cytotoxic molecules may lead to opportunistic growth of microorganism in the blood and other tissues of *P. globosa*. This situation may lead to rapid loss of this important edible species from agricultural fields of India. In this chapter, restricted practice of field burning is recommended to control smoke pollution as a measure to conserve this species in its natural habitat.

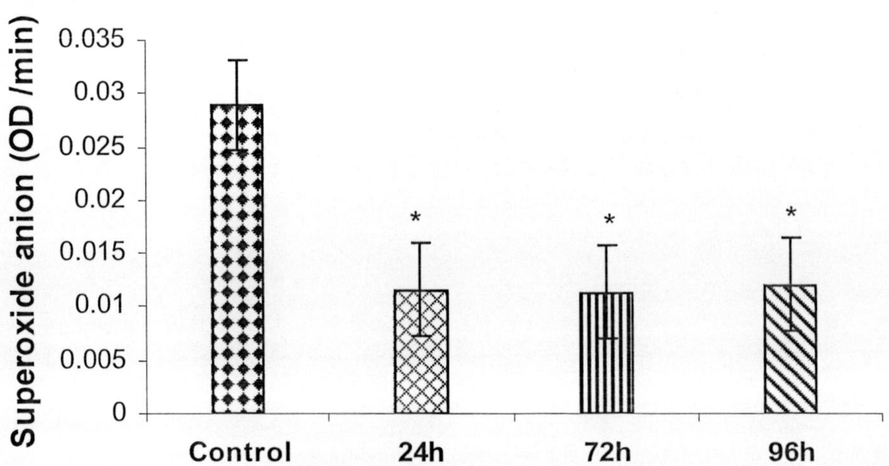

Figure X. Alteration of intrahemocyte superoxide anion (O_2^-) generation of *P. globosa* under the exposure of smoke for 24h, 72h and 96 h. Data expressed as mean ± standard error. *p< 0.05. (n =3).

REFERENCES

[1] Moore, J. (2006). *An introduction to the invertebrates* (second Ed). pp- 121. Cambridge, U.K.: Cambridge University Press.

[2] Hyman, L. H. (1967). *The invertebrates*, Mollusca I. pp-169-402. New York, London: Mc Graw Hill.

[3] Purchon, R. D. (1968). *The biology of Mollusca*. pp- 26-32. Oxford, London: Pergamon Press.

[4] Huxley, T.H. and Martin, H.N. (1892). The pulmonary sac and its associated structures. In G.B. Howes, and D. H. Scott (Eds.), *A course of elementary instruction in practical biology* (pp. 272-304). London, Mc Milian and Co.

[5] Ghiretti, F. (1966). Respiration. In K. M. Wilbur, and C. M. Yonge (Eds.), *Physiology of Mollusca* (pp. 175-193). New York, London, Academic Press.

[6] Rao, C. S. (1991). *Environmental pollution control engineering*. pp. 30- 76.New Delhi, Calcutta, London: Wiley Eastern Limited.

[7] George, W. C. and Ferguson, J. H. (1950). The blood of gastropod mollusc. *J. Morp., 86*, 315-327.

[8] Chakraborty, S., Ray, M. and Ray, S. (2008). Sodium arsenite induced alteration of hemocyte density of *Lamellidens marginalis*- an edible mollusk from India. *Clean: Soil Air Water, 36(2)*, 195-200.

[9] Bayne, C. J., Buckley, P. M. and Dewan, P. C. (1980). Macrophage-like hemocytes of resistant *Biomphalaria glabrata* are cytotoxic for sporocysts of *Schistosoma mansoni in vitro*. *J. Parasitol., 66*, 413.

[10] Burch, S. W., Fitzpatrick, L. C., Goven, A. J., Venables, B. J. and Giggleman, M. A. (1999). *In vitro* earthworm *Lumbricus terrestris* coelomocyte assay for use in terrestrial toxicity identification evaluation. *Bull. Environ. Contam. Toxicol.*, *62,* 547-554.

[11] Chen, J. and Bayne, C. J. (1995). Bivalve mollusc hemocyte behaviors: characterization of hemocyte aggregation and their inhibition in the California mussel *(Mytilus californianus*). *Biol. Bull.*, *188,* 255-266.

[12] Carballal, M. J., Lopez, C., Avzevedo, C. and Villalba, A. (1997) . *In vitro* study of phagocytic ability of *Mytilus galloprovincialis* Lmk. Hemocytes. *Fish and Shellfish Immunol., 7,* 403-416.

[13] Fryer, S. E. and Bayne, C. J. (1989) . Opsonization of yeast by the plasma of *Biomphalaria glabrata* (Gastropoda): a strain-specific, time-dependent process. *Parasite Immunol., 11,* 269-278.

[14] Sauve, S., Broussea, P., Pellerin, J., Morin, Y., Senecal, L., Goudreau, P. and Fournier, M. (2002). Phagocytic activity of marine and freshwater bivalves : *in vitro* exposure of hemocytes to metals (Ag, Cd, Hg and Zn). *Aquat. Toxicol., 508 (3-4),* 189-200.

[15] Tanner, C. A., Burnett, L. E. and Burnett, K. G. (2006). The effects of hypoxia and pH on phenoloxidase activity in the atlantic blue crab, *Callinectes sapidus*. *Comp. Biochem. Physiol., 144,* 218-223.

[16] Sung, H.H., Yang, Y.L. and Song, Y.L. (1996). Enhancement of microbicidal activity in the tiger shrimp *Penaeus monodon* via immunostimulation. *J. Crus. Biol., 6*, 278-284.

[17] Green, L. C., Wagner, D. A., Glogowski, J., Skipper, P. L., Wishnok, J.S. and Tannenbaum, S. (1982) . Analysis of nitrate, nitrite and (^{15}N) nitrate in biologic fluids. *Annal. Biochem*, *126,* 131-138.

[18] Lowry, O. H., Rosebrough, N. J., A.L., Randall, R. J. (1951). Protien measurement with the folin phenol reagent. *J. Biol. Chem., 193,* 265-275.

[19] Bayne, C.J. (1990). Phagocytosis and nonself recognition in invertebrate. *Comp. Immunol,. 40*, 723-731.

[20] Hose, J.E., Martin, G.G. and Gerand, A.S. (1990). A decapod hemocyte classification scheme integrating morphology, cytochemistry and function. *Biol. Bull., 178*, 33-45.

[21] Adamowicz, A and Balaczek, M. (2003). Blood cells morphology of the snail *Helix aspersa maxima* (Helicidae). *Zoologica Poloniac., 48,* 93-101.

[22] Scott, M.T. (1971). Recognition of foreigness in invertebrates. *In viro* studies of cockroach phagocytic hemocytes. *Immunol., 21*, 817-827.

[23] Soderhall, K. and Cerenius, L. (1998). Rate of the prophenoloxidase-activating system in invertebrate immunity. *Curr. Opin. Immunol., 10*, 23-28.

[24] Cerenius, L. and Soderhall, K. (2004). The prophenoloxidase-activating system in invertebrates. *Immunol. Rev., 198*, 116-126.

[25] Arumugan, M., Romesstand, B. and Torreilles, J. (2000). Nitrite released in hemocytes from *Mytilus galloprovincialis, Crassostrea gigas* and *Ruditapes decussaus* upon stimulation with phorbol myristate acetate. *Aqua.Living. Resour., 13,* 173-177.

[26] Takahashi, K., Akaike,T., Sato, K., Mori, K. and Maeda, H. (1993). Superoxide anion generation by pacific oyster (*Crassostrea gigas*) hemocyte: identification by electron spin resonance spin trapping and chemiluminiscence analysis. *Comp. Biochem. Physiol., 105B*, 35-41.

[27] Saha, S., Ray, M. and Ray, S. (2008). Kinetics of nonself surface adhesion and phagocytic response of hemocyte of *Scylla serrata* exposed to sodium arsenite. *Toxicol. Int., 15*, 15-19.

In: Environmental and Regional Air Pollution
Editors: D. Gallo, R. Mancini
ISBN: 978-1-60692-893-6

Chapter 15

A Spatial Econometric Approach to Measuring Damage to Health Associated to the Forest Burning in Brazilian Amazon

***Mário Jorge Mendonça*[*], *Thiago Guerrera Martins*[**] and *Paulo Roberto Loureiro*[***]**

[*]Institute for Applied Economic Research (IPEA) and
École des Hautes Études en Sciences Sociales (EHESS)
[**]IPEA and Universidade Federal do Rio de Janeiro (UFRJ)
[***]Universidade Católica de Brasília (UCB)

I. Introduction

In Brazilian Amazon forest burning is an important method used to clear and prepare land for pasture or crops. Notwithstanding, the smoke generated in this process can produce undesirable negative effects on health-induced problems by the polluted air. The most serious are respiratory ailments brought about by the inhalation of particulate material present in the smoke, and carbon-derivate gases that appear as subproducts of this practice. In this case, forest burning is a factor contributing to air pollution and, as pointed out in other studies, there is a positive correlation between the high pollution rates and the incidence of these types of ailments.

There has been a sharp increase in the incidence of respiratory diseases during periods in which this practice is most frequent (Nepstad *et alli*, 1999). Naturally, the adverse effects of this practice do translate, though not directly observable, into an increase in the cases of morbity due to respiratory ailments. Some studies point out estimates (indicating)that point towards the possibility that there may also be an atmospheric pollution impact upon the incidence of mortality due to respiratory ailments (Thomas, 1995). In this sense we also check this type of inference can be done.

Different methods may be used here, and their respective application depends on how the data are compacted. The level of aggregation normally appearing in the literature may be micro, for individual data (Ostro, 1987); as well as macro, for cities, districts, metropolitan areas, etc. (Seroa da Motta and Mendes, 1995). The basic idea here is to construct a model that associates the incidence of disease cases with the most probable determining factors. Those parameters that identify the impacts restricted to each of these causes may be individualized from the model.

Concerning the use of a model to estimate the effect of air pollution on the cases mortality and morbity rates associated with respiratory diseases some remarks need be stated. A first difficulty lies in the fact that spatial heterogeneity among cross-section units (municipalities) has to be taken into account in order to contemplate the geographic, environmental and economic diversity among those different units and the effect of their mutual interactions. In this case, we take into account the heterogeneity among land units using the panel data[1] model [(Hsiao, 1995), (Baltagi, 1995), (Arellano, 2003), etc.], mixing information concerning variation among individual units with variations taking place over time.

The object of this study is clearly the spatial process that results from the complex interplay of many phenomena occurring in a much extended spatial domain, it is obviously a spatial phenomenon and we may expect spatial autocorrelation to be present in the data. Due to geographic and climatic conditions it is possible that the emissions of smoke regard certain municipality may affect not only its own area but also the neighboring municipalities. In this sense we expect to detect transboundary effect due to the spatial nature of this phenomenon. Not to take into account spatial autocorrelation induces the bias in the model due to model misspecification. In order to account for spatial interactions in the data we used spatial regression with panel data models. We hope that the results affirm the importance of spillovers effects in determining the number of morbity and mortality cases associated to respiratory diseases in a consequence of the use of fire in Brazilian Amazon.

In this sense this study estimates a model of damage to human health due to the use of fire in the Amazon for the 1996-2000 period. This study investigates whether those effects on the region's population are statistically significant, using as variables mortality and morbity rates associated with respiratory diseases. Based on the data of municipalities of Brazilian Amazon for the period 1996-99 we will show that this forest burning practice does have significant negative health effect not only on morbity but also on mortality occurrences in this region.

This study is divided as follows. In Section 2, we present the methodology used to estimate the model. The database is described in section 3. The Section 4 contains a short description of the econometric method used in this study that take into account panel data and spatial autocorrelation in the same framework. The results are discussed in Sections 5. Finally, we offer some concluding remarks in Section 6.

2. METHODOLOGY.

In the case of damage to human health, it is necessary to previously identify the morbity and mortality cases that are exclusively associated to the incidence of fire (accidental or

[1]. Important sources of variation may be left out if the data is only pooled in a single (temporal or spatial) dimension, and more precise parameter estimates can be obtained in panel approaches that explore the variability present in the data both across counties and within counties over time.

intentional). As mentioned previously, damage to health, which is the object of this study, is damage arising from respiratory ailments. However, it is not possible to directly know the number of individuals affected by any one of these ailments due to the inhaling of smoke, for, as is the case of any other ailment, this may due to a number of causes. A classic procedure used to circumvent this type of problem is the estimation of a dose-response function (econometric model) for those morbity and mortality cases related to respiratory ailments and the incidence of fire for the same period.

This model makes it possible to associate the incidence of morbity to independent variables such as area burned, as well as important socioeconomic variables such as population and income. Variables related to local and seasonal climate factors may also be included. For example, it is reasonable to suppose that a more prolonged dry season would be associated to a higher incidence of said ailments in the Amazon region. This is due to the fact that forest burnings take place during the dry season, and if the season is a prolonged one, this considerably increases the risk of fires. The concentration of lumber mills and long stretches of unpaved roads may further contribute to the increase in the number of cases of respiratory ailments due, respectively, to the increased inhalation of particulate material such as airborne debris and dust, that may present a risk to human health.

Therefore, in order to determine the number of morbity and mortality, it is necessary to construct the model by relating said endogenous variables to the several possible exogenous variables, as mentioned earlier. With this in mind, the model proposed here is specified as follows:

$$y_{it} = F(S_{it}, Z_{it}, \rho, \alpha_i, \varepsilon_{it}) \tag{1}$$

where:

y_{it} is the number of in-patient cases of respiratory ailments or mortality in district i for period t;

S_{it} is the area burned in district i for period t;

Z_{it} is the control variable vector for district i for period t;

ρ is the effect of the neighborhoods on the district i;

α_i is the fixed effect of the district i; and

ε_{it} is the random error district i for period t.

As may be perceived, the model's data structure is a panel whose units are represented by the districts in the Amazon region, and the time frame is defined in years. Regarding the independent variable vector, it must be pointed out that the signs obtained for the parameters must obey some sort of *a priori* criterion. For example, regarding the area burned, the sign for this parameter must be positive, for the greater the area burned, the higher the production of smoke, and thus higher the probability of the appearance of respiratory ailment cases.

The independent variables to be tested in this model, as well as their respective designations, are as follows: area burned (*fire*), population (*pop*), cattle raising quantum index (*indbov*), real value of agricultural production (*vrprod*), real value of logs (*vrmad*), volume of sawed wood (*vol*), and the total extension of unpaved roads (*npav*). A brief explanation of each follows below.

The variable *fire* reflects the forest area affected by fire in each district. Naturally, our desire here is to find a positive correlation between this variable and the incidence of cases of morbity due to respiratory ailments. The inclusion of the population as an independent variable, defined here as *pop* aims at serving as a scale variable for the model. Naturally, for a particular region in which there are no great differences among the locations within its borders, a higher number of respiratory ailment cases would be found in those regions in which the population is greater. Once again, a positive sign is expected for this variable's estimated coefficient.

With regard to the *indbov, vrmad,* and *vrprod* variables, these appear as income approximations for the region. This is justified by the fact that since the districts' income is not available for the entire period of the study, it is necessary to use variable that represent these values in the model. *Indbov* is an index that represents cattle raising activities, based on the evolution of the herd, *vrmad* seeks to capture income generated by lumber activities, and *vrprod* attempts to identify the income coming from agriculture. The inclusion of more than one variable as a proxy is due to the fact that these may be able to detect the evolution in the region's municipal income, which is based on these three activities.

The variables *vol* and *npav* may have an influence upon the incidence of respiratory ailment cases. The explanation for this lies in the fact that dust and airborne debris, respectively associated to the extension of unpaved roads and the volume of sawed lumber, have a proven adverse effect upon the human respiratory apparatus, and consequently must present a positive correlation with the number of morbity cases.

There is no consensus regarding the effect of income on morbity due to respiratory ailments, the presence of these proxies aims at indicating, with regard to the signs of the parameters associated with these variables, the degree of association of these activities with the appearance of said ailments. For example, a positive correlation between the price of lumber and morbity may be related to the fact the higher the price of the lumber, the more it will be extracted from the forest. Once this is done, the subsequent stage is the secondary forest stage in which, according to specialists, burning takes place for planting or grazing purposes. Thus, according to Homma (1993), immediately after the removal of the forest, burning takes place. It then may be said that the lumber activity increases the number of morbity cases due to respiratory ailments. Technically, a second motive may be pointed out for the variable *vrmad* having a positive correlation with the incidence of morbity.

Due to the spatial nature of the observations used in our empirical study, we expect the cases of morbity or mortality in each municipality to be correlated with cases in the neighboring municipalities. It is also possible that the covariates of any given municipality may affect not only its own damage health risk but also the risks in neighboring municipalities. Given the possibility of spatial autocorrelation, it may not be appropriate to use simple linear regression models. The presence of spatial autocorrelation not only affects the estimate of the variance-covariance matrix, but can also bias the estimates due to model misspecification. In order to account for spatial interactions in the data we used spatial regression models, suggested in Anselin (1988).

3. Spatial Autocorrelation Model with Fixed Effect

The spatial heterogeneity among cross-section units (municipalities) has to be taken into account in order to contemplate the geographic, environmental and economic diversity among those different units. This heterogeneity present in the data can be controlled by using panel data methods to estimate the model. According to the panel data analysis literature [(Hsiao, 1995), (Baltagi, 1995)], the disturbance η_{it} for $i = 1,...,N$, $t = 1,...,T$ can be broken down into two stochastic components, such that, $\eta_{it} = \alpha_i + \varepsilon_{it}$. Here α_i is a fixed term specific to the observations and denoted the fixed effect. This term is introduced to model the spatial heterogeneity likely to be present among the municipalities. The term η_{it} is the noise with expected value equal to zero and variance $E(\eta_{it}^2) = \sigma_\eta^2$.

In order to assess spatial autocorrelation directly in system (3) some new elements must be introduced in our analysis. A standard choice is to include in the mean process a spatial autoregressive component that takes the spatial locations into account. This can be done by the use of a contiguity or neighborhood matrix $W = (w_{ij})$ with w_{ij} representing the neighborhood between the sites i and j, such that $w_{ij} \neq 0$ if the sites i and j are neighbors, and $w_{ij} = 0$ otherwise. This spatial weight matrix is a square matrix NxN representing the spatial context. It encodes the neighborhood relationship among the spatial units. The literature on this subject is vast, we shall only retain here that the neighborhood relationships chosen by the analyst may change the results[2]. We now present a description of the econometric model used in this paper that extends the spatial models to accommodate the presence of heterocedasticity in the residuals.

The observations can be framed in the following way. Let $Y_t = (y_{11},..., y_{1T},..., y_{N1},..., y_{NT})'$ where y_j is a $NT \times 1$ vector of dependent variables. Let $\alpha = (\alpha_1,...,\alpha_N)'$ the $N \times 1$ vector of individual or fixed effect and $NT \times 1$ the vector of noise given by $\varepsilon_t = (\varepsilon_{11},...,\varepsilon_{1T},...,\varepsilon_{N1},...,\varepsilon_{NT})'$. We assume that individual effects are fixed. In this sense the individual effect can be included into the set of parameters. Thus the error covariance matrix is defined so that $E(\varepsilon_t \varepsilon_t') = \Omega = \sigma_\varepsilon^2 I_{NT}$. In this sense the model may be rewritten in the following way:

$$
\begin{aligned}
Y_t &= \lambda W_1 Y_t + X_t \beta + \alpha + \varepsilon_t \\
\varepsilon_t &= \theta W_2 \varepsilon_t + \upsilon_t
\end{aligned}
\qquad (2)
$$

where λ and θ are parameters, and $\upsilon_t \sim N(0, \sigma_v^2 I_{NT})$, X_t is the $TN \times J$ matrix of the independent variables while $\beta = (\beta_1,...,\beta_j)'$ is the set of parameters regarding these

[2] See Anselin at al. (1988) for a detailed discussion.

variables. From the general model in (2) we can derive special models by imposing restrictions. For example, setting W_2 produces a first-order spatial autoregressive model (SAR). Letting $W_1 = 0$ produces a regression model denoted spatial autoregressive error model (SEM).

The standard estimation method for the fixed effects model is to eliminate the intercepts from the regression equation by demeaning the *Y* and *X* variables, then estimate the resulting demeaned equation by OLS and subsequently recover the intercepts (Elhost, 2003). Instead of estimating the demeaned equation by OLS, it can also be estimated by maximum likelihood (ML). The only difference is that ML estimators do not make corrections for degrees of freedom.

The Moran's $I-$ statistic can be used to check the existence of spatial dependence in the disturbances of a regression model is called. This test indicates spatial correlation in the residuals of the fixed effect panel data model estimated without introduce spatial autocorrelation, the SEM model would be an appropriate way to proceed. In order to apply Moran test the mean residuals over the all periods are required. Second, these mean residuals are further centered and standardized. In short, the residuals are transformed into:

$$\overline{e}_i^{\,s} = \frac{\overline{e}_i - \overline{\overline{e}}_i}{\sigma_{\overline{e}_i}} \tag{3}$$

with "s" standing for "standardized". Let us call $z_i = e_i^S$, given a row standardized neighborhood matrix W, then each row sums to one. The Moran's $I-$ statistic[3] is defined such as:

$$I = \left(\frac{z'Wz}{z'z} \right) \tag{4}$$

In other words it is the ratio of the covariance between neighbors to the variance of the whole series (complete graph): it is the fraction of total variance accounted by spatial neighborhood relationship. Then Moran I statistic can be easily expressed as a regression coefficient, as one can verify that if we pose: $y = Wz$, and noting that the denominator of (4) may be written as $(z'z)^{-1}$, we obtain:

$$I = (z'z)^{-1}{}'zy \tag{5}$$

This offers an easy way to compute Moran I, which is simply the slope coefficient of the regression equation (5) on the variable Wz, that is, one simply regresses $y = Wz$ on z

[3] Other tests for spatial autocorrelation will be applied in this study in section 5.

without a constant. The dependent is the spatially smoothed variable Wz and the independent is the original (standardized) variable.

4. Data Base

Regarding the health data, in this study we have used annual hospital treatment values, average in-patient treatment value, and the average in-patient treatment period related to respiratory illnesses (CID-10 X). The sample covered all districts in the Amazon region. The data also refer to the public health services, extracted from DATASUS, the Ministry of Health's information service. The estimation of the area burned used in this study was obtained from the so-called hot pixels. These hot pixels represent satellite-observed fire outbreak points. This information was processed by the Wood Hole Research Center.

The main original source of the data on population available for this study comes from Brazilian National Agriculture Census elaborated by the Brazilian Institute for Geography and Statistics (IBGE) which is usually conducted every five years. In this sense the series for population had to be constructed. This was due to the fact that there were no annual data available for the districts, since this information is only available for those periods in which the IBGE census was conducted. The procedure used in this case was to take the district population data for the 1991 – 1997 period, and determine the remaining periods by interpolation based on a linear growth rate, determined by the district population values for the two years indicated above.

Since information regarding district income was also not available for the period, the real lumber production value (logs), and the real agricultural production value (temporary planting) were tested as proxies for this variable. Another possible proxy for municipal income is the use of a cattle raising activity index, which is directly related to the growth in the region's cattle herd. All these data were constructed from information obtained from the SIDRA system of the Agricultural Census, maintained by the IBGE. The municipal income data for all municipalities were only for those years for which a demographic census was conducted.

Data regarding unpaved roads were generated from a joint study conducted by IPEA (Institute of Applied Economic Reasearch) and IBGE (IBGE, 1987). Sawed lumber production was obtained from the IPAM (Institute of Environmental Research for Amazonian) database.

Finally, In the Brazilian Amazon Basin a county can be subjected to ongoing change in its size mainly during the expansion of the agricultural frontier in Amazon. This fact obstructs the comparison between periods at county level. That is why the concept of Minimum Comparable Area (MCA) [4] was introduced, which is the smallest stable spatial unity during these five censuses that accommodates the changing county boundaries over the panel. The aggregation of counties in the later census years, in order to match the county area in 1970, is greatest in the more recently populated and sub-divided regions found in the legal Amazon.

[4] For further details about the concept of MCA see Reis at al. (2007).

5. Results

In this section, we present the analysis of econometric results for the morbity and mortality model derived from forest burning in Amazon. The panel data sample comprises 261 of the municipalities over the period 1996-1999, generating a total of 1044 observations. The districts are those for which there is available information regarding the number of sick people. The results of the morbity and mortality models are shown, respectively, in Tables 1 and 2.

Table 1. Spatial Panel Data Model for Morbity

Independent Variables	SAR Pooled (1)	SAR fixed effect (2)	SAR time period fixed effect (3)	SEM pooled (4)	SEM fixed effect (5)	SAR time period fixed effect (6)
FIRE (it)	0,4392 (0,000)	0,2969 (0,000)	0,3078 (0,000)	0,4770 (0,000)	0,2946 (0,000)	0,3061 (0,000)
VRMAD (it)	0,5601 (0,022)	0,3190 (0,049)	0,3055 (0,061)	0,6234 (0,013)	0,3152 0,052	0,2978 (0,067)
VRPROD (it)	0,0108 (0,905)	-0,0253 (0,686)	-0,0049 (0,937)	-0,0027 (0,977)	-0,0242 (0,699)	-0,0023 (0,970)
INDBOV (it)	-0,1761 (0,224)	-1,4834 (0,000)	-1,3887 (0,000)	-0,2423 (0,099)	-1,4643 (0,000)	-1,3551 (0,000)
POP (it)	0,0070 (0,000)	0,0082 (0,000)	0,0082 (0,000)	0,0070 (0,000)	0,0082 (0,000)	0,0082 (0,000)
NPAV (i)	0,6993 (0,000)	–	–	0,7037 (0,000)	–	–
VOL (i)	0,0002 (0,000)	–	–	0,0002 (0,000)	–	–
CONSTANT	110,504 (0,004)	–	–	133,507 (0,000)	–	–
W*dep.	0,0289 (0,276)	-0,0079 (0,860)	0,0119 (0,790)	0,1809 (0,000)	-0,0269 (0,557)	-0,0349 (0,447)
Observations NxT	1056	1056	1056	1056	1056	1056
R^2	0,66	0,97	0,97	0,67	0,97	0,97
The Moran's I-statistic	–	–	–	3.046 (0.002)	–	–
LM error tests	–	–	–	7.411 (0.000)	–	–
Wald test	–	–	–	5.665 (0.000)	–	–
log-likelihood	-8390,15	-6972,24	-6979,19	-8381,81	-6972,11	-6978,69

Based on the sign and statistical significance of the variable With regard to the sign found for the cattle raising index (*indbov*) it may be stated that the activities related to cattle raising tend to reduce the impact on the appearance of respiratory ailments in the Amazon region. For those activities related to lumber end up increasing the number of cases of morbity related to these maladies. Lastly, by observing Column (1) and (4), it may be noted that the non-consideration of factors that are specific to the individual units (municipalities), which is done when the pooled regression is performed, take us to overestimate the impact of the use of fire on health damage in Brazilian Amazon.

Table 2. Spatial Panel Data Model for Mortality

Independent Variables	SAR Pooled (1)	SAR fixed effect (2)	SAR time period fixed effect (3)	SEM Pooled (4)	SEM fixed effect (5)	SAR time period fixed effect (6)
FIRE (it)	-0,0166 (0,004)	0,0053 (0,037)	0,0070 (0,004)	-0,0111 (0,026)	0,0059 (0,026)	0,0081 (0,001)
VRMAD (it)	-0,0192 (0,048)	0,0010 (0,886)	0,0003 (0,965)	-0,0287 (0,001)	0,0011 (0,874)	0,0002 (0,971)
VRPROD (it)	0,0051 (0,231)	0,0011 (0,722)	0,0021 (0,518)	0,0105 (0,005)	0,0010 (0,743)	0,0023 (0,475)
INDBOV (it)	-0,0019 (0,754)	0,0041 (0,751)	0,0122 (0,330)	-0,0033 (0,582)	0,0011 (0,930)	0,0135 (0,292)
POP (it)	0,0005 (0,000)	-0,0000 (0,792)	-0,0000 (0,617)	0,0004 (0,000)	-0,0000 (0,364)	-0,0000 (0,342)
NPAV (i)	-0,0152 (0,065)	–	–	-0,0080 (0,289)	–	–
VOL (i)	-0,0000 (0,000)	–	–	-0,0000 (0,000)	–	–
CONSTANT	-4,3405 (0,005)	–	–	-8,2667 (0,000)	–	–
W*dep, Variable	-0,1829 (0,000)	0,0599 (0,264)	0,0729 (0,169)	-0,3609 (0,000)	0,1309 (0,009)	0,1029 (0,042)
Observations NxT	792	792	792	792	792	792
R^2	0,84	0,98	0,98	0,84	0,98	0,98
The Moran's I-statistic				-3.779 (0.000)		
LM error tests				15.496 (0.000)		
Wald test				18.215 (0.000)		
log-likelihood	-3669,27	-2637,17	-2642,85	-3682,77	-2636,38	-2642,46

Each table presents the results of the spatial panel data model using both specifications described in section 2. Thus the estimates for the models SAR and SEM are obtained. The columns (1) and (4) illustrates the estimation of the pooled model that does not take into account heterogeneity across the municipalities. In other words it does not consider the discrimination of the omitted individual or fixed effect. The columns (2) and (5) illustrates the model with fixed effect and the columns (3) and (6) show the results that consider not only fixed effect but also time effect. Because the variables *vol* and *npav* does not vary in time dimension the coefficients of these variables are not estimated for fixed effect models.

For all the models presented in Tables 1 and 2 there is an indication that the use of fire represented by the variable fire exerted influence upon the incidence of respiratory ailments and the cases of mortality in the Amazon region. All the tests for spatial autocorrelation[5] points out that it must be model. Results indicate the importance of spillover effect among municipalites of Brazilian Amazon in determining the air quality derived from transport of particulate material present in the smoke, and carbon-derivate gases that appear as sub-products of the use of fire. Notwithstanding our strategy to include it into the model was more successful regarding the model for mortalily which can be seen through the significance of the variable *W*dep*.

REFERENCES

Anselin, L. (1988). *Spatial Econometrics Methods and Models*. Kluwer Academic Publishers.

Arellano, M. (2003). Panel Data Econometrics. Oxford University Press.

Alves, D. and Evenson, R, 2000. *Modelling the Impacts of Climate on Health Patterns in Brazil*: A Censored Regression Analysis, mimeo.

Baltaghi, B. H., 1995. *Econometric Analysis of Panel Data*. John Wiley e Sons.

Elhost, J. P. (2003). Specification and Estimation of Spatial Panel Data Models. *International Regional Science Review*, 26(3):244-268.

Homma, A K., 1993. A Dinâmica dos Desmatamentos e das Queimadas na Amazônia: Uma Análise Microeconômica. Anais do XXXI Congresso Brasileiro de Economia e Sociologia Rural, 1, 663-676.

Hsiao, C., 1986. *Analysis of Panel Data*. Cambridge University Press.

Nepstad, D., 1999. Large-scale Impoverishment of Amazonian Forests by Logging and Fire. *Nature*, 398, 505-508.

Nepstad, D. C., Moreira, A. G. e Alencar, A. A., 1999. A Floresta em Chamas: Origens, Impactos e Prevenção de Fogo na Amazônia. Brasília: Programa Piloto para a Proteção das Florestas Tropicais do Brasil.

Ostro, D. B., 1987. Air Pollution and Morbity Revisited: A Specification Test. *Journal of Environmental Economics and Management*, 14(1), 87-98.

____________The Effects of IR Pollution on Work Loss and Morbity.,1983. *Journal of Environmental Economics and Management*, 10(2), 371-382.

Reis, E. J., Pimentel and Alvarenga, A. I. (2007). Areas Minimas Comparaveis entre os Anos de 1920 a 2000. WP NEMESIS.

Seroa da Motta, R. e Mendes, A. P. F., 1995. Custos de Saúde Associados à Poluição do Ar no Brasil. *Pesquisa e Planejamento Econômico*, 25(1), 165-198.

Seroa da Motta, R., Ortiz, R.A. e Ferreira, S.F., 2000. Health and Economic Values for Mortality Cases Associated with Air Pollution in Brazil. *Expert Workshop on Assessing The Ancillary Benefits and Costs of Greenhouse Gas Mitigation Strategies*, Washington, DC, 27-29, March.

---

[5] Moran's I-statistics, Wald and LM tests.

Thomas, V., 1985. Evaluating Pollution Control: The Case of São Paulo. *Journal Of Development Economics*, 19, 133-146.

In: Environmental and Regional Air Pollution
Editors: D. Gallo, R. Mancini

ISBN: 978-1-60692-893-6

Chapter 16

LOCAL ASSESSMENT OF VOCS AND PAHS IN THE AIR: SAMPLINGS, DISTRIBUTIONS, AND RISKS

Takeshi Ohura*

Institute for Environmental Sciences, University of Shizuoka
52-1 Yada, Shizuoka, 422-8526 Japan

ABSTRACT

When the relative magnitudes of air pollutants are arranged on a map, the local distribution of the each pollutant becomes apparent. This approach will be clues revealing the emission sources, behaviors and impacts of pollutants. Of ambient VOCs, passive air sampler will be useful device for the sampling because of highly portable, sensitive, and selective feature. The spatial distribution patterns in an industrial area were quite different from each targeted VOC, indicating that those emission sources are specific for each VOC. Similar trend has been observed in the distribution of the ambient PAHs in an industrial area. It indicates that the high-polluted distributions were apparently contributed to the industrial plants such as paper-making factories. On the other hand, the local distributions of ambient PAHs in urban city showed that the highly polluted areas were spread along the busy roads of the outer city, with an enclosed area of lower pollution in the central city. In certain industrial area, the lifetime cancer risks of exposure to ambient VOCs were significantly higher in winter than in summer, which were derived mainly from benzene, carbon tetrachloride, formaldehyde, and acetaldehyde. The carcinogenicity of ambient PAH mixture was occupied (~77%) by BaP, DBahA, and Nap in both summer and winter.

ABBREVIATIONS

VOCs	Volatile organic compounds;
TCE	Trichloroethylene;
PCE	Tetrachloroethylene;

* Tel: +81-54-264-5789, Fax: +81-54-264-5798; E-mail: ooura@smail.u-shizuoka-ken.ac.jp

PAHs	Polycyclic aromatic hydrocarbons;
Nap	Naphthalene;
Ace	Acenaphthene;
Flu;	Fluorene;
Phe	Phenanthrene;
Ant	Anthracene;
Fluor	Fluoranthene;
Py	Pyrene;
Chry	Chrysene;
BaA	Banz[*a*]anthracene;
BjF	Benzo[*j*]fluoranthene;
BbF	Benzo[*b*]fluoranthene;
BkF	Benzo[*k*]fluoranthene;
BaP	Benzo[*a*]pyrene;
IP	Indeno[1,2,3-*cd*]pyrene;
BghiP	Benzo[*ghi*]perylene;
DBahA	Dibenz[*a,h*]anthracene;
Cor	Coronene.

INTRODUCTION

The conformations of air pollutions in area could be different from the local point of view to the global point of view. Also, the difference becomes pronounced by the species of pollutants. For example, certain VOCs are emitted from only the specific source, showing that the ambient distribution is highly local. However, in handling the regional air pollutions of chemicals, we have to consider that factors affecting the levels would include various aspects at the sampling point: season, weather conditions, landform, population, industry, culture, etc.

To date, there are many reports in which the surveys of ambient levels of air pollutants are variously carried out from the local sites to country by country. The distributions in wide-range region will arrow us to show the global occurrences, whereas it is far from showed the real conditions in habitat. Therefore, the survey in local region, i.e. within a city, rather than by country could help us to evaluate the risks of exposure to air pollutants. This chapter looks at the sampling methods, distributions and risks of significant air pollutants, VOCs and PAHs in an industrial or urban local area, Japan.

SAMPLINGS

Most studies of ambient air pollutants have been performed with active samplers that require an air pump and power supply for the samplings. Although active samplers have the advantage of being able to collect a precise volume of air in a short time, it make difficult to survey the multi points and simultaneously because they require electricity. On the other hand, passive air samplers can be used where electricity is not available, and because they are small and silent, they are especially useful for measuring personal exposure and analyzing indoor air. In addition, the passive air samplers can be produced inexpensively and used to investigate many sampling points simultaneously and to locate emission sources in a particular area.

To date, a number of passive air samplers for each air pollution have been developed and commercialized: packed with activated charcoal for VOCs, and with acidified 2,4-dinitrophenylhydrazine-coated silica gel for carbonyl compounds (CCs). Kume et al. [1] suggested that passive air samplers were more effective in collecting CCs when they were covered with hoods (Figure 1). Note that applications of the passive air samplers have been limited to the pollutants at relatively high concentrations because the sampling rate of a passive air sampler is approximately 1/20 that of an active air sampler. Therefore, we may need to develop the highly sensitive analytical methods when we use the passive air samplers in the survey of air pollutants with lower concentrations. Large-volume injection of sample to HPLC or GC equipment will be one of the simple solutions.

Figure 1. Passive air samplers with hood on sampling observation.

Local Distributions of VOCs in Industrial Area

When the relative magnitudes of air pollutant concentrations are displayed on a map, the local distribution of the pollutants becomes apparent, providing clues to the emission sources [2, 3].

Our group have investigated the local distributions of ambient VOCs in an industrial city, Japan, and constructed the contour maps of the study area showing the ambient concentrations of each compound (Figure 2). In the study, the high-concentration regions of benzene were obviously spread along major roads radiating from the center of the city, indicating that traffic is the significant source. Then we have found that the high-concentration region spread several kilometers and that the concentration strongly depended on the wind direction.

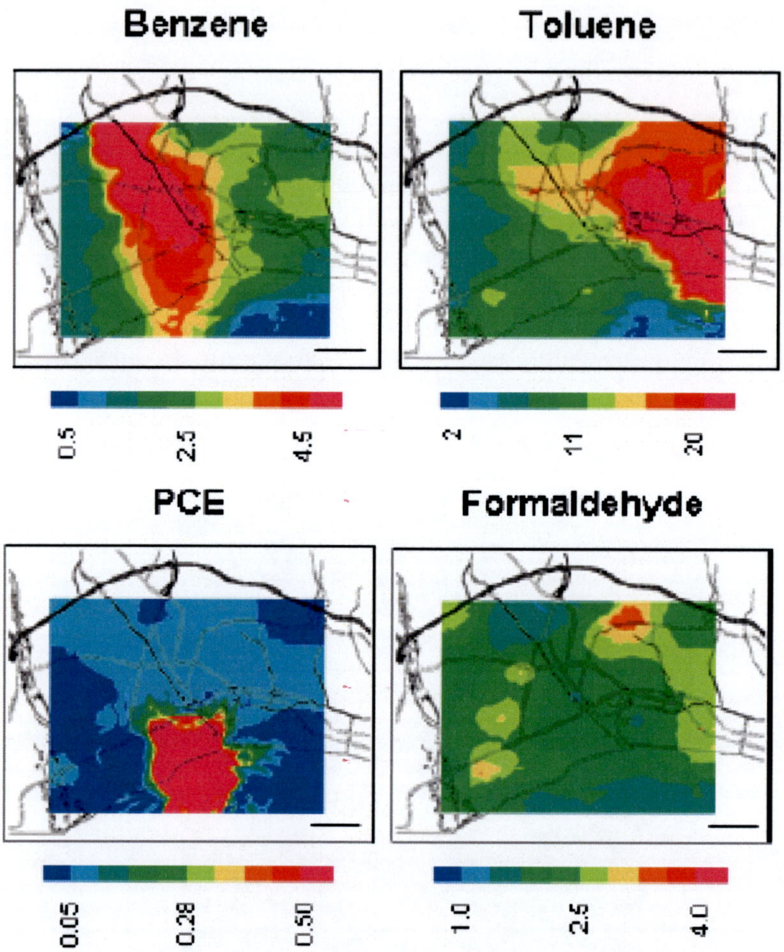

Figure 2. Contour maps of concentrations (μg/m^3) of selected VOCs in winter (March, 2004). Scale bar = 1 km.

Toluene could be one of the most abundant VOCs in the air, which is widely used as intermediate material of manufactured products, and is also included in car exhaust. Since the high-concentration regions of toluene were not spread along major roads as like benzene, the pollution cause may be due to stationary sources, rather than traffic (Figure 2). For PCE, it showed specific seasonal distribution patterns with the locally high concentrations relative to other VOCs, indicating the presence of specific emission sources (Figure 2). Note that the high-concentration region shifts only slightly (~1 km) downwind from the possible source despite of 2-3 m/s wind. For formaldehyde, the high-level regions were distributed over the center to the western side of the study area and over part of south and northwest along main roads at traffic jam points (Figure 2). They clearly indicate that traffic was the primary source of formaldehyde in the high-level regions. Traffic is believed to be a potent source of both ambient formaldehyde and benzene. However, their high-level distribution patterns were quite different in the study. This phenomenon may be due to different photochemical fates of the species in the atmosphere and the presence of other potent emission sources at nearby sites.

LOCAL DISTRIBUTIONS OF PAHS IN INDUSTRIAL AREA

By constructing the contour maps of ambient PAHs, it also reveals to show the specific distributions. Ohura et al. [2] have been investigated the regional pollution of PAHs in an industrial city, Japan, and have constructed the contour maps in which the magnitudes of individual PAHs detected were overlaid on the surveyed area (Figure 3). The PAH distributions of samples obtained in the summer showed a characteristic pattern in which the local high-level regions were polarized to the west and east areas of the city (Figure 3). Ambient low molecular weight PAHs ($\leq$4-rings), e.g. Nap and Py, were distributed at equal levels in both areas, whereas the concentrations of high molecular weight PAHs (>5-rings; BaP and Cor) were higher on the east side. This indicates that those PAH could be emitted by the specific sources. Also, the study indicated that there were no significant stationary emission sources in the west region, so that traffic could be contributed to the pollutions in the west. On the other hand, a number of large papermaking plants were located in the southeast area of the city (marked on map in Figure 3), in which high molecular weight PAHs were highly polluted. Therefore the pollutions could be due to the plants, in addition, which were distributed to north by the wind blowing from the south.

In winter, the high-level distribution for each PAH in the east region shifted as a whole to the south compared with summer (Figure 3). The wind (2–3 m/s) over the surveyed area was mostly blowing from the north when the sampling of winter was carried out. Considering the results of the distribution of PAH and the wind direction, the sources of the atmospheric PAHs in winter were also the traffic and papermaking plants, as in summer.

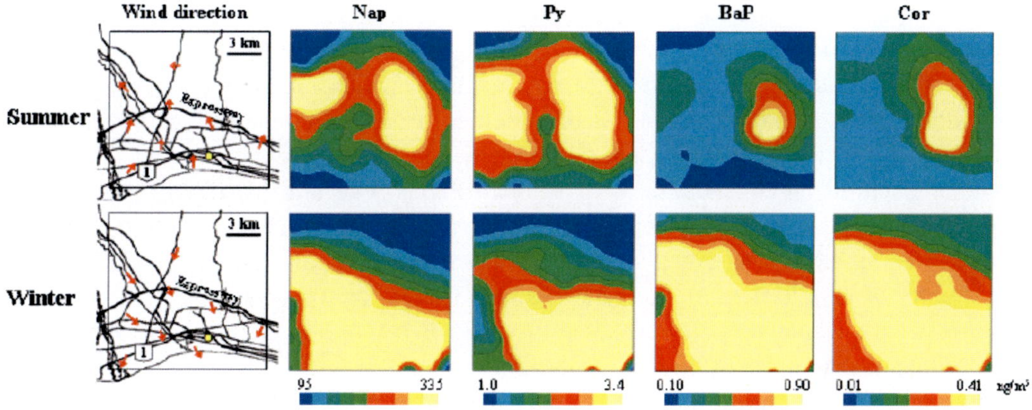

Figure 3. Seasonal wind direction and selected PAH contour maps (Nap, Py, BaP, and Cor) in an industrial city, Japan. The yellow circle in maps shows the location of paper-making plant as possible stationary emission source.

LOCAL DISTRIBUTIONS OF PAHS IN URBAN AREA

It has been carried out the surveys of particulate PAHs in an urban city, Japan, and has been constructed contour maps of the study area showing the seasonal magnitudes of PAH indicators for a gasoline engine source (Fluor and Py) and for a diesel engine source (BaP and

BghiP), and considered them taking into account the wind directions (Figure 4). During winter sampling, the wind was blowing mainly from the northeast and the areas of high concentrations of light molecular weight PAHs, Fluor and Py, were distributed along the major road in the northern area and near factories in western and southern areas. The highly polluted areas for heavy molecular weight PAHs, BaP and BghiP, were somewhat consistent with those for Fluor and Py. That is, the highly polluted areas were spread along the busy roads of the outer city, with an enclosed area of lower pollution in the central city.

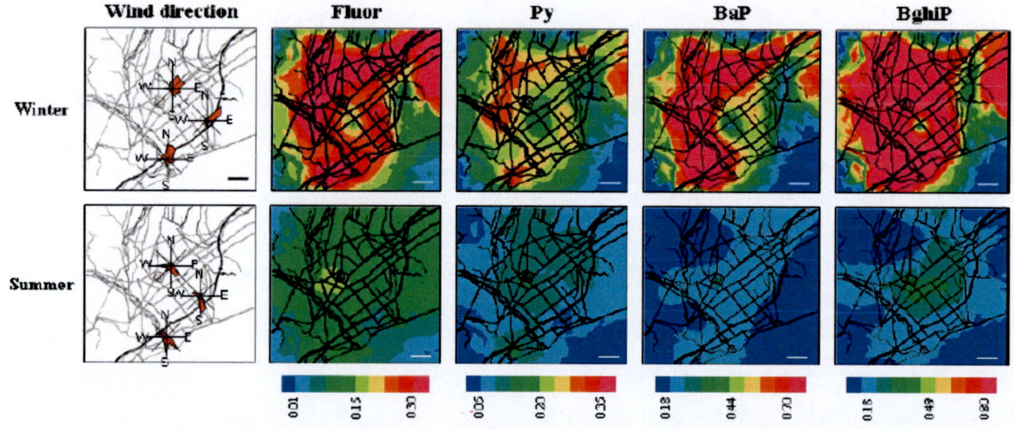

Figure 4. Seasonal wind direction and selected PAH contour maps (Fluor, Py, BaP, and BghiP) in winter (W, Feb. 2001), and summer (S, Aug. 2001) in an urban city, Japan. Scale bar = 2 km.

In summer, concentrations of all four indicator PAHs were low, but there were relatively high concentrations in the center of the study area. The wind directions in summer were mixed with northwesterly and southerly to southeasterly winds predominant. Taking wind direction into account, both the summer distributions of PAHs can be attributed to traffic on major roads and to factories in the west and south.

Contour maps of the concentrations of ambient PAHs revealed that highly polluted air was distributed along major roads outside the city center, and PAH concentrations were higher there than in the central city. This pattern may be observed in other mega-urban cities where the entry of cars to the city center is limited. That is, traffic on highways around the perimeter of a city would be potent sources of ambient PAHs and create more serious air pollution problems than those of the central area.

Exposure Risks

People spend 90% of their time indoors, but indoor air quality is strongly influenced by outdoor environments because outdoor air is easily transported indoors.

It has been reported that the concentrations of certain VOCs such as carbon tetrachloride and TCE, the primary sources of which are present in outdoors, were comparable to indoor concentrations as well as to personal exposure levels [4]. Therefore, risks in outdoor environments cannot be ignored in assessments of the human exposure risk. Here is represented the inhalation cancer risks of various ambient VOCs in terms of unit risk (UR) values for each compound in an industrial city, Japan, in which eight VOCs were selected as target carcinogens. The lifetime cancer risks (LCR) for each air pollutant and each study site were calculated using Equation 1:

$$LCR_{ij} = \sum_{ij} UR_j \times C_{ij} \quad (1)$$

where LCR_{ij} is the estimated risk from pollutant j at study site i, C_{ij} is the measured concentration of pollutant j at study site i, and UR_j is the inhalation UR for pollutant j.

Using life cycle assessment, the summed risk values for the relevant VOCs obtained in winter and spring in the study are calculated, values of which were 12×10^{-5} and 11×10^{-5}, respectively, and those values were somewhat consistent with each other (Table 1). In comparison of the values in different seasons, the values obtained in winter were significantly higher ($p<0.0001$, t-test) than in summer. The predominant risks were benzene, carbon tetrachloride, formaldehyde, and acetaldehyde in both winter and spring, which accounted for 98% of the LCR for winter and 96% for spring (Table 1).

For PAHs, assessing the toxicity (carcinogenicity, mutagenicity, Ah receptor affinity, etc.) of the individual PAHs is complicated because of a mixture of PAHs in the environment. Several approaches have been developed to evaluate the potencies of the components of a complex mixture of PAHs. One approach involves using toxicity equivalency factors (TEFs) based on BaP. The concentrations of BaP equivalent (BaP_{equiv}) for the individual PAHs were calculated using Equation 2:

$$BaP_{equivi} = TEF_i \times C_i \quad (2)$$

where C_i is the measured concentration of PAH i.

Table 2 presents the concentrations of BaP equivalent (BaP_{equiv}) together with the geometric mean concentrations of carcinogenic PAHs in an industrial city, Japan in both summer and winter. The sum of carcinogenic PAHs (ΣPAH) measured in winter tended to be slightly higher than that in summer. However, the sum of BaP_{equiv} (ΣBaP_{equiv}) in winter was approximately 3-fold higher than that in summer. The main contributor to ΣPAH was Nap in both summer (95%) and winter (98%), indicated the dominant PAH. On the other hand, in the value of ΣBaP_{equiv}, the dominant species were Nap (28%) in summer and BaP (42%) in winter (Table 2). In a previous indoor PAHs study, Nap was dominant contributor in the ΣBaP_{equiv}; the contributions are 63% in the summer and 56% in winter samplings. It indicates that there is significant sources of Nap in the indoors rather than outdoors. For assessing the health impacts of PAH mixtures, gaseous PAHs such as Nap should be taken into account, especially the samples obtained in warmer season, although high carcinogenic PAHs, e.g. BaP and DBahA, also significantly contribute to the risks. Indeed, there are reports that BaP exhibited the highest contribution to the total carcinogenic potential in ambient air, ranging from 50% to 67% [5, 6]. This could be due to measuring only particulate PAHs in air. In addition, ΣBaP_{equiv} obtained in the study was used to estimate the corresponding LCR for exposure groups. Regarding the lung cancer risk via the inhalation route, WHO suggested the unit risk of 8.7×10^{-5} $(ng/m^3)^{-1}$ for the lifetime (70 years) PAHs exposure [7].

Table 1. Geometric mean concentrations (GM) of carcinogenic VOC in an industrial city, Japan, and the lifetime cancer risks (LCR) based on the inhalation unit risks

		Winter			Spring		
	Inhalation unit risk $(\mu g/m^3)^{-1}$	GM	LCR	Contribution (%)	GM	LCR	Contribution (%)
Benzene	2.9*E-5	2.568	7.4E-05	62.4	2.053	6.0E-05	52.4
Ethylbenzene	5.0*E-7	1.030	5.2E-07	0.4	2.165	1.1E-06	1.0
Chloroform	5.3*E-6	0.249	1.3E-06	1.1	0.232	1.2E-06	1.1
Carbon tetrachloride	4.2*E-5	0.363	1.5E-05	12.8	0.409	1.7E-05	15.1
TCE	2.0*E-6	0.308	6.2E-07	0.5	0.504	1.0E-06	0.9
Tetrachloroethylene	5.9*E-6	0.061	3.6E-07	0.3	0.174	1.0E-06	0.9
Formaldehyde	6.0*E-6	2.171	1.3E-05	10.9	2.683	1.6E-05	14.2
Acetaldehyde	2.7*E-6	5.101	1.4E-05	11.5	6.072	1.6E-05	14.4
Total carcinogenic VOCs		11.852	1.2E-04	100.0	14.292	1.1E-04	100.0

Table 2. Application of TEFs to ambient PAH data in an industrial city, Japan[a]

		Summer			Winter		
Compound	TEF	Measured conc.	BaP_{equiv}	Contribution (%)	Measured conc.	BaP_{equiv}	Contribution (%)
Nap	0.001	213.006	0.213	28.05	345.003	0.345	13.33
Ace	0.001	6.415	0.006	0.84	2.870	0.003	0.11
Flu	0.001	9.841	0.010	1.30	5.767	0.006	0.22
Phe	0.001	26.274	0.026	3.46	12.568	0.013	0.49
Ant	0.01	0.422	0.004	0.56	0.929	0.009	0.36
Fluor	0.001	4.570	0.005	0.60	3.205	0.003	0.12
Py	0.001	2.998	0.003	0.39	2.860	0.003	0.11
Chry	0.01	0.409	0.004	0.54	1.604	0.016	0.62
BaA	0.1	0.120	0.012	1.58	0.983	0.098	3.80
BjF	0.1	0.196	0.020	2.58	1.010	0.101	3.90
BbF	0.1	0.527	0.053	6.94	1.542	0.154	5.96
BkF	0.1	0.175	0.018	2.31	0.675	0.067	2.61
BaP	1.0	0.178	0.178	23.51	1.075	1.075	41.54
DBahA	5.0	0.037	0.187	24.61	0.114	0.568	21.93
IP	0.1	0.177	0.018	2.34	1.141	0.114	4.41
BghiP	0.01	0.292	0.003	0.38	1.293	0.013	0.50
ΣPAH		265.638	0.759	100.00	382.636	2.588	100.00

[a]All concentrations are in unit of ng/m^3.

Therefore, the corresponding LCR were found as 6.6×10^{-5} in summer and 23×10^{-5} in winter. The level estimated considerably exceeded a health-based guideline level (10^{-5}) [8]. These risks for PAHs were somewhat same levels to those for VOCs as noted in above.

Conclusion

Atmospheric pollutions have been thought to be affected by not only possible sources such as stationary and mobile emission sources but also various factors such as climate condition, geological formation, and long-range transport. In the case that certain potent emission sources are present in the targeted area, the local high-polluted distributions, so-called "hot-spot", could be observed. As noted in this paper, the air pollution concentrations around the emission sources may be more than 10-times higher than those of the outer areas. To find a solution to such problem, creating contour map showing the local distribution of pollutants will be useful tool.

References

[1] Kume, K. Ohura, T. Amagai, T. and Fusaya, M. (2008) Field monitoring of volatile organic compounds using passive air samplers in an industrial city in Japan. *Environ. Pollut.* 153, 649-657.

[2] Ohura, T. Amagai, T. Fusaya, M. and Matsushita, H. (2004) Spatial distributions and profiles of atmospheric polycyclic aromatic hydrocarbons in two industrial cities in Japan. *Environ. Sci. Technol.* 38, 49-55.

[3] Ohura, T. Amagai, T. and Fusaya, M. (2006) Regional assessment of ambient volatile organic compounds in an industrial harbor area, Shizuoka, Japan. *Atmos. Environ.* 40, 238-248.

[4] Payne-Sturges, D.C. Burke, T.A. Breysse, P. Diener-West, M. and Buckley, T.J. (2004) Personal exposure meets risk assessment: A comparison of measured and modeled exposures and risks in an urban community. *Environ. Health Perspect.* 112, 589–598.

[5] Papageorgopoulou A. Manoli E. Touloumi E. and Samara C. (1999) Polycyclic aromatic hydrocarbons in the ambient air of Greek towns in relation to other atmospheric pollutants. *Chemosphere* 39, 2183-2199.

[6] Petry, T. Schmid, P. and Schlatter, C. (1996) The use of toxic equivalency factors in assessing occupational and environmental health risk associated with exposure to airborne mixtures of polycyclic aromatic hydrocarbons (PAHs). *Chemosphere* 32, 639-648.

[7] *Air quality guideline for Europe*. WHO European Series, World Health Organization, WHO Regional Bureau, Copenhagen, 1987.

[8] Boströn, C. E. Gerde, P. Hanberg, A. Jernström, B. Jahansson C. Kyrklund T. Rannug, A. Törnqvist, M. Victorin, K. and Westerholm, R. (2002) Cancer risk assessment, indicators, and guidelines for polycyclic aromatic hydrocarbons in the ambient air. *Environ. Health Perspect.* 110, 451-448.

In: Environmental and Regional Air Pollution
Editors: D. Gallo, R. Mancini
ISBN: 978-1-60692-893-6

Chapter 17

A RETROSPECTIVE ENVIRONMENTAL ANALYSIS, MODELING AND PREDICTION OF IMPACTS IN CONNECTION WITH INFRASTRUCTURAL DEVELOPMENTS IN SUB-SAHARAN AFRICA: A CASE STUDY OF TWO DAMS IN KUMASI, GHANA

***Isaac K. Tetteh**[*a]**, Esi Awuah**[b] **and Emmanuel Frempong**[c]*

[a] Department of Marine, Earth and Atmospheric Sciences, North Carolina State University, Campus box 8208, Raleigh, NC 27695-8208, USA.

[b] Department of Civil Engineering, Kwame Nkrumah University of Science and Technology (KNUST), Kumasi, Ghana.

[c] Department of Theoretical and Applied Biology, KNUST, Kumasi, Ghana.

ABSTRACT

In this chapter, the environmental quality (EQ) of three rural communities in connection with two dams in Kumasi, Ghana, is overhauled over four distinct years. The study is organized into two main parts: (i) an empirical analysis that aims at investigating the retrospective EQ of three riparian communities, and (ii) model development and prediction (backcasting/hindcasting) of the EQ of the communities, in one year lag. We use discontinuous, irregularly spaced, multivariate geostatistical, limited time series data consisting of sets of predictors and predictands. Missing values in the datasets are estimated using kriging technique. A multivariate empirical orthogonal function (EOF) analysis is performed on the datasets out of which the first three principal components (PCs) are retained for the study. The construction of the operational activity indices (OAIs) and EQ indices (EQIs) is based on the scores of the first three PCs of the predictor and predictand datasets, respectively. This is followed by performing dependency measures typically, zero order and lagged cross correlation analyses between

[*] Corresponding author:
Email: itetteh@ncsu.edu
Tel. No. (919) 513-0243
Fax No. (+1)919 515 7802

the two sets of indices to monitor the EQ of the communities over the four years. Generally, the results from the retrospective environmental analysis (REA) show moderate to strong relationships between the EQIs and OAIs, a signal of profound influences exerted by the dam(s). The most important signals are observed in the years when the two dams are operated in tandem. The potential forecasting (hindcasting) abilities of the OAIs determined from cross-validations, give us the confidence to incorporate the REA results in the stochastic and temporal environmental autoregressive AR (1) model developments for performing retrospective backcasts. A comparison between the parametric and non-parametric stochastic model results shows that the latter outperforms the former, and thus, better fits the data. The AR (1) model results theoretically signal progressive deteriorations in the backcast EQ for three out of the four lagged years. However, the remaining one shows no significant difference from the preceding year`s, an indication of environmental stability and non-responsiveness to the operational activities of the dam(s). Lack of observational data posed a limitation in comparing our hindcast model results with real life situation. Notwithstanding this, the model biases and forecast (hindcast) errors seem to be very reasonable. These suggest beyond doubts that the findings can be relied upon for decision support system. Remediation of the environmental impacts of the dams proposed does not differ from that of our previous studies.

1. General Introduction

The sub-Saharan Africa(SSA) is an integral part of third world economies typically characterized by political instability, relatively high unemployment and population growth rates, poverty, low life expectancy, high dependency ratio, among others (Barro 1991; Easterly et al. 1993; Barro and Lee 1994; Sahn 1994; Easterly and Levine 1997; Sachs and Warner 1997; Gyimah-Brempong and Traynor 1999; Rodrik 1999; Elbadawi and Sambanis 2000; IMF et al. 2000; Sachs 2000; Fosu 2001; Murdoch and Sandler 2002; Artardi and Sala-i-Martin 2003; Sahn and Stifel 2003; Yumkella and Vinanchiarachi 2003; Ravallion and Chen 2004; Gyimah-Brempong and Corley 2005; World Bank and IMF 2005; Verpoorten and Berlage 2007). Most governments of this subregion, in spite of the aforementioned constraints and challenges, are ambitiously committed to socioeconomic or infrastructural developments executed through economic reforms and restructuring programs enshrined in their vision statements. Their overall goals are to address socioeconomic stagnation and poverty alleviation with a view to raising the standard of living of their citizenry using various micro- and macro-economic policies (Sahn et al.1997; Fosu et al. 2003; Sahn and Younger 2004). With the United Nations Millennium Declaration on poverty alleviation and sustainable development promotions (Sahn and Stifel 2003), the visions of most SSA governments have become more enhanced. And in September 2000, 189 nations had committed themselves to a series of Millennium Development Goals (MDGs). The MDGs however, did not rule out implementation of projects or the harnessing our natural resources to solve socio-economic stagnations. For mega projects that rely substantially on all the major international multi-lateral and bi-lateral aid agencies and lending institutions such as the World Bank (WB), International Monetary Bank (IMF), African Development Bank/ African Development Fund (AfDB/AfDF), Asian Development Bank (AsDB) and, Department for International Development (DFID) for their implementation, the recipients` ambitions are regulated by the donors` stringent guidelines and conditionalities among which include adherence to project- level Environmental Impact Assessments (E IA) and/or Strategic Environmental Assessment (SEA) requirements (Goodland 1990; AsDB 1993; Canter 1996; World Bank 2000; Nielson and Tierney 2003; AfDB/AfDF 2004; ESDD 2006). These donors stress among other things, minimal damage to the ecosystem integrity, whilst maximizing the beneficial aspects of the projects for the good of the people. The donors` proactive stance on sustainable environmental development and management impinges on the ideals of the United States of America (USA) National Environmental Policy Act (NEPA) of 1969 (Sammy 1982). Donor agencies, in collaboration with the recipients` local government environmental agencies work to insure that sustainable environmental developments are not sacrificed for economic prosperity or gains, nor do they have be relegated to the background in pursuit of economic recovery and structural adjustment programs. Typically, infrastructural and economic developments of high priorities on the agenda of most SSA countries include water resources development projects (WRDPs) such as construction of dams, irrigation schemes, hydroelectric power generation, inland water transport, dredging, and channelization, as well housing projects, road construction, airports, mineral exploration and mining, manufacturing industries, and energy projects(Wathern 1988; Oehmke and Crawford 1996; Ayogu 2007; Ndulu et al. 2007) such as the West African Gas Pipeline mega project.

The huge energies that evolve from the implementation of projects during the construction, operational and post-operational phases are often accompanied by injection of environmental residues and wastes into sinks (air, water, and land). Very often, the neglected sides of such developmental projects are enormous, characterized by radical land use changes, modifications of the atmospheric constituents or compositions (typically of the troposphere and the stratosphere), aquatic resources, etc. The environmental perturbations cause changes in the biophysical environment and ultimately interfere with human welfare, ecosystem stability, and lead to degradation of EQ, extinction of species, and loss of ecosystem integrity. The perturbations, when not addressed or remedied in a timely fashion (in this case, post-development audits (PDAs)), or well in advance before projects are implemented (in this case, EIAs) may result in unprecedented environmental repercussions which are very costly, or may be irreparable due to the nature and magnitude of residual or cumulative impacts (CEQ 1987). Fortunately, the advent of EIA which originated in the USA with the enactment of the NEPA in 1969, which was promulgated on January 1, 1970 (Public Law 91-190) (US EPA 1970, Best 1972; Caldwell 1973; Kreith 1973; Bear 1993; CEQ 1993a; Gilpin 1995; Canter 1996) serving as a prototype or an archetype, and its subsequent adoption and implementation by other advanced countries (ACs) as well as less developed countries (LDCs)(Wathern 1988; Biswas and Geping 1987; Ofori 1991; Jain et al. 1993; Canter 1996; Appiah-Opoku 2001; Tran et al. 2002) has sensitized people more about the manifest and latent deleterious impacts of infrastructural developments, and the need for developers to make proper planning and decisions prior to implementation of projects. The significant role played by the NEPA earned it the "Magna Carta for the environment" in the USA (CEQ, 1993a). With the enthusiastic adoption and implementation of EIA by over a hundred countries globally, we also deem it appropriate describing the USA NEPA, the environmental torch bearer, with the following three sonorous Greek or Latin appellations, "*Primus inter pares, Primogenitus, and Prototocus*" of environmental assessments (EAs) whichever perspective one views it. Therefore, many projects whose impacts were considered "trivial" in the pre-EIA era due to ignorance, or lack of technical know-how in environmental assessment and appraisal (EAA), or due to environmental apathy of some sort, are now subject to intense scientific scrutiny empowered by governments` environmental agencies through prohibitive legislations. Interestingly, disputes relating to non-compliance are readily settled in law courts without fear or favor (Goldfarb 1999; Verghese, 2001).

One of the landmarks of the birth of EIA is the generation and evolution of countless numbers of EIA methods and techniques which have facilitated the conduct of various types of impact studies we see for almost the past four decades (Rau and Wooten 1980; Bisset 1983; Shopley and Fuggle 1984; Biswas and Geping 1987; Culhane 1993; Jain et al. 1993; Glasson et al.1994; Finsterbusch 1995; Ortolano and Shepherd 1995; Canter 1996; Tetteh et al. 2004a,c, Tetteh et al. 2006). This is a good indication of one of the tremendous benefits that has been derived from EIA legislation by countries. Lohanni and Halim(1990) describe over 15 EIA methods and techniques, whereas Canter (1979) summarizes over 100. Here, we list just a few examples of some of the common methods that have been applied to projects. These are: simple checklists (Jain et al. 1993; Canter 1996); scaling-weighting checklists eg Environmental Evaluation System (EES) (Dee et al. 1973; Goyal and Deshpande 2001); questionnaire checklists (Biswas and Geping 1987; Canter 1996); interaction matrices (Leopold et al. 1971; Rau and Wooten 1980; Glasson et al. 1994; Tetteh et al. 2004a); flow diagrams and network analysis/systems diagram (Gilliland and Risser 1977; Rau and Wooten 1980; Canter 1996; Tetteh et al. 2006); matrix-network hybrid methods (Sorensen 1971); computer-based approach egs IMPACT, and Environmental Impact Computer-aided System(Thor et al. 1978; Canter 1996); multi-attribute decision modeling methodologies (Saaty 1980; Parkin 1992; Clemen 1996; McDaniels 1996; Hokkanen et al. 1998, 1999; Bohanec and Rajkovič 1999; Triantaphyllou 2000; Turban and Aronson 2001; Kwak et al. 2002); adaptive environmental assessment and management (AEAM) (Holling 1978); resource and environmental profile analysis (REPA) (Hunt et al. 1992; Canter 1996; Tukker 2000; Brentrup et al. 2004a,b); lexicographic method (Linkov et al. 2004); conjunctive and disjunctive methods (Linkov et al. 2004); PROMETHEE methods (Brans and Vincke 1985; Brans et al.1986); analytic hierarchy process (AHP)(Saaty 1980; Ramanathan 2001; UKDTLR 2001; Tran et al. 2002; So´lnes 2003), Rapid Impact Assessment Matrix (RIAM) (Hagebro 1998; Pastakia 1998; Pastakia and Bay 1998; Pastakia and Jensen 1998; Baba 2005 ; El-Naqa 2005; Kuitunen et al. 2008); cost-benefit analysis (Petry 1995; Munda 1996; UKDTLR 2001), integrated environmental assessment (Bartlett 1993) and evidential reasoning approach (Wang, et al. 2006).

In the global environmental context, EIA has been complemented by the ratification of a series of bilateral/multilateral agreements, conventions, and international treaties, by member states to address several issues such as transboundary impacts (ECE 1991; CEQ 1993a). Examples of international treaties and conventions are seen in the Rio Earth Summit (UN 1992; Cooper and Alley 2002), the Sea and Ocean Convention (UN 1982), the Montreal Protocol(UNEP 1987), the Kyoto Protocol (Bolin 1998; Victor et al. 2001), Convention on Biological Diversity (UNCSD 1992; WCMC 1992) and its supplementary agreement, the Cartagena Protocol on Biosafety of 29th January, 2000. The overall goal includes among other things, is to address transboundary, regional, subregional and global impacts, as well as environmental sustainability and integrity. The impact of the 1972 United Nations Conference on Human Environment in Stockholm is also worth mentioning (Appiah-Opoku 2001).

In spite of the advances made in EIA methods and techniques, we all need to recognize that there are no universally accepted methods, neither, do cook books exist for them. Therefore, no method can claim all-sufficiency and be characterized as sacrosanct (see, for example, Rau and Wooten 1980; Bisset 1983; Shopley and Fuggle 1984; Biswas and Geping 1987; Jain et al. 1993; Culhane 1993; Glasson et al. 1994; Finsterbusch 1995; Ortolano and Shepherd 1995; Canter 1996). Hence, EIA professionals/experts need to be circumspect in dealing with their applications. Jain et al. (1993) have emphasized that selection of the appropriate method or technique for project-level EIA depends on its use, resources required, type of impacts, etc. Also, Clark et al.(1994) and Canter (1996) have discussed in detail comparative reviews of the utility of EIA methods and techniques, pointing out their strengths and weaknesses.

In the SSA, EIA professionals/practitioners are confronted with a great challenge regarding the development or the choice of the appropriate method(s) for the accomplishment of retrospective PDAs and hindcasting (backcasting). These challenges may be traced to the following causes: (i) the complexity of the environment displaying multidimensionality lends itself an assiduous task for deciding on the appropriateness and suitability of method(s) to do a good work since different methods cannot be efficiently amalgamated or integrated to address multifaceted environmental issues simultaneously, effectively and holistically, even though some methods such as Leopold matrix, Sorensen and integrated environmental assessment (IEA) methodologies have been utilized quite successfully in the USA (Leopold 1971; Sorensen 1971; Bartlett 1993), (ii) the environment is seen as consisting of a huge complex of interacting, competing and inextricable variables that have been modified by anthropogenic-driven disturbances (A-DDs). The degree of alterations registered by the receptor (environment) by A-DDs is not easily quantifiable, (iii) even though in project-level EIA, scoping may have attempted to focus on pertinent issues that ought to be addressed in the impact study, the problem of multicollinearity characteristic with multidimensional dependent (environmental) variables as well as independent variables (project actions (PAs)), raises some level of difficulty in drawing a sharp line between the relative contributions of the two set of variables interacting to determine the EQ, and also the dominant modes of variability governing the changes in the two categories of variables over time. Hence mitigation measures strategized without a solid knowledge of the dominant modes of variability controlling the dynamics of the EQ and PAs may not be the right and effective approach, (iv) environmental baseline and monitoring data sparsity impose a huge restriction in EAs, and rarely will EIA professionals/experts be enthusiastic about the conduct of a REA study given these conditions, and (v) the political climate or the framework under which EIA legislation operates in a given country. Sometimes politics frustrate the effective conduct of EIA. A classical example is China`s case where politics is seen to bulldoze its way through environmental regulation as can be inferred from Adams(2008) who makes a reference to Xiao (2007), and reports, "To date, many of China's dams have been built without receiving official environmental approval, often with no public knowledge, and certainly without public oversight. Even if no link between the recent earthquake and Zipingpu is conclusively drawn, public concern about dam safety in the event of earthquakes will take on a whole new life as approximately one-third of China's 88,000 dams are thought to be poorly constructed or dangerous". A situation like this will certainly be a serious set back in carrying out PDAs.

Realizing the aforementioned constraints, coupled with insufficient expertise, and the quest to improve EIA and PDAs in the subregion having sparse environmental databank, and having been spurred by the importance, the role and prospects of environmental antecedents in EQ prediction, we use a robust multivariate technique to conduct ex post evaluation of the impacts of two dams on three riparian communities in Ghana. This is followed by the development of stochastic and autoregressive (AR (1)) model(s) for backward predictions (hindcastings) of EQ in a one year lag of the affected communities.

2. A Case Study of Two Dams in Kumasi, Ghana

2. 1. Introduction

In Ghana, one of the most important infrastructural projects whose impacts cannot be overemphasized is river basin development projects (RBDPs) such dams. In this chapter, we shall use the term dams, reservoirs, man-made lakes (artificial lakes) synonymously. Reservoir constructions obstruct the natural hydrological flow regime resulting in the radical transformation of the lotic system to a lacusterine type (Baxter 1977; Beadle 1981; Petts 1984; Moss 1988). The description of environmental impacts in connection with dams just as in other projects, are very diverse: they may be good or bad; direct or indirect, beneficial or adverse; first, second or higher order; short-term or long-term; significant or insignificant; cumulative or non-cumulative; positive or negative; reversible or irreversible; reparable or irreparable; physical, biological, chemical or socioeconomic; rapidly-occurring or slowly-occurring; continuous or discontinuous; predictable or unpredictable, etc (see, for example, Rau and Wooten 1980; Biswas and Geping 1987; Culhane 1993; Jain et al. 1993; Petry 1995; Canter 1996). The benefits derived from dams include hydropower generation, potable water supply, inland water transport, flood control, fisheries development, recreation, tourism and sight-seeing (Baxter 1977; Petts 1984; Moss 1988, Biswas and Tortajada 2001). These have contributed immensely to socioeconomic developments not only in Ghana, but all other SSA countries/LDCs in the areas of industrial expansion and growth, investments (local and foreign), trade and commerce, health, and the general welfare of the people. For example in Ghana, one major dam which has contributed significantly to the nation`s socioeconomic development since 1964 is the Akosombo (Volta) Dam (Beadle 1981; Moss 1988; Kira 1995), not forgetting the several country-wide dams and weirs distributed all over the 10 administrative regions with distinct 138 metropolitan, municipal and district assemblies. Other reservoirs in the SSA/LDCs which have also made remarkable impacts are Lake Kariba (Zimbabwe/Zambia), L. Kainji (Nigeria), L. Nasser-Nubia (Egypt/Sudan) (Beadle 1981; Moss 1988) and several others (WCD 2000; Biswas and Tortajada 2001; Verghese 2001; Han et al. 2008). And more have been built in response to high demands stemming from rapid human population growth and development, urbanization, and industrialization (see, for example, Gupta 2001; Vyas 2001; Bhalla and Mookerjee 2001; Han et al. 2008).

A host of complex, complicated, and intertwined adverse impacts are however, known to be associated with reservoirs (Baxter 1977; Beadle 1981; Petts 1984; Moss 1988). Needless to say, the magnitudes of some of the adverse impacts offset some of the benefits of RBDPs. And very often riparian communities bear the brunt of unpleasant consequences from dam projects, aside from the distortion of the natural environment and ecological balance. The impacts are observed in the changes of ecology of both upstream and downstream river basin in the early and late phases of the filling of the reservoir, upswing in the prevalence of water-related diseases with serious consequences on human health with economic costs, luxuriant growth of macrophytes, socioeconomic problems, and land transformations such as the possibility of occurrence of reservoir-induced seismicities (RISs)/reservoir-triggered seimicities (RTSs) and landslides with devastating consequences (Gupta 2002). Ecological imbalances observed include biomass decomposition in the inundated catchment areas especially in the early filling stage of the newly formed man-made lake (leading to anoxia, and eutrophication problems), impact of native ichthyofauna (anadromous, catadromous, and potamodromous fishes), alteration in water quality, sedimentation, etc (Baxter 1977; Beadle 1981; Petts 1984; Moss 1988). Proliferation of water-related diseases following reservoir constructions, manifest itself in the general morbidity and mortality of affected people. In Ghana, typical reservoir-related diseases are mosquito-mediated diseases (such as malaria and yellow fever), schistosomiasis (such as vesicle/urinary and intestinal types), bancroftian (lymphatic) filariasis, dracunculiasis, onchocerciasis, and highly infective, water-borne diseases such as typhoid fever, cholera, viral hepatitis, and amebiasis (Fenwick 2006). For example, Derban(1975) reports a sharp increase in the prevalence of urinary schistosomiasis among children in communities along the Volta River banks from 5% in 1960 to 90 % in 1967 after the completion of Ghana`s Volta Dam in 1964. The health impacts of multiple infections is very serious (Tetteh et al. 2004a,c). For instance, schistosomiasis together with malaria, are often associated with chronic anemia in children especially, in the presence of low dietary availability leading to immunodeficiency problems, and reduced scholastic ability and social development in children (Pollit 1993; Soewando et al. 1989). Macrophytes are food sources and provide shelter for vectors of water-related diseases, and act as deoxyegnation agents especially along the lake shores leading to anoxia, and also impede navigation (Baxter 1977; Moss 1988). Land transformation is evidenced by alteration of geomorphology of the basin, erection of fish migration barrier, and various other physical and technological processes in the construction, operational, and post-operational phases that have the potential of causing coastal erosion, RIS/RTS, and landslides (Walter 1971; Baxter 1977; Ly 1980; Sarma 1990; WCD 2000; Gupta 2002; Xiao 2007). Socioeconomic problems manifest themselves in the expropriation of land, human displacement, conflicts, and loss of houses, indigenous culture and traditions, arable and pasture lands, livelihood, archaeological sites, among others (Beadle 1981). For example, displacements of some 84,000, 157,000, 50,000, and 100,000 people were recorded when Lake Volta, L. Kariba, L. Kainji, and L. Nasser-Nubia were formed, respectively (Beadle 1981; Moss 1988; Mills-Tettey 1989). These profound adverse impacts unequivocally, often raise a lot of social unrests, tensions, and environmental concerns among the communities and other stakeholders proactive in environmental protection and conservations, whenever and wherever RBDPs/WRDPs have been envisioned for their implementation in any part of the world (Biswas and Tortajada 2001). Here, we briefly present an overview on few examples. India for instance, has over 3,000 dams in Narmada River Valley, which have been constructed for multipurpose uses. Nonetheless, there is evidence of turbulent but nonviolent confrontations between pro-dam and anti-dam activists (Bhalla and Mookerjee 2001). In this country, the construction of the Narmada Dam had been designed to generate hydropower for promotion and sustenance of industrialization, and supply of clean water in areas where these utilities are lacking (Bhalla and Mookerjee 2001; Verghese 2001; Vyas 2001). This was the stance of pro-dam activists who describe the project as "Gujarat`s lifeline". However, anti-dam protesters describing it as an "environmental disaster", demanded a complete stop to the project (Gupta 2001;Verghese 2001). Central to the stiff, but nonviolent opposition by these protesters, was the displacement and relocation of people, which highly incurred their displeasure. Biswas and Tortajada

(2001) report a similar case involving the construction of the Sardar Sardar Dam in India which was opposed by the Narmada Bachao Andolan (NBA), a democratic movement and a mouthpiece of the *adivasis* (a tribal group), which was severely going to be affected. Other interesting and parallel studies have been reported elsewhere (see, Nandy 2001).

Ghana is not an exception to these multifarious reservoir-related problems. Major projects that have contributed substantially to the socioeconomic growth and development of the country were implemented in the pre-EIA era, and therefore did not have to pass any EIA test, and that any latent adverse and complex impacts could have been transmitted over time to present through cumulative or residual effects (CEQ 1987). Following the NEPA of 1969 and the 1972 Stockholm Conference on Human Environment, Ghana became more environmentally conscious. And in March 1988, the government of Ghana prepared the National Environmental Action Plan (NEAP) in response to deal with a myriad of environmental problems that had been connected to the exploitation of Ghana`s rich natural resources. The final adoption of NEAP in 1991 sought to deal with issues relating to land management, human settlement, mining, forestry and wildlife, marine and coastal ecosystems, industrial pollution, water management, hazardous chemicals and control (Workshop on Environmental Statistics for the ECOWAS Region, Dakar – Senegal 2005). With the adoption and implementation of EIA legislation following the enactment of the Environmental Protection Agency (EPA) Act 1994 Act 490 (Ghana EPA 1994), the previously existing Environmental Protection Council (EPC) was transformed into EPA, receiving an enhanced constitutional backing charged with environmental stewardship. The framework under which EIA operates has been described in detail by Ofori (1991) and Appiah-Opoku (2001). And now, all projects that have the potential of disturbing the EQ, are under obligation, to undergo a thorough EIA procedure before permits are granted. Another positive impact of the EIA legislation relates to dealing with all projects that existed prior to the EIA implementation, and had had the potential of causing environmental disturbances. The environmental regulation stipulates the submission of periodic Environmental Management Plans (EMPs) to the EPA on such projects as a way of handling post-development impacts (PDIs). Unquestionably, the Barekese and Owabi dams fall perfectly under this category of projects that require EMP preparations. However, sparse attention has been given to retrospective PDAs, monitoring, and evaluation (Culhane 1993; Dipper et al. 1998; Wood 1999) of these dams with special reference to downstream communities (Tetteh et al. 2004a,c; Tetteh et al. 2006). Among the very few studies that have addressed retrospective impact analysis (RIA) of these dams are that of Tetteh et al.(2004a,c), Tetteh et al.(2006), and Tetteh et al. (unpublished) . Our reconnaissance study conducted in 1998 indicated that five water-related diseases endemic in the communities downstream were malaria, vesicle (urinary) schistosomiasis, infectious hepatitis, diarrheal and skin diseases. This motivated us to conduct a full-scale retrospective environmental health impact assessment (REHIA) involving these multiple infections in the communities (Tetteh et al. 2004a). In this study, we utilized a weighted matrix technique for assessing the health impacts of the Barekese Dam. It was concluded that the dam had exerted an influence on the health of the downstream communities. In a similar study, we developed a weighting system for a comparative evaluation of the environmental health status at the local and national levels (Tetteh et al. 2004c), and the results were consistent with our previous findings (Tetteh et al. 2004a). Also, we used a network diagram that incorporated a probabilistic model to conduct post-project analysis (PPA) of the Barekese Dam (Tetteh et al. 2006). The expected environmental impacts (EEIs) computed showed negative values substantiating the adverse impacts of the dam. Other field and experimental studies as part of our monitoring program conducted from 1999 to 2004 continued to show endemicity of vesicle (urinary) schistosomiasis (Tetteh et al. 2004a,b,c; Tetteh et al. unpublished). We noted that well over 85 % of the children under 16 years had the characteristic hematuria in the urine samples collected, and worm burden evidenced through the isolation of copious quantities of *Schistosoma haematobium* eggs with their unique terminal spines, using urine filtration techniques (Jordan and Webb 1982; WHO 1983b; Tetteh et al. 2004b). Community diagnostics conducted on this disease also showed that the prevalence rates had doubled within three decades. We concluded that there had not been any comprehensive program of action to control the disease, and that occasional chemotherapeutic treatment of disease administered by the Ministry of Health as a control strategy, was not a lasting solution since treatment did not debar the children from re-entering the rivers to get re-infected with the disease. The communities argued that even though they are very close to the dams, they did not have access to clean water since the Owabi and Barekse Dams were commissioned in 1932 and 1972, respectively. They sharply drew a case and buttressed their argument against the municipal areas far removed from the treatment plants and are schistosomiasis-free, enjoy clean water whereas they the brunt bearers, live in total deprivation or neglect of this indispensable commodity. Their arguments however, are not peculiar to Ghana (see, Lerer and Scudder 1999). Nonetheless, several questions intuitively revolve around the clean water issue in these communities. But the prominent one that readily comes to mind is: Who is responsible for providing potable water for the downstream communities? The Ghana Water Company (GWC) Ltd., formerly, Ghana Water and Sewerage Corporation (GWSC) has the responsibility of providing potable water for the country. However, the Community Water and Sanitation Agency (CWSA), now an autonomous body, was created primarily with the responsibility of providing clean water in areas and communities, especially the rural areas beyond the GWC distribution system (Mensah 1998). As at the latter part of the 1990s when we conducted our study, nothing concrete had been done for the villagers. Interestingly, it was during that time that the GWC Ltd. had undertaken a refurbishment of the Barekese treatment plant (BTP) by duplicating raw water pipelines. This expansion project was intended to increase water production capacity by 50% in response to increase in demand for more clean water resulting from population growth and urbanization occurring in the metropolis. This initiative has made water more accessible, thus increasing the quality of life (QOL) in the metropolis and beyond. Associated with this expansion however, is a corresponding increase in the generation of wastewater which is discharged downstream and eventually mixes with the basin water flowing through the villages. Even though " Dams have been used as a political tool to deprive certain groups of water"(The Economist 1992) elsewhere, we cannot assume that the same applies the communities under study. Therefore, the answer to the question we posed unfortunately is beyond the scope of our study. However, the plight of the villagers therefore appeared to be in the domain of environmental and social injustice (Goldfarb 1999). Our monitoring program issued us enough signal to undertake a comprehensive study to ascertain the historical impacts of the dams. This further indicated that it appears REA is either missing, or not given the due recognition in the country`s environmental regulation in contradistinction to new projects which tend to receive better recognition and intense scrutiny to pass the EIA requirements. For example, the construction of the Bui Hydroelectric Dam in Ghana, which is currently underway, attracted a lot of attention from well-intentioned environmental pressure groups, anti-dam activists, conservationists, and communities who would suffer from the implementation of the project. Their opposition can be connected to lessons learned from analogous projects such as the extent to which the operations of the mining sectors have distorted the ecosystem integrity and stability, and have had profound environmental degradation, contamination, and pollution (Acquah and Boateng 2000; Nyamekye 2000; Babut et al. 2003; Bonzongo et al. 2003; Golow and Mingle 2003; Donkor et al. 2005; Donkor et al. 2006; Hilson and Pardie 2006; Obiri et al. 2006; Asante et al. 2007; Hilson and Pardie 2007), for which the country is struggling to grapple with managing these myriad of problems. In contrast, infinitesimal voice has been added to the plight of the communities under study. Needless to say, the role of environmental antecedents in helping us to understand the present better and to re-shape our future environmental policies is indispensable. And this forms the justification, foundation and motivation of our research.

In spite of all these Ghana has made a giant leap step forward internationally in the area of environmental sustainability given the fact that EIA legislation is barely less than 15 years old. It ranked the 49th best out of 146 countries that met the inclusion criteria, based on 2005 Environmental Sustainability Index (ESI) (http://epi.yale.edu). In the area of environmental performance, it ranked 72th based on the Pilot 2006 Environmental Performance Index (EPI). However, in April 2008, it occupied the 86th position out of 149 countries that met the inclusion criteria, a position which was below the midpoint rank. The results testify the need for the country to be infused with environmental preservation, conservation and rehabilitation with more scientific rigor. And this can be achieved if it becomes more committed to REA as well. The rationale is that with about 70% of the population dwelling in rural areas whose economic base is agriculture, the bedrock of national economy, environmental sustainability and performance problems can be tackled expertly at the grass root levels. As our objectives, we set forth the mantle by overhauling the past EQ of three riparian communities downstream of the two dams, using empirical orthogonal function (EOF) analysis, making use of limited data for four distinct years. We then move a step beyond this by developing stochastic and temporal environmental AR (1) models to theoretically backcast EQ of the communities in one year lead time from the baseline environmental states.

3. Materials and Methods

3. 1. Brief Description of the Study Area

Our study area comprises two dams, and three riparian communities namely, Barekuma, Aninkroma, and Hiawo Besease located in the Atwima District of the Ashanti Region of Ghana (Figure 1). Geographically, the Owabi and Barekese Dams are located at 6° 44′,1° 42′ W and 6° 50′,1° 44′ W, respectively. The former, a much smaller dam compared with the latter, was constructed between1928-1932, whereas the latter was constructed between 1967-72. They lie across River Owabi and R. Offin, respectively (Figure 1). River Offin is much bigger than the R. Owabi, and two rivers form a confluence beyond Hiawo Besease, continuing as the major Offin River. Both dams are the composite type with central concrete gravity spillway and earth-fill embankments. The Barekese and Owabi Dams control catchment areas of 906 km^2 and 69 km^2, respectively. Other characteristics of the dams have been in described in detail by Wassink(2004). These dams constitute the principal sources of water supply for the Kumasi Metropolis, which has an estimated population of 1,517,000 and an urban annual population growth rate of 3.81 % between 2000- 2005 (UNPD 2005).

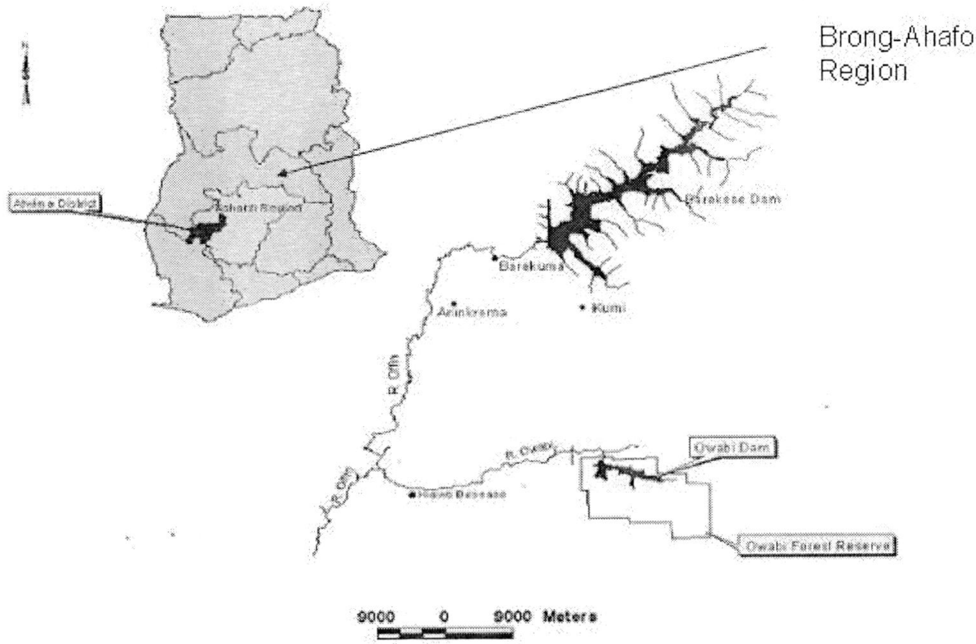

Figure 1. Map of Ghana showing the study areas in the Atwima District in the Ashanti Region of Ghana.

The communities are typically peasant farmers. Barekuma and Aninkroma are located downstream of R. Offin, and Hiawo Besease downstream of R. Owabi. Their respective geographic coordinates are 6° 49′, 1° 46′ W (Barekuma), 6° 48′, 1° 46′ W (Aninkroma), and 6° 44′, 1° 47′ W (Hiawo Besease). As at 1999, their respective estimated populations stood at 2000, 1200, and 2500 (Tetteh et al. 2004a,c; Tetteh et al. 2006). Lacking access to clean water, the two rivers constitute their primary source of water for domestic, economic, agricultural, recreational, and other activities. Whereas Hiawo Besease uses on both rivers, the other two communities use R. Offin. Rain harvesting is an extra source of water during the rainy season. Brief community characteristics have been reported by Tetteh et al.(2004a,c) and Tetteh et al. (2006).

3.2. Research Methodology

3.2.1. Predictor and Predictand Determinants

The primary objective was to select predictors and predictands for the construction of two sets of indices over four distinct time periods for REAs as well as EQ prediction backward in time for the communities. The predictors (independent or explanatory variables) were generated from the OAs of the dams from which we constructed operational activity indices (OAIs). OAIs were intended to give a description of the combined effects of the most important aspects of the multidimensional OAs of the dams, and for the analysis of changes in the EQ of the communities over some specified time periods. The predictands (dependent or response variables) are the environmental baseline conditions used to construct the EQ indices (EQIs). The EQIs similarly, were intended to give a description of the most important aspects of the EQ of the communities from a multidimensional dataset over the same time periods. Selection of appropriate variables for the construction of these indices were achieved through multidisciplinary team discussions, reconnaissance study, extensive literature search (eg in Environmental Science, EIA documents, etc), the use of professional or expert judgment, and analogous information elsewhere (Appiah-Opoku 2001; Jain et al. 1993; Canter 1996).

A reconnaissance study involving visits to both water treatment plants and the communities in 1997 and 1998 was undertaken to guide us in determining pertinent issues to be incorporated in the terms of reference (TOR). In our visit to the treatment plants, we had the opportunity of getting first hand information on clean water production cycle from raw water (untreated) from the dam to the end-product (treated water) released into the main distribution pipeline. The effluents (sludge) produced and discharged downstream were well noted. Exploratory visits to the treatment plants were followed with visits to communities. After seeking permission from the chiefs, elders, and Village Committee members and leaders, the whole communities were invited to meetings that were scheduled at different times in which the rationale of the study was conveyed to them explicitly. Through this community involvement (Canter 1996) we had the opportunity to know more about them, and their grievances. In the well attended meetings, water-related health issues, drinking water problems, and wastewater from the treatment plants were the communities' concerns and were accorded valuable inputs for incorporating into the TOR for the study (Tetteh et al. 2004a,b,c; Tetteh et al. 2006). We then explored their general environments and made visits to their water collection sites from the rivers.

After the completion of the pilot study, we decided on the specific variables for creating predictor and predictand data sets. Based on the OAs, guided by the criteria suggested by Jain et al. (1993) and Canter (1996), we arrived at three major activities that were perceived to have impacted the EQ of the communities (Tetteh et al. 2006), and deemed them important ingredients for the construction of OAIs. These are: (i) the presence of the earth-filled embankments acting as physical barriers to the flow of the basin water. Even though the embankments *per se* are static engineering structures, their manifest and latent adverse impacts cannot be overemphasized. Their presence constitutes huge barriers dividing the river basins asunder, culminating into long-lasting environmental changes both up and downstream, which have been reviewed in the previous sections, (ii) discharge of wastewater resulting from water treatment and its impacts on the downstream water quality (WQ), and (iii) occasional backwashing of clogged filters and their associated effluents released into the downstream river basin. On the basis of the physical and chemical inputs used during water treatment process and with reference to Ghana`s effluent guidelines for discharges into water bodies (Ghana EPA Effluent Guidelines 1999), 11 physico-chemical parameters of the effluents were thought as important in impacting downstream water quality (WQ). The OAs were generally categorized as physical process I, II, and III. In all, we had 13 distinct dam-related variables that continuously impacted the downstream EQ, and constituted predictor sets used to construct the OAIs shown in the framework (Figure 2). For the predictands, 44 potential variables were initially listed. Through a rigorous screening, any dependent variable(s) which had not been impacted upon continuously by any of the OAs (ie, finding of no significant impact (FONSI) (Canter 1996)) over all the specified time periods were dropped. This was achieved partly by using a rating scale developed for qualitative data, relying on the premise that a rating scale is developed when there is lack of uniform standards (Hickie and Wade 1989; Tzoumis and Finegold 2000). The selection of the most relevant predictands was further enhanced by relying on the results of the probability of occurrence from our previous study (Tetteh et al. 2006). Through this process, 20 out of the 44 variables met the selection criteria. The 20 predictands were broadly classified under: (i) environmental health, (ii) water quality, (iii) biodiversity, (iv) social psychological factors, and (v) natural hazards. The predictands were then used to construct the EQIs whose framework is shown in Figure 3.

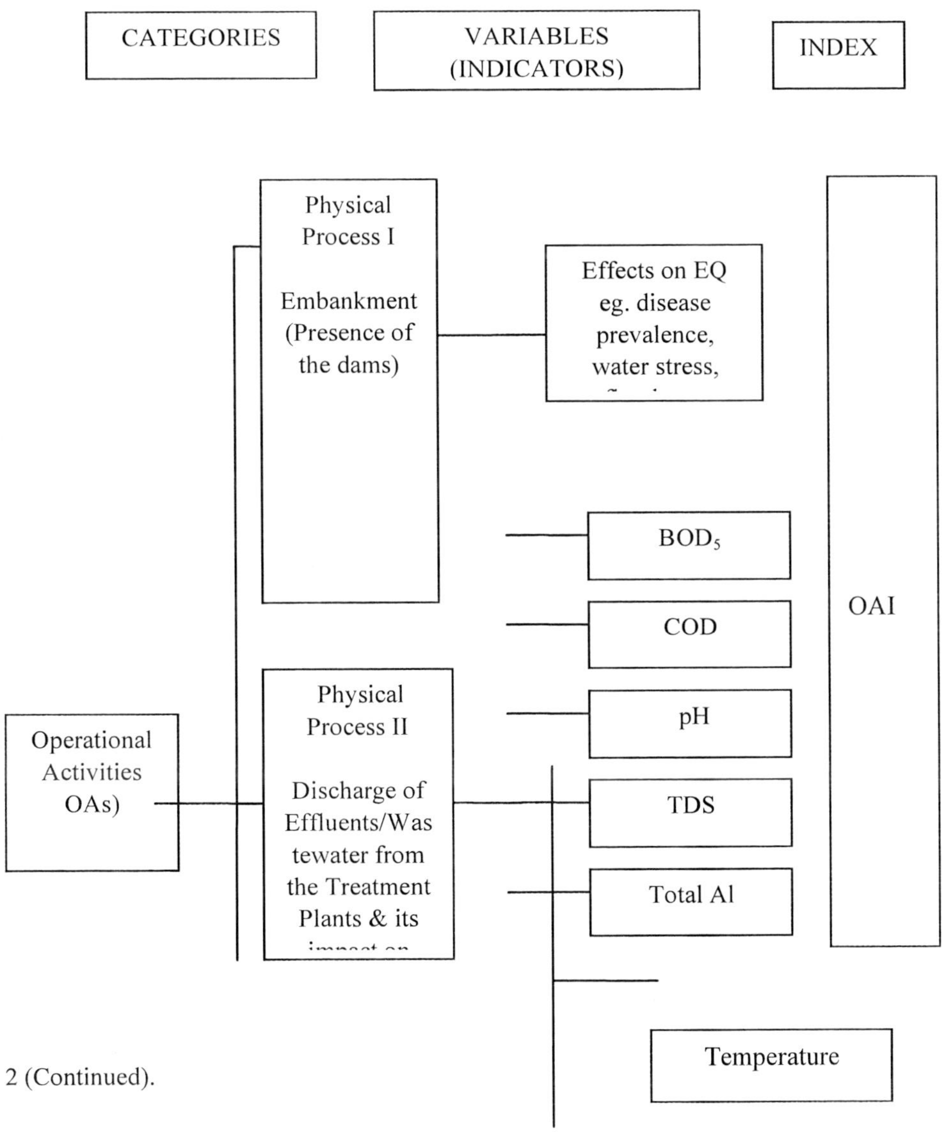

Figure 2 (Continued).

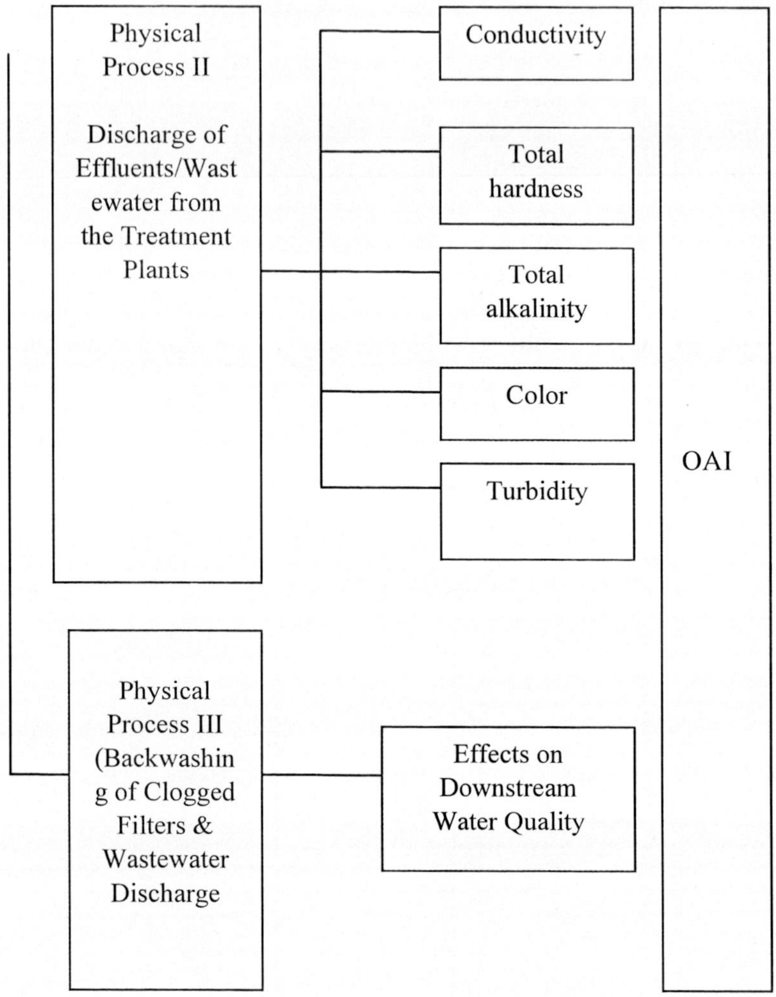

Figure 2. Construction of the OAI.

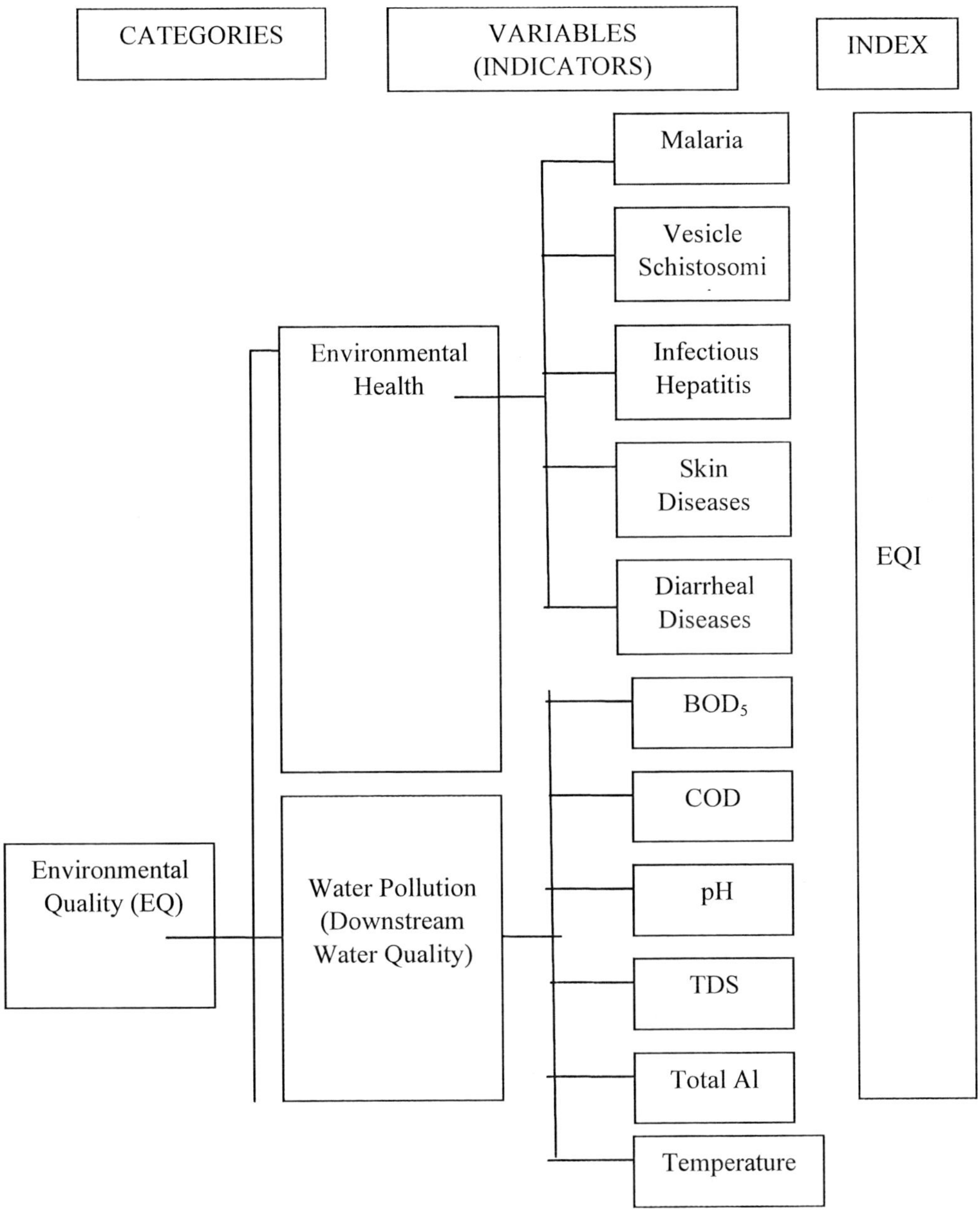

Figure 3 (Continued).

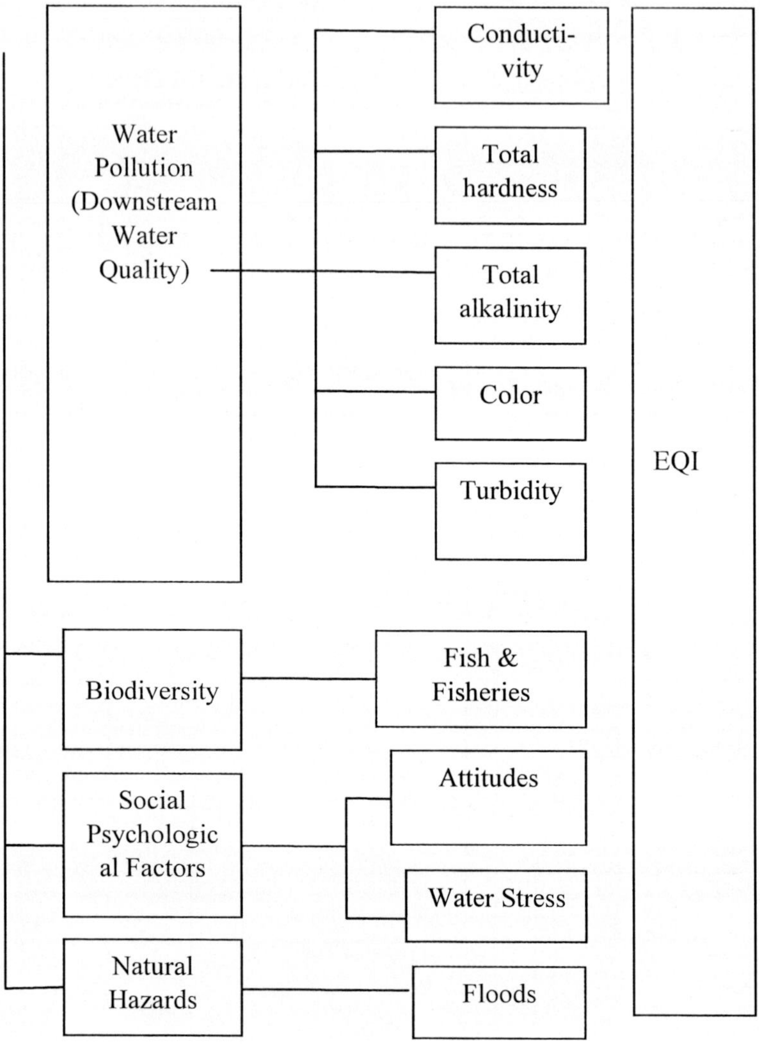

Figure 3. Construction of the EQI.

3.2.2. Experimental Design, Data Collection and Measurements

The paucity of environmental baseline and monitoring data necessitated obtaining primary data. The data collection was organized data along two lines. These are field survey, and laboratory study. The field survey needed to satisfy a minimum sample size requirement that would be a representation of the total populations under study. To achieve this, we used vesicle (urinary) schistosomiasis, one of the most recognizable indicators of EQ incorporated in the TOR to compute the minimum number. In a cross-sectional survey, a simple random sampling was conducted to estimate the proportion of people who had the disease in an initial sample size of 50 in each of the communities. The sample size was determined for each community using a formula in Sanders (1990) which is given as follows:

$$E = K\left[\sqrt{P(1-P)\frac{1}{n}}\right]$$

where,

E= desired error; in this case our E=$\pm$ 0.05

K=confidence multiplier which depends on alpha (α); in this case K=1.96

P= proportion of a variable in a sample; in this case *P* refers to the proportion of people with disease

n= sample size; in this case n is number of subjects required for the full-scale study

Based on this, we needed a total sample size of 618 (207, 226, and 185 from Barekuma, Aninkroma, and Hiawo Besease, respectively). To increase the efficiency of the data collection, we divided the respective numbers by the average household sizes (Tetteh et al. 2004a) which indicated that we needed at least 41 households in all (14, 15, and 12 for Barekuma, Aninkroma, and Hiawo Bease, respectively). However, we decided to use 54 households (18 per community) to increase the power of our analysis. One of the greatest challenges we had had to contend with was the reliability of field data we had planned to collect (not for the year 1999 we started the study where everything remained relatively fresh in the respondents` memories) but rather, the information of the past years since we were going to rely on test of memory. This is because reliance on primary data in a situation like this (ie, for REA), could be undermined if historical information is dominated by oral tradition, and the right approach and target group for information retrieval are not used. Nonetheless, the quality of data obtained from oral tradition or undocumented (informal) depends to a large extent on the type of study being undertaken. More so, cultures, traditions and historic events in African societies that had exited over hundreds of centuries ago before the advent of computer or electronics age, or the invention of printing press, have been preserved and handed down from generation to generation accurately via oral tradition to present. And this has been considered as one of the most powerful and trustworthy information dissemination technique (Horton 1967; Cohen 1989; Hale 1997). This gave us the assurance and credibility of using such resources for obtaining appropriate information. In our case, we were guided by the importance and the role of traditional ecological knowledge (TEK) and 'griots' in making effective EA studies (Cohen 1989; Johannes 1993; Kuhn and Duerden 1996). Our pilot study also helped us in deciding which time periods in the country`s history when connected to the study, would insure efficient and reliable data collection.. We had also realized that time series data are indispensable for the study, and any attempts made to acquire historical facts/data in a regularly-spaced time periods would be tantamount to incurring huge error and make the data worthless, since it would be extremely difficult if not impossible, for the interviewees/respondents to systematically and sequentially recollect historical facts over decades. An approach to handle such a situation is to use a limited data, this justifying four distinct years used for our study. The years selected were 1957, 1966, 1972, and 1999 which formed important parts in the life cycles of the dams. We deemed it very important to limit ourselves to on four years for the following reasons. First, to avoid putting a huge burden on the informants in overstretching their memories to retrieve past events for generating regularly-spaced time series data from 1957-1999 (ie, 43 continuous years), which would not be a possibility in any way. Second, these years constituted parts of the life cycles of the dams which, overlapped with landmarks in the political history of the country (Boafo-Arthur 2000; Tetteh et al. 2004a). And any historical information on the communities' EQ needed, when connected to the dams` history (existence and operation) and historic political events of the country, was perceived to be an appropriate circumstance and a powerful catalyst that could facilitate respondents` memory recall (Bonsu and Thatcher 2007).

We present here a brief recap of Ghana`s political history (Boafo-Arthur 2000; www.ghanaweb.com/GhanaHomePage/republic/polit_hist.php) vis-à-vis the life cycles of the dams and the environmental states of the communities. On 6th March 1957, Ghana became the first country in the SSA to become politically independent after gaining her independence from the British colony, with the first republic commencing on 1st July 1960. On 6th March 1957 when the first president, Dr. Kwame Nkrumah publicly declared Ghana`s independence to the people of the motherland, the Owabi Dam had been in operation for about 25 years, and any operational impacts could have well been transmitted to Hiawo Besease (see Figure 1) over these years through several pathways. However, the Barekese Dam had not been constructed by then. So, any impacts at Barekuma and Aninkroma could have been attributed to natural variability. Dr. Nkrumah`s government, the Convention Peoples Party (CPP) grew less successful with time. His government was overthrown by the military and the police on 24th February 1966. And this was the maiden coup d`etat in the annals of Ghana`s political history. The nation lost its political virginity, and became more vulnerable to political instabilities. At that time the Owabi Dam was still in operation. However, the Barekese Dam had been on the drawing board before the political upheaval. Dr. Nkrumah`s government had intended to build this dam for a dual purpose (Amoah, B. personal communication 1997): water supply for the Kumasi Metropolis and hydroelectricity for both the Ashanti and Brong Ahafo Regions (see Figure 1 for the location of the Brong Ahafo Region). The dam`s foundation stone was later laid in 1967 (Kumah and Baah 1971), but the hydropower generation component was suspended by Dr. Nkrumah`s successor, and is still missing in the political agenda of various governments which have come into power since then. Ghana returned to a constitutional rule in 1969 under Dr. K. A. Busia, the Prime Minister of the then ruling Progress Party (PP) government, that commenced the second republic. Dr. Busia was known for his anti-Nkrumaist stance and adopted a liberalized economic system. In his regime, there was a mass repatriation of 500,000 Nigerian citizens living in the country. His era was also characterized by 44% devaluation of the cedi in 1971 which met a public outcry. The PP government ruled from 1969-1972 when it was toppled down on 13th January 1972 by the military led by the then Col. I. K. Acheampong who established National Redemption Council (NRC), which was later transformed into Supreme Military Council (SMC) on 9th October 1975. It was in 1972 that the Barekese Dam was commissioned and they city could pride themselves with two functional dams serving the metropolis with clean water. The EQ at Barekuma and Aninkroma at least from that time could have started experiencing the operational impacts of the new dam. Col. Acheampong, the head of state was promoted to the rank of a General. Interestingly, he hailed from the Ashanti Region, specifically, Atwima District, where the two dams are sited in our study area. He was popularly known for his ideology on Union Government with the motto "One People, One Nation, One Destiny", which he introduced, and the radical agriculture philosophy, "Operation Feed Yourself", which was embraced by the nation as a way of stimulating every person to become self-reliant. During his era, the country shifted from the imperial system to the metric system of measurements. The country also witnessed "Operation Keep Right" on 4th August 1974, where there was a change from driving on the left to driving on the right. On 5th July 1978, a palace coup was successfully staged to overthrow the SMC that led to the establishment of SMC II headed by Lt. Gen. F. W. K. Akuffo with a view to restructuring the military government, and continue with the preparation to return the country to a constitutional rule. His rule was cut short when on 4th June 1979 the SMC II was overthrown by junior military officers led by Flt. J. J. Rawlings, who declared himself as the chairman of the Armed Forces Revolutionary Council (AFRC). During his regime, the country witnessed a heinous execution by firing squad, most senior military officials including the former heads of states of SMC I and SMC II. The third republic was ushered in 1979 after Dr. H. Limann and his party, the Peoples National Party (PNP) had won the general election. The third republic was overthrown on 31st December 1981 by the military junta led by Flt Lt. J. J. Rawlings, again, and became the chairman of Provisional National Defense Council (PNDC) and the head of state of ruling from 1982-1992. During his regime, the 31st December Women`s Movement was created with his wife, Nana Konadu Agyemang-Rawlings as the President of the movement. Eventually, the fourth republic began on 7th January 1993 when Flt. Lt. J.J. Rawlings and his party, National Democratic Party (NDC) won the general elections in December 1992. He was re-elected in the December 1996 general elections. He ruled from 1993-2000, and he is known for being the longest ruler (at least 19 years) in Ghana; as a military junta (11 years) and a democrat (8 years). It was in 1999, a year prior to his last term in office, when we conducted our study. At that time, the Barekese Dam had operated for nearly three decades and Owabi Dam over six decades, and EIA legislation in the country had been only 5 years old. The EQ of the communities could have also been altered by the OAs of the dams for these numbers of years. With this information, we deemed it important to design the study to focus only on 4 years (1957, 1966, 1972, and 1999): 1957-1999 marks post-operational phase of the Owabi Dam; 1957-1966 and 1972-1999 mark pre-construction and post-operational phases of the Barekese Dam, respectively. Data collection was strictly tied to the four distinct years, which marked some of the important events in the country`s political history.

Using the OAI and EQI frameworks (Figures 2, 3), data collection methods were organized into two: quantitative and qualitative. Quantitative methods focused on water quality and environmental health measurements. All others were determined qualitatively. In the field work, a simple random sample was performed to select 18 households per community as experimental units. In the longitudinal studies, we started with 1999 and retrogressed to 1957. The rationale was to let the respondents get familiarized with the questions addressing issues in 1999, which were comparatively more current in contrast to the other years, and then retrogressed to those years using the same questions. Questionnaire checklists (Canter 1996) were used as measurement tools in the field survey. The questions were designed based on level zero, and occasionally, level one, as seen in the mode of formulating instructional objectives in pedagogy and didactics in Educational Psychology, Educational Philosophy, Sociology, and in the Behavioral Sciences (Anderson and Glew 2002; Gosh 2007). The questions were connected to 1957, 1966, 1972, and 1999 historic political events in the country. The advantage of these techniques was to insure efficiency in data collection without loss of any vital information. Another advantage was to insure that questions designed were not cognitive-driven (VanLehn 1996; Johnson et al. 2006). A violation of this, could lead to difficulty in translating the questions exactly to their dialect that could easily be comprehended by communities, and this could trigger off poor responses/feedbacks. The questions were directed to the elderly, parents, and adults. Based on the responses, the qualitative information obtained was transformed into weights using a rating scale (Hickie and Wade 1989; Tzoumis and Finegold 2000) developed for study. The scale theoretically ranges from 0-10, and has five distinct ranges with their corresponding comments indicating the level/magnitude of any environmental variable measured against the OAs of the dam(s). The same scale was used to assign weights to all qualitative OAs. The scale is presented as follows:

- 8 - 10 very high
- 6 – 8 high
- 4 - 6 moderate
- 2 - 4 low
- 0 - 2 insignificant

Effluents from the BTP were sampled fortnightly from January to March, from behind the pumping stations (Tetteh, et al. 2006), and were analyzed using standard methods techniques (APHA/AWWA/WEF 1995), focusing on the WQ parameters listed in Figure 2. Three replications were made for each parameter analyzed per sampling day. Sampling and analyses were done in the dry season to remove the influence of allochthonous materials that otherwise, could have masked the effects of the effluents from the treatment plants had we sampled in the rainy season. We were constrained by repeating the analysis for the OTP since its effluents were inaccessible for sampling. So, we assumed that the data obtained from the effluent analysis from BTP, a far bigger treatment plant than OTP, would be an adequate representation of the composite action (effluents) from the two dams for the construction of the variance-covariance matrix needed for other parts of the studies.

The next issue was to estimate the effluent quality for 1972, 1966, and 1957 where no information exists. The effluent data for these years were treated as "missing values". In this sense, the variance-covariance matrices with missing values cannot permit principal component analysis (PCA) to be performed easily. Several methods are available for handling missing values. These are mean substitution, and various types of extrapolations, and interpolations such as smoothing, spline interpolation, bicubic interpolation, kriging, linear interpolation and universal kriging, etc (see, for example, Dempster et al. 1977; Little and Rubin 1987; Cressie 1993; Ghahramani and Jordan 1994; Kitanidis 1997). Also, pairwise and casewise (listwise) deletion of missing data as well as expectation-maximization (EM) algorithm can be used. For instance, an application of the EM algorithm in EOF analysis is probabilistic PCA (PPCA)(Tipping and Bishop 1999). This approach has been used by Houseago-Stokes and Challenor (2004) to perform EOF analysis of North Atlantic sea surface temperature (SST) in the presence of missing values. In our case, we used kriging (Cressie 1993; Kitanidis 1997) to estimate the effluent quality of 1972, based on the 1999 station (BTP) data. The values obtained were used to represent the effluent quality from the two treatment plants. In a stepwise and backward moving fashion, the 1972 values were used to estimate the effluent quality for 1966, and finally, moved to 1957 using that of 1966. The values obtained for 1957 and 1966, were however used to represent the effluent quality of OTP only, since the BTP was not in existence by then. In this regard, the covariance matrices for 1957 and 1966 for the OAs had some zero values (assessed on our rating scale), an indication of no effluents discharged into R. Offin, and consequently, an insignificant impact assigned.

The monitoring of the communities` downstream river quality in 1999 as part of determining the EQ, was similarly done in the dry season from January to March. Sampling was done in the dry season fortnightly at different sites along the river basins: Barekuma and Aninkroma (R. Offin); Hiawo Besease (R. Owabi), and beyond the confluence (R. Offin and R. Owabi)(see Figure 1). Analysis was done using standard methods techniques (APHA/AWWA/WEF 1995) using the same number of replicates as before. Kriging technique (Cressie 1993; Kitanidis 1997) was similarly adopted to estimate the WQ of the communities for 1972, 1966, and 1957. In all, the standard deviations for our kriging estimates ranged between 1.08-1.14, giving us the confidence in the reliability of estimates.

4. Analysis Techniques

4.1. Data Structure and Exploratory Data Analysis

The data collected form discontinuous, irregularly spaced, multivariate geostatistical, limited time series data (Cressie 1993) consisting of sets of predictors and predictands with the following breakdown: four datasets each for dependent variables (predictands) and independent variables (predictors) for the construction of EQIs and OAIs for the years (1957, 1966, 1972, and 1999). Each of the predictand and predictor datasets has 54 x 20 and 54 x 13 dimensionality, respectively. Exploratory data analysis (EDA) was carried to test for multivariate normality and multicollinearity (Johnson and Wichern 2007). The rationale behind the multivariate test was that even though we had satisfied the minimum requirement for sample size for the fieldwork, implicitly suggesting normality of the population, the same could not be implied for the actual data collected. Test of normality can be done either quantitatively or graphically (ibid). Here, we used the graphical approach for the analysis. The tests revealed lack normality in the datasets exemplified by two normal quantile-quantile(Q-Q) plots of their marginal distributions (Figures 4a,b). The variance-covariance matrices of the datasets also showed matrix singularity (which is not abnormal) characterized by zero determinant, and noninvertible matrices (only the generalized inverse can be computed), and existence of strong mutual correlations (multicollinearity). The test for multivariate normality in singular matrices like these was achieved by using the upper triangles of the variance-covariance matrices. The chi-square plots exemplified demonstrate non-normality (Figures 4c,d). The existence of multicollinearity is also an indication of data redundancy. Moreover, strong mutual correlations and data redundancy "lead to poor estimates (high-variance sampling distributions) for the parameters" (Wilks 2006). We solved the non-normality problem by using data transformation technique carried out by finding the square root of all the observations in data frames (Johnson and Wichern 2007). The multicollinearity problem was solved through empirical orthogonal function (EOF) analysis.

In this chapter, we shall use PCA and EOF analysis interchangeably (Preisendorfer 1988; Peixoto and Oort 1992). This is one of the main analytical tools used in the study.

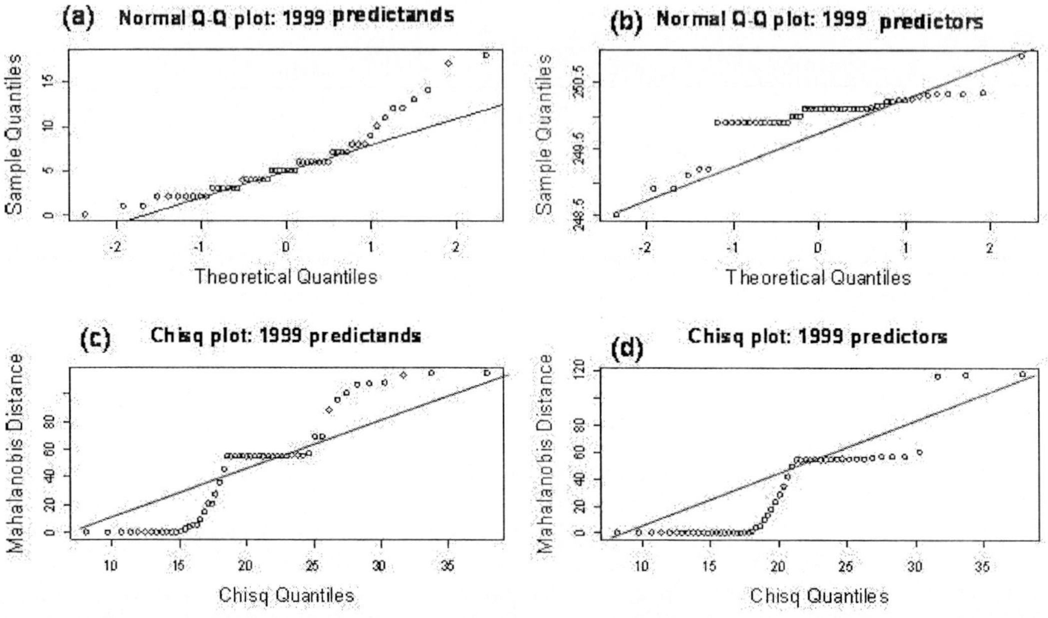

Figure 4. Test of marginal and multivariate normality.

4.2. EOF Analysis and Dependency Measures

Analysis of multivariate data is often fraught with difficulty because of multi-dimensionality of datasets. However, readily available analytical techniques for dealing with multivariate data analysis are EOF analysis, factor analysis, canonical correlation analysis, cluster analysis, discriminant analysis, logistic regression, etc (Wood 2000; Wilks 2006; Johnson and Wichern 2007). EOF/PC analysis derives its roots in the Social Sciences (Pearson 1902), where there is evidence of its substantial applications. However, its utility in Geophysical Sciences such as Atmospheric Sciences is relatively recent. It was used in the late 1940s by Obukhov (1947), followed by Fukuoka (1951), Lorenz (1956), and Kutzbach (1967). Since then it has gained a lot of popularity as a robust analysis tool in climate research. Rarely, has this tool been explored specifically, in the field of EIAs/EAs with particular reference to Ghana`s environmental system. Among the few studies that have used PCA as analysis tool in EIA, is that of Ying and Liu (1995). In the global context, PCA has been used for the development of the Environmental Sustainability Index (ESI) such as the 2005 ESI, and Environmental Performance Index (EPI) such as the Pilot 2006 EPI and 2008 EPI, where 21, 16, and 25 variables(indicators), respectively, were used to construct a set of indices for the global initiative conducted by Yale Center for Environmental Law and Policy (YCELP), Yale University and Center for International Earth Science Information Network (CIESIN) in collaboration with World Economic Forum, Geneva, Switzerland and Joint Center of the European Commission, Ispra, italy(http://epi.yale.edu).The types of PCA are traditional/standard/regular/conventional EOF analysis, extended EOF analysis, combined EOF analysis, simplified EOF analysis, functional PCA, Bayesian PCA, rotated EOF analysis, multivariate EOF analysis, single (univariate) empirical orthogonal function(sEOF) analysis, cyclostationary EOF analysis, probabilistic PCA, real vector EOF analysis, and complex EOF analysis (Jolliffe 1972; Weare and Nasstrom 1982; Horel 1984; Branstator 1987; Chen and Hart 1993; Wang et al. 1995; Kaihatu et al. 1998; Kim and Wu 1999; Houseago-Stokes and Challenor 2004; Buongiorno Nardelli and Santoleri 2005). However, they share the basic underlying principles. EOF/PC analysis is a vector space orthogonal linear transformation that converts a variance-covariance matrix (correlation matrix) dataset of n x p dimensions into a new coordinate system represented by a few linear combinations that explains or carries a very large proportion of information contained in the original data. The new components, representative of the original data set n x p dimensions are known as principal components (PCs) (Jolliffe 2002; Wilks 2006; Johnson and Wichern 2007). In this case, PCA is also a dimension reduction technique in which the new coordinate system has PCs with a lower dimensionality, but at the same time retains those important characteristics of the original data that contributed a greater proportion of the variance in the transformation. The first PC has the highest (maximum) variance, followed by the second, third, etc. But overall, the variance of the original data and the newly transformed PCs are conserved. The computation leads to three important products namely, the eigenvalues, eigenvectors, and scores, the dominant modes representing the most important information that is captured from the original data. The three products are a depiction of the variance, spatial loadings/spatial patterns (loadings) of variability, and the time series, respectively. EOF analysis has the ability to orthogonalize a matrix of vectors making the variables in the new coordinate system uncorrelated with each other in space. Thus, the multicollinearity problem observed in our datasets was solved with this technique. The multidimensionality of our datasets necessitated the adoption of the multivariate EOF analysis technique (Buongiorno Nardelli and Santoleri 2005). Prior to doing the analysis, all the data were non-dimensionalized via standardization so that all the data are put on the same platform (Wilks 2006; Johnson and Wichern 2007). Another advantage of this is that it eliminates the undue advantage of variables having large values in the n x p variance-covariance matrix to be weighted heavily in the total variation as compared to variables with smaller values (Buongiorno Nardelli and Santoleri 2005; Johnson and Wichern 2007). This standardized variance -covariance matrix now becomes a correlation matrix which is more refined, better, and robust. For each year, we generated a 54 x 20 and 54 x 13 correlation matrices, and their corresponding PCA generated 24 and 13 PCs for the predictands and predictors, respectively. In this study, the scores fixed in time, consisting of 54 data points were used as the indices. So, we had 24 and 13 indices which represent 24 EQIs and 13 OAIs, respectively. The contributions of the PCs of the predictors and predictands in explaining the variances were determined from their eigenvectors.

On the basis of the results obtained from the EOF analysis, we decided to retain the first three dominant modes of the PCs- the eigenvalues, eigenvalues, and scores (now the EQIs and OAIs) for further analysis. Following that, dependency measures (concurrent and lagged cross correlations), typically zero-order parametric and nonparametric correlations, were carried out to determine the degree of the predictand-predictor (EQI-OAI) relationships, and for monitoring the EQ over the study periods. For the parametric correlation, we used Pearson product-moment correlation, whereas Spearman rank correlation (rho), and Kendall's tau were used for nonparametric correlations (Helsel and Hirsch 1992; Sheskin 2004). The moderate to strong lagged correlations suggested a good forecasting potential for the predictors. The first three scores of the EOF analysis (ie, the first three OAIs and EQIs) were used as inputs for the stochastic and AR (1) model developments (Spall 2003; Sprent and Smeeten 2007).

5. Stochastic and Autoregressive AR (1) Model Developments and Analysis Techniques

5.1. A Brief Overview of Forecast Models

Predictive (forecast) models are very diverse and have several applications both in the Social and Geophysical Sciences. Forecast models have been classified differently by different authors. Examples of these models include: the linear trend model, the constant mean model, seasonal models, regression models, probabilistic models, stochastic models, growth curves, dynamical models, and multivariate models (Gilchrist 1976; Wilks 2006). Sparse attention has been to the utility of these models in EIA neither have audit methodologies been adequately developed (Wood 2000). Tomlinson and Atkinson (1987) however, have reviewed a substantial literature on the applications of predictive technique audits (PTAs) studies in the U.S.A. PTAs are recognized as a general class of techniques known for their important roles in EIA audits. Of the forecast models, those which have received little attention in PTAs include stochastic and AR (1) models, which are dealt with in the next subsection.

5.2. Development of Parametric, Nonparametric and AR (1) Models

This section is concerned with the development of parametric, nonparametric stochastic, and AR (1) models with the overall aim of theoretically, making retrospective predictions (hindcasts or backcasts) with a positive one year lag (t +1 time step) from the baseline for each of the four years using inputs from the EOF analysis results. Eventually, an attempt would be made to decide which of the stochastic models better fits the data. Prior to the development of these models, the potential backcasting (hindcasting) ability of predictors were ascertained by performing cross-validation techniques (Sheskin 2004; Shumway and Stoffer 2006; Wilks 2006).

The parametric model construction made use of the EOF 1, 2, and 3 predictor indices of the 4 years (hereafter OAI 1_{1957}, OAI 2_{1957}, OAI 3_{1957}, OAI 1_{1966}, OAI 2_{1966}, OAI 3_{1966}, OAI 1_{1972}, OAI 2_{1972} OAI 3_{1972}, OAI 1_{1999}, OAI 2_{1999}, and OAI 3_{1999}) and EOF 1 of the predictand indices of the 4 years (hereafter EQI_{1957}, EQI_{1966}, EQI_{1972}, and EQI_{1999}). The underlying model construction is multiple regression analysis. Parametric models assume Gaussian, Poisson, and Binomial distributions underlying them (Spall 2003; Sheskin 2004; Wilks 2006). In our study, we assumed that our data follow the Gaussian distribution (Wilks 2006) as they had undergone square root transformation to eliminate the non-normality (Johnson and Wichern 2007) prior to the PCA (see, Section 4.1). On the other hand, nonparametric models are distribution-free and follow resampling procedures (Sheskin 2004; Wilks 2006; Sprent and Smeeton 2007). In this case, we used the EOF analysis results generated from the original untransformed datasets which lacked multivariate normality, and then used the dominant three predictor EOF indices (hereafter OAI $1_{nsrt1957}$, OAI $2_{nsrt1957}$,OAI $3_{nsrt1957}$,OAI $1_{nsrt1966}$, OAI $2_{nsrt1966}$,OAI $3_{nsrt1966}$,OAI $1_{nsrt1972}$, OAI $2_{nsrt1972}$,OAI $3_{nsrt1972}$, OAI $1_{nsrt1999}$, OAI $2_{nsrt1999}$, and OAI $3_{nsrt1999}$) and first dominant predictand EOF indices (hereafter $EQI_{nsrt1957}$, $EQI_{nsrt1966}$, $EQI_{nsrt1972}$, and $EQI_{nsrt1999}$) where the subscript "*nsrt*" denotes no square root transformation. The development of models such as parametric and nonparametric stochastic models is also underpinned by model selection and verification which address bias-variance tradeoff (Spall 2003). The rationale in addressing the bias-variance tradeoff is to minimize either implicitly or explicitly, the fitting error and model complexity (Linhart and Zucchini 1986; Shao 1993; McQuarrie and Tsai 1998; Spall 2003). Spall (2003) lists some of the renowned model selection methods. These are: an information criterion (AIC) (Akaike 1974), the principle of minimum description (Rissanen 1978; Wei 1992); bootstrap model selection (Efron and Tibshiranni 1997; Wilks 2006), Bayesian model selection (Akaike 1977; George and McCulloch 1997; Wilks 2006), Vapnik-Chervonenkis (V-C) dimension (Vapnik and Chervonenkis 1971; Cherkassky and Mulier 1998), the Fisher information criterion (Wei 1992), and cross-validation (Allen 1974; Stone 1974; Geisser 1975; Michaelson 1987; Efron and Tibshiranni 1993). Among these methods, the most familiar is cross-validation because of its simplicity, directness and other advantages, even though it has its own limitations (Spall 2003).

Our model selection for the study which is based on split sample cross-validation technique uses a regression equation with multiple predictors against a vector of predictands. Cross-validation is a technique in which the dataset is repeatedly partitioned into training/fitting/developmental and test/verification/prediction subsets (Stone 1974; Efron and Tibshiranni 1993; Spall 2003; Wilks 2006), with the partitions satisfying the general combination formula (Riordan 2002; Wilks 2006; Kumar et al. 2007). For the parametric stochastic model, we used a computer-assisted and intensive K-fold cross-validation technique (also known as rotation estimation) in which the data for each year were partitioned into K roughly equal-sized subsets (Efron and Tibshiranni 1993), wherein K-1 subset data (training) are used to predict the remaining one subset data (test) (Stone 1974; Efron and Tibshiranni 1993). The process is repeated until all the subset data are fully captured in the "fitting-testing recycling exercise". In the model, nine subset data were generated for each of the years, with K-1 developmental and one test subsets for model fitting and testing, respectively. The model building also incorporated ensemble members (Toth and Kalnay 1993, 1997; Molteni et al. 1996; Buizza 1997; Ehrendorfer 1997; Kalnay 2003) to generate a cumulative distribution function (CDF) to integrate them backward in time and to achieve compactness (Wilks 2006). Although there is no general agreement on a specific method for choosing ensemble members, subjective approach is also useful (ibid), and this is one adopted for the study. With the introduction of ensemble members into the model building, the model subsumes in essential respects a "Monte Carlo-like resampling test", which is an approximation to "stochastic dynamic model" with a phase space of high dimensionality (Wang et al. 1995; Wilks 2006). An ensemble of 100,000 members was used for each year for the computer simulation (Gilchrist 1976; Ripley 1987), with their associated prediction (forecast) errors generated (Gilchrist 1976; Wilks 2006).

For the nonparametric model selection, we used the leave-one-out cross-validation (LOOCV) technique (Spall 2003; Wilks 2006). LOOCV is a cross-validation technique which is similar to the K-fold cross validation utilized for our first model construction. But it differs from the first in that, if *n* is the sample size (data points) of the full dataset, it repeatedly splits the data into *n*-1 fitting subset, and the remaining one (a single data point) is the test. At any given time, the subset *n-1* is used to predict the remaining one that has been excluded from the full dataset. Similarly, the process is repeated until all the possible combinations of fitting and test subsets have gone through the cycle. This results in the production of *n* model fits which are also associated with prediction errors (Gilchrist 1976; Wilks 2006). Here too, we used 100,000 ensemble members for the computer simulation. With this approach, the LOOCV being a resampling test subsumes the Monte Carlo approximation to a stochastic dynamic model. Another feature incorporated into the model is K-nearest neighbor (K-NN) based approach with the generalized *kth* nearest neighbor estimate (Silvermann 1986) given as follows:

$$f-\text{hat}(t)=\frac{1}{nd_k(t)}\sum_{i=1}^{n}K\left(\frac{t-X_i}{d_k(t)}\right) \tag{1}$$

where,

f-hat (t) = the kernel estimate evaluated at t having a smoothing parameter(bandwidth or window width) $d_k(t)$

K= kernel function

n= sample size

X_i= real observations

The rationale behind its inclusion was to guide us in the selection of window (bandwith) that best fits the data for subsequent model development and verification. Using this approach a bandwidth which had the minimum root mean square error (RMSE) (Gilchrist 1976; Spall 2003; Wilks 2006) was selected. After training the models, we used 1966, 1972, and 1999 predictor EOF 1, 2, and 3 (ie, the respective OAIs for the parametric and nonparametric) which lag positively 1957, 1966, and 1972 predictand EOF 1 (ie, the EQIs for the parametric and nonparametric), respectively, to theoretically gauge the EQ of 1958, 1967, and 1973, a year lead time from their baselines. Since we did not have data beyond 1999 (especially OAIs) to backcast EQ in 2000, 1957 EOF 1, 2, and 3 (OAI 1, 2, and 3) predictors were used as substitutes. The use of 1957 predictor dataset is underpinned by exhibition of a good potential predictability in the lagged cross correlation (see Table 1), giving us the confidence and credibility to use it as an approximation to 2000 OAIs (predictors). In summary, using computer simulations, below are the various combinations employed for backcasting the EQ:

- 1966 (predictors) vs 1957 (predictand) to backcast 1958
- 1972 (predictors) vs 1966 (predictand) to backcast 1967
- 1999 (predictors) vs 1972 (predictand) to backcast 1973
- 1957(predictors) vs 1999 (predictand) to backcast 2000

Table 1. Cross correlations of the first three EOFs of predictands and predictors (EQIs vs OAIs)

		Predictand 1957			Predictand 1966			Predictand 1972			Predictand 1999		
		EOF 1	EOF2	EOF 3	EOF 1	EOF2	EOF 3	EOF 1	EOF2	EOF 3	EOF 1	EOF2	EOF 3
Predictor 1957	EOF 1	-0.69 (5.9e-09)	-0.02 (0.89)	-0.16 (0.25)	0.10 (0.48)	-0.70 (3.5e-09)	-0.083 (0.55)	-0.70 (3.6e-09)	-0.02 (0.87)	0.07 (0.62)	0.69 (7.8e-09	-0.14 (0.33)	-0.02 (0.90)
	EOF 2	0.58 (4.2e-06)	-0.15 (0.27)	0.27 (0.05)	-0.04 (0.79)	0.55 (1.4e-05)	0.42 (0.002)	0.62 (7.4e-07)	0.22 (0.11)	-0.17 (0.21)	-0.67 (2.6e-08)	-0.11 (0.43)	0.03 (0.84)
	EOF 3	-0.01 (0.94)	-0.03 (0.83)	0.09 (0.51)	-0.05 (0.75)	-0.01 (0.94)	0.08 (0.54)	0.002 (0.99)	-0.04 (0.76)	-0.20 (0.15)	-0.04 (0.76)	-0.17 (0.21)	-0.21 (0.13)
Predictor 1966	EOF 1	0.83 (1.1e-14)	-0.11 (0.45)	0.22 (0.11)	-0.08 (0.58)	0.81 (1.7e-13)	0.32 (0.02)	0.85 (4.e-16)	0.20 (0.16)	-0.08 (0.58)	-0.87 (2.2e-16)	0.06 (0.64)	0.06 (0.66)
	EOF 2	0.32 (0.018)	0.04 (0.79)	0.28 (0.04)	-0.15 (0.28)	0.33 (0.016)	0.20 (0.15)	0.35 (0.009)	0.01 (0.92)	-0.30 (0.02)	-0.41 (0.002)	-0.20 (0.16)	-0.15 (0.26)
	EOF 3	-0.06 (0.69)	-0.09 (0.51)	0.12 (0.38)	0.06 (0.68)	-0.04 (0.79)	0.18 (0.20)	-0.04 (0.79)	0.20 (0.15)	-0.07 (0.60)	-0.01 (0.92)	-0.17 (0.22)	0.11 (0.43)

Predictor 1972	EOF 1	0.88 (2.2e-16)	-0.12 (0.37)	0.30 (0.025)	-0.04 (0.78)	0.86 (2.2e-16)	0.40 (0.003)	0.91 (2.2e-16)	0.19 (0.16)	-0.16 (0.26)	-0.95 (2.2e-16)	-0.03 (0.85)	0.02 (0.91)
	EOF 2	-0.13 (0.33)	-0.04 (0.76)	-0.25 (0.07)	0.08 (0.59)	-0.29 (0.036)	0.27 (0.047)	-0.147 (0.29)	-0.19 (0.16)	0.09 (0.52)	0.19 (0.18)	0.04 (0.78)	-0.18 (0.19)
	EOF 3	-0.003 (0.98)	-0.089 (0.53)	0.05 (0.70)	0.13 (0.326)	-0.013 (0.925)	0.21 (0.126)	0.003 (0.98)	0.24 (0.08)	0.05 (0.70)	-0.04 (0.79)	-0.07 (0.63)	0.10 (0.46)
Predictor 1999	EOF 1	0.16 (0.23)	-0.22 (0.11)	-0.15 (0.29)	0.21 (0.13)	0.11 (0.45)	0.08 (0.55)	0.14 (0.40)	0.09 (0.52)	0.24 (0.09)	-0.13 (0.36)	0.15 (0.28)	-0.11 (0.44)
	EOF 2	-0.86 (2.2e-16)	0.05 (0.73)	-0.31 (0.02)	0.09 (0.50)	-0.85 (8.7e-16)	-0.35 (0.009)	-0.89 (2.2e-16)	-0.21 (0.12)	0.14 (0.31)	0.92 (2.2e-16)	-0.01 (0.94)	-0.06 (0.67)
	EOF 3	0.26 (0.06)	0.13 (0.34)	0.06 (0.64)	-0.25 (0.07)	0.19 (0.16)	0.10 (0.47)	0.27 (0.05)	-0.16 (0.24)	-0.18 (0.18)	-0.26 (0.06)	-0.05 (0.72)	-0.26 (0.06)

Figures in parenthesis are p-values; bold-faced figures are correlation coefficients significant at least at 95 % confidence level (all others not significant).

For both models, the relationships between the environmental states (1957 vs 1958; 1966 vs 1967; 1972 vs 1973; and 1999 vs 2000) with a positive year lag were established using model AR (1) (Gilchrist 1976; Wilks 2006; Metcalf and Koons 2007). Under these stochastic environments, we modeled the AR (1) which is essentially a linear simple regression model given by the following equation:

$$X_{t+1} = \Phi X_t + \varepsilon_{t+1}, \tag{2}$$

where ε is a random variable (quantity) which is independent and identically distributed (iid) with zeo mean and variance σ^2, and Φ, the autocorrelation (autoregressive) parameter (Gilchrist 1976; Schwager et al. 2006; Wilks 2006; Johnson and Wichern 2007; Metcalf and Koons 2007).

6. Results and Discussion

6.1. Retrospective Environmental Quality Analysis of the Communities

The first three dominant modes of the PCA- eigenvalues, eigenvectors, and scores were retained for discussion since they carried the most important information in the original datasets. The scree plots and the scores (indices) showing the percent variances (eigenvalues) and variability in EQIs and OAIs, respectively, for 1957, 1966, 1972, and 1999 are presented in Figures 5-14. The EQIs and OAIs are analogous to the typical time series of EOF analysis that exhibit amplitude fluctuations (variabilities). The eigenvector analysis results (not shown) indicate the relative contributions (weights) of the variables in explaining the observed variances. Particular attention has also been given to the interpretation of sign of the weights (coefficients) of the eigenvectors (loadings). This was based essentially on literature review and expert judgment. For instance, a positive (negative) weight of a WQ parameter means high (low) level. Also, a positive (negative) weight associated with a disease means an increase/enhancement (a decrease/suppression/decline) in disease prevalence. A positive (negative) weight of fisheries means high (low) fish biodiversity. A positive weight for an OA, for example, embankment is adverse when the weight of a disease is also positive, suggesting the impoundment effects promote disease proliferation. However, a negative weight of embankment is adverse when the weight for downstream fisheries is also negative, suggesting that impoundment effects suppress downstream fisheries development. In the nutshell, the sign of the weight of each variable is interpreted based on its own merit, and the discussion will be based on these criteria. Interpretation of the EQIs and OAIs was based on the nature and direction of the phase in the amplitude fluctuations of the OAIs since cause-effect relation established in our previous study (Tetteh et al. 2006) attributed the communities` EQ to presence of the dam and its related operations. Thus, a synchronization in the phase between an OAI and an EQI above the mean zero level, is an indication of a poor EQ since the environmental state is driven by the OAs. Conversely, a synchronization in the phase between an OAI and an EQI below the mean zero level is an indication of a better EQ since the thresholds of the OAs have subsided. An inverse relationship between any of the phases (OAI vs EQI), depicts antagonism. Therefore, a positive (negative) phase of a point in the OAI means adverse (beneficial) impacts. A positive (negative) phase of a point in the EQI means poor (good) EQ.

A comparative analysis of the cross correlations using the first three EOF analysis of the OAIs and EQIs shows that Pearson product-moment correlation best describes predictor-predictand relationships among the three dependency measures. The results shown only for Pearson product-moment correlation (Table 1) further suggest that the relationships are more of a linear function than nonlinear. The percent total variation in EQ of the communities expressed by the first three dominant modes of 1957 predictands was about 87% (Figure 5a).The corresponding eigenvector analysis showed that the EOF 1 loadings were essentially dominated by WQ. Specifically, the EQ was due to the weighted sum of nine WQ parameters (all weighted positively) namely, color, turbidity, pH, total hardness, total alkalinity, TDS, BOD_5, conductivity, and total aluminum (Al) concentration, an indication of poor environmental state based on the criteria we established earlier on. Water quality parameters are known to exhibit some relationships. For instance, the relationship between conductivity, pH, and total alkalinity of natural water has been reported (Peavy et al. 1985). Also, pH is partly influenced by the levels of TDS. Waters of high TDS have displeasing taste and are potentially unhealthy (Berger 1982). The detection of color and turbidity in EOF loadings 1, corroborated the indigenous visual pollution criteria used by the communities in assessing WQ (Tetteh et al. 2006). Color, turbidity, and TDS are indicators of the presence of microbes, and on the whole, consumption of water with these characteristics is certainly injurious to human health (Berger 1982). Aluminum (Al) is naturally present in water (Srinivasan et al. 1999). It has been found that the solubility of Al and residual Al are affected by temperature, pH and turbidity of the water (Jekel 1991; Van Benschoten 1994; Srinivasan et al. 1999). The communities, especially, Hiawo Besease ran a risk of exposure to high levels of ingested Al because of operations of the OTP. Aluminum is a powerful neurotoxicant, which is associated with neurological disorders, Alzheimer's disease and prehensile dementia, and other pathological diseases (Crapper and Boni 1980; Davidson et al. 1982; Schecher and Driscoll 1988; Martyn et al.1989; Gardner and Gunn 1991; Jekel 1991; Flaten 2001). Unacceptable limits of pH affect mucous membrane, imparts bitter taste, affect aquatic life, and can be corrosive. BOD_5 is an indicator of organic pollution. Conductivity, a measure of the ability of water to conduct electric current is directly related to TDS. Total hardness may reduce the toxicity of other compounds such as zinc, lead, and copper which are more toxic in soft water. It is also known to reduce the risk of heart conditions. The EOF 2 loadings were attributed to the weighted sum of infectious hepatitis and skin diseases. Infectious hepatitis may be fatal, and skin diseases are debilitating and lead to secondary infections. The EOF 3 loadings were governed by weighted sum of COD (positive) and fisheries (positive). The relationship between COD and biomass production has been reported by Kawabe and Kawabe (1997). They report that anomalously high COD has been attributed to phytoplankton in Tokyo Bay. Biomass production mainly through photosynthesis of phytoplankton has been observed in highly eutrophic Tokyo Bay (Hattori et al. 1983; Ogawa and Ogura 1990a). Kawabe and Kawabe (1997) have found that in Tokyo Bay, COD correlates highly with the level of chlorophyll *a,* which is a measure of phytoplankton biomass production. In our situation, we guess a similar event but perhaps, on a smaller scale occurred where the phytoplankton provided enough food to support fisheries development in their natural environment with a concomitant increase in COD, especially when R. Offin had not been impounded (Tetteh et al. 2006). The 1957 EQIs (Figure 7) show interesting observations. Whereas EQIs 2 and 3 show typical amplitude fluctuations about the mean environment equal to zero, EQI 1 is described as a "staircase index" with three distinct levels of variabilities- a constant state below the environmental mean (zero), at zero, and above zero. Thus, this observation shows stability of the EQ (Hollings 1978) in three different phases characterized by enhancement (a good EQ), non-responsiveness (environmental neutrality), and hostility (a poor EQ), respectively.

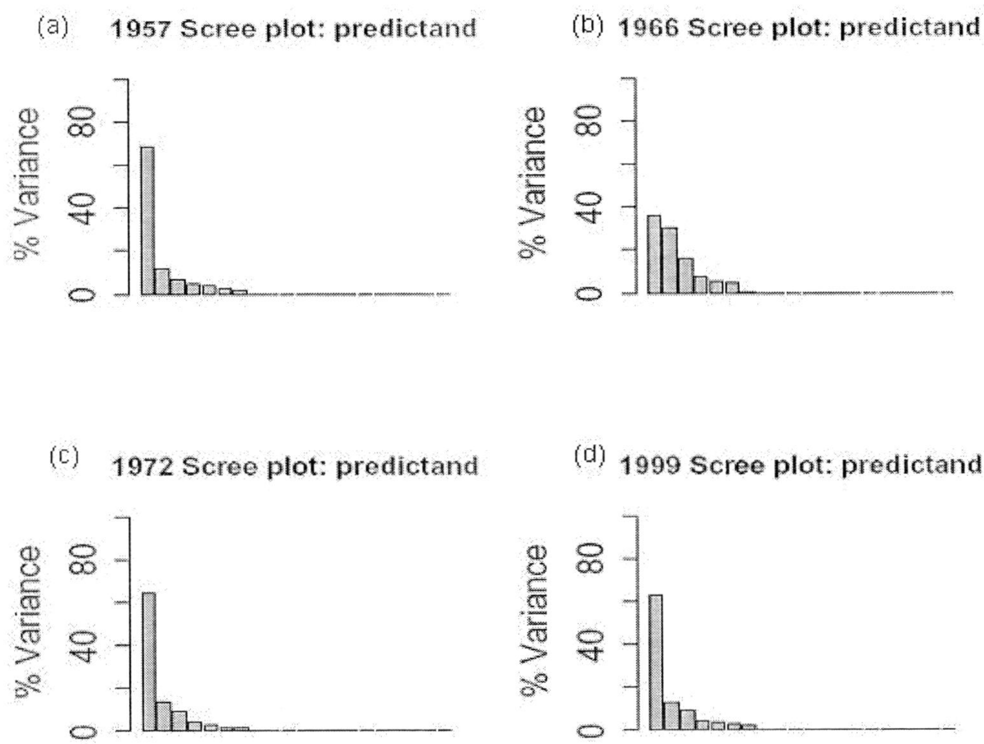

Figure 5. Percent variance of predicatand EOF modes: (a) 1957, (b) 1966, (c) 1972, (d) 1999.

In 1957, the percent variance explained by the first three dominant EOF modes for the OAs shown in the scree plot (Figure 6a) was 45.8 %. The EOF 1 loadings due to weighted sum of pH (positive), TDS (positive), and total Al (positive) negatively influenced embankment (negative), and backwashing of clogged filters (negative). The positives (pH, TDS, and total Al) were apparently, the result of conventional water treatment (from the OTP) that introduced effluents into the downstream river water (ie R. Owabi), which impacted the WQ at Hiawo Besease. Thus, they could be linked to the adverse impacts of pH, TDS, and Al on human welfare summarized already. The EOF 2 loadings were attributed to the weighted sum all of which were positives for the following variables of the OAs: pH, total Al, embankment, and backwashing of clogged filters. By matching these with their corresponding environmental states, the embankment (impoundment) effects synchronized the weighted sum infectious hepatitis (positive) and skin diseases (positive), suggesting that embankment positively influenced the two diseases which is consistent with Petts(1984). The EOF 3 loadings were due to weighted sum of TDS (positive) and COD (positive) that negatively influenced pH (negative) and BOD_5 (negative). The COD was connected to the EQ COD that insured the promotion of fisheries via phytoplankton biomass production, the chief energy base of aquatic ecosystems (Moss 1988). The first three OAIs (ie, EOF 1, 2, and 3 scores) of 1957 (Figure 11) generally, show some level of resemblance and whose amplitudes are dominated by fluctuations centered around the environmental mean (zero), suggestive no significant changes in the thresholds of OAs. Cross correlations between the EQIs and OAIs (Table 1) show that EQI 1 correlated significantly with OAI 1 and OAI 2 ($r = -0.69$; $p=5.9e\text{-}09$, and $r= 0.58$; $p= 4.2e\text{-}06$, respectively), a signal of the influence of the OAs on the EQ of the communities.

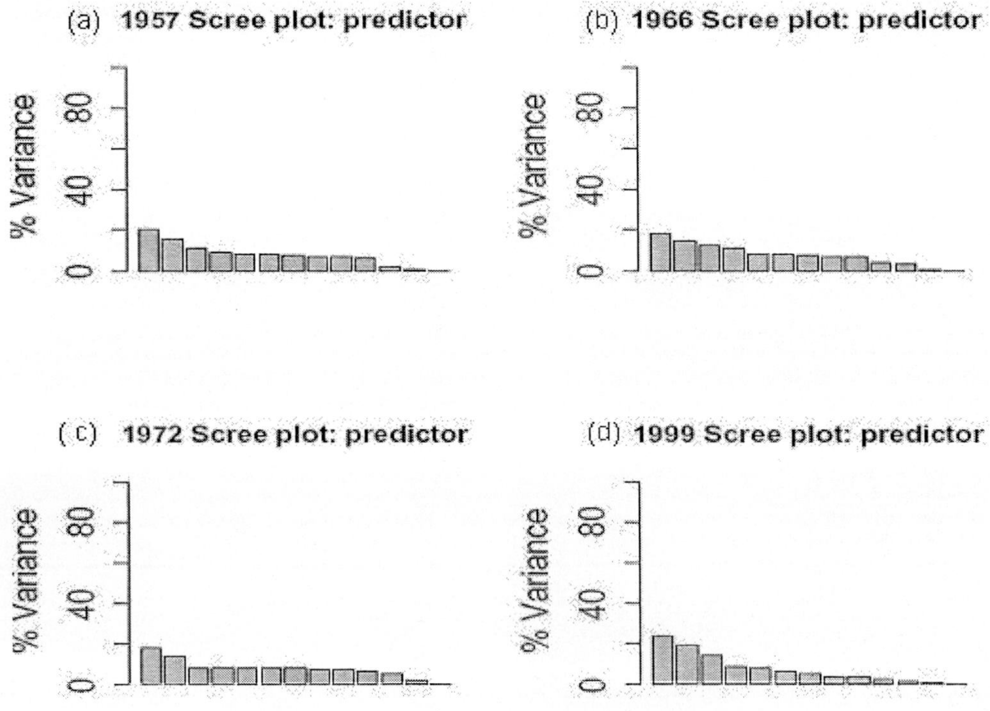

Figure 6. As in Figure 5 but for the predictors.

The eigenvalue analysis shows that in 1966, the percent variation explained by EOF 1, 2, and 3 predictands was 89.0 % (Figure 5b). The corresponding EOF 1 loadings were determined and controlled by the weighted sum of turbidity (positive), total hardness (positive), total alkalinity (positive), BOD_5 (positive), and total aluminum concentration (positive). Just as in the EOF 2 loadings of 1957, the EOF 2 loadings of 1966 were also controlled by weighted sum of infectious hepatitis and skin diseases, an indication of the maintenance of the same level of herd immunity by the communities over the two years (Lilienfeld and Lilienfeld 1980; Tetteh et al. 2004a). Also, the relative contributions of the dominant factors in the EOF 3 loadings of 1966 were the same as that of 1957, which was tied to COD and fisheries. Figure 8 shows amplitude fluctuations of the EQI 1, 2, and 3 which demonstrate variability and instability. However, the fluctuations in EQI 1 were more pronounced than in EQI 2 and 3.

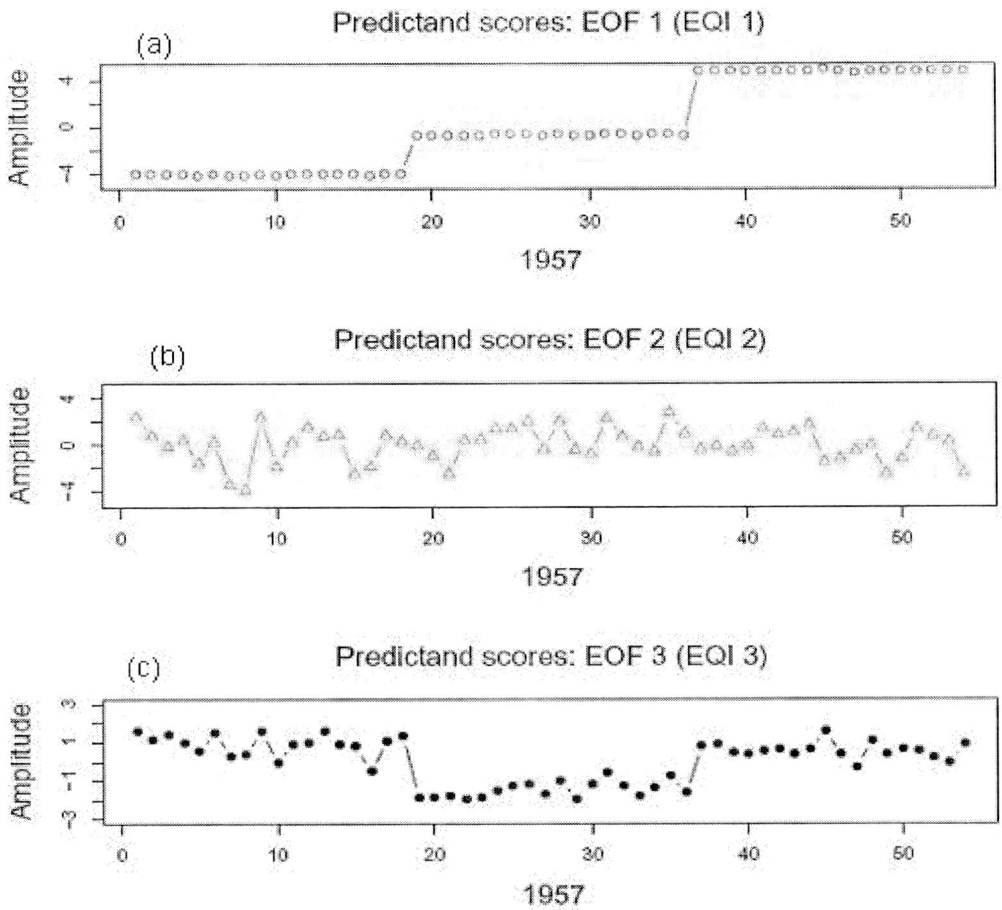

Figure 7. EQIs of 1957 based on the first three EOF scores.

The scree plot presented in Figure 6b shows the percent variance explained by the OAs in 1966. The EOF 1, 2, and 3 accounted for 45.3%. The corresponding EOF 1 loadings were governed by the weighted sum of embankment effects (positive) and backwashing of clogged filters (positive). These could be linked to the environmental WQ dynamics observed previously. The EOF 2 loadings were due to the weighted sum of TDS (positive) and BOD_5 (positive). High TDS and BOD_5 positively influenced infectious hepatitis and skin diseases which is concordance with Berger (1982). Also, the EOF 3 loadings were due to weighted sum of color (positive) and turbidity (positive). Colored and turbid waters rich in phytoplankton biomass (Beadle 1981; Peavy et al. 1985) could positively be related to COD and fish biodiversity observed in the environmental state. The three OAIs generally show fairly constant fluctuations centered around the mean (zero) except for some few localized spikes as well as constant positive phase (above zero) for OAI 1 (Figure 12). The cross correlations among three EOF scores (EQIs and OAIs) showed that OAI 1 correlated very well with the EQI 1 ($r = 0.81$; $p = 1.7e\text{-}13$), whereas a lower but significant correlation was found between OAI 2 and EQI 2 ($r = 0.33$; $p = 0.016$) (Table 1), this suggests that the influence of the Owabi Dam and the construction phase impacts of the Barekese Dam are consistent with literature (Baxter 1977; Rau and Wooten 1980; Beadle 1981; Petts 1984; Moss 1988).

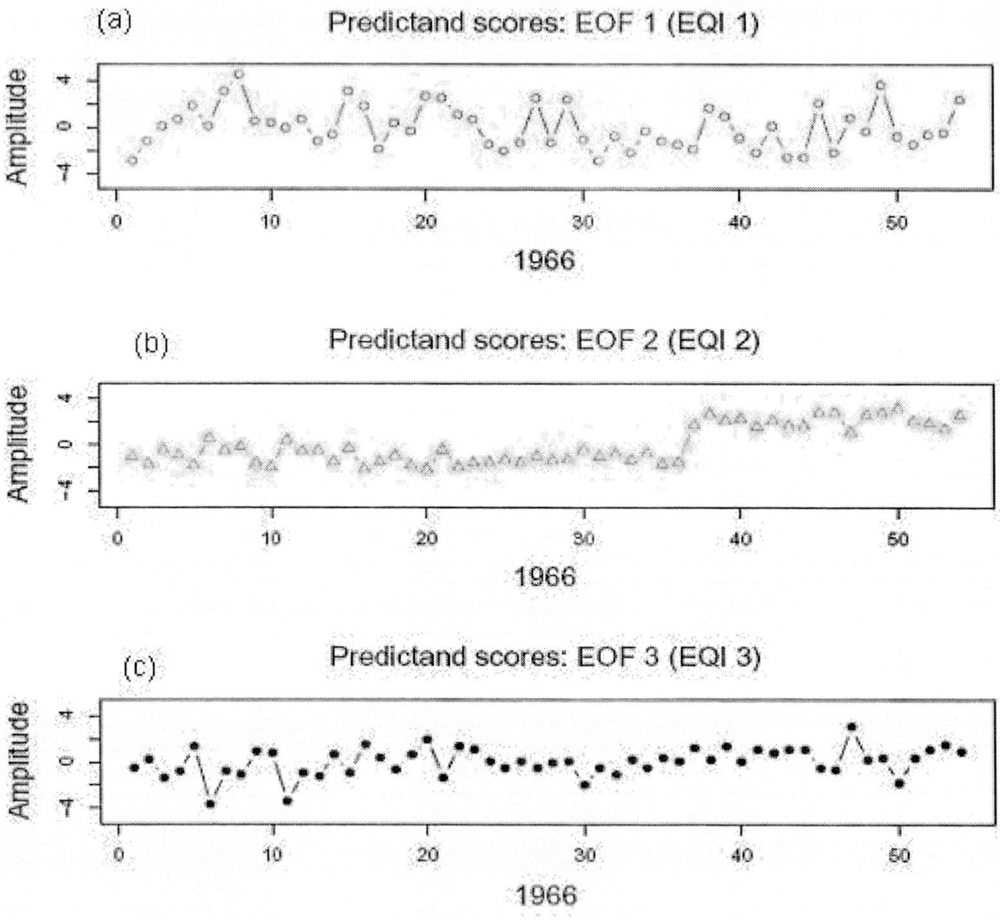

Figure 8. As in Figure 7 but for 1966.

With the commissioning of the Barekese Dam in 1972, the two dams started operating in tandem. It was observed that the percent variance explained by the baseline environmental state in 1972 EOF 1, 2, and 3 was 88.0% (Figure 5c). Their corresponding EOF loadings are as follows. The EOF 1 loadings were essentially governed by the weighted sum of six WQ parameters, namely, turbidity, pH, total hardness, total alkalinity, BOD_5, and conductivity, an increase over the 1966 environmental state. The EOF 2 loadings were attributed largely to weighted sum of vesicle schistosomiasis, infectious hepatitis, and skin diseases. And EOF 3 loadings were due to weighted difference between diarrheal diseases (positive) and COD (negative). We notice that aside from the WQ issues, vesicle schistosomiasis and diarrheal diseases have for the first time emerged as part of the dominant diseases contributing to the disease dynamics. These two diseases are acknowledged as well known water-related diseases associated with non-potable water sources and dams in the tropics (Jordan and Webb 1982; Ukoli 1984). Vesicle schistosomiasis is well known for its reduced scholastic ability in children, impairment of growth in children, reproductive health problems, etc (Jordan and Web 1982). Diarrheal diseases are known for their malady and fatality, and claim over 2 million people annually and world-wide (Parashar et al. 2003). Most of the victims are children in the developing countries who are under 5 years old. The EOF 1, 2, and 3 scores (EQIs) are shown in Figure 9. We observe that the amplitude of fluctuations in EOF 1 scores (EQI 1) here too, mimics the EOF 1 (EQI 1) of 1957, with its characteristic "staircase index" having the three distinct phases. The EOF 2 and 3 scores (EQI 2 and 3) also displayed typical amplitude fluctuations (variabilities) about the mean environment (zero).

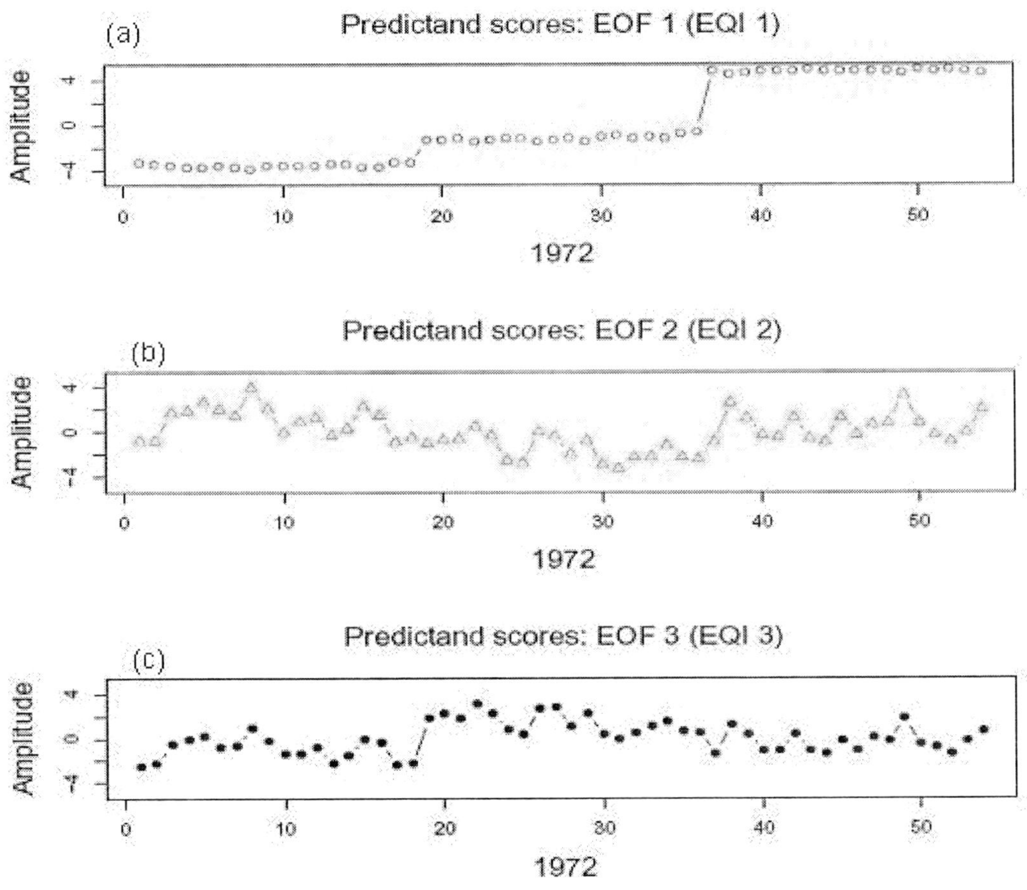

Figure 9. As in Figure 7 but for 1972.

The percent variance explained by 1972 EOF 1, 2, and 3 predictors (OAs) as shown in Figure 6c was 40.2%. The eigenvector analysis revealed that the EOF 1 loadings were contributed by the weighted sum of the embankment (positive) and backwashing of clogged filters (positive) which were associated with the downstream WQ issues in the environmental state. The impacts of these post-operational activities on the downstream EQ are therefore, in concordance with Rau and Wooten (1980), Jain et al.(1993) and Canter (1996). The EOF 2 loadings were determined by the weighted sum of pH (negative) and total hardness (negative). The most significant environmental impact of pH involves synergism. And it appears that pH and total hardness operated in concert to create favorable conditions for the perpetuation of vesicle schistosomiasis, infectious hepatitis, and skin diseases observed in the environmental state. On the other hand, the EOF 3 loadings were governed chiefly by the weighted difference between TDS (positive) and total alkalinity (negative). TDS is an indicator of microbial pollution by providing sites for harboring highly infective gastro-intestinal diseases (Berger 1982; Gerba et al. 1996), and this parameter could be connected to diarrheal diseases observed in the corresponding eigenvector analysis of the environmental state. The fluctuations in 1972 EOF 1, 2, and 3 scores (OAIs) were heavily concentrated on the environmental mean (zero) (Figure 13), an indication of stability of OAs (Hollings 1978). Correlation analysis shows that the most significant correlation exited between EOF 1 predictand (EQI 1) and EOF 1 predictor (OAI 1)($r=0.91$; $p=2.2e-16$)(Table 1). Here too, this observation provides a signal of profound influence exerted by the dams.

In 1999, the percent variance for the EOF 1, 2, and 3 of the predictands (EQI 1, 2, and 3) was 85.5% (Figure 5d). The corresponding EOF 1 loadings were ascribed to the weighted difference between water stress (scarcity) during the dry season (positive) that negatively influenced the composite effects of pH (negative), total hardness (negative), total alkalinity (negative), attitudinal behavior towards the dam (negative). Water stress during the dry season was observed in our field study at Hiawo Besease where the site the villagers abstract the water from the river had dried up completely. The negative attitudinal behavior towards the dams, a social psychological factor, was a commonality of the riparian communities because of the environmental impacts associated with the dams. Awakul and Ogunlana (2002) recognize two categories of conflicts- internal conflicts and interface conflicts characteristic of large construction projects. Internal conflicts occur among project participants, whereas interface conflicts are experienced between construction project and people outside the project. The negative attitudinal behavior of communities is a clear manifestation of an interface but "nonviolent conflict" between the dams (in other words, the GWC Ltd.) and the communities. Our field study revealed that invariably, any ill omen that befell the communities was blamed on the dams, a perception that would take time to correct because they have been peeved for a long time. The EOF 2 loadings were ascribed to weighted sum of COD (negative) and fisheries (negative) which contrast the findings of EOF 3 loadings of 1957 and 1966, where the latter two years were all weighted positively. This explains why there has been significant loss in fish biodiversity (Tetteh et al. 2006; Tetteh et al. unpublished). The EOF 3 loadings were attributed to the weighted sum of vesicle schistosomiasis and diarrheal diseases, and these two have maintained their importance since the two dams began operated jointly, a warning signal of a poor environmental health state of the communities. The corresponding 1999 predictand EOF 1, 2, and 3 scores (EQI 1, 2, and 3) are presented in Figure 10. It is seen that EQI 1 (Figure 10a) was characterized by two distinct phases: i. two-thirds of the phase, constant and above the environmental mean (zero) which indicate hostility, and ii. one-third of phase below the environmental mean (zero), an indication of a good EQ. EQI 2 (Figure 10b) is characterized by three has distinct phases from the left to right of the year axis as enhancement, hostility, and neutrality, respectively, whereas the EQI 3 displays irregular fluctuations, characteristic of instabilities (Figure 10c).

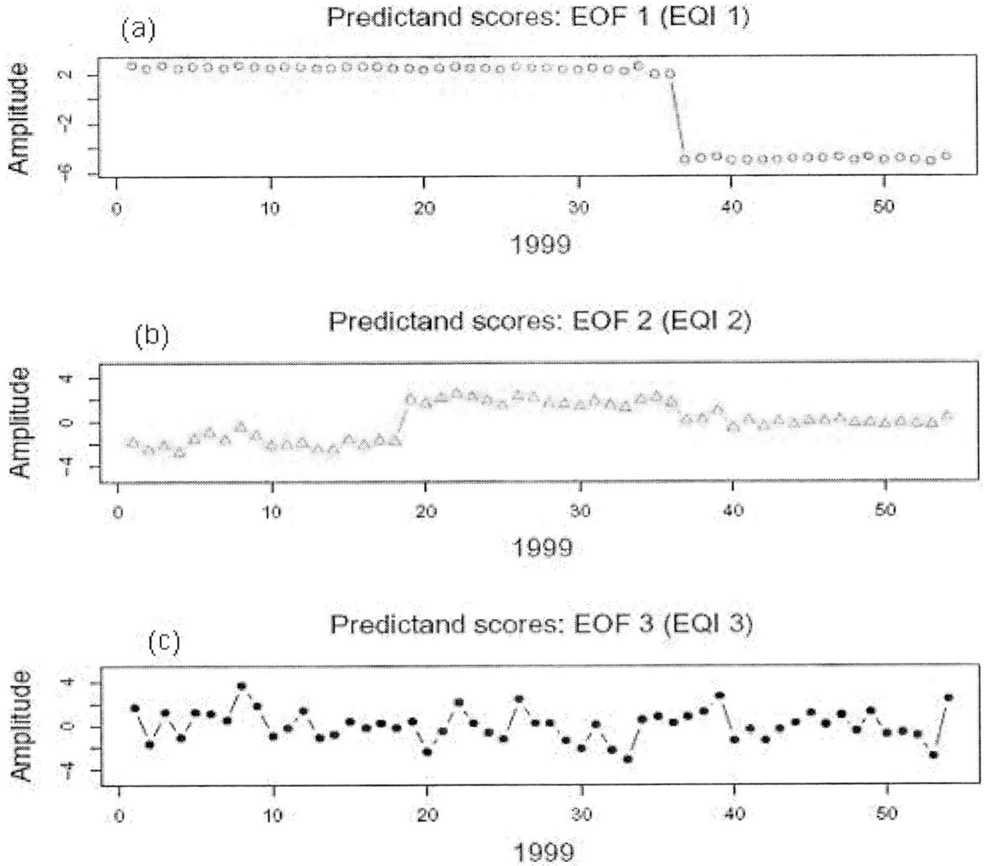

Figure 10. As in Figure 7 but for 1999.

The scree plot (Figure 6d) that shows the percent variance for the three most important PCs (EOF 1, 2, and 3) of predictors (OAs) in 1999 was 57.6%, the highest so far. The EOF 1 loadings were attributed to weighted sum of pH (negative) and total alkalinity (negative), which synchronized pH and alkalinity observed in the 1999 EQ. The EOF 2 loadings which were governed by the weighted sum of embankment (negative) and occasional backwashing of clogged filters (negative) which were in contrast to the 1972 EOF 1 loadings of the OAs. The embankment effect (negative) was responsible for the decline in fish biodiversity and abundance, which is consistent with our previous findings (Tetteh et al. 2006). In one of our previous studies (Tetteh et al. unpublished), analysis of the 1999 fisheries resources of the rivers based on Frempong and Nijjhar`s(1973) baseline study (19 species reported) showed that 42.1% of the species appeared to have become endangered or threatened following the construction of the Barekese Dam. The effects of the pre-existing Owabi Dam magnified by the new dam (Barekese Dam) are an evidence of cumulative impacts (CEQ 1987). The 57.9% surviving species, which are relatively fewer in number as compared to at least the past 30 years (Tetteh et al, unpublished), are represented only by three dominant genera (Tetteh et al. 2006). We argue that these fundamental changes could be attributed to the dams whose presence has interfered with the migration and biology of potamodromous fishes. A similar observation was made where there was a decline in fish and fisheries in the estuaries, following the constructions of the Akosombo (Volta) Dam (Dankwa and Gordon 2002) and the Kpong Dam (Moss 1988; Zakhary 1997) in 1964 and 1981, respectively. Among the fishing industries that was heavily affected was the *Eigeria radiata* (a bivalve) industry, which was due to salinity changes that hindered the survival, development and growth of the veliger larvae (Moss 1988). The collapse in the sardine industry in the Nile Delta following the construction of the Aswan High Dam in 1964 (Nixon 2004), is also worth mentioning. Petry (1995) has also reported that the expected losses in pre-project (existing) riverine species could be caused by: (i) impoundment of the river, (ii) reduction in the downstream inundation because of reservoir storage, (iii) reduction in downstream fisheries in the river zone below the dam where changes in the hydrologic regime is expected to be large, and (iv) reduction in downstream fisheries due to trapping of nutrients in the reservoir. These factors or a combination of them had probably resulted in the decline of fish and fisheries resources in the communities. The impoundment effects of the Owabi Dam appeared to have been responsible for interrupting the hydrological flow downstream, creating water shortage at the abstraction sites at Hiawo Besease in the dry season, which was enhanced by high evapotranspiration. Backwashing of clogged filters could be connected to low COD and fisheries observed in the downstream. The EOF 3 loadings indicated a weighted difference between temperature (positive) and Al (negative), an inverse relation which is consistent with the findings of Vesely´ et al.(2003). In their study, they noted that increasing temperature in strongly acidified Central Europe Lakes declined the total Al concentration, and this was found to be driven by climate warming. In RBDPs, it has been found that impoundment effects, apart from creating thermally stratified lake in the reservoir area, also reduce the discharge downstream (Rau and Wooten 1980; Canter 1996), making it readily warmed up by radiative heating. In the present study, we therefore guess that the impoundment effects similarly and indirectly affected the downstream water temperature, and this possibly played an important role in epidemiology of vesicle schistosomiasis (Mousa and el-Hassan 1972) and diarrheal diseases captured in the EOF 3 loadings. For instance, the intermediate host, *Bulinus sp* (a snail) of *Schistosoma haematobium*, requires ecologically warm water to thrive and proliferate, where the mean water temperature is about 30°C, but crashes at 35 °C because it cannot survive (Jordan and Web 1982; Mangal et al. 2008). The mean surface temperature of the river water in the communities was approximately 30° C, which possibly supported the snail vector. Cross correlations show that the most significant correlation existed between predictor EOF 2 (OAI 2) and predictand EOF 1 (EQI 1) ($r=0.92$; $p= 2.2e-16$) (Table 1), also a prominent signal of the dams` impacts on the communities` environmental states.

A monitoring of the OAIs and EQIs involving test of hypothesis (Johnson and Wichern 2007) based only on their respective EOF 1 scores over the four years, is shown in the correlation matrices (Tables 2, 3). No significance difference was observed between EQI of 1957 and 1966, an indication of similar EQ. On the other hand, a statistically significant difference occurred between 1957 and 1972 ($r=0.99$; $p=2.2e-16$) which gives a signal of a significant change in EQ after completion the second dam (Barekese Dam), and the two dams started operating jointly. The EQ of these two years graphically displayed similar characteristic "staircase index" (Figure 7a vs Figure 9a). A comparison between the EQ of 1957 and 1999 showed very significant inverse relationship ($r= -0.93$; $p=2.2e-16$). The correlation matrix shows that the EQ of 1966 correlated poorly with that of 1972, and 1999.

The correlation between the EQ of 1972 and 1999 also showed a significantly high inverse relation ship (r=-0.96; p=2.2e-16). We conclude that the EQs of 1957, 1972, and 1999 were distinct. In Table 3, it is observed that the 1957 OAI 1 anti-correlated significantly with that of 1966 (r=-0.67; p= 2.3e-08), and similarly with 1972 (r=-0.68; p=2.1e-08). We therefore reject the null hypothesis, and conclude that the thresholds (magnitudes) of the OAs for these years were different. However, non-significance was found for 1957 vs 1999, 1966 vs 1999, and 1972 vs 1999, an indication of similar thresholds. On the other hand the 1966 OAs correlated positively with that of 1972 (r= 0.82; p=5.4e-14).

Table 2. Correlation matrix of the first EOFs of the predictands (EQIs)

Year	1957	1966	1972	1999
1957	1.00	0.16	0.99	-0.93
1966		1.00	-0.17	0.11
1972			1.00	-0.96
1999				1.00

Table 3. Correlation matrix of the first EOFs of the predictors (OAIs)

Year	1957	1966	1972	1999
1957	1.00	0.16	0.99	-0.93
1966		1.00	-0.17	0.11
1972			1.00	-0.96
1999				1.00

6.2. Parametric, Nonparametric and AR (1) Model Analyses

The goal of this section is to evaluate the performance of the parametric and nonparametric models, and use the results of the superior model as inputs for AR (1) model construction under stochastic environments for backcasting the EQ of the communities. Table 4 shows the results from the parametric model construction, the serial correlation coefficients, and other summary statistics. The observed and cross-validation correlation coefficients for 1957, 1966, 1972 and 1999 were 0.84, -0.15, 0.86, 0.95, respectively. Their respective root mean square errors (RMSEs) and model biases were 2.0 (-0.0006), 2.74 (0.24), 1.83 (-0.03), and 1.11(0.033), respectively, where the figures without parentheses are the RMSEs, and those enclosed in parentheses are the model biases. The observed and cross-validation negative correlation coefficients obtained for 1966 (Table 4) depicts an error, which inherently denies the parametric model as a better fit for the data, although those for the other years yielded desirable results of varying degrees of fitness.

Table 4. Parametric and AR (1) model analyses showing the correlation coefficients, RMSEs, and the biases between the observed and cross-validated hindcasts

Model	Correlation Coef.	P-Value	RMSE	Bias
1957: observed & cross-valid	0.84	2.00E-15	2.00	-0.0006
1- year lead of 1957(1958): (AR 1)	0.64	1.40E-07	3.70	-0.32
1966: observed & cross-valid	-0.15	0.29	2.74	0.24
1-year lead of 1966 (1967): (AR1)	-0.06	0.67	2.03	-0.05
1972: observed & cross-valid	0.86	0.00	1.83	-0.03
1-year lead of 1972 (1973): (AR 1)	0.89	0.00	1.60	-0.02
1999: observed & cross-valid	0.95	0.00	1.11	0.033
1-year lead of 1999 (2000): (AR 1)	0.95	0.00	1.11	-0.01

The nonparametric results, together with the serial correlation coefficients, and other summary statistics are presented in Table 5. Of paramount importance are K-NNs which decided the bandwidth that would optimize the potential predictability of the model. Among the 54 predictor data points for each year (see Figures 11-14), the K-NNs with the minimum RMSE, an indication of having a good forecasting ability (Gilchrist 1976; Wilks 2006) was chosen. The most desirable K-NNs for 1957, 1966, 1972, and 1999 were K-7, K-4, K-11, and K-9, respectively. Their corresponding observed and cross-validation correlation coefficients were 0.91, 0.90, 0.95, and 0.99. These results demonstrate the superiority of the nonparametric model over the parametric one.

Table 5.

Model	Correlation Coef.	P-Value	RMSE	Bias
1957: observed & cross-valid using K-NN =7	0.91	0.00	1.58	-0.36
1- year lead of 1957(1958): (AR 1)	0.89	0.00	1.67	-0.06

1966: observed & cross-valid using K-NN = 4	0.90	0.00	1.67	-0.38
1-year lead of 1966 (1967): (AR1)	0.013	0.92	1.92	-0.03
1972: observed & cross-valid using K-NN = 11	0.95	0.00	1.07	0.015
1-year lead of 1972 (1973): (AR 1)	0.94	0.00	1.26	-0.40
1999: observed & cross-valid using K-NN = 9	0.99	0.00	0.60	0.05
1-year lead of 1999 (2000): (AR 1)	0.997	0.00	0.51	0.28

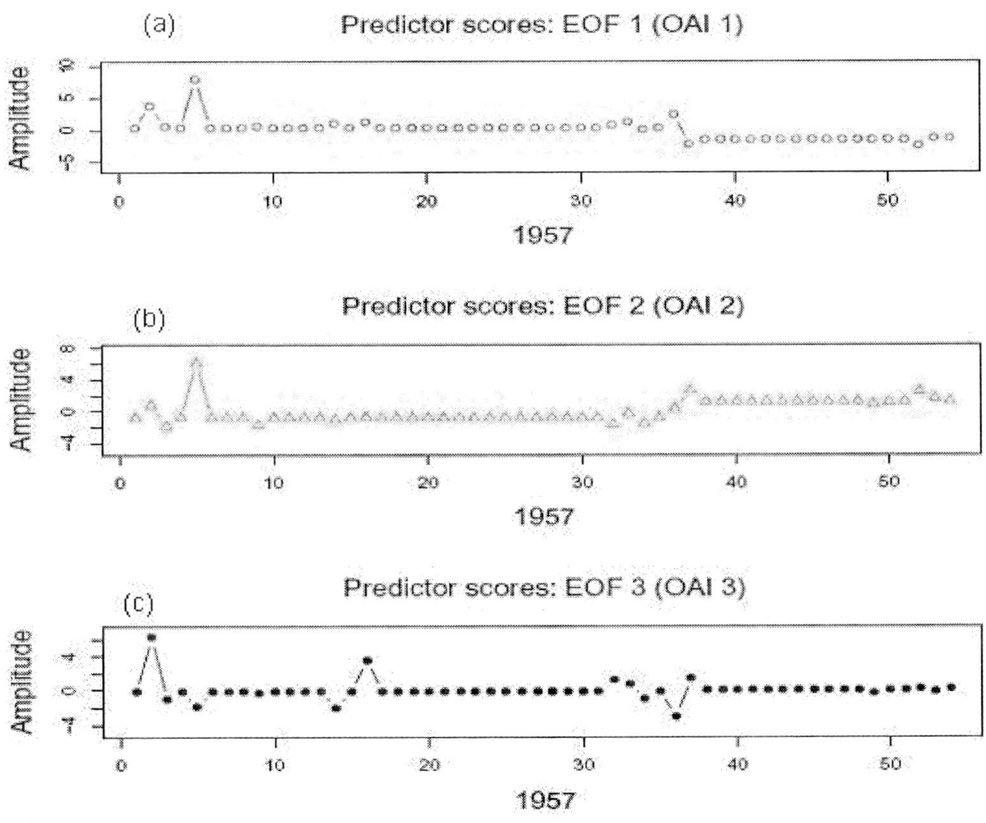

Figure 11. OAIs of 1957 based on the first three EOF scores.

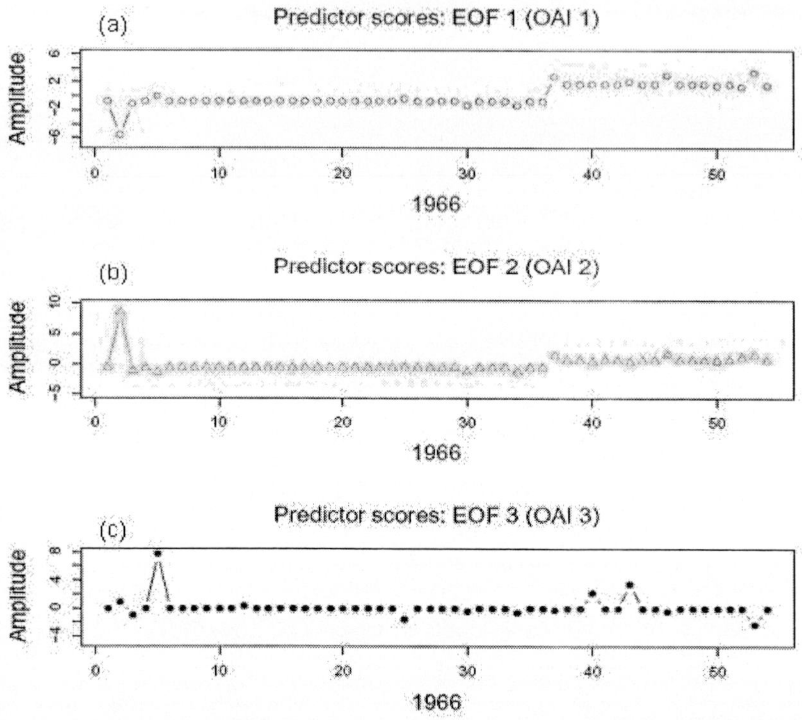

Figure 12. As in Figure 11 but for 1966.

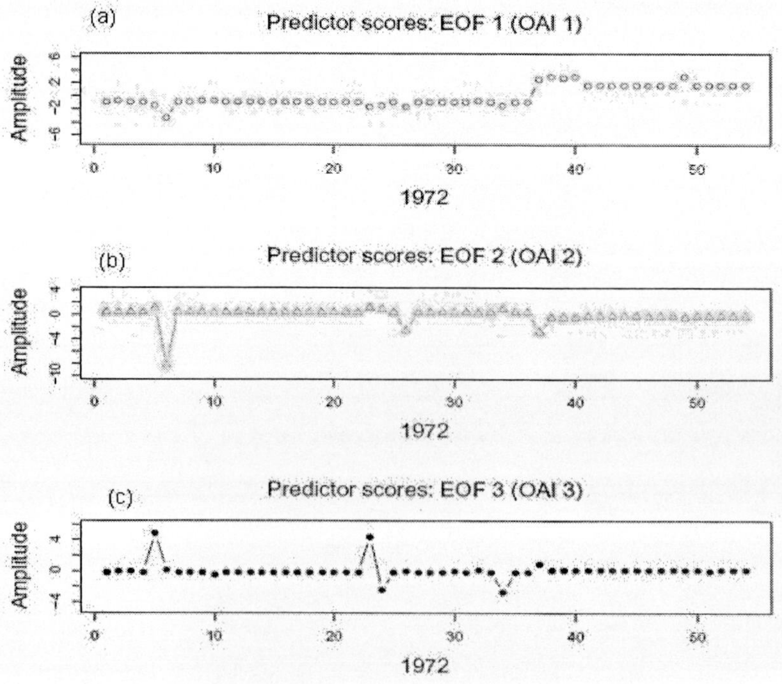

Figure 13. As in Figure 11 but for 1972.

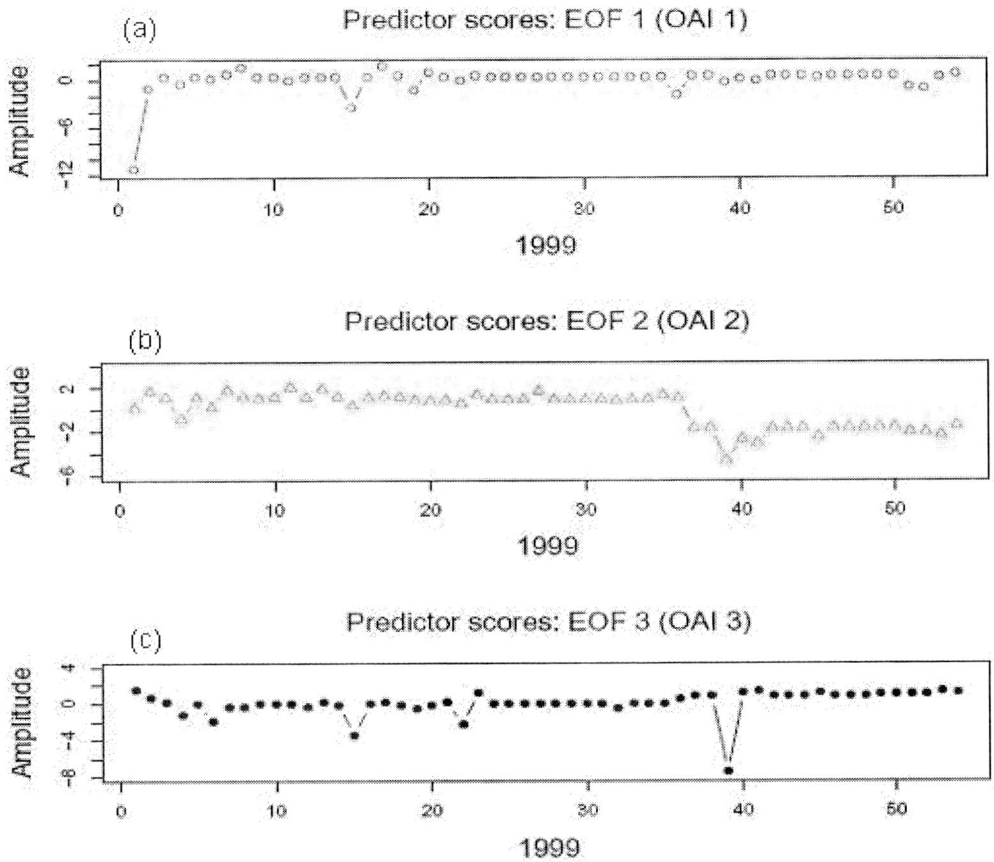

Figure 14. As in Figure 11 but for 1999.

In the final part of this section, we discuss the AR (1) results based on the inputs from the nonparametric stochastic model. The temporal environmental autocorrelation coefficients for 1957 vs 1958; 1966 vs 1967; 1972 vs 1973; and 1999 vs 2000 were 0.89, 0.013, 0.94, and 0.997, respectively (Table 5). The limitation of the discussion centers on the lack of 1958, 1967, 1973, and 2000 EQ observational data for comparison with the predicted. However, the values of the model biases coupled with low RMSEs demonstrate acceptable model performance. The importance of environmental autocorrelation, both spatial and temporal in helping to understand the dynamics of EQ has been addressed abundantly in population ecology (Goodman 1987; Mode and Jacobson 1987; Rotenberg 1987; Caswell and Cohen 1995; Halley 1996; Johst and Wissel 1997; Petchey et al. 1997; Halley and Kunin 1999; Morales 1999; Petchey 2000; Gonzalez and Holt 2002; Holt et al. 2003; Schiegg 2003; Pike et al. 2004). Generally, it has been proven that positive environmental autocorrelation is associated with decline in population species (promotes extinction risk) in contrast to negative autocorrelations that are linked to increase in population species (reduce extinction risk) (Mode and Jacobson 1987; Rotenberg 1987; Haccou and Vatutin 2003; Wichmann et al. 2003; Pike et al. 2004). The theoretical grounds of this assertion reported by Schiegg (2003) is that "correlated environments can cause populations to fluctuate synchronously", 'the

Moran effect' (Royama 1992) which promotes extinction risk (Petchey et al. 1997; Halley and Kunin 1999; Morales 1999; Petchey 2000; Gonzalez and Holt 2002; Holt et al. 2003; Schiegg 2003; Pike et al. 2004). Since literature is sparse on the application of autocorrelation or AR (1) model in addressing stochastic issues as is the case of our study, we adapt the theory of population dynamics based on environmental autocorrelation concept to shed some light on the dynamics of the communities' EQ at irregularly spaced t+1 time step (1957 - 1958; 1966-1967; 1972-1973; 1999-2000).

In an attempt to explain the factors underlying the AR(1) or temporal environmental autocorrelation for the t +1 time step, special reference would be made to the summary of EOF loadings of the predictors (OAs) and predictands (environmental variables), and their corresponding scores/indices (as inputs for the model developments) of 1957, 1966, 1972, and 1999 (see Section 6.1) as they give more insights to appreciate the dynamics the EQ of 1957-1958, 1966-1967, 1972-1973, and 1999-2000. We recall that the variability of the EQ in 1957, 1966, 1972, and 1999 were determined and controlled by WQ issues, environmental health, fish biodiversity, water stress during the dry season, and negative attitudinal behavior towards the dams. Similarly, the thresholds of OAs of the four years were governed by effluent quality, the presence of the embankment, and backwashing of clogged filters and concomitant discharge of wastewater. In Table 5, the temporal environmental autocorrelation coefficients show a high value of (r =0.84, p= 0.0) for 1957 vs 1958, followed by a poor correlation (r =0.013, p=0.92) for 1966 vs 1967. A higher positive autocorrelation was registered for 1972 vs 1973 (r =0.94; p=0.0), and peaked to r =0.997, p= 0.0 for 1999 vs 2000. The good temporal coherence between 1957 vs 1958, 1972 vs 1973, and 1999 vs 2000 theoretically suggests unacceptable EQ on the basis of environmental autocorrelation (Pike et al. 2004). Thus, a positive autocorrelation between 1957 and 1958 environmental states suggests an unacceptable EQ baseline in 1957, which effectively and correspondingly predicted the 1958 EQ. In line with the results obtained for 1957 vs 1958, as opposed to increase in extinction of species in population dynamics arising from increasing autocorrelation (Johst and Wissel 1997; Petchey et al. 1997; Halley and Kunin 1999; Morales 1999; Petchey 2000; Gonzalez and Holt 2002; Holt et al. 2003; Schiegg 2003), we explicitly or implicitly suggest that the high correlation signified a high deterioration in EQ in 1958 mostly dominated by WQ issues. Pike et al.(2004) have reported that a situation where practically, no correlation exits between successive environmental states is an indication of various components of the environment exerting equal influence because of equality in their magnitudes (thresholds) . In our study, the fact that the 1967 EQ did not differ significantly from that of 1966 possibly due to similarity of their magnitudes, is suggestive of environmental stability (Holling 1978) and non-responsiveness to the OAs of the dam(s). And the pattern of fluctuations in 1967 is therefore, expected to be similar to Figure 8a. The temporal autocorrelations for 1972-1973 and 1999-2000 revealed the worst cases in environmental deteriorations. Consistent with 1957-1958 findings, the 1973 EQ appeared to have been controlled to a large extent by WQ issues. The 1972-1973 period marked the phase where the two treatment plants started to operate synchronously. It is therefore expected that the 1972 environmental state would respond accordingly to the OAs of the dams. In other words, the OAs appeared to have created the condition optimal for 1972 EQ to predict 1973 EQ, and that their scores (indices) would tend to fluctuate with close patterns. A similar explanation is deemed appropriate for the 1999-2000 scenario. However, the 1999 EOF 1 loadings, unlike the other years, were governed by water weighted difference between water stress (positive) and WQ issues (negative) and attitudinal behavior (negative). And these gave the basis for near perfect prediction of the 2000 EQ.

CONCLUSION

A thorough retrospective analysis and the development of stochastic and AR (1) models for gauging the environmental states and making backward prediction of EQ, one year positive lag in three communities in the vicinity of the dams, have been carried out. Using discontinuous, irregularly spaced, multivariate geostatistical, limited time series data consisting of sets of predictors and predictands, we have analyzed the dynamics of the EQ (of the communities) and the OAs (of the dams) for the four years. The multivariate EOF analysis has proved to be a robust analysis tool and efficient in data dimensionality and redundancy reductions. The most important information was captured by the three PCs of the eigenvalues, eigenvectors, and scores (indices). The eigenvalues adequately have explained the variations of the EQ and OAs of the communities and the dams, respectively. The eigenvector analysis has shed some light in understanding the contributions of variables to the variances observed. This, together with the eigenvalues and the scores (indices), have elucidated the dynamics of the EQ and OAs. The results have shown that the EQ was governed by WQ issues, environmental health (water-related diseases), fish biodiversity, dry season water stress, and negative attitudinal behavior towards the dam (from the communities` perspective). The OAs were essentially governed by effluent quality, the impoundment effects, and backwashing of clogged filters and associated wastewater discharge. The construction of important indices (EQIs and OAIs) from the scores have shed some light on the relationships between the EQIs and OAIs graphically and the use of dependency measures. The results from the correlation analyses have shown that the relationship between the two indices generally ranged from moderate to high significance, demonstrating the influence of the dams on the communities` EQ. Also, the relationship between the two sets of indices is best described by Pearson product- moment correlation. Thus, the relationship between the dams and the communities` EQ is more of a linear function than nonlinear. The stochastic models developed have shown that the nonparametric model, which incorporated LOOCV and K-NNs outperforms the parametric one, and hence, better fits the study. The AR (1) backcasting model results theoretically, have shown serious progressive deterioration in the EQ from 1957-1958, 1972-1973, and finally from 1999-2000. However, the EQ of 1967 did not significantly differ from that of 1966, a sign of environmental stability, neutrality and non-responsiveness to the OAs of the dam(s).

The limitations of the study were due to lack of observational data for (i) the downstream river WQ and effluent quality prior to 1999, and (ii) comparing with the predicted EQ/backcasts (model results). However, the study has shown beyond doubts that the EQ needs remediation. And this is important to prevent any undesirable impacts to escalate since cumulative impacts are real. Moreover, the role of environmental antecedents in predicting the future is vital in forecasting schemes in all fields, and should be given the due regard. Therefore, we should not view the study merely as historical, but rather a resource valuable for environmental decision support system, especially for WRDP/RBDP policies. If the country will like to achieve its long term vision for example the MDGs, the environment that provides resources for the maintenance and sustenance of humanity and other forms of biota should receive a better stewardship to prevent it from spewing us out. In the country, whole economy which is agriculture-driven is centered on rural communities which constitute about 70% of the total population and human resource, and any environmental policy that is well organized and effectively and efficiently implemented at the grass roots level will shine on the pinnacle, which will be emulated by other countries in the SSA. Our remediation of the environmental impacts of the dams is not different from what we have previously indicated (Tetteh et al. 2004a; Tetteh et al. 2006). Finally, it our hope that this piece of work will stimulate governmental agencies, the ministries, research institutions, universities, the private sectors and environmentally conscious non-governmental organizations (NGOs) to be more proactive and be infused with more scientific vigor for the creation and maintenance of environmental database readily accessible for public use, since proper decision-making to a very large extent is data-driven. Also, this piece of work should propel the government to be more stringent on environmental regulation in the country.

Future Work

This will involve canonical correlation analysis (CCA) for EQ prediction. This multivariate approach will use multiple dependent variables against multiple independent variables in the prediction scheme. In this case, we will use all of the EOF 1, 2, and 3 predictands in addition to the EOF 1, 2, and 3 predictors. The results will be compared with the nonparametric stochastic model which used one dependent (EOF 1 predictand) against three dependents (EOF 1, 2, and 3 predictors) augmented by the AR (1) model for the prediction. This will help us to assess the strengths and weaknesses of various models for the purpose of making better environmental decisions.

References

Acquah, P.C. and Boateng, A. Planning for Mine Closure: Case Studies in Ghana. In ENVIRONMENTAL POLICY IN MINING, Corporate Strategy and Planning for Closure; Warhurst, A.; Noronha, L.;Eds.; Lewis Publishers: NY, 2000, pp 383-395.

Adams, P. U. S. experts echoes Chinese concerns about dam-induced earthquakes. The Three Gorges Probe Feature, Probe International, 5thJune 2008. http://www.probeinternational.org/three-gorges-probe/news-and-opinion/u-s-expert-echoes-chinese-concerns-about-dam-induced-earthquakes.

Adjei, R. O.; Sasu, S. and Appiah-Kwkye, L. Index of Potential Contamination: Schistosoma haematobium Infections in School Children in the Ashanti Region of Ghana. *East. Afr. Med. J.* October 2004b, Vol. 81, No.10, pp.520-523.

African Development Bank (AfDB)/African Development Fund (AfDF). African Development Bank Group's Policy on the Environment, February 2004, 42 pp.

Akaike, H. A. New Look at the Statistical Model Identification, *IEEE Transactions on Automatic Control,* 1974; Vol. AC-19, pp.716-723.

Akaike, H. An Objective Use of Bayesian Models, *Annals of the Institute Statistical Mathematics*, 1977; Vol. 29, pp. 9-20.

Allen, D. M. The Relationship between Variable Selection and Data Augmentation and a Method for Prediction. *Technometrics*. 1974; Vol. 16, pp. 125-127.

American Public Health Association/American Water Works Association/Water Environment Federation (APHA/AWWA/WEF). 18th Edition. Standard Methods for the Examination of Water and Wastewater; Eaton, A. D.; Clesceri, L. S.; Greenberg, A. E. (Eds.), Washington, DC 20005, 1995.

Anderson, W. L. and Glew, R. H.(2002). Support of problem-based learning curriculum by basic science faculty. Med. Educ. Online [serial online], 7:10. Available from URL http://www.med-ed-online.org.

Appiah-Opoku, S. Environmental impact assessment in developing countries: The case of Ghana. Environ. Impact. Assess. Rev. 2001; Vol. 21, pp. 59–71.

Artardi, E.V. and Sala-i-Martin, X. *The Economic Tragedy of the XXth Century: Growth in Africa,*: NBER Working Paper No. 9865, National Bureau of Economic Research, Cambridge, MA, July 2003; 33pp.

Asante, K.A, Agusa, T. Subramanian, A., Ansa-Asare, O. D., Biney, C.A. and Tanabe, S. Contamination status of arsenic and other trace elements in drinking water and residents from Tarkwa, a historic mining township in Ghana. *Chemosphere*. January 2007; Vol. 66, Issue 8, pp. 1513-1522.

Asian Development Bank(AsDB). Environmental Assessment Requirements and Environmental Review Procedures of the Asian Development Bank. Office of the Environment, Asian Development Bank. March 1993; 44 pp.

Awakul, P. and Ogunlana, S. O. The effect of attitudinal differences on interface conflict on large construction projects: The case of the Pak Mun Dam project. *Environ. Impact. Assess. Rev*. 2002; Vol. 22, Issue 4, pp. 311-335.

Awuah, E. and Frempong, E. (2004c). Development of a weighting system for use in Environmental Health Impact Assessment associated with water impoundment projects in Ghana and its application 1. *Journal of the Ghana Science Association.* Vol.6, No. 2, pp. 59-69.

Ayogu, M. Infrastructure and Economic Development in Africa. *J. Afr. Econ.* 2007; Vol. 16 Supplement No.1, pp. 75-126.

Baba, A. Rapid Impact Assessment Matrix (RIAM) Method for the Tuzla GEOTHERMAL and CAN THERMAL Power Plant in Canakkale, Turkey, *Fresenius Environmental Bulletin*, 2005; Vol.14, No. 2, pp. 113-119.

Babut, M., Sekyi, R., Rambaud, A., Potin-Gautier, M., Tellier, S., Bannerman, W. and Beinhoff, C. Improving the environmental management of small- scale gold mining in Ghana: a case study of Dumasi. *Journal of Cleaner Production*. March 2003; Vol. 11, Issue 2, pp. 215-221.

Barro, R. Economic Growth in a Cross-section of Countries, *Quarterly Journal of Economics*. 1991; Vol. 106, Issue 2, pp.407-443.

Barro, R.J. and Lee, J. W. Sources of Economic Growth. *Carnegie-Rochester Conference Series on Public Policy*, 1994; Vol. 40, pp. 1-46.

Bartlett, R. V. (1993). Integrated impact assessment as environmental policy. *Policy Studies Review*, 1993; Vol. 12, pp.162-177.

Baxter .R. M. Environmental Effects of Dams and Impoundments *Annual Review of Ecology and Systematics,* 1977; Vol. 8, pp. 255-283.

Beadle, F. C. The Inland Waters of Tropical Africa. An Introduction to Tropical Limnology. Longman Group Ltd: NY, 1981; pp.362-406

Bear, D. NEPA: Substance or Merely Process. *Forum for Applied Research and Public Policy*. Summer 1993; Vol. 8, Issue 2, pp. 85-88.

Berger, B. B. Water and wastewater quality control and the public health. *Annual Review of Public Health*.1992; Vol. 3, pp. 339-92.

Best, J. A. The National Environmental Policy act as a Full Disclosure Law, NTIS Rep. PB-227-809, U.S. Department of Commerce, National Technical Information Service, Springfield, Va., December 1972.

Bhalla, S. S. and Mookerjee, A. Big Dam Development: Facts, Figures and Pending Issues. *Water Resources Development,* March 2001; Vol. 17, Issue 1, pp. 89–98.

Bisset, R. A critical survey of methods of Environmental Impact Assessment; In *An annotated Reader in Environmental Planning and Management*; O`Riorden, T.; Turner, R. K. Eds.; Pergamon Press: England, 1983; pp. 168-186.

Biswas, A. K. and Geping, Q. (Eds.); *Environmental Impact Assessment for Developing Countries,* Tycooly International for U.N. University, London, 1987; 232 pp.

Biswas, A. K. and Tortajada, C. Development and Large Dams: A Global Perspective. *Water Resources Development*, 2001; Vol.17, Issue 1, pp. 9–21.

Boafo-Arthur, K. The political economy of Ghana's foreign policy. Past, present and the future, Ghana. *Soc. Sci. J.* January 2000, Vol.1, Issue 1, pp. 1-30.

Bohanec, M. and Rajkovič, V. Multi-Attribute Decision Modeling: Industrial Applications of DEX. *Informatica*, 1999; Vol. 23, Issue 4, pp. 487-491.

Bolin, B. The Kyoto Negotiations on Climate Change: A Science Perspective', *Science, 1998;* Vol. 279, pp.330–331.

Bonsu, O.A. and Thatcher, M. P. African Oral Genealogy: Collecting and Preserving Yesterday for Tomorrow. World Library and Information Congress: 73rd IFLA General Conference and Council, 19-23 Aug, Durban, South Africa, 2007; 9pp, available at http://www.ifla.org/iv/ifla73/index.htm.

Bonzongo, J.C.J.; Donkor, A. K. and Nartey, V. K. Environmental impacts of mercury related to artisanal gold mining in Ghana. *Journal De Physique Iv (Proceedings).*XIIth International Conference on Heavy Metals in the Environment - Volume I and II Grenoble, France, May 26-30, 2003; Vol. 107, pp. 217-220.

Brans J.P. and Vincke, P. A preference ranking organization method: the PROMETHEE method for multiple criteria decision-making, *Management Science*, 1985; Vol. 31, Issue 6, pp. 647- 665.

Brans, J.P.; Vincke, P. and Marechal, B. How to select and how to rank projects: The PROMETHEE method", *European Journal of Operational Research*, 1986; Vol. 24, pp. 228- 238.

Branstator, G.. A Striking Example of the Atmosphere's Leading Traveling Pattern. Journal of the Atmospheric Sciences*. 15 August 1987; Vol. 44, Issue 16, pp. 2310-2323.*

Brentrup, F.; Ku¨sters, J.; Kuhlmann, H.; and Lammel, J. Environmental impact assessment of agricultural production systems using the life cycle assessment methodology I. Theoretical concept of a LCA method tailored to crop production. *European Journal of Agronomy*. 2004a,Vol. 20, pp. 247–264.

Brentrup, F.; Ku¨sters, J.; Kuhlmann, H; and Lammel, J. Environmental impact assessment of agricultural production systems using the life cycle assessment (LCA) methodology II. The application to N fertilizer use in winter wheat production systems. *European Journal of Agronomy*, 2004b, Vol. 20, pp. 265–279.

Buizza, R. Potential forecast skill of ensemble prediction and ensemble spread and skill distributions of the ECMWF Ensemble Prediction System. *Mon. Wea. Rev.* 1997, Vol. 125, pp. 99-119.

Buongiorno Nardelli, B. and Santoleri, R. Methods for the Reconstruction of Vertical Profiles from Surface Data: Multivariate Analyses, Residual GEM, and Variable Temporal Signals in the North Pacific Ocean .*Journal of Atmospheric and Oceanic Technology*. November 2005, Vol. 22 , Issue 11, pp.1762–1781.

Caldwell, L. K. The National Environmental Policy Act: Status and Accomplishments, Proceedings of the 38th North American Wildlife Natural Resources Conference, Wildlife Management Institute, Washington, D.C. 1973.

Canter, L. W. Water Resources Assessment- Methodology and Technology Sourcebook, Ann Arbor Science Publishers, Ann Arbor, MI., 1979; pp. 203.

Canter, L W. Environmental Impact Assessment, 2nd ed. McGraw-Hill, Inc, NY, 1996; 660 pp.

Caswell, H. Matrix population models: construction, analysis, and interpretation. Second Edition. Sinauer Associates, Sunderland, MA, 2001; 772 pp.

Chen, J. M. and Hart, P.A. Interpretation of extended empirical orthogonal function (EEOF) analysis. *Mon. Weather Rev.* 1993; Vol. 121, pp. 2631–2636.

Cherkassky, V. and Mulier, F. Learning from Data: Concepts, Theory and Methods, Wiley & Sons, NY.1998; 441 pp.

Clark, B.D.; Gilad, A.; Bisset, R. and Tomlinson, P. (Eds.); Perspectives on Environmental Impact Assessment.Proceedings of the Annual WHO Training Courses on Environmental Impact Assessment, Centre for Environmental Management and Planning, University of Aberdeen, Scotland, 1980-1983, 1984; 540 pp.

Clemen, R. T. Making Hard Decisions: An Introduction to Decision Analysis, Second Edition, Duxbury Press, Belmont, California, 1996; 664 pp.

Cohen, D. W. The Undefining of Oral Tradition, *Ethnohistory*. Winter 1989, Vol. 36, No. 1, pp. 9-18.

Cohen, J. E. Unexpected dominance of high frequencies in chaotic nonlinear population models. *Nature*, 1995, Vol. 378, pp. 610–612.

Cooper, C.D. and Alley, F. C. Air Pollution Control: A Design Approach, Third Edition. Waveland Press, Inc: IL, 2002; pp.14-15.

Council on Environmental Quality (CEQ). Regulations for Implementing the Procedural Provisions of the National Environmental Policy Act, 1987; 40 CFR Parts 1500 - 1508.

Council on Environmental Quality (CEQ). Environmental Quality, Twenty- third Annual Rep., U.S. Government Printing Office, Washington, D.C., January 1993a; pp. 151-172.

Crapper, M.D.R. and Boni, D. Aluminum in human brains disease-an overview. *Neurotoxicol.* 1980; Vol.1, pp. 3-16.

Cressie, N. A. C. Statistics for Spatial Data. Revised Edition. Wiley Series in Probability and Mathematical Statistics; John Wiley and Sons, Inc: NY,1993; 900 pp.

Culhane, P. J. Post-EIS environmental auditing: a first step to making rational environmental assessment a reality. *Environ. Prof.* 1993, Vol.15, Issue 1, pp. 66-75.

Dankwa, H.R. and Gordon, C. The Fish and Fisheries of the Lower Volta Mangrove Swamps in Ghana. *African Journal of Science and Technology (AJST) Science and Engineering Series,* 2002, Vol. 3, No. 1, pp. 25-32.

Davidson, A. M.; Walker, G. S. and Lewins, A. M.. Water supply aluminum concentrations, dialysis dementia, and effects of reverse osmosis water treatment. *Lancet.*1982, Vol. 2, pp. 785-787.

Dee, N.; Baker, J.; Drobny, N.; Duke, K.; Whitman, I.; and Fahringer, D. An environmental evaluation system for water resource planning. *Water Resources Research,* 1973, Vol. 9, pp.523-535.

Dempster, A. P.; Laird, N. M. and Rubin, D. B. Maximum likelihood from incomplete data via the EM algorithm. *J. Roy. Stat. Soc.,* 1977, Vol. 39B, pp. 1–38.

Derban, L. K. A. Some environmental health problems associated with industrial development in Ghana. Institute of Aquatic Biology, Council for Scientific and Industrial Research, Accra, Ghana, 1975; 22 pp.

Dipper B.; Jones, C. and Wood, C. Monitoring and Post-auditing in Environmental Impact Assessment: A Review. *Journal of Environmental Planning and Management.* November 1998, Vol. 41, No. 6, pp. 731-747.

Donkor, A. K.; Bonzongo, J-C.; Nartey, V. K.; and Adotey, D. K. Heavy metals in Sediments of the gold mining impacted Pra River Basin, Ghana, West Africa. *Soil & Sediment Contamination*, 2005; Vol. 14, Issue 6, pp. 479-503.

Donkor, A. K.; Bonzongo, J. C.; Nartey, V. K.; and Adotey, D. K. Mercury in different environmental compartments of the Pra River Basin, Ghana. *Science of The Total Environment*, September 2006, Vol. 368, Issue 1, pp.164-176.

Easterly, W. and Levine , R. Africa`s Growth Tragedy: Policies and Ethnic Divisions. *Quarterly Journal of Economics*, 1997, Vol.112, pp. 1203-1250.

Easterly, W.; Kremer, M. and Pritchett, L.; Summers, L. Good Policy or Good Luck? Country Growth Performance and Temporary Shocks. *Journal of Monetary Economics*, 1993, Vol. 32, Issue 3, pp: 459-483.

Economic Commission for Europe (ECE). Convention on Environmental Impact Assessment in a Transboundary Context, E/ECE/1250, United Nations, NY, February 1991.

Efron, B. and Tibshirani, R.J. An introduction to the bootstrap. Monographs in Statistics and Applied Probability 57. Chapman & Hall: London, 1993.

Efron, B. and Tibshirani, R. Improvements on Cross-Validation: The 0.632 + Bootstrap Method, *Journal of the American Statistical Association*, 1997; Vol. 92, pp. 548-560.

Ehrendorfer, M. Predicting the uncertainty of numerical weather forecasts: a review. *Meteorol. Zeitschrift*. 1997, Vol. 6, pp. 147-183.

Elbadawi, I. and Sambanis, N. Why Are There So Many Civil Wars in Africa? Understanding and Preventing Violent Conflict. *J. Afr. Econ*. December 2000, Vol. 9, Issue 3, pp. 244-69.

El-Naqa, A. Environmental impact assessment using rapid impact assessment matrix (RIAM) for Russeifa landfill, Jordan, *Environ. Geol.* 2005, Vol. 47, pp. 632–639.

Environmental Performance Index (EPI). Yale Center for Environmental Law and Policy, Yale University; Center for International Earth Science Information Network (CIESIN), Columbia University; In collaboration with World Economic Forum Geneva, Switzerland; Joint Research Centre of the European Commission Ispra, Italy, 2008. Full report and data available at http://epi.yale.edu.

Environment and Social Development Department (ESDD). Environmental Impact Assessment Regulations and Strategic Environmental Assessment Requirements Practices and Lessons Learned in East and Southeast Asia, SAFEGUARD DISSEMINATION NOTE NO. 2, April 2006.

Environmental Sustainability Index (ESI). Yale Center for Environmental Law and Policy, Yale University; Center for International Earth Science Information Network (CIESIN), Columbia University; In collaboration with World Economic Forum Geneva, Switzerland; Joint Research Centre of the European Commission Ispra, Italy, 2005. Full report and data available at available at http://epi.yale.edu.

Fenwick, A. Waterborne Infectious Diseases—Could They Be Consigned to History? *Science* 25 August 2006, Vol. 313. No. 5790, pp. 1077 – 1081.

Finsterbusch, K. In Praise of SIA: A Personal Review of the Field of Social Impact Assessment. *Impact Assessment*, 1995, Vol. 13, Issue, 3, pp 229-252.

Flaten,T.P. Aluminium as a risk factor in Alzheimer's disease, with emphasis on drinking water. *Brain Research Bulletin*. 2001, Vol. 55, Issue 2, pp. 187–196.

Fosu, A .K. Political instability and economic growth in developing economies: some specification empirics. *Economic Letters*, February 2001, Vol. 70, Issue 2, pp. 289-294.

Fosu, A. K.; Kimenyi, M.,S. and Ndung'u, N.S. Economic Reforms and Restructuring in Sub-Saharan Africa: An Overview. *J. Afr. Econ.* September 2003; Vol. 12, AERC Supplement 2, ppii1-ii11.

Frempong, E. and Nijjhar, B. Some preliminary observations on the fauna and flora of Barekese Lake, Ghana. *Bulletin de l'Institut Fondamental d'Afrique noire*, 1973, Series A, Vol. 35, pp.67-78.

Fukuoka. A. A Study of 10-day Forecast (A Synthetic Report), Vol. XXII. *The Geophysical Magazine*. Tokyo, 1951, pp. 177–218.

Gardner, M.J. and Gunn, A.M. Bioavailability of aluminum from food and drinking water. In Alzheimer's disease and the environment. Proceedings of the conference. Walton, L.; (Ed.); *Proc. Royal Soc. Med. Round Table Series*, London, UK, 1991; No. 26, 79 pp.

Geisser, S. The Predictive Sample Reuse Method with Applications. *Journal of the American Statistical Association*. 1975, Vol. 70, pp. 320-328.

George, E. I. and McCulloch, R. E. Approaches for Bayesian Variable Selection, *Statistica Sinica*, 1997, Vol. 7, pp. 339-373.

Gerba, C. P.; Rose, J. B., Haas, C. N.; and Crabtree, K. D. Waterborne rotavirus: A risk assessment. *Water Research*, December 1996, Vol. 30, Issue 12, pp. 2929-2940.

Ghahramani, Z. and Jordan, M. I. Supervised learning from incomplete data via an EM approach. In Cowan, J.D.; Tesauro, G.; Alspector, J.; Eds.; Advances in Neural Information Processing Systems 6, Morgan Kaufman: San Mateo, CA, 1994; pp.120-127.

Ghana Environmental Protection Agency (EPA). General Effluent Quality Guidelines for Discharges into Natural Bodies of Water, 1999.

Ghana Environmental Protection Agency Act (EPA). Government of Ghana, 1994.

Ghana. *J. Environ. Manage*. 2004a, Vol. 72, pp.189-194.

Gilchrist, W. Statistical Forecasting, John Wiley and Sons: NY, 1976; 308 pp.

Gilliland, M. W. and Risser, P.G. The Use of Systems Diagrams for Environmental Impact Assessment: Procedures and an Application. *Ecological Modeling,* August 1977, Vol. 3, No. 3, pp 183-199.

Gilpin, A. Environmental impact assessment. Cambridge University Press, London, 1995.

Glasson, J.; Therivel, R. and Chadwick, A. Introduction to environmental impact assessment. Principles and procedures, process, practice and prospects. Second Edition. Spon Press :London, 1994; 496 pp.

Goldfarb, T. D. (Ed.); Taking Sides. Clashing Views on Environmental Issues. Eighth Edition. Duskin/McGraw-Hill: CT, 1999; 353 pp.

Golow, A. A.; L. C. Mingle. Mercury in river water and Sediments in some rivers near Dunkwa-On-Offin, an alluvial goldmine, Ghana. *Bulletin of Environmental Contamination and Toxicology*, 2003, Vol. 70, Issue 2, pp. 379-384.

Gonzalez, A. and Holt, R. D. The inflationary effects of environmental fluctuations in source–sink systems. *Proc. Natl Acad. Sci.* USA 99, 2002, pp.14 872–14 877.

Goodland, R. The world bank's new environmental policy for dams and reservoirs.*International Journal of Water Resources Development*. 1990, Vol. 6, Issue 4, pp. 226-239.

Goodman, D. The demography of chance extinction. In Soule, M.E.; Ed; Viable populations for conservation , Cambridge University Press: UK,1987; pp. 11–34.

Gosh, S. Combination of didactic lectures and case-oriented problem-solving tutorials toward better learning: perceptions of students from a conventional medical curriculum. *Advan. Physiol. Educ.* January 2007, Vol. 31, pp. 193-197.

Goyal, S. K. and Deshpande, V.A. Comparison of weight assignment procedures in evaluation of environmental impacts. *Environ. Impact. Assess. Rev.* 2001,Vol. 21, pp.553–563.

Gupta, R. K. (2001). Human rights dimension of regional water transfer. Experience of the Sardar Sarovar Project, *International Journal of Water Resources Development*, 2001, Vol. 17, No.1, pp. 125–147.

Gupta, H. K. A review of recent studies of triggered earthquakes by artificial water reservoirs with special emphasis on earthquakes in Koyna, India. *Earth- Science Reviews*, 2002, Vol.58, pp. 279–310.

Gyimah-Brempong, K. and Corley, M.E. Civil Wars and Economic Growth in Sub-Saharan Africa. *J. Afr. Econ.* .June 2005, Vol.14, Issue 2, pp.270-311.

Gyimah-Brempong, K. and Traynor, T. L. Political Instability, Investment, and Economic Growth in Sub-Saharan Africa. *J. Afr. Econ.* 1999, Vol. 8, Issue 1, pp.52-86.

Haccou, P. and Vatutin, V. Establishment success and extinction risk in autocorrelated environments. *Theor. Popul. Biol.* 2003, Vol. 64, pp.303–314.

Hagebro, C. Flood damage assessment in Dac Lac Province, Vietnam. In Environmental Impact Assessment Using the Rapid Impact Assessment Matrix (RIAM); Jensen, K.; (Ed.); Olsen & Olsen, Fredensborg, Denmark, 1998; 69pp.

Hale, T.A .From the Griot of *Roots* to the Roots of *Griot*:A New Look at the Origins of a Controversial African Term for Bard, *Oral Tradition*, 1997, Vol. 12, Issue 2, pp.249-278.

Halley, J. M. Ecology, evolution and 1/f-noise. *Trends. Ecol. Evol.* 1996, Vol. 52, pp.91–100.

Halley, J. M. and Kunin, W. E. 1999 Extinction risk and the 1/f family of noise models. *Theor. Popul. Biol.* 1999, Vol. 56, pp. 215–230.

Han, S.Y,; Kwak, S.J. and Yoo, S.H. Valuing environmental impacts of large dam construction in Korea: An application of choice experiments. *Environ. Impact. Assess. Rev.* 2008, Vol. 28, pp. 256–266.

Harin, L. Introduction: The Search for Grounds in African Oral Tradition, *Oral Tradition, 1994, Vol.* 9, Issue 1, pp. 3-22.

Hattori, A.; Ohtsu, M. and Koike, I. Distribution, metabolism and budget of nitrogen in Tokyo Bay. *Chikyu Kagaku (Geochemistry).*1983, Vol.17, pp.33-41 (in Japanese).

Helsel, D.R. and Hirsch, R. M. Statistical Methods in Water Resources; Studies in Environmental Science 49; Elsevier: NY, 1992; 522pp.

Hickie D. and Wade M. Development of guidelines for improving the effectiveness of environmental assessment. *Environ. Impact. Assess.* Rev. 1998, Vol.18, Issue 3, pp 267–287.

Hilson, G..; Hilson, C. J. and Pardie, S. Improving awareness of mercury pollution in small-scale gold mining communities: Challenges and ways forward in rural Ghana. *Environmental Research*. February 2007, Vol.103, Issue 2, pp. 275-287.

Hilson, G.. and Pardie, S. Mercury: An agent of poverty in Ghana's small-scale gold-mining sector? *Resources Policy*. June 2006, Vol 31, Issue 2, pp106-116.

Holt, R. D., Barfield, M. and Gonzalez, A. Impacts of environmental variability in open populations and communities: inflation' in sink environments. *Theor. Popul. Biol.* 2003, Vol. 64, pp.315–330.

Hokkanen, J.; Lahdelma, R.; Miettinen, K.; and Salminen, P. Determining the implementation order of a general plan by using a multicriteria method. *Journal of Multi-criteria Decision Analysis*. 1998, Vol. 7, pp. 273–284.

Hokkanen, J.; Lahdelma, R. and Salminen, P. A multiple criteria decision model for analyzing and choosing among different development patterns for the Helsinki cargo harbor. *Socio-Economic Planning Sciences*. 1999, Vol. 33, pp.1–23.

Holling, C.S. (Ed.); Adaptive environmental assessment and Management, John Wiley & Sons: London, 1978.

Horel, J.D. Complex Principal Component Analysis: Theory and Examples. Journal of Climate and Applied Meteorology. *December 1984, Vol. 23, pp.1660-1673.*

Horton, R. African Traditional Thought and Western Science, *Journal of the International African Institute*, April 1967, Vol. 37, No. 2, pp.155-187.

Houseago-Stokes, R. E. Using optimal interpolation and EOF analysis on North Atlantic satellite data. *International WOCE Newsletter,* WOCE International Project Office, Southampton, United Kingdom, 2000, No. 28, pp. 26–28.

Houseago-Stokes, R.F. and Challenor, P.G. Using PPCA to Estimate EOFs in the Presence of Missing Values. *Journal of Atmospheric and Oceanic Technology*. September 2004, Vol. 21, pp. 1471-1480.

Hunt, R. G.; Sellers, J.D. and Franklin, W. E.. Resource and Environmental Profile Analysis: A Life Cycle Assessment for Products and Procedures, *Environmental Impact Assessment Review*, 1992, Vol. 12, pp. 245-269.

International Monetary Fund (IMF), Organization for Economic Co-operation and Development, United Nations, and World Bank Group. A Better World for All, 2000; 28 pp; available at http://www.paris21.org/betterworld.

Jain, R. K; Urban, L. V.; Stacey, G. S.; and Balbach, H. E. Environmental Assessment. McGraw-Hill, Inc: NY, 1993; 526pp.

Jekel, M. R. Aluminum in water: How it can be removed? Use of aluminum salts in treatment. *Proc. of the Int. Water Supply Ass.*, Copenhagen, Denmark, May 25-31 1991.

Johannes R.E. Integrating traditional ecological knowledge and management with environmental impact assessment. In Inglis J.T.; Ed;. Traditional Ecological Knowledge: Concepts and cases. Ottawa, Canada: International Development Research Centre (IDRC) Press, 1993.

Johnson, R. E.; Chang, C.H. and Lord, R. G. (2006). *Moving From Cognition to Behavior: What the Research Says. Psychological Bulletin*, 2006, Vol. 132, Issue 3, pp 381-415.

Johnson, R. A. and Wichern, D. W. Applied Multivariate Statistical Analysis. Sixth Edition. Pearson Education, INC: N. J, 2007; 773pp.

Johst, K. and Wissel, C. Extinction risk in a temporally correlated fluctuating environment. *Theor. Popul. Biol.* 1997, Vol. 52, pp.91–100.

Jolliffe, I. T. Discarding variables in a principal component analysis. I: Artificial data. *Appl. Stat.*1972, Vol. 21, pp. 160–173.

Jolliffe I. T. *Principal Component Analysis*. Second Edition. Springer Series in Statistics.Springer: NY, 2002; 487 pp.

Jordan, P. and Webbe, G. (Eds.), Schistosomiasis, Epidemiology, Treatment and Control. William Heinemann Medical Books, London, 1982; pp. 376 pp.

Kaihatu, J. M.; Handler, R. A.; Marmorino, G. O.; and Shay, L. K. (1998). Empirical Orthogonal Function Analysis of Ocean Surface Currents Using Complex and Real-Vector Methods. *Journal of Atmospheric and Oceanic Technology*. August 1998, Vol. 15, pp. 927-941.

Kalnay, E. Atmospheric Modeling, Data Assimilation and Predictability. Cambridge University Press: N.Y., 2003; 341 pp.

Kawabe, M. and Kawabe, M. Factors Determining Chemical Oxygen Demand in Tokyo Bay. *Journal of Oceanography*, 1997, Vol. 53, pp. 443- 453.

Kim, K.W. and Wu, Q. 1999. A comparison study of EOF techniques:analysis of nonstationary data with periodic statistics. *Journal of Climate,* 1999, Vol. 12, pp. 185–199.

Kira, T. Lake Volta Utilization. In International Lake Environment Committee Foundation (ILEC) ôtsushigyô Photoprinting Co. Ltd, Otsu, Japan, 1995, pp132-133.

Kitanidis, P.K. Introduction to Geostatistics. Applications in Hydrogeology, Cambridge University Press: N.Y., 1997; 249 pp.

Kreith, F. Lack of Impact. *Environment.*1973, Vol. 15, No. 1, pp. 26-33.

Kuhn, R. and Duerden, F. A review of traditional ecological knowledge:an interdisciplinary canadian perspective. 1996, Culture XVI(1), 71–84.

Kuitunen, M.; Jalava, K. and Hirvonen., K. Testing the usability of the Rapid Impact Assessment Matrix (RIAM) method for comparison of EIA and SEA results, *Environ. Impact. Assess. Rev.* 2008, Vol. 28, pp. 312–320.

Kumah, J.F and Baah, E.Y. (1971), Asuofia Resettlement Project. Town and Country Planning Department. Ashanti Regional Office, Kumasi, Ghana, 1971; 99 pp.

Kumar, A. P.; Reddy, P. R. and Tagore, R. Permutation and Combinations Approach to Program Evaluation and Review Technique, *ARPN Journal of Engineering and Applied Sciences*. December 2007, Vol. 2, No. 6, pp. 20-26.

Kung, H-T.; Ying, L. G. and Liu, Y. C. Fuzzy clustering analysis in environmental impact assessment — A complement tool to environmental quality index. *Environmental Monitoring and Assessment*, October 1993, Vol 28, No.1, pp.1-14.

Kutzbach J. E. 1967. Empirical eigenvectors of sea-level pressure, surface temperature and precipitation complexes over North America. *Journal of Applied Meteorology* 1967, Vol. 6, pp.791–802.

Kwak, S.J.;Yoo, S. H. and Shin, C.O. A multiattribute index for assessing environmental impacts of regional development projects: A case study of Korea. *Environ. Manage.*, 2002, Vol. 29, No. 2, pp.301–309.

Leopold, L.; Clarke, F. E.; Hanshaw, B. B.; and Balslay, J. R. A procedure for evaluating environmental impact. US Geological Survey (USGS) circular 645. Washington, DC: USGS, 1971.

Lerer, L. B. and Scudder, T. Health impacts of large dams. *Environ. Impact. Assess. Rev.* March 1999, Vol. 19, Issue 2, pp.113-123.

Lilienfeld, A.M. and Lilienfeld, D.E. Foundations of Epidemiology. Second Edition. Oxford University Presss: Oxford, 1980; pp. 61–62.

Linhart, H. and Zucchini, W. Model Selection, Wiley: NY, 1986.

Linkov. I.; Varghese, A.; Jamil, S.; Seager, T. P.; Kiker, G.; and Bridges, T. Multi-criteria Decision Analysis: a Framework for Structuring Remedial Decisions at Contaminated Sites. In Comparative Risk Assessment and Environmental Decision Making; Linkov, I.; Ramadan, A.; Eds; Kluwer Academic Publishers: The Netherlands, 2004; pp.15-54.

Little, R. J. A. and D. B. Rubin. *Statistical Analysis with Missing Data.* Wiley: NY, 1987; 278 pp.

Lohani, B. N. and Halim, N. Environmental Impact Identification and Prediction: Methodologies and Resource Requirements, background papers for course. Environmental Impact Assessment of Hydropower and Irrigation Projects. International Center for Water Resources Management and Training (REFIGURE), Bangkok, Thailand, 13-31 August 1990; pp.152-182.

Lorenz, E. N. Empirical Orthogonal Functions and Statistical Weather Prediction. Technical report, Statistical Forecast Project Report 1, Dep of Meteor, MIT: 49, 1956.

Ly, C. K. The Role of the Akosombo Dam on the Volta River in Causing Coastal Erosion in Central and Eastern Ghana (West-Africa). *Marine Geology*, 1980, Vol. 37, Issue (3-4), pp. 323-332.

Mangal, T. D.; Paterson, S. and Fenton, A. Predicting the Impact of Long-Term Temperature Changes on the Epidemiology and Control of Schistosomiasis: A Mechanistic Model. *PLoS ONE.* January 2008, Vol.3, No.1, pp1-9.

Maritz, J. S. Distribution-free statistical methods: Monographs in Statistics and Applied Probability 17. Second Edition. Chapman & Hall: London, 1995; 255pp.

Martyn, C. N.; Osmand, C.; Edwardson, J. A.; Barker, D. J. P.; Harris, E. C.; and Lacey, R. F. C. Geographical relation between Alzheimer's disease and aluminum in drinking water. *The Lancet.* 1989, I (8629), pp. 59-62.

McDaniels, T. L. A multiattribute index for evaluating environmental impacts of electric utilities. *J. Environ. Manage.* 1996, Vol. 46, pp. 57–66.

McQuarrie, A. D. R. and Tsui, C.L. Regression and Time Series Model Selection, World Scientific: Singapore, 1998.

Metcalf, C. J. E.; D. N. and Koons, D. N. Environmental uncertainty, autocorrelation and the evolution of survival. *Proc. R. Soc. B*. 2007, Vol. 274, pp. 2153–2160.

Mensah, K. B. Restructuring the delivery of clean water to rural communities in Ghana: the institutional and regulatory issues. *Water Policy*.1998, Vol.1, pp. 383-395.

Mills-Tettey R. African resettlement housing: a revisit to the Volta and Kainji schemes. *Habitat Int*. 1989, Vol.13, No. 4, pp.71- 81.

Mode, C. and Jacobson, M. E.. On estimating critical population size for an endangered species in the presence of environmental stochasticity. *Math. Biosci*. 1987, Vol. 85, pp.185–209.

Molteni, F.; Buizza, R.; Palmer, T. N.; and Petroliagis, T. The new ECMWF Ensemble Prediction System: methodology and validation. *Quarterly Journal of the Royal Meteorological Society* 1996, Vol. 122, pp. 73-119.

Morales, J. M. Viability in a pink environment: why 'white noise' models can be dangerous. *Ecol. Lett*. 1999, Vol. 2, pp. 228–232.

Moss, B. Ecology of Freshwater. . Second Edition. Blackwell Scientific Publications: Oxford, 1988; pp. 258-272.

Mousa A.H. and el-Hassan A.A. The effect of water movements on the snail intermediate hosts of Schistosomiasis in Egypt. *J. Egypt. Med. Assoc*. 1972, Vol. 55, No.4, pp. 255-260.

Munda, G. Cost-benefit analysis in integrated environmental assessment: some methodological issues. *Ecological Economics* 1996, Vol. 19, pp.157-168.

Murdoch, J.C and Sandler, T. Economic Growth, Civil Wars, and Spatial Spillovers. *Journal of Conflict Resolution*. 2002, Vol. 46, No.1, pp. 91-110.

Nandy, A. Dams and dissent: India's first modern environmental activist and his critique of the DVC project. *Futures*. 2001, Vol. 33, pp.709–731.

Ndulu, B.; Chakraborti, L.; Lijane, L.; Ramachandran, V.; and Wolgin, J. Challenges of African Growth; Opportunities, Constraints and Strategic Directions .The International Bank for Reconstruction and Development / The World Bank, Washington, D.C., 2007.

Nielson, D. L. and Tierney, M. J. Delegation to International Organizations: Agency Theory and World Bank Environmental Reform. *International Organization* 57, Spring 2003, pp 241–276.

Nixon, S.W. The Artificial Nile. American Scientist, March-April 2004, Vol. 92, pp. 158-165.

Nyamekye, E. Socioeconomic Impact of Mine Closures: A Case Study of Ghana: In Environmental Policy I Mining, Corporate Strategy and Planning for Closure; Warhurst, A.; Noronha, L.; Ed.; Lewis Publishers: NY, 2000, pp. 397-413.

Obiri, S.; Dodoo, D. K.; Okai-Sam, F.; and Essumang, D. K. Non-cancer health risk assessment from exposure to cyanide by resident adults from the mining operations of Bogoso gold limited in Ghana. *Environmental Monitoring and Assessment*. 2006 Vol.118, Issue (1-3), pp.51-63.

Obukhov A. M. Statistically homogeneous fields on a sphere. *Uspethi Mathematicheskikh Nauk* 1947, Vol. 2, pp. 196–198.

Oehmke, J. F. and Crawford, E. W. The Impact of Agricultural Technology in Sub-Saharan Africa. *J. Afr. Econ.* 1996, Vol. 5, No. 2, pp. 271-292.

Ofori, S.C. Environmental impact assessment in Ghana: current administration and procedures- towards appropriate methodology. *Environmentalist.* March 1991, Vol. 11, No.1, pp 45-54.

Ogawa, H.; N. Ogura. Source and behavior of organic carbon of seawater in Tokyo Bay. *Geochemistry*, 1990a, Vol. 24, pp. 27–41 (in Japanese).

Ortolano, L. and Shepherd, A. Environmental impact assessment:challenges and opportunities. *Impact Assessment.* IAIA Spring 1995, Vol. 13, No. 1, pp 8.

Parashar, U. D.; Hummelman, E. G. ; Bresee, J. S.; and Miller, M. A.; R. I. Glass, R.I. Global Illness and Deaths Caused by Rotavirus Disease in Children. *Emerging Infectious Diseases.* 2003, Vol. 9, pp. 565–72.

Parkin, J. A philosophy for multiattribute evaluation in environmental impact assessments. *Geoforum* 1992, Vol. 23, No. 4, pp.467–475.

Pastakia, C.M.R. The rapid impact assessment matrix (RIAM)—a new tool for environmental impact assessment. In Environmental Impact Assessment Using the Rapid Impact Assessment Matrix (RIAM); Jensen, K.; Ed.; Olsen & Olsen, Fredensborg: Denmark, 1998.

Pastakia, C.M.R. and Bay, J. Initial environmental evaluation of alternative methods to conserve the Rupa Tal, Nepal. In Environmental Impact Assessment Using the Rapid Impact Assessment Matrix (RIAM); Jensen, K. Ed.; Olsen and Olsen, Fredensborg: Denmark, 1998.

Pastakia C .M. R. and Jensen A. The Rapid Impact Assessment Matrix (RIAM) for EIA. *Environ. Impact. Asses. Rev.* 1998, Vol. 18, pp.461–482.

Pearson, K. On lines and planes of closest fit to systems of points
in space. *Philosophical Magazine* 1902, Vol. 2, pp. 559–572.

Peavy, H. S., Rowe, D.R. and Tchobanoglous, G.. Environmental Engineering. McGraw-Hill Book Company: London, 1985.

Peixoto, J. P.; Oort, A.H. Physics of Climate. American Institute of Physics, NY, 1992.

Petchey, O. L. Environmental colour affects aspects of single-species population dynamics. *Proc. R. Soc. Lond.* B 2000, Vol. 267, pp.747–754.

Petchey, O. L.; Gonzalez, A. and Wilson, H. B. Effects on population persistence: the interaction between environmental noise colour, intraspecific competition and space. *Proc. R. Soc. Lond.* B 1997, Vol. 264, pp.1841–1847.

Petry, B. Lecture Notes and References. Refresher course on environmental impacts: Assessment, Monitoring and Control. Sponsored by: Ministry of Foreign Affairs Directorate –General for International Co-operation, the Netherlands. In collaboration with U.S.T., Kumasi, Ghana. International Institute for Infrastructural , Hydraulic and Environmental Engineering(IHE), the Netherlands, 18-29 September 1995.

Petts, G. E. Impounded Rivers. Perspectives of Ecological Management. John Wiley and Sons: NY, 1984.

Pike, N.; Tully, T.; Haccou, P.; and Ferrière, R. The effect of autocorrelation in environmental variability on the persistence of populations: an experimental test, *Proc. R. Soc. Lond.* B 2004, Vol. 271, pp.2143–2148.

Pilot Environmental Performance Index (EPI) 2006 . Yale Center for Environmental Law and Policy ,Yale University; Center for International Earth Science Information Network

(CIESIN), Columbia University; In collaboration with World Economic Forum Geneva, Switzerland; Joint Research Centre of the European Commission Ispra, Italy. Full report and data available at http://epi.yale.edu.

Pollit, E. Iron Deficiency and Cognitive Function. Annal Review of Nutrition. *1993, Vol. 13, pp.521-37.*

Preisendorfer, R. W. *Principal Component Analysis in Meteorology and Oceanography.* Elsevier: NY, 1988.

Ramanathan, R., 2001. A note on the use of the Analytical Hierarchy Process for Environmental Impact Assessment. *J. Environ. Manage.* 2001, Vol. 63, pp.27–35.

Rau, J. K. and Wooten, D.C. (Eds.); *Environmental Impact Analysis Handbook* , McGraw-Hill Book Company: NY, 1980.

Ravallion, S. and Chen, M. How have the world's poorest fared since the early 1980s? *The World Bank Research Observer* 2004, Vol. 19, No. 2, pp.141–169.

Riordan, J. *Introduction to Combinatorial Analysis*. Rover Publications Inc.: NY, 2002; pp.4.

Ripley, B. D. *Stochastic Simulation*. John Wiley and Sons: NY, 1987.

Rissanen, J. *Modeling by Shortest Data Description. Automatica.* 1978; Vol.14, pp. 465-471.

Rodrick, D. Where Did All the Growth Go? External Shocks, Social Conflict, and Growth Collapses . *Journal of Economic Growth.* 1999, Vol. 4pp 385-412.

Rotenberg, M. Effect of certain stochastic parameters on extinction and harvested populations. *J. Theor. Biol.* 1987, Vol. 124, pp. 455–471.

Roweis, S. EM algorithms for PCA and SPCA. In Advances in Neutral Information Processing Systems 10; Jordan, M. I.; Kearns, M. J.; Solla, S. A.; Eds.; MIT Press: MA, 1997, pp. 626–632.

Royama, T. *Analytical Population Dynamics*, Chapman & Hall: London ,1992.

Saaty, T. L. *The Analytic Hierarchy Process*, McGraw-Hill: NY, 1980.

Sachs, J. and Warner, *A. Sources of Slow Growth in African Economies. J. Afr. Econ.* 1997, Vol. 6, pp. 335-76.

Sachs, J.D. *Tropical Underdevelopment*, CID Working Paper No. 57, Center for International Development, Harvard University. Cambridge, MA, 2000.

Sahn, D.E. (Ed.); *Adjusting to Policy Failure in African Economies*, Cornell University Press, Ithaca NY, 1994.

Sahn, D.E. (Ed.); *Economic Reform and the Poor in Africa*, Oxford University Press: Oxford, 1996.

Sahn, D.E.; Dorosh, P., A. and Younger, S.D. *Structural Adjustment Reconsidered: Economic Policy and Poverty in Africa,* Cambridge University Press: Cambridge, 1997.

Sahn, D.E. and Stifel, D.C. *Progress Toward the Millennium Development Goals in Africa.* World Development.2003, Vol. 31, No. 1, pp. 23-52.

Sahn, D. E. and Younger, S.D. *Growth and Poverty Reduction in Sub-Saharan Africa: Macroeconomic Adjustment and* Beyond. *J. Afr. Econ.* July 2004, Vol.13, No.1, pp. i66-i95.

Sammy, G.K. *Environmental Impact Assessment in Developing Countrie*s, Ph.D. dissertation, University of Oklahoma, Norman, 1982.

Sanders, D. H. *Statistics—A Fresh Approach. Fourth Edition*. McGraw-Hill: NY, 1990; pp. 253–288.

Sarma, N. K. *Environmental impacts of water resources projects: the Indian experience, in UNESCO/UNEP*. Impacts of Large Water Projects on the Environment. UNEP,

EMINWA, UNESCO. Proceedings of an International Symposium, UNESCO, UNEP, IIASA and IAHS 21-31 October 1990; pp. 239-248.

Schecher, W. D. and Driscoll, C.T. An evaluation of equilibrium calculations within acidification models: The effect of uncertainty in measured chemical components. *Water Resour. Res*. 1988, Vol. 24, pp. 533-542.

Schiegg, K. Environmental autocorrelation: curse or blessing? *Trends. Ecol. Evol.* May 2003, Vol.18, No.5, pp. 212-214.

Schneider, T. Analysis of incomplete climate data: Estimation of mean values and covariance matrices and imputation of missing values. *J. Climate.* 2001, Vol. 14, pp. 853–871.

Schwager, M.; Johst, K. and Jeltsch, F. Does red noise increase or decrease extinction rist? Single extreme events versus series of unfavorable conditions. *Am. Nat.* 2006, Vol.167, pp.879–888.

Schwartz, G. *Estimating the Dimension of a Model. Annals of Statistics.* 1978, Vol. 6, pp. 461-464.

Shao, J. Linear Model Selection by Cross-Validation. *Journal of the American Statistical Association*. 1993, Vol. 88, pp. 486-494.

Sheskin, D.J. *Handbook of Parametric and Nonparametric Statistical Procedures*. Third Edition. Chapman and Hall/CRC: NY, 2004.

Shopley, J.; Fuggle, R. A comprehensive review of current environmental Impact assessment methods and techniques. *J. Environ. Manage*. 1984, Vol. 18, pp.25-47.

Shumway, R. H.; Stoffer, D.S. *Time Series Analysis and Its Applications*.. With R Examples. Second Edition. Springer, NY, 2006; 575pp.

Silverman, B. W. *Density Estimation for Statistics and Data Analysis.* Chapman and Hall/CRC: NY, 1986; 175pp.

Soewando, S., Husaini, M. and Pollit, E. Effects of iron deficiency on attention and learning processes in pre-school children: Bandung, Indonesia. *American Journal of Clinical Nutrition*, 1989, Vol. 50, pp. 667-674.

So´lnes, J. Environmental quality indexing of large industrial development alternatives using AHP. *Environ. Impact. Assess. Rev*. Vol. 23, pp. 283–303.

Sorensen, J. C. *A Framework for Identification and Control of Resource Degradation and Conflict in the Multiple Use of the Coastal Zone*. Masters Thesis, June 1971.

Spall, J. C. Introduction to Stochastic Search and Optimization: Estimation, Simulation, and Control. *Wiley-InterScience Series in Discrete Mathematics and Optimization*; John Wiley and Sons: NJ, 2003; 595pp.

Sprent, P. and Smeeton, N. C. *Applied Nonparametric Statistical Methods*. Fourth Edition.Chapman and Hall /CRC: NY, 2007; 530 pp.

Srinivasan, P. T., Viraraghavan, T. and Subramanian, K. S. Aluminium in drinking water: An overview, *Water SA*. January 1999, Vol. 25, No. 1, pp.47-56.

Stone, M. Cross-validation choice and assessment of statistical predictions. *Journal of the Royal Statistical Society*.1974; Vol. B-36, pp.111–147.

Tetteh, I. K.; Frempong, E. and Awuah, E. *An Analysis of the Environmental Health Impact of the Barekese Dam in Kumasi*,

Tetteh, I. K.; Awuah, E. and Frempong, E. Post-project analysis: The use of a network diagram for environmental evaluation of the Barekese Dam, Kumasi, Ghana. *Environmental Modeling and Assessment*. 2006, Vol.11, pp. 235–242.

Tetteh, I. K. Awuah, E, and Frempong, E. (Unpublished): *Post-project analysis: An Application of a Weighted Matrix Model for Environmental Evaluation of the Barekese Dam, Kumasi, Ghana.*

The Economist (US). The beautiful and the dammed. March 28, 1992. Available at http://www.highbeam.com/doc/1G1-12086595.html.

Thor, E. C.; Elsner, G. H; Travis, M.R.; and O'Loughlin, K. M. Forest Environmental Impact Analysis--A New Approach. *Journal of Forestry*. November 1978, Vol. 76, No.11, pp. 723-725.

Tipping, M. E. and C. M. Bishop. Probabilistic principal component analysis. *J. Roy. Stat. Soc.* 1999, Vol. 61B, pp.611–622.

Tomlinson, P. and Atkinson, S.F. Environmental Audits: A Literature Review. Journal of Environmental Monitoring. *May 1987, Vol. 8, Issue 3, pp.239-261.*

Toth, Z. and Kalnay, E. Ensemble Forecasting at NMC: the generation of perturbations. *Bulletin of the American Meteorological Society*. 1993, Vol.74, pp.2317-2330.

Toth, Z. and Kalnay, E. Ensemble Forecasting at NCEP the breeding method. *Mon. Wea. Rev.* 1997, Vol. 125, pp.3297-3318.

Tran, L.; Knight, C. G.; O'Neill, R.; Smith, E.; Ritters, K.; and Wickham, J. Environmental Assessment Fuzzy Decision Analysis for Integrated Environmental Vulnerability Assessment of the Mid-Atlantic Region. *Environ. Manage.* 2002, Vol. 29, No. 6, pp. 845-859.

Triantaphyllou, E. *Multi-Criteria Decision Making Methods: A Comparative Study*. Kluwer Academic Publishers: The Netherlands, 2000.

Tukker, A. Life cycle assessment as a tool in environmental impact assessment. *Environ. Impact. Assess. Rev.* 2000, Vol. 20, pp. 435–456.

Turban, E. and Aronson, J. E. *Decision Support Systems and Intelligent Systems*, Prentice Hall: NJ, 2001.

Tzoumis, K. and Finegold, L. Looking at the quality of draft environmental impact statements over time in the United States: Have ratings improved? *Environ. Impact. Assess. Rev.* 2000, Vol. 20, pp. 557–578.

UKDTLR. *Multi Criteria Analysis: A Manual*, Department for Transport, Local Government and the Regions, United Kingdom, 2001.

Ukoli, F.M.A. Introduction to Parasitology in Tropical Africa, Wiley: Great Britain, 1984.

United Nations(UN). Agenda 21. NY, 1992.

United Nations (UN). United Nations Convention on the Law of the Sea of 10 December 1982.

United Nations Conference on Sustainable Development (UNCSD). Convention on Biological Diversity, artt. 1-8, Rio de Janeiro, 1992.

United Nations Environment Program (UNEP). Montreal Protocol on Substances that
Deplete the Ozone Layer, 1987.

United Nation Population Division (UNPD). World Urbanization Prospects: The 2005 Revision Population Database, 2005.

United States (US) Environmental Protection Agency (EPA). Washington, D.C., 2nd December 1970.

Van Benschoten, J. E. Effects of Temperature and pH on Residual Aluminum in Alkaline-Treated Waters. *Journal of Environmental Engineering.* May/June 1994, Vol. 120, No. 3, pp. 543-559.

VanLehn, K. Cognitive Skill Acquisition. *Annu. Rev. Psychol.*1996. Vol. 47, pp. 513–539.

Vapnik, V. and Chervonenkis, A.Y. On the Uniform Convergence of Relative Frequencies of Events and Their Probabilities. *Theory of Probability and Its Applications*. 1971 Vol. 16, pp. 264-280.

Victor, D.G,; Naki´cenovi´c, N. and Victor, N. The Kyoto Protocol Emission Allocations: Windfall Surpluses for Russia and Ukraine. *Climatic Change*. 2001. Vol.49, pp.263–277.

Verghese, B. G. Sardar Sarovar Project Revalidated by Supreme Court. *Water Resources Development,* 2001, Vol. 17, No. 1, pp.79–88.

Verpoorten, M. and Berlage, L. Economic Mobility in Rural Rwanda: A Study of the Effects of War and Genocide at the Household Level. *J. Afr. Econ.* June 2007, Vol. 16, No. 3, pp. 349-392.

Vesely´, J.; Majer, V.; Kopa´cˇek, J.; and Norton, S. A. Increasing temperature decreases aluminum concentrations in Central European lakes recovering from acidification *Limnol. Oceanogr.* 2003, Vol. 48, No. 6, pp. 2346–2354.

Vyas, J. N. Water and Energy for Development in Gujarat with special focus on the Sardar Sarovar Project, *International Journal of Water Resources Development,* 2001, Vol. 17, No. 1, pp.37–54.

Walters , R.C.S. Dam Geology. Second Edition. Wiley: Texas, 1971, pp.50-62.

Wang, R. , Fraedrich, K. and Pawson, S. Phase-Space Characteristics of the Tropical Stratospheric Quasi-Biennial Oscillation. *Journal of the Atmospheric Sciences*. 1995, Vol. 52, No. 24, pp. 4482-4500.

Wang, Y.M.; Yang, J.B. *and* Xu, DL. Environmental impact assessment using the evidential reasoning approach, *European Journal of Operational Research.* 2006, Vol. 174, pp.1885–1913.

Wassink, H. Ghana Water Restructuring Project. Dam Safety Assessment. Draft final report 9P2503, April 2004; Vol. 1, 96pp.

Wathern, P. (Ed.); Environmental Impact Assessment: Theory and Practice. Winchester, Mass: Allen and Unwin: London, 1988.

Wichmann, M. C.; Johst, K.; Moloney, K. A.; Wissel, C.; *and* Jeltsch, F. Extinction risk in periodically fluctuating environments. *Ecol. Model.* 2003, Vol. 167, pp. 221–231.

Weare, B.C. and Nasstrom, J.S. Examples of extended empirical orthogonal function analyses. Mon. Wea. Rev. 1982, Vol. 110, pp.481-485.

Wei, C. Z. On Predictive Least Squares Principles. Annals of Statistics, 1992, Vol. 20, pp. 1-42.

Wilks D.S. 2006. Statistical Methods in the Atmospheric Sciences. Second Edition. Academic Press: Amsterdam, 2006; 627 pp.

Wood. G. Post-development Auditing of EIA Predictive Techniques: A Spatial Analytical Approach. Journal of Environmental Planning and Management.September 1999, Vol. 42, No. 5, pp. 671-689.

Wood, G. Is what you see what you get? Post-development auditing of methods used for predicting the zone of visual influence in EIA. *Environ. Impact. Assess. Rev.* October 2000, Vol. 20, Issue 5, pp. 537-556.

Workshop on Environmental Statistics for the ECOWAS Region, Dakar – Senegal. Country Paper – GHANA. Environmental Statistics in Perspective. 28th February – 4th March, 2005.

World Bank. The World Bank's Approach to the Environment. Washington, D.C. World Bank, 2000.

World Bank and IMF. Heavily Indebted Poor Countries (HIPC) Initiative—Statistical update. International Development Association and International Monetary Fund, 2005; 39 pp.

World Commission on Dams(WCD).*Dams and development: a new framework for decision-making. The Report of world commission on Dams*, Earthscan Publications : London, UK, 2000.

World Conservation Monitoring Center (WCMC). *Global Biodiversity: Status of the Earth's Living Resources.* Chapman & Hall: London, 1992; 544 pp.

World Health Organization (WHO). *Urine filtration technique for S. haematobium infection.* Document PDP/83.4, 1983b.

www.ghanaweb.com/GhanaHomePage/republic/polit_hist.php. Political History of Ghana.

Xiao, F. China`s big dams: Are they safe? Probe International. 29 January 2007.

Ying, L.G.. and Liu, Y. C. A model of objective weighting for EIA. *Environmental Monitoring and Assessment.* 1995, Vol 36, No.2, pp.169-182.

Yumkella, K. and Vinanchiarachi, J. Leading Issues on Africa`s Path to Industrialization; The Role of Support Systems and Instruments. *J. Afr. Econ.* March 2003, Vol 12, No.1, pp.30-40.

Zakhary, K. *Factors Affecting the Prevalence of Schistosomiasis in the Volta Region of Ghana.* McGill J. Med. 1997, Vol. 3, pp. 93-101.

In: Environmental and Regional Air Pollution
Editors: D. Gallo, R. Mancini
ISBN: 978-1-60692-893-6

Short Communication

BIOSYNTHESIS OF POLY(γ-L-GLUTAMIC ACID) AND COMPARATIVE STUDIES ON THE BIODEGRADABILITY OF THE γ- AND α-ENANTIOMERIC FORMS OF THE POLY(GLUTAMIC ACID) BY *BACILLUS LICHENIFORMIS* NCIMB 11709

M. Soledad Marqués-Calvo[1*], Jordi Bou[3] and Marta Cerdà-Cuéllar[2]
[1] Departament d'Òptica i Optometria, EUOOT,
Universitat Politècnica de Catalunya, E-08222 Terrassa, Barcelona, Spain
[2] Centre de Recerca en Sanitat Animal (CReSA), Esfera UAB,
Campus de Bellaterra, UAB, E-08193 Bellatera, Barcelona, Spain
[3] Departament d'Enginyeria Química, ETSEIB,
Universitat Politècnica de Catalunya, E-08028 Barcelona, Spain

ABSTRACT

Poly-γ-glutamic acid (γ-PGA) is a water soluble, polyanionic, extracellular polymer produced by several members of the genus *Bacillus*, the most notable among them *Bacillus subtillis* and *Bacillus licheniformis.* This polymer consists of D- and L-glutamic acid repeating units that are linked between the α-amino and γ-carboxylic acid functionalities. A peculiar characteristic of γ-PGA formed by *B. licheniformis* is that the polymer stereochemistry can be controlled by the composition of the medium used in the polymer production in the laboratory. γ-PGA is water soluble, biodegradable, edible and non-toxic to humans and the environment. Therefore, the potential applications of this polymer and its derivatives have been of growing interest in the past few years in a broad range of industrial fields such as food, cosmetics, medicine and water treatment.

* To whom correspondence should be addressed: M. Soledad Marqués-Calvo, Departament d'Òptica i Optometria, Universitat Politècnica de Catalunya, C/Violinista Vellsolà, 37.08222 Terrassa, Spain. vre.marques@upc.edu and marques@oo.upc.edu

Most of the reports published on the microbial or enzymatic biodegradation of microbial γ-PGA are only concerned with γ-D-PGA. In the present work, γ-L-PGA was biosynthesized using *B. licheniformis* NCIMB 11709, and the biodegradation of the different enantiomeric forms of PGA, as well as the α and γ forms (γ-L-PGA, γ-D-PGA, γ-D,L-PGA, α-L-PGA and α-D-PGA) of this bacterial strain, were studied. γ-PGA with L-units content of 70% was produced with a Mw of 1 million g/mole. *B. licheniformis* NCIMB 11709 did not grow when the α-forms of the polymer were used as sole carbon source, independently of the D or L contents. However, a decrease in the molecular weight of the α-L-PGA in the bacterial culture was observed, which was not due to hydrolytic degradation. On the other hand, this strain could degrade and use the polymer when it was in the γ-form, regardless of the D or L contents.

This finding can have an implication in defining bacterial factors involved in the degradation of the polymer. Moreover, the generated knowledge of the degradation of the different forms of the poly(glutamic acid) will facilitate selection of the better form according to the intended application.

INTRODUCTION

Poly-γ-glutamic acid (γ-PGA) is a naturally-occurring homo-polyamide made of D- and L-glutamic acid units connected by amide linkages between α-amino and γ-carboxylic acid groups (Figure 1). Several microorganisms produce this polymer (Table 1), but it is mainly synthesized by various members of the genus *Bacillus*.

Table 1. Organisms reported to produce PGA

Organism	Configuration	Content (%)		References
		D-Enantiomer	L-Enantiomer	
Bacillus anthracis	D	100	0	Hanby and Rydon (1946) [40]
Bacillus licheniformis	D and L	10–100	0–90	Pérez-Camero et al. (1999) [26]
Bacillus megaterium	D and L	30	70	Ashiuchi et al. (2003) [19]
Bacillus pumilus	D and L	60	40	Schneerson et al. (2003) [41]
Bacillus subtilis (*chungkookjang*)	D and L	60–70	30–40	Ashiuchi and Misono (2002) [27]
Bacillus subtilis (*natto*)	D and L	50–80	20–50	Ashiushi and Misono (2002) [27]
Bacillus halodurans	L	0	100	Aono (1987) [42]
Planococcus halophilus	D	ND	ND	Kandler et al. (1983) [43]
Sporosarcina halophila	D	ND	ND	Kandler et al. (1983) [43]
Staphylococcus epidermidis	D and L	40	60	Kocianova et al. (2005) [44]
Natrialba aegyptiaca	L	0	100	Hezayen et al. (2001) [45]
Hydra	L	0	100	Weber (1990) [46]

ND, not determined.

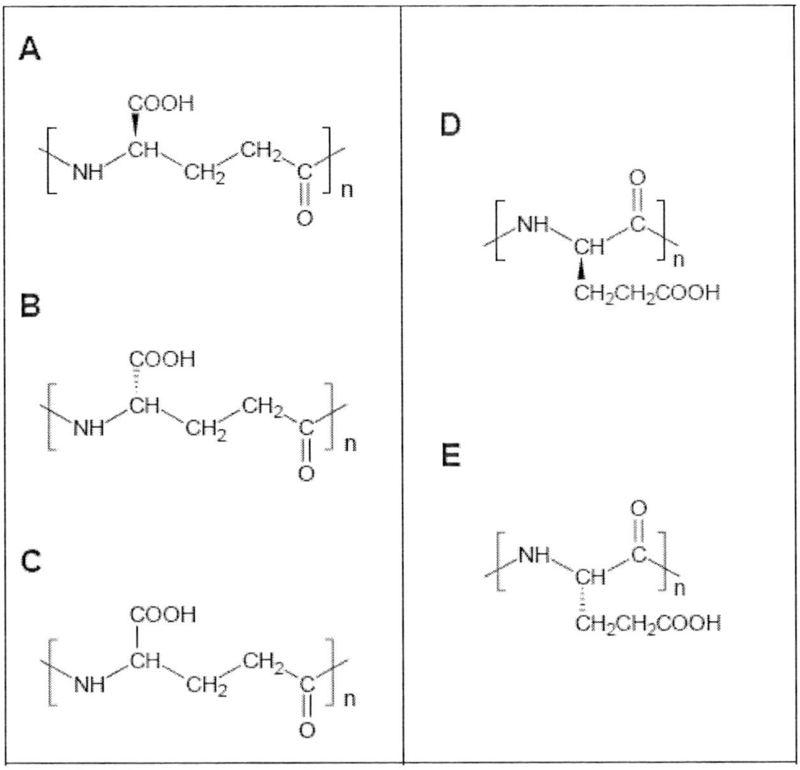

Figure 1. Chemical structures of PGA. A: Poly-γ-D-glutamate (γ-D-PGA), B: poly-γ-L-glutamate (γ-L-PGA), C: poly-γ-DL-glutamate (γ-DL-PGA), D: poly-α-D-glutamate and E: poly-α-L-glutamate. The term *n* indicates the number of glutamyl linkages in PGA, and the range of the *n* value is estimated to be 100 to over 10,000.

The specific biological functions of the polymer are due to its poly(anionic) nature and to its water solubility. This poly(amino acid) is present in many different environments and protects the producing organisms against adverse environmental conditions like dehydration or high salt concentration of the medium, against the effects of heavy metal ions or against the response of the human immune system during infections [1]. The role of PGA depends on whether it is anchored or released to the medium. Thus, anchored PGA is a virulence factor, whereas released PGA is a storage element or a persistence factor.

Because γ-PGA is water soluble, biodegradable, edible and non-toxic to humans and the environment, it could act as a substitute for synthetic polymers with similar characteristics. Therefore, it has been of growing interest in the past few years due to the potential applications of this polymer and its derivatives in a broad range of industrial, medical, pharmaceutical, and personal care applications, as well as in agriculture and environmental fields. γ-PGA and its derivatives are used, among other applications, as a thickener, moisturizer in cosmetics, cryoprotectant, sustained-release material, drug carrier, curable biological adhesive, biodegradable fibre, biopolymer flocculant, heavy metal absorber from water, dispersant, animal feed additive, intestinal Ca^{2+} solubility agent, in antibody production, and as an antitumor agent [2-4]. Biodegradability is one important advantage of naturally-occurring poly(amino acid)s over many synthetic polymers.

The minimal gene sets required for γ-PGA synthesis and release were recently defined. Four genes are involved in the synthesis and only one more is necessary for the release. The synthesis complex is probably responsible for the stereochemical specificity of γ-PGA composition [1]. γ-PGA is optically active and its specific rotation depends on the different proportion of the two enantiomers, L and D.

Polymer production in the laboratory and its stereochemistry can be controlled by the composition of the medium used. Various factors such as inorganic salts, glutamic acid, citric acid, glycerol and size of inoculum affect the production of γ-PGA by *B. licheniformis* in static and shaken cultures. In general, bacteria producing γ-PGA are L-glutamic acid dependent, meaning that exogenous L-glutamic acid is necessary for γ-PGA production. The production of γ-PGA increases with the addition of L-glutamic acid to the medium. Nevertheless, some genetically stable strains of *B. subtilis* and *B. licheniformis* are L-glutamic acid-independent bacteria and can synthesize γ-PGA without the addition of L-glutamic acid to the medium [2]. Poly(glutamic)acid is present as poly(acid) or poly(salt) form, depending on the pH of the aqueous solution.

Molecular weight and polydispersity are important features of microbial PGA for the effect that molecular size has on polymer properties. γ-PGA can have molecular weights ranging from 100 to over 1,000 KDa [5]. Final molecular weight depends on many factors, including medium composition, mode of feeding and fermentation conditions [6-11].

The γ-PGA produced by *B. licheniformis* ATCC 9945a is a homopolymer of D- and L-glutamic acid repeat units. The proportion of D-glutamic acid grows gradually as the concentration of Mn^{+2} increases. Synthesis and assembly of bacterial poly(γ–D-glutamic acid) occur through the function of a membranous multienzyme complex, or polyglutamyl synthetase complex, acting in a concerted fashion [12]. This complex catalyzes the activation, racemization, and polymerization of L-glutamic acid to form a high molecular weight polymer of γ–D–glutamic acid [13]. The extracellular synthetase enzyme activity was measured in *B. licheniformis* but was not detectable in cultures of *B. subtilis* [14]. Ashiuchi et al. [15] showed evidence that a multienzyme system, PgsBCA, was the sole machinery of γ-PGA synthesis in *B. subtilis (chungkookjang)*.

Few reports exist on the microbial and enzymatic biodegradation of microbial γ-PGA. Already in the early studies with *B. licheniformis* a depolymerase degrading γ-PGA was described [16]. Later, several authors have described enzymes from *B. licheniformis* and *B. subtilis* that can hydrolize the γ-glutamyl linkage. These studies deal with the kind of cleavage that takes place (between L- and L-, D- and D-, or D- and L- glutamic acid units) and if the enzyme acted in and *endo-* or *exo*-type manner [17-22]. Also, Oppermann et al. [23] investigated the ability of a non-adapted microbial sewage sludge community to grow with γ-D-PGA as sole carbon and nitrogen source, and several bacteria were isolated and identified. However, the existing studies of PGA degradation by *B. licheniformis* are mainly concerned only with γ-D-PGA and none of them compare the degradation of the different enantiomeric forms of PGA, and its α and γ forms. The present study is concerned with the ability of *B. licheniformis* NCIMB 11709 to degrade all of them. The biosynthesis of γ-L-PGA was also carried out as it was not possible to obtain the commercialized product.

At present, our studies are focussed to develop new block copolymers with controlled-structure in which the bacterial γ-PGA is one of the components. These block copolymers have a clear industrial applicability. To deepen the knowledge of the biodegradation of the different forms of γ-PGA will facilitate the selection of the most suitable form of the polymer for the synthesis of the block copolymers, according to the physical and the chemical characteristics of the final polymer, and to the biodegradation response.

Bacteria and Polymers

The bacteria *B. licheniformis* NCIMB 11709 (ATCC 9945a) from the National Collection of Industrial and Marine Bacteria (Great Britain) was used to synthesize the poly(γ-L-glutamic acid) and to study its ability to degrade the different forms of PGA. The following powdered enantiomeric forms of poly(glutamic acid) were used in this study: i) poly(γ-70%L-glutamic acid), produced in the present study by *Bacillus licheniformis* NCIMB 11709 as described below; ii) poly(γ-77%D-glutamic acid), previously synthesized in our laboratory from *Bacillus licheniformis* NCIMB 11709; iii) poly(γ-D,L-glutamic acid), kindly provided by Dr. Kubota of Meiji Co. (Japan); iv) poly(α-D-glutamic acid) and poly(α-L-glutamic acid), purchased from Sigma–Aldrich, Inc. (Madrid, Spain) and used without further purification.

Polymer Biosynthesis and Characterization

Highly mucoid colonies of the bacterium grown on modified Sauton's agar (Thorne et al. [24]) were selected and grown overnight aerobically at 37°C in a reciprocally shake flask containing 10 ml of modified Sauton's medium (Table 2). Two ml of this culture was used to inoculate two 1L Erlenmeyer flasks each containing 250 ml of the fermentation medium, which was medium E (Table 3) without $MnSO_4$. The pH of the medium was adjusted to 6.5 since it is the optimum for a maximum production of γ-PGA [8]. The cultures were incubated at 37°C in shake flasks at 250 rpm, for 74 h. Studies from Cromwick and Gross [7] and Pérez-Camero et al. [26] demonstrated that both the yield and the enantiomeric content of γ-PGA obtained in cultures of *B. licheniformis* were modulated by the concentration of Mn^{2+} in the culture medium. The L-glutamic acid in γ-PGA varied inversely with the content in $MnSO_4$ over a range of 80% to 10% for $[Mn^{2+}]$ increasing from 0 to 1230 $\mu mol \cdot L^{-1}$. The γ-PGA yield varied from 1 to 6 g L^{-1} over such an interval of manganese concentrations. Since we wanted to obtain a γ-PGA with the maximum content of L-glutamic acid units, we used medium E with no $MnSO_4$, although the yield was going to be low. Optimal γ-PGA production has been determined for both *B. subtilis* and *B. licheniformis* [10,27-30], but the polymer obtained is mainly formed of D-units. As γ-PGA produced by *B. licheniformis* is an exocellular polymer that is freely excreted into the fermentation broth, its recovery is relatively easy. The broth was first centrifuged to remove the cells (9000 rpm, 30 min). The γ-PGA in the supernatant was then precipitated using cold ethanol (3:1 volume ratio) to concentrate and partially purify the polymer. The resulting crude γ-PGA was then redisolved in 100 volumes of distilled water and filtered to remove any remaining solids. Next, the PGA solution was dialyzed (to remove salts and low-molecular weight impurities) and concentrated with a S-Minitan equipment and Prep/Scale-TFF cartridge (Millipore). The polymer was then rotational evaporated to finally obtain a film, which was then redisolved to a final volume of 10 ml in distilled water. This PGA solution was acidified with 12M HCl and precipitated with 100 ml of cold 1-propanol. The polymer was finally washed with ethanol and air dried.

Table 2. Composition of modified Sauton's medium

Content [a]	Concentration (g/L)
L-glutamic acid	20
Glycerol	80
Citric acid	12
NH_4Cl	7
$MgSO_4$	0.5
$FeCl_3 \cdot 6H_2O$	0.04
K_2HPO_4	0.5
pH is adjusted to 7.4 with NaOH	

[a] 15 g/L of agar are added to the medium to prepare agar plates.

Table 3. Composition of the medium E for production of γ-L-PGA by *B. licheniformis* NCIMB11709 (adapted from Leonard et al., 1958 [25])

Content	Concentration (g/L)
L-glutamic acid	20
Glycerol Citric acid	80
Citric acid	12
NH_4Cl	7
$MgSO_4 \cdot 7H_2O$	0.5 (2.03×10^{-3} M)
$FeCl_3 \cdot 6H_2O$	0.04 (1.44×10^{-4} M)
K_2HPO_4	0.5 (2.88×10^{-3} M)
$CaCl_3 \cdot 2H_2O$	0.15 (1.02×10^{-3} M)
$MnSO_4 \cdot H_2O$ [a]	0
pH is adjusted to 6.5 with NaOH	

[a] The concentration of Mn^{+2} can be varied from 1.54×10^{-7} M to 2.46×10^{-3} M to give the desired enantiomeric content in γ- PGA. In this study we used the medium without $MnSO_4$, to obtain polymer with maximum L content.

The number- and weight-average molecular weights (Mn and Mw, respectively) of the synthesized polymer, chemical structure, and stereochemical composition were performed as described by Pérez-Camero et al. [26]. Mn and Mw were determined by gel permeation chromatography (GPC). Aliquots of synthesized PGA were filtered through a 0.45 μm cellulose acetate syringe filter unit (Millipore) to remove the cells, and then directly injected into the GPC column using a Perkin-Elmer LC 2500 system equipped with two Ultrahydrogel™ linear columns (Waters) and a UV detector (Applied Biosystems 785 A) operating at 220nm [26]. Chemical structure of the γ-PGA was evaluated by NMR (Figure 2).

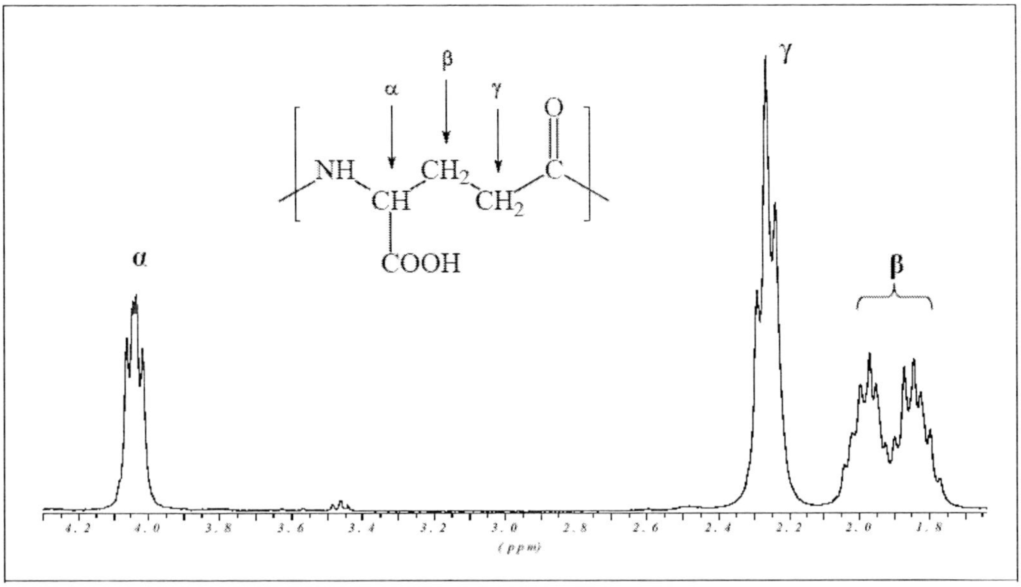

Figure 2. ^{1}H-NMR spectrum of pure γ-D-PGA obtained by fermentation. Solvent was D_2O. Doublet assigned to β position is due to effect of neighbor chiral carbon. Signals of hydrogen atoms in COOH and NH are not present because they interchange rapidly with D_2O.

^{1}H and ^{13}C NMR spectra were recorded on a Bruker AMX-300 spectrometer operating at 300.13 and 70.48 MHz, respectively [26]. The stereochemical composition of γ-PGA was determined by HPLC as described by Pérez-Camero et al. [26].

The γ-PGA obtained in this study had a L-glutamic acid content of 70% (γ-70% L-PGA), higher than that obtained by biosynthesis previously reported by other authors with *B. licheniformis* [8] and similar to that obtained by Pérez-Camero et al. [26]. The final yield of the polymer was 90 mg/L, similar to that obtained by these authors when no $MnSO_4$ was added to the fermentation medium. Low volumetric yields are obtained in cultures with no or very low $MnSO_4$ concentrations, since Mn^{2+} affects cell viability [7].

The Mn and Mw of poly (γ-L-glutamic acid) are shown in Table 4. Taking into account that Mn^{2+} was not present in the fermentation medium, high molecular weight for γ-70%L-PGA has been obtained in this work, which is relatively higher than that previously described [26]. Culture conditions, like incubation time, might have determined the final molecular weight of the polymer.

Table 4. Molecular weight of the polymers before and after biodegradation assays

Polymer	Origin	% D:L	Mn		Mw	
			Initial	Final	Initial	Final
γ-70%L-PGA	biosynthesis	30:70	193000	61000	986000	228000
γ-77%D-PGA	biosynthesis	23:77	462100	81700	1954000	272000
γ-D,L-PGA	Meiji Co	59:41	147000	912000	388000	288000
α-L-PGA	Sigma-Aldrich	0:100	53000	15000	154000	33000
α-D-PGA	Sigma-Aldrich	100:0	36000	34000	83000	81000

Biodegradation

The ability of *B. licheniformis* NCIMB 11709 to degrade the different enantiomeric forms of PGA, as well as its α and γ forms (γ-L-PGA, γ-D-PGA, γ-D,L-PGA, α-L-PGA and α-D-PGA) was investigated. For this purpose, a basal medium supplemented with the corresponding polymer solution at a final concentration of 0.25% was used. The composition of the basal medium is shown in Table 5. The pH was adjusted to 7.0. For each culture, a 100 ml Erlenmeyer flask containing 30 ml of the sterilized basal medium and 0.25% (w/v) of filter sterilized polymer solution was inoculated with 300 μl of bacteria and was incubated at 37°C with shaking. The inocula was obtained from an overnight culture at 37°C in tryptone soy broth (TSB) of *B. licheniformis* NCIMB 11709; the cells were washed twice with 6 ml of basal medium and finally resuspended in 5 ml of basal medium before use. Degradation assays were carried out for over 260 h. As control, cultures without polymer or bacteria were also performed. Bacterial growth in culture medium, polymers concentration, and their average molecular weights were routinely measured at various incubation times. Bacterial growth was monitored by the level of absorbance at 660 nm using a Pharmacia Biotech Novaspec II UV spectrophotometer. Mn and Mw were measured removing the cells by filtration [17] and carrying out GPC analyses using the same above described chromatographic conditions for the determination of the molecular weight of poly (γ-L-PGA). PGA concentration was also measured by GPC.

Table 5. Composition of the basal medium for biodegradation assays

Content	Concentration (g/L)
KH_2PO_4	0.7
K_2HPO_4	0.7
$MgSO_4 \cdot 7\ H_2O$	0.7
NH_4NO_3	1.0
NaCl	0.005
$FeSO_4 \cdot 7\ H_2O$	0.002
$ZnSO_4 \cdot 7\ H_2O$	0.002
$MnSO_4 \cdot 1\ H_2O$	0.001
pH is adjusted to 7.0 with NaOH	

Control runs for hydrolytic degradation, where polymer samples (γ-L-PGA, γ-D-PGA, γ-D,L-PGA, α-L-PGA and α-D-PGA) were incubated over identical exposure conditions in the absence of bacteria, showed negligible molecular weight loss. Also, no bacteria could grow in the basal medium without polymer (Figures 3 and 4).

B. licheniformis could grow using poly-γ-glutamic acid as carbon and nitrogen source, regardless of its D- or L- composition (Figure 3). The biodegradation studies of PGA have usually been performed with the polymer composed mainly of D- units, as it is the form of the polymer obtained when culturing the bacteria in the optimal conditions to achieve high yield of PGA. Here, we analyzed the three enantiomeric forms of the polymer (D-, L,- and racemic), and a similar behaviour of the bacteria was observed.

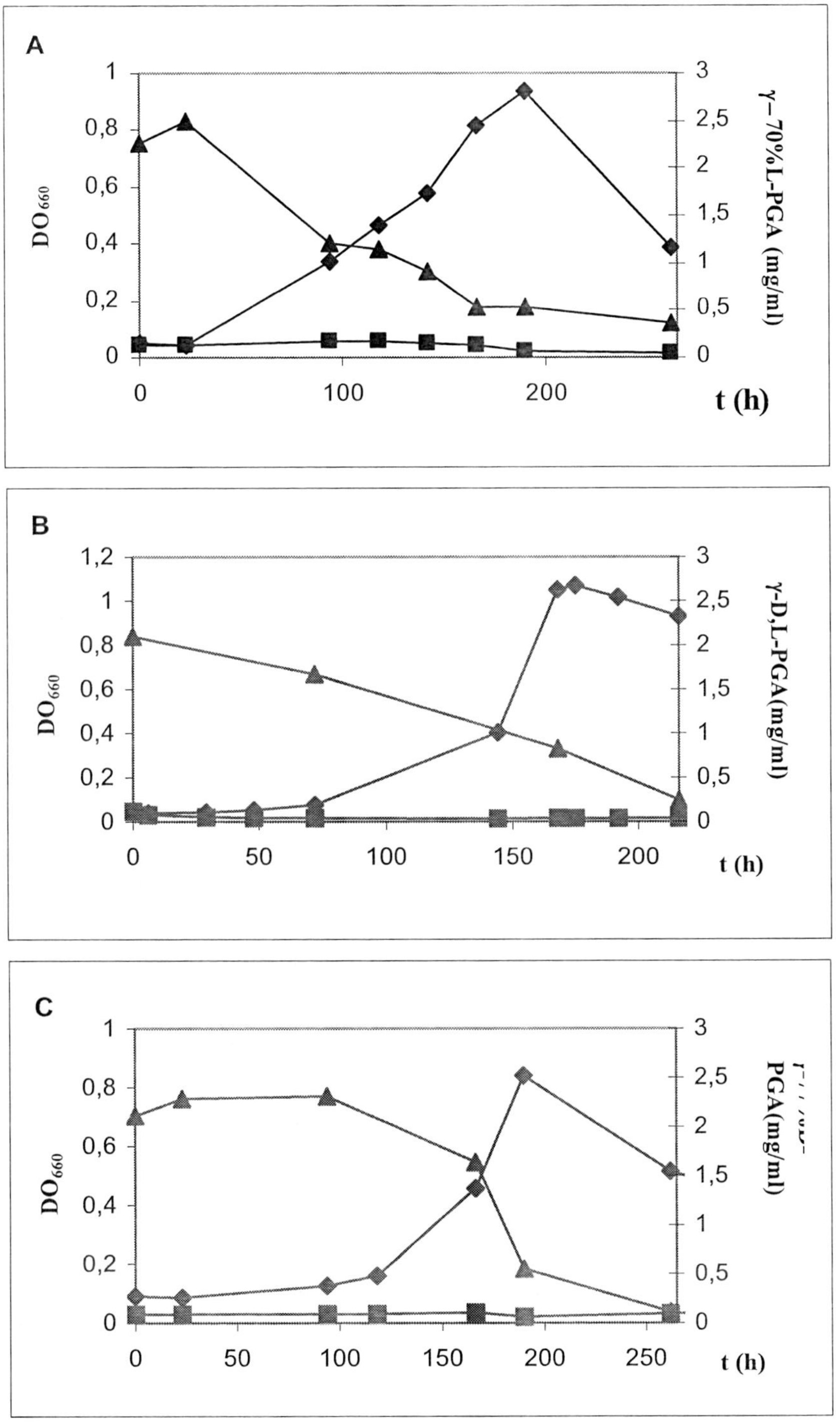

Figure 3. Biodegradation of γ-PGA by *B. licheniformis* NCIMB 11709. A: biodegradation of γ–L-PGA; B: biodegradation of γ-D,L-PGA; C: biodegradation of γ-D-PGA. Polymer concentration is indicated by triangles (▲); growth in minimal medium is indicated by squares (■); growth in minimal medium with γ-PGA is indicated by diamonds (♦).

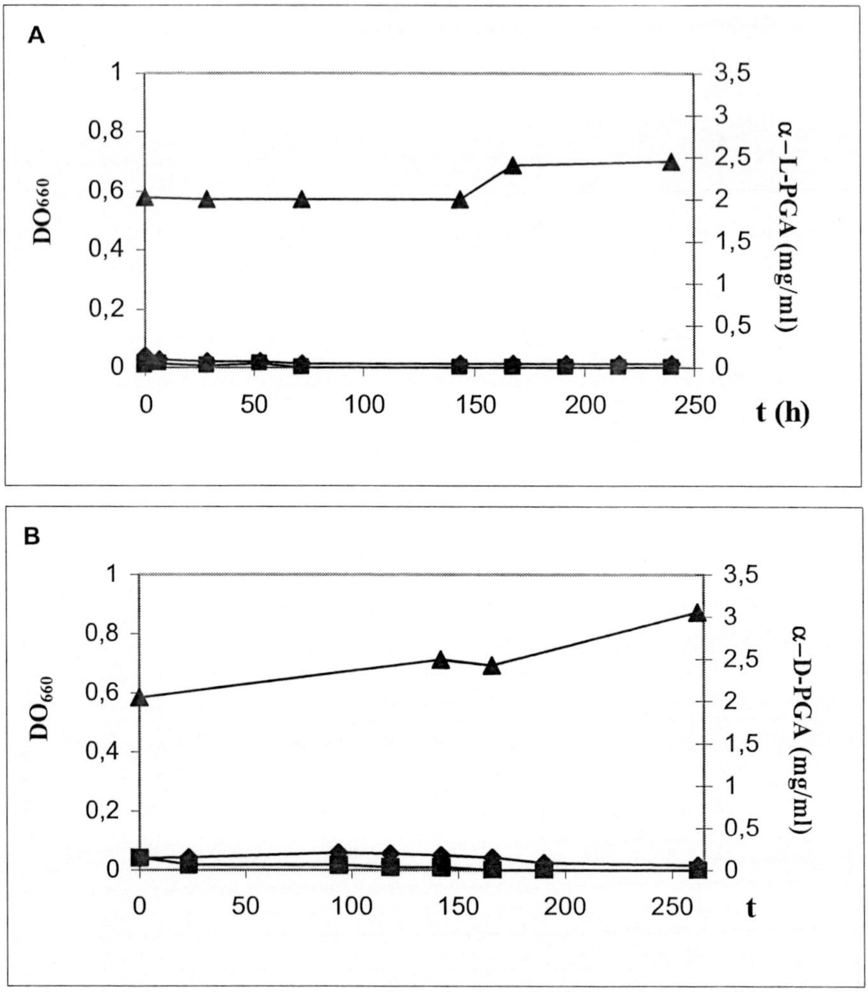

Figure 4. Biodegradation of α-PGA by *B. licheniformis* NCIMB 11709. A: biodegradation of α–L-PGA; B: biodegradation of α–D-PGA. Polymer concentration is indicated by triangles (▲); growth in minimal medium is indicated by squares (■); growth in minimal medium with α–L-PGA or α–D-PGA is indicated by diamonds (♦).

The three polymers were degraded by *B. licheniformis* to a similar final molecular weight, ranging from 227 kDa to 288 kDa. However, the higher decrease of molecular weight was observed in γ-D and γ-L-PGA (Table 4). Since *B. licheniformis* NCIMB 11709 is able to degrade both D- and L-PGA, as well as the racemic PGA, it might secrete different depolymerases which cleave the γ-glutamyl bonds between D- and D-, L- and L-, and D- and L-glutamic acid units. Several degrading γ-PGA depolymerases have been described, mainly from *B. subtilis*, but also from *B. licheniformis*. King et al. [18] characterized a γ-PGA depolymerase from *B. licheniformis* ATCC 9945a, which was tightly, but not covalently, associated with γ-PGA and showed an *endo*-D type activity. The enzyme was activated by Zn^{2+} and Ca^{2+} ions and was more active at high pH (pH= 8-10). Based on the studies from King et al. [18] with γ-D-PGA and the γ-PGA *endo*-D-depolymerase, Ashiuchi and Misono [27] suggested the biodegradation mechanism of γ-PGA: highly elongated γ-PGA is first digested by γ-PGA *endo*-L- and *endo*-D-depolymerases to yield γ-PGA with low molecular masses, which are further hydrolyzed to oligo-γ-glutamic acids or the amino acid monomer by γ-PGA *exo*-depolymerase. The results from our study strongly support this hypothesis, since all γ-PGA were degraded by *B. licheniformis* NCIMB 11709, independently of the L- or D- contents of the polymer.

Lately, several studies have been published regarding the isolation and characterization of PGA-degrading enzymes from *Bacillus* species, with different stereospecificities and *exo*- or *endo*-activities. Kimura et al. [31] characterized a γ-glutamyltransferase (GGT) from *B. subtilis* NAFM5, involved in the degradation of capsular poly(γ-glutamic acid), which contains D- and L-glutamate. The enzyme was synthesized during the stationary phase and the capsule γ-PGA was first internally degraded by an *endo*-type of γ-PGA hydrolase into 1×10^5 Da intermediates, then externally into glutamates via GGT. This GGT had a powerful *exo*-γ-glutamyl hydrolase activity. GGTs from *B. subtilis* (natto) seem functioning as γ-PGA *exo*-depolymerases. These enzymes hydrolyze more effectively the fragmentized γ-PGA than the highly elongated γ-PGAs [32]. Ashiuchi et al. [19] described the production and purification of an *endo*-D-depolymerase from *B. subtilis* (*chungkookjang*). It was produced by cloning the *pgdS* gene in *E. coli*. The enzyme digested PGAs from *B. subtilis* (D-glutamate content, 70%) and from *B. megaterium* (D-glutamate content, 30%), but not from *Natrialba aegyptiaca*, which consists only of L-glutamate. Further studies from Ashiuchi et al. [20] described a novel *endo*-peptidase that cleaves the γ-amide linkage between two D-glutamate residues in DL-PGA. Another hydrolase, the YwtD hydrolase cloned from *B. subtilis* IFO 16449, was shown to be a γ-DL-glutamyl hydrolase, cleaving only the γ-glutamyl bond between D- and L-glutamic acids of PGA. The enzyme acts in an *endo*-type manner [33]. This hydrolase has recently been further analyzed and it seems to act in a different way: it cleaves the γ-glutamyl bond between two consecutive D-glutamic acid residues, recognizing adjacent L-glutamic acid toward the N-terminal region of PGA [22]. Also, a non-PGA producer, *Bacillus* sp. CMU29, has been isolated from a natto-like product that produces an *endo*-type PGA-degrading enzyme [21]. This γ-glutamyl hydrolase cleaves the γ-glutamyl linkage between two contiguous L-glutamic acids in polymer chain of PGA. On the other hand, the PGA-depolymerase from *B. anthracis* catalyses the hydrolysis of the poly-D-glutamic acid capsule in an *exo*-peptidase manner [34]. Apart from all of these depolymerases from *Bacillus* species, it should be mentioned that the first characterized γ-PGA degrading exoenzyme was from *Flavobacterim polyglutamicum* [35]. The crude extract obtained from a culture of this bacterium completely degraded γ-L-PGA to free L-glutamic acid. Finally, a γ-PGA hydrolase from a bacteriophage that infects *B. subtilis* host cells was described [36]. The enzyme might contribute to the phage infection of *B. subtilis* strains that produce γ-PGA by providing a better accessibility of phage receptors on the cell surface.

On the other hand, information regarding the utilization and degradation by bacteria of α-PGA, both the D- and L- forms, is scarce. No *Bacillus* species producing α-PGA have been described. Only α-L-PGA polymer has been found in the cell walls of pathogenic mycobacteria [37], and naturally-occurring α-D-PGA has not yet been identified. Therefore, in this study we have also explored the possible degradation and utilization of α-PGA by *B. licheniformis*. The α-PGAs (D- and L-) used in this study are of synthetic origin, and it should be pointed out that the molecular sizes of synthetic polymers are low compared with those of naturally-occurring γ-PGA (Table 4). Interestingly, *B. licheniformis* could not grow when the medium was supplemented with α-PGA, regardless of the D or L contents (Figure 4), indicating that α-PGA (carbon source) cannot be utilized by this bacterium for its growth. In the medium with α-D-PGA, there was no bacterial growth, no decrease in the molecular weight of the polymer (Table 4) and no significant decrease of polymer concentration. Furthermore, in the medium supplemented with α-L-PGA, although there was no bacterial growth and no decrease of polymer concentration, a decrease in the molecular weight was observed (Table 4).

Since there was no hydrolytic degradation of the polymer, this degradation was enzymatic. This indicates that in the culture supernatant there are one or perhaps more enzymes secreted by the bacterium that degrade the polymer, although the bacterium cannot utilize the degradation products. These enzymes are stereospecific, since this behaviour was not observed with the α-D-PGA. Early studies by Troy [38] also detected an *endo*-type activity in the culture supernatant of *B. licheniformis*, which showed no activity against α-D-glutamyl polymers.

More recent studies have described depolymerases from fungal or bacterial origin with no activity against α-PGA. Tanaka et al. [39] described an *endo*-depolymerase from the fungus *Myrothecium* sp. TM-4222 which acted on γ-DL-PGA of *B. subtilis* (*natto*), but did not show any activity on α-L-PGA or α-D-PGA. Also, the purified YwtD hydrolase from *B. subtilis* IFO 16449 (a γ-DL-glutamyl hydrolase) did not have any in vitro activity, neither on α-L-PGA nor on α-D-PGA [33].

Conclusion

Culture conditions used in this work have allowed us to produce γ-PGA with a high percentage of L units (γ-70%L-PGA). The assays of biodegradation reveal that *B. licheniformis* NCIMB 11709 cannot grow with α-PGA as C source, regardless of its enantiomeric form. This incapability of use by the bacteria can be related to the fact that there is no *Bacillus* species producing α–PGA. However, the bacteria can degrade all forms of γ-PGA, independently of the D or L contents. Results obtained here can be useful in future investigations to easily select the better form of the poly-glutamic acid according to their broad range of applications.

Acknowledgements

The authors are grateful for the financial support received from the Comisión Interministerial de Ciencia y Tecnología (MAT2002-04600-CO2-O2 and MAT2006-05979). We also wish to thank Dr. Kubota for kindly providing us the poly(γ-D,L-glutamic acid) used in this study.

REFERENCES

[1] Candela T, Fouet A. Poly-gamma-glutamate in bacteria. *Mol. Microbiol.* 2006; 60: 1091-1098.

[2] Shih IL, Van YT. The production of poly-(gamma-glutamic acid) from microorganisms and its various applications. *Biores. Technol.* 2001;79:207-225.

[3] Richard A, Margaritis A. Poly(glutamic acid) for biomedical applications. *Critical Rev Biotechnol.* 2001;21:219-232.

[4] Sung M-H, Park C, Kim C-J, Poo H, Soda K, Ashiuchi M. Natural and Edible Biopolymer Poly-γ-glutamic Acid: Synthesis, Production, and Applications. *Chem. Rec.* 2005;5:352-366.

[5] Kubota H, Nambu Y, Endo T. Alkaline hydrolysis of poly(γ–glutamic acid) produced by microorganisms. *J. Pol. Sci: Part A: Polymer Chemistry* 1996; 34:1347-1351.

[6] Giannos S, Gross A, Kaplan D, Mayer J. Poly(glutamic acid) produced by Bacterial Fermentation. In: *Novel Biodegradable Microbial Polymers*. Dordrecht: Kluwer Academic Publishers; 1990. p. 457-460.

[7] Cromwick AM, Gross RA. Effects of manganese (II) on *Bacillus licheniformis* ATCC9945A physiology and gamma-poly(glutamic acid) formation. *Int. J. Biol. Macromol.* 1995;17:259-267.

[8] Cromwick AM, Birrer GA, Gross RA. Effects of pH and aeration of gamma-poly(glutamic acid) formation by *Bacillus licheniformis* in controlled batch fermentor cultures. *Biotechnol. Bioeng.* 1996;50:222-227.

[9] Ko YH, Gross RA. Effects of glucose and glycerol on gamma-poly(glutamic acid) formation by *Bacillus licheniformis* ATCC 9945a. *Biotechnol. Bioeng* 1998;57:430-437.

[10] Yoon SH, Do JH, Lee SY, Chang HN. Production of poly-gamma-glutamic acid by fed-batch culture of *Bacillus licheniformis*. *Biotechnol. Lett.* 2000;22:585-588.

[11] Morikawa M, Kagihiro S, Haruki M, Tanako K, Branda S,Kolter R, Kanaka S. Biofilm formation by *Bacillus subtilis* strain that produces gamma-polyglutamate. *Microbiology* 2006;152:2801-2807.

[12] Troy F A. Chemistry and biosynthesis of the poly(γ–d-glutamyl) capsule in *Bacillus licheniformis*. In: *Peptide Antibiotics-Biosynthesis and Functions*. New York: Walter de Gruyter; 1982. p. 49-83.

[13] Gardner JM, Troy FA. Chemistry and Biosynthesis of the Poly(γ–D–glutamyl) Capsule in *Bacillus licheniformis*. *J. Biol. Chem.* 1979;254:6262-6269.

[14] Kambourova M, Tagney M, Priest F. Regulation of polyglutamic acid synthesis by glutamate in *Bacillus licheniformis* and *Bacillus subtilis*. *Appl. Environ. Microbiol.* 2001;67:1004-1007.

[15] Ashiushi M, Kamei T, Baek DH, Shin SY, Sung M-H, Soda K, Misono H. Isolation of *Bacillus subtilis* (chungkookjang), a poly-γ-glutamate producer with high genetic competence. *Appl. Microbiol. Biotechnol.* 2001;57:764-769.

[16] Troy FA. Chemistry and biosynthesis of the poly(γ–d-glutamyl) capsule in *Bacillus licheniformis*. I. Properties of the membrane-mediated biosynthesis reaction. *J. Biol. Chem.* 1973;248:305-315.

[17] Birrer GA, Cromwick AM, Gross RA. Gamma-poly(glutamic acid) formation by *Bacillus licheniformis* 9945a: physiological and biochemical studies. *Int. J. Biol. Macromol.* 1994;16:265-275.

[18] King EC, Blacker AJ, Bugg TDH. Enzymatic breakdown of poly-gamma-D-glutamic acid in *Bacillus licheniformis*: Identification of a polyglutamyl gamma-hydrolase enzyme. *Biomacromolecules* 2000;1:75-83.

[19] Ashiuchi M, Nakamura H, Yamamoto T, Kamei T, Soda K, Park C, Sung M-H, Yagi T, Misono H. Poly depolymerase of *Bacillus subtilis*: production, simple purification and substrate selectivity. *J. Molec Catal B: Enzym.* 2003;23:249-255.

[20] Ashiuchi M, Nakamura H, Yamamoto M, Misono H. Novel poly-gamma-glutamate-processing enzyme catalyzing gamma-glutamyl-DD-amidohydrolysis. *J. Biosci. Bioeng.* 2006;102:60-65.

[21] Chunhachart O, Hanayama T, Hidesaki M, Tanimoto H, Tahara Y. Structure of the hydrolyzed product (F-2) released from gamma-polyglutamic acid by gamma-glutamyl hydrolase YwtD of *Bacillus subtilis*. *Biosci. Biotechnol. Biochem.* 2006;70:2289-91.

[22] Chunhachart O, Itoh T, Sukchotiratana M, Tanimoto H, Tahara Y. Characterization of gamma-glutamyl hydrolase produced by Bacillus sp. Isolated from Thai Thua-nao. *Biosci. Biotechnol. Biochem.* 2006;70:2779-2782.

[23] Oppermann FB, Pickartz S, Steinbuchel A. Biodegradation of polyamides. *Polym. Degrad. Stabil.* 1998;59:337-344.

[24] Thorne CB, Gómez CG, Noyes HE, Housewright RD. Production of glutamyl polypeptide by *Bacillus subtilis*. *J. Bacteriol.* 1954;68:307-315.

[25] Leonard CG, Housewright RD, Thorne CB. Effects of some metallic ions on glutamic polypeptide synthesis by *Bacillus subtilis*. *J. Bacteriol.* 1958;76:499-503.

[26] Pérez-Camero G, Congregado F, Bou JJ, Muñoz-Guerra S. Biosynthesis and ultrasonic degradation of bacterial poly(gamma-glutamic acid). *Biotechnol. Bioeng.* 1999;63:110-115.

[27] Ashiuchi M, Misono H. Poly-γ-glutamic acid. In Fahnestock SR and Steinbüchel A (Ed.), *Biopolymers*, vol. 7. Weinheim: Wiley-VCH; 2002. p. 123–174.

[28] Richard A, Margaritis A. Optimization of cell growth and poly(glutamic acid) production in batch fermentation by *Bacillus subtilis*. *Biotechnol. Lett.* 2003;25:465-468.

[29] Chen X, Chen S, Sun M, Yu Z. High yield of poly- γ -glutamic acid from *Bacillus subtilis* by solid-state fermentation using swine manure as the basis of a solid substrate. *Biores. Technol.* 2005;96:1872–1879.

[30] Soliman NA, Berekaa MM, Abdel-Fattah YR. Polyglutamic acid (PGA) production by *Bacillus* sp. SAB-26: application of Plackett–Burman experimental design to evaluate culture requirements. *Appl. Microbiol. Biotechnol.* 2005;69: 259–267.

[31] Kimura K, Tran LSP, Uchida I, Itoh Y. Characterization of *Bacillus subtilis* gamma-glutamyltransferase and its involvement in the degradation of capsule poly-gamma-glutamate. *Microbiology-Sgm* 2004;150:4115–4123.

[32] Abe K, Ito Y, Ohmachi T, Asada Y. Purification and properties of two isozymes of γ-glutamyltranspeptidase from *Bacillus subtilis* TAM-4. *Biosci. Biotechnol. Biochem.* 1997;61:1621–1625.

[33] Suzuki T, Tahara Y. Characterization of the *Bacillus subtilis ywtD* Gene, Whose Product Is Involved in γ-Polyglutamic Acid Degradation. *J. Bacteriol.* 2003;185:2379-2382.

[34] Uchida I, Makino S, Sasakawa C, Yoshikawa M, Sugimoto C, Terakado N. Identification of a novel gene, dep, associated with depolymerization of the capsular polymer in *Bacillus anthracis*. *Mol. Microbiol.* 1993;9:487-496.

[35] Volcani BE, Margalith P. A new species (*Flavobacterium polyglutamicum*) which hydrolyzes the gamma-L-glutamyl bond in polypeptides. *J. Bacteriol.* 1957;74:646–655.

[36] Kimura K, Itoh Y. Characterization of poly-c-glutamate hydrolase encoded by a bacteriophage genome: possible role in phage infection of *Bacillus subtilis* encapsulated with poly-c-glutamate. *Appl. Environ. Microbiol.* 2003;69:2491–2497.

[37] Wietzerbin J, Lederer F, Petit JF. Structural study of the poly-L-glutamic acid of the cell wall of *Mycobacterium tuberculosis* var *hominis*, strain Brevannes. *Biochem. Biophys. Res. Commun.* 1975;62:246-52.

[38] Troy FA. Chemistry and biosynthesis of the poly(γ–d-glutamyl) capsule in *Bacillus licheniformis*. II. Characterization and structural properties of the enzymatically synthesized polymer. *J. Biol. Chem.* 1973;248:316-324.

[39] Tanaka T, Hiruta O, Futamura T, Uotani K. Purification and characterization of poly(γ-glutamic acid) hydrolase from a filamentous fungus, *Myrothecium* sp. TM-4222. *Biosci. Biotechnol. Biochem.* 1993;57:2148-2153.

[40] Hanby WE, Rydon HN. The capsular substance of *Bacillus anthracis*. *Biochem. J.* 1946;40: 297–309.

[41] Schneerson R, Kubler-Kielb J, Liu TY, Dai ZD, Leppla SH, Yergey A, Backlund P, Shiloach J, Majadly F, Robbins JB. Poly(gamma-D-glutamic acid) protein conjugates induce IgG antibodies in mice to the capsule of *Bacillus anthracis*: a potential addition to the anthrax vaccine. *Proc. Natl. Acad. Sci. USA* 2003;100:8945-50.

[42] Aono R. Characterization of structural component of cell walls of alkalophilic strain of *Bacillus* sp. C-125. *Biochem. J.* 1987;245:467.

[43] Kandler O, König H, Wiegel J, Claus D. Occurrence of poly-γ-D-glutamic acid and poly-α-Lglutamine in the genera *Xanthobacter*, *Flexithrix*, *Sporosarcina* and *Planococcus*. *Syst. Appl. Microbiol.* 1983;4:34–41.

[44] Kocianova S, Vuong C, Yao Y, Voyich JM, Fischer ER, DeLeo FR, Otto M. Key role of poly-gamma-DL-glutamic acid in immune evasion and virulence of *Staphylococcus epidermidis*. *J. Clin. Invest* 2005;115: 688–694.

[45] Hezayen FF, Rehm BHA, Tindall BJ, Steinbüchel A. Transfer of *Natrialba asiatica* B1T to *Natrialba taiwanensis* sp. nov. and description of *Natrialba aegyptiaca* sp. nov., a novel extremely halophilic, aerobic, non-pigmented member of the *Archaea* from Egypt that produces extracellular poly (glutamic acid). Polymer production by two newly isolated extremely halophilic archaea: application of a novel corrosion-resistant bioreactor. *Int. J. Syst. Evol. Microbiol.* 2001;51:1133-1142.

[46] Weber J. Poly(γ-glutamic acid)s are the major constituents of nematocysts in *Hydra* (*Hydrozoa, Cnidaria*). *J. Biol. Chem.* 1990;265:9664-9669.

INDEX

A

B

C

D

E

F

G

H

I

J

K

L

M

N

O

P

Q

R

S

T

U

V

W

X

Y

Z